INFINITE-DIMENSIONAL DYNAMICAL SYSTEMS IN ATMOSPHERIC AND OCEANIC SCIENCE

INFINITE-DIMENSIONAL DYNAMICAL SYSTEMS IN ATMOSPHERIC AND OCEANIC SCIENCE

Guo Boling
Huang Daiwen
Institute of Applied Physics and Computational Mathematics, Beijing, China

World Scientific
Zhejiang Science and Technology Publishing House

Published by

World Scientific Publishing Co. Pte. Ltd.
5 Toh Tuck Link, Singapore 596224
USA office: 27 Warren Street, Suite 401-402, Hackensack, NJ 07601
UK office: 57 Shelton Street, Covent Garden, London WC2H 9HE

and

Zhejiang Science and Technology Publishing House
No. 347 Tiyuchang Road
Hangzhou, China

Library of Congress Cataloging-in-Publication Data
Guo, Boling.
Infinite-dimensional dynamical systems in atmospheric and oceanic science / by Boling Guo & Daiwen Huang (Institute of Applied Physics and Computational Mathematics, China).
pages cm
Translated from Chinese.
Includes bibliographical references.
ISBN 978-981-4590-37-2 (hardcover : alk. paper)
1. Atmospheric circulation. 2. Dynamic meteorology. 3. Marine sciences. I. Huang, Daiwen. II. Title.
QC880.4.A8G86 2014
551.4601'185--dc23

2013050767

British Library Cataloguing-in-Publication Data
A catalogue record for this book is available from the British Library.

The ISBN for China market is 978-7-5341-5935-0

Printed in Singapore

Preface

The intent of the book is to introduce some results in the study of partial differential equations and infinite-dimensional dynamical systems in geophysical fluid dynamics, which has been mainly focused on the dynamics of large-scale phenomena in the atmosphere and the oceans. In the past several decades, there are many research works in the field. In 1979, Zeng Qingcun made some pioneering research on the theories of some mathematical models governing the atmospheric and oceanic motions. His works have aroused many mathematicians' interest in the study of the theories about partial differential equations of the atmosphere and the oceans. From the 1980s, Chou Jifan and his collaborators made many works on the global analysis theory about the dissipative primitive equations of the atmosphere. In 1992, Jacques Louis Lions, Roger Temam and Wang Shouhong introduced a new formulation of the dissipative primitive equations of the atmosphere, and proved the global existence of weak solutions of these equations. Later, they also made many works on partial differential equations in the atmospheric and oceanic dynamics. Peter Constantin, Andrew Majda and Esteban G. Tabak studied the formation of strong fronts in the surface quasi-geostrophic equations. Andrew Majda and his collaborators contribute many theoretical and numerical works on PDEs and waves for the atmosphere and oceans. Mu Mu and Li Jianping both studied extensively on PDEs in the atmospheric and oceanic dynamics. In 2005, Cao Chongsheng and Edriss S. Titi proved the global well-posedness for the 3D viscous primitive equations. Recently, Zhou Xiuji presented the necessity and great meaning of the atmospheric random dynamics research. From 2006, the authors obtained some results about the primitive equations and some stochastic PDEs in the atmospheric and oceanic science.

This book consists of five chapters. In Chapter 1, we briefly recall some partial differential equations of the atmosphere and oceans. In Chapter 2, the quasi-geostrophic models of the atmospheric and the oceanic motions are introduced. In Chapter 3, we consider the initial boundary value problem for the three-dimensional viscous primitive equations of the large-scale moist atmosphere and oceans. In Chapter 4, we consider some stochastic models in the atmospheric and oceanic science. Chapter 5 is reserved for stability and instability theory of waves for the atmosphere and oceans.

The authors would like to thank Academician Zeng Qingcun, Academician Mu Mu, Prof. Liu Shishi, Prof. Li Jianping and so on. We acknowledge the generous support of both National Natural Science Foundation and Ministry of Science and Technology of China.

Contents

Chapter 1

Nonlinear Equations of the Atmospheric and the Oceanic Motions

There are usually two methods for predicting long-term weather and climate. First, by statistical methods, we can use the current climate, the historical record and numerical analysis to predict the future climate and the possible global climatic changes. Second, because air is compressible, and seawater is incompressible, by dynamical methods, we consider that the future status of climate is a consequence determined by the current status and the physical principles dominating these changes, thus we study equations and models describing the atmospheric and oceanic motions. Regarding weather prediction as an initial-boundary value problem in mathematical physics, we can establish numerical weather prediction models based on mathematical physical equation.

Numerical weather prediction is an outstanding applied research achievement of atmospheric science in the 20th century, of which theoretical foundation is the atmospheric dynamics. In 1922, Richardson introduced the concept of numerical weather prediction for the first time ([183]). His idea is that through solving the complete primitive equations governing the atmosphere motions numerically, one can simulate the evolution process of atmosphere, thus may predict weather quantitatively. Due to the weak calculation ability at that time, the dream of numerical weather prediction did not exist. Applying the long-wave theory and the scale-analysis theory established by Rossby and others, Charney set up a two-dimensional geostrophic model. With this model, he and his collaborators successfully made true 24-hour numerical weather prediction on the ENIAV computer of the Institute for Advanced Studies in Princeton for the first time. Along with the boom of atmosphere science and the enhancing of data dealing ability and numerical calculation ability of computer, researchers turn to numerical weather prediction by the primitive equation models from 1960s

([112,147,181,218]), greatly extend the time-range of numerical weather prediction. Afterward researchers started to make long-term numerical weather prediction, climate forecasting and numerical simulation of atmospheric circulation by some primitive equation models of the atmosphere and oceans.

To actualize long-term numerical weather prediction, climate prediction and numerical simulation of atmospheric circulation based on physical methods, the first thing is to establish some atmospheric and oceanic dynamical models, which are the nonlinear partial differential equations with initial-boundary value conditions which govern the atmospheric and oceanic motion. In this chapter, we mainly present basic and primitive equations and their boundary conditions which govern the atmospheric and oceanic motion. For more detail see [220], and also [84,145,162,205,211].

1.1 Basic Equations of the Atmospheric and the Oceanic Motions

1.1.1 *Basic Equations of the Atmosphere*

Regarding air and seawater as continuous media, one can use the Euler method to describe the atmospheric and oceanic motions. In the inertial coordinate frame (the coordinate axis is fixed with respect to the stellar), according to the Newton second law, the momentum conservation equation of the atmosphere is given by

$$\frac{\mathrm{d}_I \boldsymbol{V}_I}{\mathrm{d}t} = -\frac{1}{\rho}\mathrm{grad}_3 p + g_I + D,$$

where $\boldsymbol{V}_I$ is the absolute velocity of the atmosphere (velocity in the inertial coordinate frame), $\dfrac{\mathrm{d}_I \boldsymbol{V}_I}{\mathrm{d}t} = \dfrac{\partial \boldsymbol{V}_I}{\partial t} + (\boldsymbol{V}_I \cdot \boldsymbol{\nabla}_3)\boldsymbol{V}_I$ is the absolute acceleration (acceleration in the inertial coordinate frame), ρ is the density of air, p is the atmospheric pressure, $-\dfrac{1}{\rho}\mathrm{grad}_3 p$ is the pressure-gradient force, g_I is the gravity, and D is a molecular viscous force (molecular friction force, dissipative force), which is a dissipative force caused by air internal friction or turbulent momentum transmission.

In general, researchers are concerned with the relative motions of the atmosphere to the earth. So taking a coordinate frame rotating together with the earth as a reference frame, researchers can observe atmospheric relative motions. Suppose that the angular velocity of rotation in the rotating coordinate frame is $\boldsymbol{\Omega}$ (that is the rotational angular velocity of the earth), $\boldsymbol{V}$ is the atmospheric relative velocity, $\frac{\mathrm{d}\boldsymbol{V}}{\mathrm{d}t}$ is the atmospheric relative acceleration in the rotating coordinate frame, then

$$\boldsymbol{V}_I = \boldsymbol{V} + \boldsymbol{\Omega} \times \mathbf{r},$$

$$\frac{\mathrm{d}_I \boldsymbol{V}_I}{\mathrm{d}t} = \frac{\mathrm{d}\boldsymbol{V}_I}{\mathrm{d}t} + \boldsymbol{\Omega} \times \boldsymbol{V}_I,$$

where $\mathbf{r}$ is the radius vector. The proof of the second equation above appears in section 1.5 in [172]. According to the previous three equations, we get in the rotating coordinate frame **the atmospheric momentum conservation equation**

$$\frac{\mathrm{d}\boldsymbol{V}}{\mathrm{d}t} = -\frac{1}{\rho}\mathrm{grad}_3 p + g - 2\boldsymbol{\Omega} \times \boldsymbol{V} + D, \tag{1.1.1}$$

where $g = g_I + \Omega^2 \mathbf{r}$ is commonly referred to gravity (Ω is the value of the earth rotation angular velocity), $-2\boldsymbol{\Omega} \times \boldsymbol{V}$ is the Coriolis force, $\Omega^2 \mathbf{r}$ is the inertial centrifugal force,

$$\frac{\mathrm{d}}{\mathrm{d}t} = \frac{\partial}{\partial t} + \boldsymbol{V} \cdot \boldsymbol{\nabla}_3$$

is the substantial derivative (often called the total derivative).

According to the mass conservation law, **the continuity equation** is given by

$$\frac{\mathrm{d}\rho}{\mathrm{d}t} + \rho \mathrm{div}_3 \boldsymbol{V} = 0. \tag{1.1.2}$$

In general, when describing large-scale motions of the troposphere and the stratosphere, one may consider dry air as ideal gas, and can get the **atmospheric state equation**

$$p = R\rho T, \tag{1.1.3}$$

where the vaporation in the atmosphere is negligible, T means the temperature absolute term of the atmosphere, and $R = 287\ \mathrm{J{\cdot}kg^{-1}K^{-1}}$ is a gas constant of dry air.

According to the first law of thermodynamics, the **atmospheric thermodynamic equation** is given by

$$c_v \frac{\mathrm{d}T}{\mathrm{d}t} + p\frac{\mathrm{d}\frac{1}{\rho}}{\mathrm{d}t} = \frac{\mathrm{d}Q}{\mathrm{d}t},$$

where $c_v = 718$ J·kg^{-1}K^{-1}, and $\frac{\mathrm{d}Q}{\mathrm{d}t}$ is the quantity of heat per unit mass of air obtained from external environment per unit time. Applying (1.1.3), we have

$$R\frac{\mathrm{d}T}{\mathrm{d}t} = \frac{\mathrm{d}\frac{p}{\rho}}{\mathrm{d}t} = \frac{1}{\rho}\frac{\mathrm{d}p}{\mathrm{d}t} + p\frac{\mathrm{d}\frac{1}{\rho}}{\mathrm{d}t} = \frac{RT}{p}\frac{\mathrm{d}p}{\mathrm{d}t} + p\frac{\mathrm{d}\frac{1}{\rho}}{\mathrm{d}t}.$$

Combining the above two equations together, we get

$$c_p\frac{\mathrm{d}T}{\mathrm{d}t} - \frac{RT}{p}\frac{\mathrm{d}p}{\mathrm{d}t} = \frac{\mathrm{d}Q}{\mathrm{d}t}, \tag{1.1.4}$$

where $c_p = c_v + R$ is specific heat at constant pressure.

Equations (1.1.1)-(1.1.4) are called the **fundamental equations of dry air**, where the unknown functions are $\boldsymbol{V}$, ρ, p, and T in these equations. If D and $\frac{\mathrm{d}Q}{\mathrm{d}t}$ are fixed, equations (1.1.1)-(1.1.4) are self-closed.

When one has to consider vaporation in the air, the moist air state equation is

$$p = R\rho T(1 + cq), \tag{1.1.5}$$

where $q = \frac{\rho_1}{\rho}$ is the mixing ratio of water vapor in the air, and ρ_1 is the density of water vapor in the air. Here, c represents positive constant varying with context. $c = 0.618$ in (1.1.5). The thermodynamic equation of the moist atmosphere is

$$c_p\frac{\mathrm{d}T}{\mathrm{d}t} - \frac{RT(1+cq)}{p}\frac{\mathrm{d}p}{\mathrm{d}t} = \frac{\mathrm{d}Q}{\mathrm{d}t}, \tag{1.1.6}$$

the conservation equation of the water vapor in the air is

$$\frac{\mathrm{d}q}{\mathrm{d}t} = \frac{1}{\rho}W_1 + W_2, \tag{1.1.7}$$

where W_1 is the condensation ratio of steam per unit volume, and W_2 is the volume change ratio of unit mass steam due to horizontal and vertical diffusions. Equations (1.1.1), (1.1.2) and (1.1.5)-(1.1.7) are called the **equations of the moist atmospheric**.

1.1.2 *Basic Equations of the Oceans*

Suppose that there are massless source-sinks within the oceans. In the rotating coordinate frame, the equations of oceans consist of the following equations:

the momentum conservation equation

$$\rho \frac{\mathrm{d}\boldsymbol{V}}{\mathrm{d}t} = -\mathrm{grad}_3 p + \rho g - 2\rho \boldsymbol{\Omega} \times \boldsymbol{V} + D,$$

the continuity equation

$$\frac{\mathrm{d}\rho}{\mathrm{d}t} + \rho \mathrm{div}_3 \boldsymbol{V} = 0,$$

the state equation

$$\rho = f(T, S, p),$$

the thermodynamic equation

$$\frac{\mathrm{d}T}{\mathrm{d}t} = Q_1,$$

and the salinity conservation equation

$$\frac{\mathrm{d}S}{\mathrm{d}t} = Q_2,$$

where S is salinity, Q_1 is the heat source per unit mass seawater derive from the external environment in unit time, and Q_2 is the salt source per unit mass seawater derive from the external environment in unit time.

Since the equations above are too complex, one has to do some simplification. Generally, one takes **Boussinesq approximation**, that is, consider ρ in ρg and the state equation as unknown function, but ρ in other position as constant ρ_0. Moreover, we use the following approximation equation to replace the above state equation

$$\rho = \rho_0[1 - \beta_T(T - T_0) + \beta_S(S - S_0)],$$

where β_T and β_S are positive constants, and T_0, S_0 are the reference values of temperature and salinity, respectively. Thus, we get the equations of oceans as

$$\rho_0 \frac{\mathrm{d}\boldsymbol{V}}{\mathrm{d}t} = -\mathrm{grad}_3 p + \rho g - 2\rho_0 \boldsymbol{\Omega} \times \boldsymbol{V} + D, \tag{1.1.8}$$

$$\mathrm{div}_3 \boldsymbol{V} = 0, \tag{1.1.9}$$

$$\rho = \rho_0[1 - \beta_T(T - T_0) + \beta_S(S - S_0)], \tag{1.1.10}$$

$$\frac{\mathrm{d}T}{\mathrm{d}t} = Q_1, \tag{1.1.11}$$

$$\frac{\mathrm{d}S}{\mathrm{d}t} = Q_2. \tag{1.1.12}$$

Remark 1.1.1. State equation (1.1.10) is an empirical equation, which appears in [212]. The more general form is

$$\rho = \rho_0 \left[1 - \beta_T(T - T_0) + \beta_S(S - S_0) + \frac{p}{\rho_0 c_s^2}\right],$$

where c_s is a positive constant, and this equation appears in section 2.4.1 of [205].

1.2 Equations of the Atmosphere and the Oceans in the Sphere Coordinate Frame

1.2.1 *Equations of the Atmosphere in the Sphere Coordinate Frame*

The atmosphere moves on the rotating earth surface. To study the relative motion of the atmosphere, assuming that the earth surface is simulated by a sphere surface, we discuss the atmosphere motion in spherical coordinate system.

Let's deduce the basic atmospheric equations in spherical coordinate frame. Setting the center of earth as the origin of the spherical coordinate, $\theta(0 \le \theta \le \pi)$ denotes the co-latitude of earth (it mutually complement to latitude), $\varphi(0 \le \varphi \le 2\pi)$ denotes the longitude of earth, r denotes the distance between the center and point on the surface of the earth, $\boldsymbol{e}_\theta$, $\boldsymbol{e}_\varphi$ and $\boldsymbol{e}_r$ are the unit vectors in the directions of θ, φ, r respectively, $\boldsymbol{e}_\theta$ tends to the south along the longitude, $\boldsymbol{e}_\varphi$ tends to the east along the latitude, and $\boldsymbol{e}_r$ tends outward along the radius. Using differential geometry symbols, we have

$$\boldsymbol{e}_\theta = \frac{1}{r}\frac{\partial}{\partial \theta}, \quad \boldsymbol{e}_\varphi = \frac{1}{r\sin\theta}\frac{\partial}{\partial \varphi}, \quad \boldsymbol{e}_r = \frac{\partial}{\partial r}.$$

According to the definition of velocity, the air velocity $\boldsymbol{V}$ is expressed as

$$\boldsymbol{V} = v_\theta \boldsymbol{e}_\theta + v_\varphi \boldsymbol{e}_\varphi + v_r \boldsymbol{e}_r,$$

where

$$v_\theta = r\frac{\mathrm{d}\theta}{\mathrm{d}t} = r\dot{\theta}, \ v_\varphi = r\sin\theta\frac{\mathrm{d}\varphi}{\mathrm{d}t} = r\sin\theta\dot{\varphi}, \ v_r = \frac{\mathrm{d}r}{\mathrm{d}t} = \dot{r}.$$

In spherical coordinate frame, the substantial derivative of any vector F is given by

$$\begin{aligned}\frac{\mathrm{d}F}{\mathrm{d}t} &= \lim_{\Delta t\to 0}\frac{1}{\Delta t}[F(t+\Delta t, \theta(t+\Delta t), \varphi(t+\Delta t), r(t+\Delta t)) \\ &\quad -F(t,\theta(t),\varphi(t),r(t))] \\ &= \left(\frac{\partial}{\partial t} + \dot{\theta}\frac{\partial}{\partial \theta} + \dot{\varphi}\frac{\partial}{\partial \varphi} + \dot{r}\frac{\partial}{\partial r}\right)F \\ &= \left(\frac{\partial}{\partial t} + \frac{v_\theta}{r}\frac{\partial}{\partial \theta} + \frac{v_\varphi}{r\sin\theta}\frac{\partial}{\partial \varphi} + v_r\frac{\partial}{\partial r}\right)F.\end{aligned}$$

Since $\boldsymbol{\nabla}_3 = \boldsymbol{e}_\theta\frac{1}{r}\frac{\partial}{\partial \theta} + \boldsymbol{e}_\varphi\frac{1}{r\sin\theta}\frac{\partial}{\partial \varphi} + \boldsymbol{e}_r\frac{\partial}{\partial r}$ in spherical coordinate frame, the substantial derivative in spherical coordinate frame is

$$\frac{\mathrm{d}}{\mathrm{d}t} = \frac{\partial}{\partial t} + \boldsymbol{V}\cdot\boldsymbol{\nabla}_3.$$

By direct calculation, we have

$$\frac{\mathrm{d}\boldsymbol{e}_\theta}{\mathrm{d}t} = \frac{v_\varphi \cot\theta}{r}\boldsymbol{e}_\varphi - \frac{v_\theta}{r}\boldsymbol{e}_r,$$

$$\frac{\mathrm{d}\boldsymbol{e}_\varphi}{\mathrm{d}t} = -\frac{v_\varphi \cot\theta}{r}\boldsymbol{e}_\theta - \frac{v_\varphi}{r}\boldsymbol{e}_r,$$

$$\frac{\mathrm{d}\boldsymbol{e}_r}{\mathrm{d}t} = -\frac{v_\varphi}{r}\boldsymbol{e}_\varphi + \frac{v_\theta}{r}\boldsymbol{e}_\theta.$$

Angular velocity of earth rotation is given by $\boldsymbol{\Omega} = -\Omega\sin\theta\boldsymbol{e}_\theta + \Omega\cos\theta\boldsymbol{e}_r$, so

$$-2\boldsymbol{\Omega}\times\boldsymbol{V} = 2\Omega\cos\theta v_\varphi\boldsymbol{e}_\theta + (-2\Omega\cos\theta v_\theta - 2\Omega\sin\theta v_r)\boldsymbol{e}_\varphi + 2\Omega\sin\theta v_\varphi\boldsymbol{e}_r.$$

Using

$$\mathrm{div}_3\boldsymbol{V} = \boldsymbol{\nabla}_3\cdot\boldsymbol{V} = \frac{1}{r\sin\theta}\frac{\partial v_\theta\sin\theta}{\partial\theta} + \frac{1}{r\sin\theta}\frac{\partial v_\varphi}{\partial\varphi} + \frac{1}{r^2}\frac{\partial r^2 v_r}{\partial r},$$

and

$$\begin{aligned}\frac{\mathrm{d}\boldsymbol{V}}{\mathrm{d}t} &= \frac{\mathrm{d}(v_\theta\boldsymbol{e}_\theta + v_\varphi\boldsymbol{e}_\varphi + v_r\boldsymbol{e}_r)}{\mathrm{d}t}\\ &= \boldsymbol{e}_\theta\frac{\mathrm{d}v_\theta}{\mathrm{d}t} + \boldsymbol{e}_\varphi\frac{\mathrm{d}v_\varphi}{\mathrm{d}t} + \boldsymbol{e}_r\frac{\mathrm{d}v_r}{\mathrm{d}t} + v_\theta\frac{\mathrm{d}\boldsymbol{e}_\theta}{\mathrm{d}t} + v_\varphi\frac{\mathrm{d}\boldsymbol{e}_\varphi}{\mathrm{d}t} + v_r\frac{\mathrm{d}\boldsymbol{e}_r}{\mathrm{d}t}.\end{aligned}$$

We rewrite equations (1.1.1)-(1.1.4) following the basic equations of atmosphere in spherical coordinate frame

$$\begin{aligned}&\frac{dv_\theta}{dt} + \frac{1}{r}(v_r v_\theta - v_\varphi^2\cot\theta) = -\frac{1}{\rho r}\frac{\partial p}{\partial\theta} + 2\Omega\cos\theta v_\varphi + D_\theta,\\ &\frac{dv_\varphi}{dt} + \frac{1}{r}(v_r v_\varphi + v_\theta v_\varphi\cot\theta) = -\frac{1}{\rho r\sin\theta}\frac{\partial p}{\partial\varphi} - 2\Omega\cos\theta v_\theta - 2\Omega\sin\theta v_r + D_\varphi,\\ &\frac{dv_r}{dt} - \frac{1}{r}(v_\theta^2 + v_\varphi^2) = -\frac{1}{\rho}\frac{\partial p}{\partial r} - g + 2\Omega\sin\theta v_\varphi + D_r,\\ &\frac{\mathrm{d}\rho}{\mathrm{d}t} + \rho\left(\frac{1}{r\sin\theta}\frac{\partial v_\theta\sin\theta}{\partial\theta} + \frac{1}{r\sin\theta}\frac{\partial v_\varphi}{\partial\varphi} + \frac{1}{r^2}\frac{\partial r^2 v_r}{\partial r}\right) = 0,\\ &c_p\frac{\mathrm{d}T}{\mathrm{d}t} - \frac{RT}{p}\frac{\mathrm{d}p}{\mathrm{d}t} = \frac{\mathrm{d}Q}{\mathrm{d}t},\\ &p = R\rho T,\end{aligned}$$

where $D = (D_\theta, D_\varphi, D_r)$ is viscosity term.

Because the thickness of the atmospheric layer to be studied (about 120 kilometers) is far less than the radius of earth $a \approx 6,371$ kilometers, we use a instead of previous r which appears as coefficient in the above equations.

For the large-scale motions, the item $\frac{2v_r}{r}$ in the mass conservation equation can be omitted. Thus we simplify the above equations as

$$\frac{dv_\theta}{dt} + \frac{1}{a}(v_r v_\theta - v_\varphi^2 \cot\theta) = -\frac{1}{\rho a}\frac{\partial p}{\partial \theta} + 2\Omega\cos\theta v_\varphi + D_\theta, \tag{1.2.1}$$

$$\frac{dv_\varphi}{dt} + \frac{1}{a}(v_r v_\varphi + v_\theta v_\varphi \cot\theta) = -\frac{1}{\rho a \sin\theta}\frac{\partial p}{\partial \varphi} - 2\Omega\cos\theta v_\theta - 2\Omega\sin\theta v_r + D_\varphi, \tag{1.2.2}$$

$$\frac{dv_r}{dt} - \frac{1}{a}(v_\theta^2 + v_\varphi^2) = -\frac{1}{\rho}\frac{\partial p}{\partial r} - g + 2\Omega\sin\theta v_\varphi + D_r, \tag{1.2.3}$$

$$\frac{d\rho}{dt} + \rho\left(\frac{1}{a\sin\theta}\frac{\partial v_\theta \sin\theta}{\partial \theta} + \frac{1}{a\sin\theta}\frac{\partial v_\varphi}{\partial \varphi} + \frac{\partial v_r}{\partial r}\right) = 0, \tag{1.2.4}$$

$$c_p\frac{dT}{dt} - \frac{RT}{p}\frac{dp}{dt} = \frac{dQ}{dt}, \tag{1.2.5}$$

$$p = R\rho T, \tag{1.2.6}$$

where $\frac{d}{dt} = \frac{\partial}{\partial t} + \frac{v_\theta}{a}\frac{\partial}{\partial \theta} + \frac{v_\varphi}{a\sin\theta}\frac{\partial}{\partial \varphi} + v_r\frac{\partial}{\partial r}$.

1.2.2 *Equations of the Oceans in the Sphere Coordinate Frame*

Suppose that the velocity of seawater is $\boldsymbol{V} = (u, v, w)$, and u, v, w are the velocity of seawater respectively in the direction of θ, φ, r. In spherical coordinate frame, the equations of the oceans under Boussinesq approximation are

$$\frac{du}{dt} + \frac{1}{a}(wu - v^2\cot\theta) = -\frac{1}{\rho_0 a}\frac{\partial p}{\partial \theta} + 2\Omega\cos\theta v + D_u, \tag{1.2.7}$$

$$\frac{dv}{dt} + \frac{1}{a}(wv + uv\cot\theta) = -\frac{1}{\rho_0 a\sin\theta}\frac{\partial p}{\partial \varphi} - 2\Omega\cos\theta u - 2\Omega\sin\theta w + D_v, \tag{1.2.8}$$

$$\frac{dw}{dt} - \frac{1}{a}(u^2 + v^2) = -\frac{1}{\rho_0}\frac{\partial p}{\partial r} - \frac{\rho}{\rho_0}g + 2\Omega\sin\theta v + D_w, \tag{1.2.9}$$

$$\frac{1}{a\sin\theta}\frac{\partial u\sin\theta}{\partial \theta} + \frac{1}{a\sin\theta}\frac{\partial v}{\partial \varphi} + \frac{\partial w}{\partial r} = 0, \tag{1.2.10}$$

$$\rho = \rho_0[1 - \beta_T(T - T_0) + \beta_S(S - S_0)], \tag{1.2.11}$$

$$\frac{dT}{dt} = Q_1, \tag{1.2.12}$$

$$\frac{dS}{dt} = Q_2, \tag{1.2.13}$$

where $D = (D_u, D_v, D_w)$ is viscosity term.

$$\frac{\mathrm{d}}{\mathrm{d}t} = \frac{\partial}{\partial t} + \frac{u}{a}\frac{\partial}{\partial \theta} + \frac{v}{a \sin\theta}\frac{\partial}{\partial \varphi} + w\frac{\partial}{\partial r}.$$

1.3 Equations of the Atmosphere in Atmospheric Pressure Coordinate Frame

The basic equations of atmosphere motions are so complicated that researcher is not able to solve them numerically or theoretically at present. Therefore, researchers have to omit the minor and medium scale factors, and simplify the basic equations of atmosphere motions reasonably in order to achieve numerical weather prediction. As the vertical scale of the atmosphere is far smaller than horizontal scale, the most natural simplification method is to adopt the **hydrostatic approximation**, that is, substituting the hydrostatic equilibrium equation

$$\frac{\partial p}{\partial r} = -\rho g$$

for the vertical momentum conservation equation. The hydrostatic equilibrium equation demonstrates the equilibrium relationship between the vertical pressure-gradient force and the gravity. It's in conformity with the weather observation data of the large-scale atmosphere, and also the theoretical analysis.

Here we use the scale analysis to interpret briefly the rationality of the hydrostatic approximation. For large-scale atmosphere motions, the horizontal characteristic length scale of the motion is $L \approx O(10^6)$, the vertical characteristic length scale of the motion is $D \approx O(10^4)$, the characteristic scale of horizontal velocity is $U \approx O(10^1)$, the characteristic scale of vertical velocity is $W \approx O(10^{-2})$, $\Omega \approx O(10^{-4})$, and the characteristic scale of atmospheric pressure is $P \approx O(10^5)$. Thus we know, in the vertical momentum equation, except $-\frac{1}{\rho}\frac{\partial p}{\partial r} \approx O(10^1)$, $-g \approx O(10^1)$, scale of other terms is all less than $O(10^{-3})$. So we can replace the vertical momentum conservation equation by the hydrostatic equilibrium equation.

According to the hydrostatic equilibrium equation, we know that the pressure p is a monotonic decreasing function of r, that is, the mapping $(\theta, \varphi, r; t) \to (\theta, \varphi, p; t)$ is one-to-one. Thus, we substitute a pressure coordinate system $(\theta, \varphi, p; t)$ (also called the isobaric surface coordinate frame) for the coordinate frame $(\theta, \varphi, r; t)$. Introducing a new pressure coordinate frame $(\theta^*, \varphi^*, p; t^*)$, we have

$$t^* = t,\ \theta^* = \theta,\ \varphi^* = \varphi,\ p = p(\theta, \varphi, r; t).$$

Now, let's deduce the form of the atmospheric equations in the new pressure coordinate frame $(\theta^*, \varphi^*, p; t^*)$. Firstly, in the pressure coordinate frame, the substantial derivative of any vector F is

$$\begin{aligned}\frac{\mathrm{d}F}{\mathrm{d}t^*} &= \left(\frac{\partial}{\partial t^*} + \dot{\theta}^*\frac{\partial}{\partial \theta^*} + \dot{\varphi}^*\frac{\partial}{\partial \varphi^*} + \dot{p}\frac{\partial}{\partial p}\right)F \\ &= \left(\frac{\partial}{\partial t^*} + \dot{\theta}\frac{\partial}{\partial \theta^*} + \dot{\varphi}\frac{\partial}{\partial \varphi^*} + \dot{p}\frac{\partial}{\partial p}\right)F.\end{aligned}$$

To obtain a new form of the momentum equation in the new pressure coordinate frame, we only compute a new form of the force here. In meteorology, usually substitute height $z = r - a$ for r, thus, the original coordinate is expressed as a function of the new coordinates $t = t^*,\ \theta = \theta^*,\ \varphi = \varphi^*,\ z = r - a = z(\theta^*, \varphi^*, p; t^*)$. So we have

$$p = p(\theta, \varphi, a + z(\theta^*, \varphi^*, p; t^*); t).$$

Differentiating the above function with respect to the variable p, we have

$$1 = \frac{\tilde{\partial} p}{\tilde{\partial} r}\frac{\partial r}{\partial p} = \frac{\tilde{\partial} p}{\tilde{\partial} r}\frac{\partial z}{\partial p},$$

where, to distinguish derivatives in the two coordinate frames, we use $\frac{\tilde{\partial} p}{\tilde{\partial} r}$ to indicate the derivative of p with respect to r in the original coordinate frame, $\frac{\partial z}{\partial p}$ to indicate the derivative of z with respect of p in the new coordinates frame. The following symbols in this section are defined similarly. Taking differential quotient of θ^* and φ^* in the above relationship of p, we have

$$0 = \frac{\tilde{\partial} p}{\tilde{\partial} \theta}\frac{\partial \theta}{\partial \theta^*} + \frac{\tilde{\partial} p}{\tilde{\partial} r}\frac{\partial r}{\partial \theta^*} = \frac{\tilde{\partial} p}{\tilde{\partial} \theta} + \frac{\tilde{\partial} p}{\tilde{\partial} r}\frac{\partial r}{\partial \theta^*},$$

$$0 = \frac{\tilde{\partial} p}{\tilde{\partial} \varphi}\frac{\partial \varphi}{\partial \varphi^*} + \frac{\tilde{\partial} p}{\tilde{\partial} r}\frac{\partial r}{\partial \varphi^*} = \frac{\tilde{\partial} p}{\tilde{\partial} \varphi} + \frac{\tilde{\partial} p}{\tilde{\partial} r}\frac{\partial r}{\partial \varphi^*}.$$

Combining the above two equations with the hydrostatic equilibrium equation, we obtain

$$-\frac{1}{\rho a}\frac{\tilde{\partial} p}{\tilde{\partial} \theta} = -\frac{1}{a}\frac{\partial \Phi}{\partial \theta^*},\quad -\frac{1}{\rho a\sin\theta}\frac{\tilde{\partial} p}{\tilde{\partial} \varphi} = -\frac{1}{a\sin\theta^*}\frac{\partial \Phi}{\partial \varphi^*},$$

where $\Phi = gz$ is generally called the geopotential. With the equation $1 = \frac{\tilde{\partial} p}{\tilde{\partial} r}\frac{\partial z}{\partial p}$ and the hydrostatic equilibrium equation, we have

$$p\frac{\partial \Phi}{\partial p} = -RT.$$

Therefore, we get the new form of momentum equations in the new pressure coordinate frame as follows

$$\frac{\mathrm{d}v_\theta}{\mathrm{d}t^*} - \frac{1}{a}v_\varphi^2\cot\theta^* = -\frac{1}{a}\frac{\partial \Phi}{\partial \theta^*} + 2\Omega\cos\theta^* v_\varphi + D_\theta, \tag{1.3.1}$$

$$\frac{\mathrm{d}v_\varphi}{\mathrm{d}t^*} + \frac{1}{a}v_\theta v_\varphi\cot\theta^* = -\frac{1}{a\sin\theta^*}\frac{\partial \Phi}{\partial \varphi^*} - 2\Omega\cos\theta^* v_\theta + D_\varphi, \tag{1.3.2}$$

$$p\frac{\partial \Phi}{\partial p} = -RT. \tag{1.3.3}$$

According to the principle that Coriolis force does no work, we omit the term $-2\Omega\sin\theta^* v_r$. Similarly, because the scale of v_r is very small to the large scale atmospheric motions, we also omit $\frac{1}{a}v_r v_\theta$, $\frac{1}{a}v_r v_\varphi$.

Next, let's deduce a new form of the mass conservation equation in the new pressure coordinate frame. According to the hydrostatic equilibrium equation, we have $\rho = -\frac{1}{g}\frac{\tilde{\partial} p}{\tilde{\partial} r}$. Substituting this equality in (1.2.4), we have

$$\frac{\tilde{\mathrm{d}}\frac{\tilde{\partial} p}{\tilde{\partial} r}}{\tilde{\mathrm{d}}t} + \frac{\tilde{\partial} p}{\tilde{\partial} r}\left(\frac{1}{a\sin\theta}\frac{\tilde{\partial} v_\theta\sin\theta}{\tilde{\partial}\theta} + \frac{1}{a\sin\theta}\frac{\tilde{\partial} v_\varphi}{\tilde{\partial}\varphi} + \frac{\tilde{\partial} v_r}{\tilde{\partial} r}\right) = 0. \tag{1.3.4}$$

With the definition of substantial derivative in the original coordinate

$$\frac{\tilde{\mathrm{d}}}{\tilde{\mathrm{d}}t} = \frac{\tilde{\partial}}{\tilde{\partial} t} + \frac{v_\theta}{a}\frac{\tilde{\partial}}{\tilde{\partial}\theta} + \frac{v_\varphi}{a\sin\theta}\frac{\tilde{\partial}}{\tilde{\partial}\varphi} + v_r\frac{\tilde{\partial}}{\tilde{\partial} r},$$

and

$$\frac{\tilde{\mathrm{d}}p}{\tilde{\mathrm{d}}t} = \frac{\tilde{\partial} p}{\tilde{\partial} t} + \frac{v_\theta}{a}\frac{\tilde{\partial} p}{\tilde{\partial}\theta} + \frac{v_\varphi}{a\sin\theta}\frac{\tilde{\partial} p}{\tilde{\partial}\varphi} + v_r\frac{\tilde{\partial} p}{\tilde{\partial} r},$$

we obtain

$$\frac{\tilde{\mathrm{d}}\frac{\tilde{\partial} p}{\tilde{\partial} r}}{\tilde{\mathrm{d}}t} = \frac{\tilde{\partial}\frac{\tilde{\mathrm{d}}p}{\tilde{\mathrm{d}}t}}{\tilde{\partial} r} - \frac{\tilde{\partial} v_\theta}{\tilde{\partial} r}\frac{\tilde{\partial} p}{a\tilde{\partial}\theta} - \frac{\tilde{\partial} v_\varphi}{\tilde{\partial} r}\frac{\tilde{\partial} p}{a\sin\theta\tilde{\partial}\varphi} - \frac{\tilde{\partial} v_r}{\tilde{\partial} r}\frac{\tilde{\partial} p}{\tilde{\partial} r}. \tag{1.3.5}$$

With the relationship between the pressure coordinate frame and the original coordinate frame, we have

$$\frac{\tilde{\partial}}{\tilde{\partial} r} = \frac{\tilde{\partial} p}{\tilde{\partial} r}\frac{\partial}{\partial p}, \quad \frac{\tilde{\partial}}{\tilde{\partial}\theta} = \frac{\partial}{\partial\theta^*} + \frac{\tilde{\partial} p}{\tilde{\partial}\theta}\frac{\partial}{\partial p}, \quad \frac{\tilde{\partial}}{\tilde{\partial}\varphi} = \frac{\partial}{\partial\varphi^*} + \frac{\tilde{\partial} p}{\tilde{\partial}\varphi}\frac{\partial}{\partial p}.$$

Thus,

$$\frac{\tilde{\partial}\frac{\tilde{\mathrm{d}}p}{\tilde{\mathrm{d}}t}}{\tilde{\partial}r} = \frac{\tilde{\partial}\dot{p}}{\tilde{\partial}r} = \frac{\tilde{\partial}p}{\tilde{\partial}r}\frac{\partial\dot{p}}{\partial p},$$

$$\begin{aligned}
&-\frac{\tilde{\partial}v_\theta}{\tilde{\partial}r}\frac{\tilde{\partial}p}{a\tilde{\partial}\theta} + \frac{\tilde{\partial}p}{\tilde{\partial}r}\frac{\tilde{\partial}v_\theta \sin\theta}{a\sin\theta\tilde{\partial}\theta} \\
&= -\frac{\tilde{\partial}p}{\tilde{\partial}r}\left(\frac{\tilde{\partial}p}{a\tilde{\partial}\theta}\frac{\partial v_\theta}{\partial p} - \frac{\tilde{\partial}v_\theta \sin\theta}{a\sin\theta\tilde{\partial}\theta}\right) \\
&= -\frac{\tilde{\partial}p}{\tilde{\partial}r}\left[\frac{\tilde{\partial}p}{a\tilde{\partial}\theta}\frac{\partial v_\theta}{\partial p} - \frac{1}{a\sin\theta}\left(\frac{\partial}{\partial\theta^*} + \frac{\tilde{\partial}p}{\tilde{\partial}\theta}\frac{\partial}{\partial p}\right)v_\theta\sin\theta\right] \\
&= \frac{\tilde{\partial}p}{\tilde{\partial}r}\left(\frac{\partial v_\theta \sin\theta^*}{a\sin\theta^*\partial\theta^*}\right),
\end{aligned}$$

$$\begin{aligned}
&-\frac{\tilde{\partial}v_\varphi}{\tilde{\partial}r}\frac{\tilde{\partial}p}{a\sin\theta\tilde{\partial}\varphi} + \frac{\tilde{\partial}p}{\tilde{\partial}r}\frac{\tilde{\partial}v_\varphi}{a\sin\theta\tilde{\partial}\varphi} \\
&= -\frac{\tilde{\partial}p}{\tilde{\partial}r}\left(\frac{\tilde{\partial}p}{a\sin\theta\tilde{\partial}\varphi}\frac{\partial v_\varphi}{\partial p} - \frac{\tilde{\partial}v_\varphi}{a\sin\theta\tilde{\partial}\varphi}\right) \\
&= -\frac{\tilde{\partial}p}{\tilde{\partial}r}\left[\frac{\tilde{\partial}p}{a\sin\theta\tilde{\partial}\varphi}\frac{\partial v_\varphi}{\partial p} - \frac{1}{a\sin\theta}\left(\frac{\partial}{\partial\varphi^*} + \frac{\tilde{\partial}p}{\tilde{\partial}\varphi}\frac{\partial}{\partial p}\right)v_\varphi\right] \\
&= \frac{\tilde{\partial}p}{\tilde{\partial}r}\left(\frac{\partial v_\varphi}{a\sin\theta^*\partial\varphi^*}\right).
\end{aligned}$$

In the process of verifying the first equality, we have used the relationship $p = \frac{\tilde{\mathrm{d}}p}{\tilde{\mathrm{d}}t} = \frac{\mathrm{d}p}{\mathrm{d}t^*}$. Combining the above three equations together, we deduce the continuity equation in the pressure coordinate frame from (1.3.4) and (1.3.5)

$$\frac{\partial\dot{p}}{\partial p} + \frac{1}{a\sin\theta^*}\left(\frac{\partial v_\theta \sin\theta^*}{\partial\theta^*} + \frac{\partial v_\varphi}{\partial\varphi^*}\right) = 0, \tag{1.3.6}$$

the thermodynamic equation in the pressure coordinate frame is

$$c_p\frac{\mathrm{d}T}{\mathrm{d}t^*} - \frac{RT}{p}\dot{p} = \frac{\mathrm{d}Q}{\mathrm{d}t^*}. \tag{1.3.7}$$

The equations (1.3.1)-(1.3.3), (1.3.6) and (1.3.7) are known as **the dry atmospheric equations in pressure coordinate frame,** where

$$\frac{\mathrm{d}}{\mathrm{d}t^*} = \frac{\partial}{\partial t^*} + \dot{\theta}^*\frac{\partial}{\partial\theta^*} + \dot{\varphi}^*\frac{\partial}{\partial\varphi^*} + \dot{p}\frac{\partial}{\partial p}.$$

With definitions of the substantial derivative and the hydrostatic equilibrium equation in pressure coordinate frame, we get the vertical velocity given by

$$\begin{aligned} v_r &= \frac{\mathrm{d}r}{\mathrm{d}t} = \frac{\mathrm{d}z}{\mathrm{d}t} = \frac{\partial z}{\partial t^*} + \dot{\theta}^* \frac{\partial z}{\partial \theta^*} + \dot{\varphi}^* \frac{\partial z}{\partial \varphi^*} + \dot{p}\frac{\partial z}{\partial p} \\ &= \frac{\partial z}{\partial t^*} + \frac{v_\theta}{a}\frac{\partial z}{\partial \theta^*} + \frac{v_\varphi}{a\sin\varphi^*}\frac{\partial z}{\partial \varphi^*} + \dot{p}\frac{\partial z}{\partial p} \\ &= \frac{\partial z}{\partial t^*} + \frac{v_\theta}{a}\frac{\partial z}{\partial \theta^*} + \frac{v_\varphi}{a\sin\varphi^*}\frac{\partial z}{\partial \varphi^*} - \frac{\dot{p}}{\rho g}. \end{aligned}$$

1.4 Equations of the Atmosphere in the Topography Coordinate Frame

In the practical case, researchers sometimes need to consider the variation of topography of the earth surface. As the earth surface is not an isobaric surface in this situation, we usually can't take the pressure coordinate frame, otherwise it's difficult to give a reasonable lower boundary condition. Therefore, we take the following topography coordinate $(\theta, \varphi, \pi; t)$, that is

$$t = t^*,\ \theta = \theta^*,\ \varphi = \varphi^*,\ \pi = \pi\left(\frac{p}{p_s}\right),$$

where $p_s(\theta^*, \varphi^*; t^*)$ denotes pressure of the earth surface, and π is a strictly monotonic function of $\frac{p}{p_s}$. Here $\pi(1)$ denotes the earth surface, and $\pi(0)$ denotes the upper boundary of atmosphere. Here, we suppose $\pi = \zeta = \frac{p}{p_s}$.

Then, let's deduce the form in the new topography coordinate $(\theta, \varphi, \zeta; t)$ of the equations of the atmosphere in the pressure coordinate frame $(\theta^*, \varphi^*, p; t^*)$, which appear in the above section. First, in the topography coordinate frame, the substantial derivative of any vector F is

$$\frac{\mathrm{d}F}{\mathrm{d}t} = \left(\frac{\partial}{\partial t} + \dot{\theta}\frac{\partial}{\partial \theta} + \dot{\varphi}\frac{\partial}{\partial \varphi} + \dot{\zeta}\frac{\partial}{\partial \zeta}\right)F = \left(\frac{\partial}{\partial t} + \dot{\theta}^*\frac{\partial}{\partial \theta} + \dot{\varphi}^*\frac{\partial}{\partial \varphi} + \dot{\zeta}\frac{\partial}{\partial \zeta}\right)F,$$

where

$$\dot{\zeta} = \frac{p_s\dot{p} - p\dot{p}_s}{p_s^2}.$$

Applying the relationship between the pressure coordinate frame and the topography coordinate frame, we have

$$\frac{\bar{\partial}}{\bar{\partial} p} = \frac{\bar{\partial}\zeta}{\bar{\partial} p}\frac{\partial}{\partial \zeta} = \frac{1}{p_s}\frac{\partial}{\partial \zeta}, \tag{1.4.1}$$

$$\frac{\bar{\partial}}{\bar{\partial}\theta^*} = \frac{\partial}{\partial\theta} + \frac{\bar{\partial}\zeta}{\bar{\partial}\theta^*}\frac{\partial}{\partial\zeta} = \frac{\partial}{\partial\theta} - \frac{\zeta}{p_s}\frac{\partial p_s}{\partial\theta}\frac{\partial}{\partial\zeta}, \tag{1.4.2}$$

$$\frac{\bar{\partial}}{\bar{\partial}\varphi^*} = \frac{\partial}{\partial\varphi} + \frac{\bar{\partial}\zeta}{\bar{\partial}\varphi^*}\frac{\partial}{\partial\zeta} = \frac{\partial}{\partial\varphi} - \frac{\zeta}{p_s}\frac{\partial p_s}{\partial\varphi}\frac{\partial}{\partial\zeta}. \tag{1.4.3}$$

To distinguish derivatives in the two coordinate frames, we use here $\frac{\bar{\partial}\zeta}{\bar{\partial}p}$ to denote the derivative of ζ with respect to p in the original coordinate, $\frac{\partial}{\partial\zeta}$ to denote in the new coordinate. The following symbols are defined in a similar way. With (1.4.1) and equation $p\frac{\partial\Phi}{\partial p} = -RT$, we get

$$\zeta\frac{\partial\Phi}{\partial\zeta} = -RT. \tag{1.4.4}$$

We obtain the momentum equation in the new topography coordinate frame. Here we only give the form of the pressure-gradient force. Using (1.4.2), (1.4.3) and (1.4.4), we have

$$-\frac{1}{a}\frac{\bar{\partial}\Phi}{\bar{\partial}\theta^*} = -\frac{\partial\Phi}{a\partial\theta} + \frac{\zeta}{ap_s}\frac{\partial p_s}{\partial\theta}\frac{\partial\Phi}{\partial\zeta} = -\frac{\partial\Phi}{a\partial\theta} - \frac{RT}{ap_s}\frac{\partial p_s}{\partial\theta},$$

$$-\frac{1}{a\sin\theta^*}\frac{\bar{\partial}\Phi}{\bar{\partial}\varphi^*} = -\frac{\partial\Phi}{a\sin\theta\partial\varphi} + \frac{\zeta}{a\sin\theta p_s}\frac{\partial p_s}{\partial\varphi}\frac{\partial\Phi}{\partial\varphi} = -\frac{\partial\Phi}{a\sin\theta\partial\varphi} - \frac{RT}{a\sin\theta p_s}\frac{\partial p_s}{\partial\varphi}.$$

According to the above two equations, we get the horizontal momentum equation in the new topography coordinate frame as

$$\frac{\mathrm{d}v_\theta}{\mathrm{d}t} - \frac{1}{a}v_\varphi^2\cot\theta = -\left(\frac{\partial\Phi}{a\partial\theta} + \frac{RT}{ap_s}\frac{\partial p_s}{\partial\theta}\right) + 2\Omega\cos\theta v_\varphi + D_\theta, \tag{1.4.5}$$

$$\frac{\mathrm{d}v_\varphi}{\mathrm{d}t} + \frac{1}{a}v_\theta v_\varphi\cot\theta = -\left(\frac{\partial\Phi}{a\sin\theta\partial\varphi} + \frac{RT}{a\sin\theta p_s}\frac{\partial p_s}{\partial\varphi}\right) - 2\Omega\cos\theta v_\theta + D_\varphi. \tag{1.4.6}$$

Next, let's deduce the mass conservation equation in the new topography coordinate frame. With $\dot{\zeta} = \frac{p_s\dot{p} - p\dot{p}_s}{p_s^2}$, we have

$$\dot{p} = \dot{\zeta}p_s + \zeta\dot{p}_s.$$

Applying (1.4.1) and the above equation, we get

$$\frac{\bar{\partial}\dot{p}}{\bar{\partial}p} = \frac{\bar{\partial}\dot{\zeta}p_s}{\bar{\partial}p} + \frac{\bar{\partial}\zeta\dot{p}_s}{\bar{\partial}p} = \frac{\partial\dot{\zeta}}{\partial\zeta} + \frac{\dot{p}_s}{p_s} + \frac{\zeta}{p_s}\frac{\partial\dot{p}_s}{\partial\zeta}. \tag{1.4.7}$$

With (1.4.2) and (1.4.3), we have

$$\begin{aligned}
&\frac{1}{a\sin\theta^*}\left(\frac{\partial v_\theta \sin\theta^*}{\partial\theta^*}+\frac{\partial v_\varphi}{\partial\varphi^*}\right)\\
=&\frac{1}{a\sin\theta}\left[\left(\frac{\partial v_\theta\sin\theta}{\partial\theta}-\frac{\zeta}{p_s}\frac{\partial p_s}{\partial\theta}\frac{\partial v_\theta\sin\theta}{\partial\zeta}\right)+\left(\frac{\partial v_\varphi}{\partial\varphi}-\frac{\zeta}{p_s}\frac{\partial p_s}{\partial\varphi}\frac{\partial v_\varphi}{\partial\zeta}\right)\right]\\
=&\frac{1}{a\sin\theta}\left(\frac{\partial v_\theta\sin\theta}{\partial\theta}+\frac{\partial v_\varphi}{\partial\varphi}\right)-\frac{\zeta}{a\sin\theta p_s}\left(\frac{\partial p_s}{\partial\theta}\frac{\partial v_\theta\sin\theta}{\partial\zeta}+\frac{\partial p_s}{\partial\varphi}\frac{\partial v_\varphi}{\partial\zeta}\right).
\end{aligned}\tag{1.4.8}$$

Since $p_s(\theta^*,\varphi^*;t^*)=p_s(\theta,\varphi;t)$, we get the following system by the definition of substantial derivative in the topography coordinate frame

$$\dot{p_s}=\frac{\partial p_s}{\partial t}+\frac{v_\theta}{a}\frac{\partial p_s}{\partial\theta}+\frac{v_\varphi}{a\sin\theta}\frac{\partial p_s}{\partial\varphi},\tag{1.4.9}$$

$$\frac{\zeta}{p_s}\frac{\partial\dot{p_s}}{\partial\zeta}=\frac{\zeta}{p_s}\left(\frac{\partial p_s}{a\partial\theta}\frac{\partial v_\theta}{\partial\zeta}+\frac{\partial p_s}{a\sin\theta\partial\varphi}\frac{\partial v_\varphi}{\partial\zeta}\right).\tag{1.4.10}$$

Substituting (1.4.7) and (1.4.8) into (1.3.6), and applying (1.4.9) and (1.4.10), we deduce the continuity equation in the new topography coordinate frame as

$$\frac{\partial p_s}{\partial t}=-\frac{\partial p_s\dot{\zeta}}{\partial\zeta}-\frac{1}{a\sin\theta}\left(\frac{\partial p_s v_\theta\sin\theta}{\partial\theta}+\frac{\partial p_s v_\varphi}{\partial\varphi}\right).\tag{1.4.11}$$

According to $\dot{p}=\dot{\zeta}p_s+\zeta\dot{p_s}$, we obtain the thermodynamic equation in the new topography coordinate frame

$$c_p\frac{\mathrm{d}T}{\mathrm{d}t}-\frac{RT}{\zeta p_s}(\dot{\zeta}p_s+\zeta\dot{p_s})=\frac{\mathrm{d}Q}{\mathrm{d}t}.\tag{1.4.12}$$

We denote equations (1.4.4)-(1.4.6), (1.4.11) and (1.4.12) as **the dry atmospheric equations in topography coordinate frame,** where

$$\frac{\mathrm{d}}{\mathrm{d}t}=\frac{\partial}{\partial t}+\dot{\theta}\frac{\partial}{\partial\theta}+\dot{\varphi}\frac{\partial}{\partial\varphi}+\dot{\zeta}\frac{\partial}{\partial\zeta}=\frac{\partial}{\partial t}+\frac{v_\theta}{a}\frac{\partial}{\partial\theta}+\frac{v_\varphi}{a\sin\theta}\frac{\partial}{\partial\varphi}+\dot{\zeta}\frac{\partial}{\partial\zeta}.$$

When studying the atmospheric motions below specific barometric altitude (supposing the upper bound of atmospheric pressure is $p=p_0$, where p_0 is a positive constant), researchers use the modified topography coordinate frame $(\theta,\varphi,\pi;t)$, where

$$t=t^*,\ \theta=\theta^*,\ \varphi=\varphi^*,\ \pi=\pi(\eta),\ \eta=\frac{p-p_0}{p_s-p_0}.$$

With the above methods, we can also deduce the atmospheric equations in the modified topography coordinate frame.

1.5 Equations of the Atmosphere and the Oceans in Local Rectangular Coordinate Frame under β-Plane Approximation

For the oceanic motions, researchers sometimes are concerned with some local areas containing no polar region. One can simplify the local earth surface into a flat plane, then choose the local rectangular coordinate frame. In the same way, studying the properties of local atmospheric motions, we can also choose local rectangular coordinate frame. Now let's discuss the oceanic equations in local rectangular coordinate frame.

Let O as a point on the sea level (its co-latitude is θ_0), the forward direction of x-axis point to the south, y-axis to the east, z-axis vertically upward, $\boldsymbol{e}_x$, $\boldsymbol{e}_y$ and $\boldsymbol{e}_z$ in the three-coordinate direction of local rectangular coordinate frame $\{O; x, y, z\}$ are respectively unit constant vectors in x, y, z axis directions. According to the definition of velocity, the velocity $\boldsymbol{V}$ is expressed as

$$\boldsymbol{V} = u\boldsymbol{e}_x + v\boldsymbol{e}_y + w\boldsymbol{e}_z,$$

where

$$u = \frac{\mathrm{d}x}{\mathrm{d}t}, \quad v = \frac{\mathrm{d}y}{\mathrm{d}t}, \quad w = \frac{\mathrm{d}z}{\mathrm{d}t}.$$

In the local rectangular coordinate frame, the total derivative of any vector F is

$$\begin{aligned}&\frac{\mathrm{d}F}{\mathrm{d}t}\\ &= \lim_{\Delta t \to 0} \frac{1}{\Delta t}[F(t+\Delta t, x(t+\Delta t), y(t+\Delta t), z(t+\Delta t)) - F(t, x(t), y(t), z(t))]\\ &= \left(\frac{\partial}{\partial t} + \dot{x}\frac{\partial}{\partial x} + \dot{y}\frac{\partial}{\partial y} + \dot{z}\frac{\partial}{\partial z}\right)F = \left(\frac{\partial}{\partial t} + u\frac{\partial}{\partial x} + v\frac{\partial}{\partial y} + w\frac{\partial}{\partial z}\right)F.\end{aligned}$$

As in the local rectangular coordinate frame, the gradient operator $\boldsymbol{\nabla}_3 = \boldsymbol{e}_x\frac{\partial}{\partial x} + \boldsymbol{e}_y\frac{\partial}{\partial y} + \boldsymbol{e}_z\frac{\partial}{\partial z}$, the total derivative in the local rectangular coordinate frame is

$$\frac{\mathrm{d}}{\mathrm{d}t} = \frac{\partial}{\partial t} + \boldsymbol{V} \cdot \boldsymbol{\nabla}_3.$$

To acquire the oceanic equations in the local rectangular coordinate system, we have to find the approximate form of Coriolis force $-2\boldsymbol{\Omega} \times \boldsymbol{V}$ in the local rectangular coordinate frame. In the sphere coordinate frame,

$$-2\boldsymbol{\Omega} \times \boldsymbol{V} = 2\Omega \cos\theta v \boldsymbol{e}_\theta + (-2\Omega\cos\theta u - 2\Omega \sin\theta w)\boldsymbol{e}_\varphi + 2\Omega \sin\theta v \boldsymbol{e}_r.$$

Because $2\Omega\sin\theta w \ll 2\Omega\cos\theta u$, $2\Omega\sin\theta v \ll g$, in general situations, one can omit $2\Omega\sin\theta w$, $2\Omega\sin\theta v$. Thus, we just have to find the approximate forms of $2\Omega\cos\theta v$, $2\Omega\cos\theta u$ in the local rectangular coordinate frame, that is, we should deal with the Coriolis parameter $f = 2\Omega\cos\theta$. Using the Taylor expansion of $f = 2\Omega\cos\theta$ at θ_0, we get

$$f = 2\Omega\cos\theta = 2\Omega[\cos\theta_0-(\theta-\theta_0)\sin\theta_0-(\theta-\theta_0)^2\frac{\cos\theta_0}{2!}+(\theta-\theta_0)^3\frac{\sin\theta_0}{3!}+\cdots].$$

When $\theta-\theta_0$ is small enough, there is

$$\theta-\theta_0 \approx \frac{x}{a},$$

where a is the earth radius. When $\frac{y}{a} \ll 1$, we take

$$f \approx f_0 = 2\Omega\cos\theta_0,$$

which means, the Coriolis parameter f can be considered as a constant. So, when $\frac{y}{a} < 1$, and $(\frac{y}{a})^2 \ll 1$. So we can take

$$f \approx 2\Omega\cos\theta_0 - x\frac{2\Omega\sin\theta_0}{a} = f_0 - \beta_0 x.$$

Here $\beta_0 = \dfrac{2\Omega\sin\theta_0}{a}$ is the origin value of Rossby parameter $\beta = \dfrac{2\Omega\sin\theta}{a}$ in the local rectangular coordinate frame. The above equation is usually called β-plane approximation. The Coriolis parameter can be considered as a linear function of x in the local rectangular coordinate frame.

Remark 1.5.1. If O is a point on the sea level (its latitude is $\phi_0 = \frac{\pi}{2} - \theta_0$), the forward direction of x-axis point to the east, y-axis to the north, z-axis vertically upward, then the β-plane approximation is given by

$$f = 2\Omega\sin\phi \approx 2\Omega\sin\phi_0 + y\frac{2\Omega\cos\phi_0}{a} = f_0 + \beta_0 y.$$

After taking Boussinesq approximation (except buoyancy term and density in the state equation, densities of other position are all considered constant) and β-plane approximation, we get oceanic equations in the local rectangular coordinate frame

$$\frac{\mathrm{d}u}{\mathrm{d}t} = -\frac{1}{\rho_0}\frac{\partial p}{\partial x} + f_\beta v + D_u, \tag{1.5.1}$$

$$\frac{\mathrm{d}v}{\mathrm{d}t} = -\frac{1}{\rho_0}\frac{\partial p}{\partial y} - f_\beta u + D_v, \tag{1.5.2}$$

$$\frac{\mathrm{d}w}{\mathrm{d}t} = -\frac{1}{\rho_0}\frac{\partial p}{\partial z} - \frac{\rho}{\rho_0}g + D_w, \tag{1.5.3}$$

$$\frac{\partial u}{\partial x} + \frac{\partial v}{\partial y} + \frac{\partial w}{\partial z} = 0, \tag{1.5.4}$$

$$\rho = \rho_0[1 - \beta_T(T - T_0) + \beta_S(S - S_0)], \tag{1.5.5}$$

$$\frac{\mathrm{d}T}{\mathrm{d}t} = Q_1, \tag{1.5.6}$$

$$\frac{\mathrm{d}S}{\mathrm{d}t} = Q_2, \tag{1.5.7}$$

where $f_\beta = f_0 - \beta_0 x$, $D = (D_u, D_v, D_w)$ is the viscosity term, and

$$\frac{\mathrm{d}}{\mathrm{d}t} = \frac{\partial}{\partial t} + u\frac{\partial}{\partial x} + v\frac{\partial}{\partial y} + w\frac{\partial}{\partial z}.$$

In the same way, one can obtain the equations of the dry atmosphere (without hydrostatic approximation) in the local rectangular coordinate frame

$$\frac{\mathrm{d}u}{\mathrm{d}t} = -\frac{1}{\rho}\frac{\partial p}{\partial x} + f_\beta v + D_u, \tag{1.5.8}$$

$$\frac{\mathrm{d}v}{\mathrm{d}t} = -\frac{1}{\rho}\frac{\partial p}{\partial y} - f_\beta u + D_v, \tag{1.5.9}$$

$$\frac{\mathrm{d}w}{\mathrm{d}t} = -\frac{1}{\rho}\frac{\partial p}{\partial z} - g + D_w, \tag{1.5.10}$$

$$\frac{\mathrm{d}\rho}{\mathrm{d}t} + \rho(\frac{\partial u}{\partial x} + \frac{\partial v}{\partial y} + \frac{\partial w}{\partial z}) = 0, \tag{1.5.11}$$

$$p = R\rho T, \tag{1.5.12}$$

$$c_p\frac{\mathrm{d}T}{\mathrm{d}t} - \frac{RT}{p}\frac{\mathrm{d}p}{\mathrm{d}t} = \frac{\mathrm{d}Q}{\mathrm{d}t}, \tag{1.5.13}$$

where $f_\beta = f_0 - \beta_0 x$, $D = (D_u, D_v, D_w)$ is the viscosity term.

1.6 Equations of the Atmosphere and the Oceans under Satification Approximation

The inhomogeneous heating of sun to earth causes noticeable density variance of the atmosphere and oceans, thus the density statification in the atmosphere and oceans. One observational property of statification is that, in general, the large-scale atmosphere and oceans are always gravity stable, that is, the lighter fluids are always above the heavier fluids. The atmospheric density is almost monotonic decreasing with the height, and the ocean density is almost monotonic increasing with the depth. The existence of stable statification in the atmosphere and oceans, which results in the inhibition of vertical motion, contributes to the formation of the almost horizontal motions.

Basing on the existence of stable statification in the atmosphere, researchers can think that the large-scale atmospheric motions are that the oscillations occur near the average state $((v_\theta, v_\varphi, \dot{p}, \varPhi, T) = (0,0,0, \bar{\varPhi}(p), \bar{T}(p))$, also called standard state), where $\bar{\varPhi}(p)$ and $\bar{T}(p)$ satisfy the hydrostatic equilibrium relationship, that is,

$$p\frac{\partial \bar{\varPhi}(p)}{\partial p} = -R\bar{T}(p).$$

Next, let's deduce the equations for perturbation states near the average state $(0,0,0, \bar{\varPhi}(p), \bar{T}(p))$ from (1.3.1)–(1.3.3), (1.3.6) and (1.3.7). Suppose

$$(v_\theta, v_\varphi, \dot{p}) = (0 + v'_\theta, 0 + v'_\varphi, 0 + \dot{p}'),$$

$$\varPhi = \bar{\varPhi}(p) + \varPhi', \quad T = \bar{T}(p) + T'.$$

The equations (1.3.1)-(1.3.3) and (1.3.6) are written as

$$\frac{\mathrm{d}v'_\theta}{\mathrm{d}t} - \frac{1}{a}v'^2_\varphi \cot\theta^* = -\frac{1}{a}\frac{\partial \varPhi'}{\partial \theta^*} + 2\varOmega\cos\theta^* v'_\varphi + D_\theta, \tag{1.6.1}$$

$$\frac{\mathrm{d}v'_\varphi}{\mathrm{d}t} + \frac{1}{a}v'_\theta v'_\varphi \cot\theta^* = -\frac{1}{a\sin\theta^*}\frac{\partial \varPhi'}{\partial \varphi^*} - 2\varOmega\cos\theta^* v'_\theta + D_\varphi, \tag{1.6.2}$$

$$p\frac{\partial \varPhi'}{\partial p} = -RT', \tag{1.6.3}$$

$$\frac{\partial \dot{p}'}{\partial p} + \frac{1}{a\sin\theta^*}\left(\frac{\partial v'_\theta \sin\theta^*}{\partial \theta^*} + \frac{\partial v'_\varphi}{\partial \varphi^*}\right) = 0. \tag{1.6.4}$$

Substituting $T = \bar{T}(p) + T'$ into equation (1.3.7), we have

$$c_p\frac{\mathrm{d}T'}{\mathrm{d}t} + c_p\frac{\partial \bar{T}(p)}{\partial p}\dot{p} - \frac{R(\bar{T}(p) + T')}{p}\dot{p} = \frac{\mathrm{d}Q}{\mathrm{d}t}.$$

As $|T - T'| \ll \frac{p}{R\dot{p}}$, we take $\frac{R(\bar{T}(p)+T')}{p}\dot{p} \approx \frac{R\bar{T}(p)}{p}\dot{p}$. Thus

$$c_p\frac{\mathrm{d}T'}{\mathrm{d}t} + \left(c_p\frac{\partial \bar{T}(p)}{\partial p} - \frac{R\bar{T}(p)}{p}\right)\dot{p} = \frac{\mathrm{d}Q}{\mathrm{d}t}.$$

If $\bar{T}(p)$ satisfies $R\left(\frac{R\bar{T}(p)}{C_p} - p\frac{\partial \bar{T}(p)}{\partial p}\right) = C^2$, where C is a positive constant, then

$$c_p\frac{\mathrm{d}T'}{\mathrm{d}t} - \frac{c_pC^2}{pR}\dot{p} = \frac{\mathrm{d}Q}{\mathrm{d}t}. \tag{1.6.5}$$

Getting rid of "′" and "*", introducing viscosity to the equation, and letting $\frac{\mathrm{d}Q}{\mathrm{d}t} = \mu_2 \Delta T + \nu_2 \frac{\partial}{\partial p}[(\frac{gp}{R\bar{T}})^2 \frac{\partial T}{\partial p}] + F(\theta, \varphi, p)$, we write equations (1.6.1)-(1.6.5) as

$$\frac{\partial \boldsymbol{v}}{\partial t} + \boldsymbol{\nabla}_{\boldsymbol{v}} \boldsymbol{v} + \omega \frac{\partial \boldsymbol{v}}{\partial p} + f \boldsymbol{k} \times \boldsymbol{v} + \mathrm{grad}\, \varPhi - \mu_1 \Delta \boldsymbol{v} - \nu_1 \frac{\partial}{\partial p}\left[\left(\frac{gp}{R\bar{T}}\right)^2 \frac{\partial \boldsymbol{v}}{\partial p}\right] = 0, \tag{1.6.6}$$

$$\mathrm{div} \boldsymbol{v} + \frac{\partial \omega}{\partial p} = 0, \tag{1.6.7}$$

$$\frac{\partial \varPhi}{\partial p} + \frac{bP}{p} T = 0, \tag{1.6.8}$$

$$\frac{R^2}{C^2}\left(\frac{\partial T}{\partial t} + \boldsymbol{\nabla}_{\boldsymbol{v}} T + \omega \frac{\partial T}{\partial p}\right) - \frac{R}{p}\omega - \mu_2 \Delta T - \nu_2 \frac{\partial}{\partial p}\left[\left(\frac{gp}{R\bar{T}}\right)^2 \frac{\partial T}{\partial p}\right] = F, \tag{1.6.9}$$

where $\omega = \dot{p}$,

$$\begin{aligned}
\boldsymbol{\nabla}_{\boldsymbol{v}} \widetilde{\boldsymbol{v}} = &\left(\frac{v_\theta}{a} \frac{\partial \widetilde{v_\theta}}{\partial \theta} + \frac{v_\varphi}{a \sin\theta} \frac{\partial \widetilde{v_\theta}}{\partial \varphi} - \frac{v_\varphi \widetilde{v_\varphi}}{a} \cot\theta\right) \boldsymbol{e}_\theta \\
&+ \left(\frac{v_\theta}{a} \frac{\partial \widetilde{v_\varphi}}{\partial \theta} + \frac{v_\varphi}{a \sin\theta} \frac{\partial \widetilde{v_\varphi}}{\partial \varphi} + \frac{v_\varphi \widetilde{v_\theta}}{a} \cot\theta\right) \boldsymbol{e}_\varphi, \\
\mathrm{grad}\, \varPhi = &\frac{\partial \varPhi}{a \partial \theta} \boldsymbol{e}_\theta + \frac{1}{a \sin\theta} \frac{\partial \varPhi}{\partial \varphi} \boldsymbol{e}_\varphi, \\
\Delta \boldsymbol{v} = &\left(\Delta v_\theta - \frac{2\cos\theta}{a^2 \sin^2\theta} \frac{\partial v_\varphi}{\partial \varphi} - \frac{v_\theta}{a^2 \sin^2\theta}\right) \boldsymbol{e}_\theta \\
&+ \left(\Delta v_\varphi + \frac{2\cos\theta}{a^2 \sin^2\theta} \frac{\partial v_\theta}{\partial \varphi} - \frac{v_\varphi}{a^2 \sin^2\theta}\right) \boldsymbol{e}_\varphi, \\
\mathrm{div} \boldsymbol{v} = &\mathrm{div}\,(v_\theta \boldsymbol{e}_\theta + v_\varphi \boldsymbol{e}_\varphi) = \frac{1}{a \sin\theta}\left(\frac{\partial v_\theta \sin\theta}{\partial \theta} + \frac{\partial v_\varphi}{\partial \varphi}\right), \\
\boldsymbol{\nabla}_{\boldsymbol{v}} T = &\frac{v_\theta}{a} \frac{\partial T}{\partial \theta} + \frac{v_\varphi}{a \sin\theta} \frac{\partial T}{\partial \varphi}, \\
\Delta T = &\frac{1}{a^2 \sin\theta}\left[\frac{\partial}{\partial \theta}\left(\sin\theta \frac{\partial T}{\partial \theta}\right) + \frac{1}{\sin\theta} \frac{\partial^2 T}{\partial \varphi^2}\right],
\end{aligned}$$

where $\boldsymbol{v} = v_\theta \boldsymbol{e}_\theta + v_\varphi \boldsymbol{e}_\varphi$, and $\widetilde{\boldsymbol{v}} = \widetilde{v_\theta} \boldsymbol{e}_\theta + \widetilde{v_\varphi} \boldsymbol{e}_\varphi$. In the atmosphere science, the equations (1.6.6)–(1.6.9) are also called the dry atmospheric primitive equations.

For the ocean equations (1.5.1)-(1.5.7) in Boussinesq approximation, ignoring salinity, taking $\rho = \rho_0[1 - \beta_T(T - T_0)]$, and introducing viscosity, we have the following equations

$$\begin{aligned}
\frac{\mathrm{d}u}{\mathrm{d}t} &= -\frac{1}{\rho_0}\frac{\partial p}{\partial x} + f_\beta v + \nu\Delta u,\\
\frac{\mathrm{d}v}{\mathrm{d}t} &= -\frac{1}{\rho_0}\frac{\partial p}{\partial y} - f_\beta u + \nu\Delta v,\\
\frac{\mathrm{d}w}{\mathrm{d}t} &= -\frac{1}{\rho_0}\frac{\partial p}{\partial z} - \frac{\rho}{\rho_0}g + \nu\Delta w,\\
\frac{\partial u}{\partial x} &+ \frac{\partial v}{\partial y} + \frac{\partial w}{\partial z} = 0,\\
\frac{\mathrm{d}\rho}{\mathrm{d}t} &= \kappa\Delta\rho.
\end{aligned}$$

$(u, v, w, p, \rho) = (0, 0, 0, \bar{p}(z), \bar{\rho}(z))$ is a solution of the above equations, which is always denoted as an average state (or a standard state). Where $\bar{p}(z)$ and $\bar{\rho}(z)$ satisfy the hydrostatic equilibrium relationship, that is,

$$\frac{\partial\bar{p}(z)}{\partial z} = -g\bar{\rho}(z).$$

Researchers can consider motions near the average state $(0, 0, 0, \bar{p}(z), \bar{\rho}(z))$. Next, let's deduce equations for motions perturbed near the average state $(0, 0, 0, \bar{p}(z), \bar{\rho}(z))$. Supposing

$$(u, v, w) = (0 + u', 0 + v', 0 + w'), \quad p = \bar{p}(z) + p', \quad \rho = \bar{\rho}(z) + \rho',$$

we get the equations of oceans as follows

$$\frac{\mathrm{d}u'}{\mathrm{d}t} = -\frac{1}{\rho_0}\frac{\partial p'}{\partial x} + f_\beta v' + \nu\Delta u', \tag{1.6.10}$$

$$\frac{\mathrm{d}v'}{\mathrm{d}t} = -\frac{1}{\rho_0}\frac{\partial p'}{\partial y} - f_\beta u' + \nu\Delta v', \tag{1.6.11}$$

$$\frac{\mathrm{d}w'}{\mathrm{d}t} = -\frac{1}{\rho_0}\frac{\partial p'}{\partial z} - \frac{\rho'}{\rho_0}g + \nu\Delta w', \tag{1.6.12}$$

$$\frac{\partial u'}{\partial x} + \frac{\partial v'}{\partial y} + \frac{\partial w'}{\partial z} = 0, \tag{1.6.13}$$

$$\frac{\mathrm{d}\rho'}{\mathrm{d}t} + \frac{\mathrm{d}\bar{\rho}}{\mathrm{d}z}w = \kappa\Delta\rho' + \kappa\frac{\mathrm{d}^2\bar{\rho}}{\mathrm{d}z^2}. \tag{1.6.14}$$

If we take the static approximation, that is, substituting $\frac{\partial P'}{\partial Z} = \rho' g$ for (1.6.12), and the equations (1.6.10)–(1.6.14) are called the ocean primitive equations in the rectangular coordinate frame.

1.7 Boundary Conditions

The atmospheric and oceanic motions can not only be described by the equations listed above, but also are under the influence of boundary. So, in this section we discuss the boundary conditions of the atmosphere and oceans.

1.7.1 *The Lower Boundary Conditions of the Atmosphere*

In the pressure coordinate frame, if the lower interface of the atmosphere $p = P$ (the approximate value of earth surface, seen as a constant) is considered as ideal rigid body, and it is a material surface, then the normal velocity of the air is zero, that is,

$$\dot{p}|_{p=P} = 0. \tag{1.7.1a}$$

In the same way, taking the spherical coordinate frame, and setting $z = r - a$, z standing for the elevation, where a is the earth radius, we can assume that the lower boundary condition of the atmosphere is

$$v_r|_{z=0} = 0. \tag{1.7.1b}$$

In the topography coordinate frame, the lower boundary condition of the atmosphere is

$$\dot{\zeta}|_{\zeta=1} = 0, \tag{1.7.1c}$$

where $\zeta = \frac{p}{p_s}$. If the viscosity of the lower boundary surface $p = P$ is considered, then the velocity is zero, that is,

$$v_\theta|_{p=P} = 0, \ v_\varphi|_{p=P} = 0, \ \dot{p}|_{p=P} = 0. \tag{1.7.2a}$$

In the spherical coordinate frame, if the viscosity of the lower boundary surface $z = 0$ is considered, then the lower boundary conditions of the atmosphere are

$$v_\theta|_{z=0} = 0, \ v_\varphi|_{z=0} = 0, \ v_r|_{z=0} = 0. \tag{1.7.2b}$$

In the same way, in the topography coordinate frame, the lower boundary conditions of the atmosphere are

$$v_\theta|_{\zeta=1} = 0, \ v_\varphi|_{\zeta=1} = 0, \ \dot{\zeta}|_{\zeta=1} = 0. \tag{1.7.2c}$$

(1.7.1a)-(1.7.2c) are generally called **kinematic boundary conditions**.

In the local rectangular coordinate frame, if the vertical motion of the lower boundary surface is caused by the force of landform $z = h(x, y)$, then the lower boundary condition is given by

$$w|_{z=h(x,y)} = \frac{\mathrm{d}h}{\mathrm{d}t} = u\frac{\partial h}{\partial x} + v\frac{\partial h}{\partial y}.$$

In the spherical coordinate frame, if the lower boundary surface is the surface of the ocean, of which pressure is $p_0(\theta, \varphi, t)$, then the lower boundary condition of pressure p is given by

$$p(\theta, \varphi, z, t)|_{z=0} = p_0(\theta, \varphi, t). \tag{1.7.3a}$$

If the lower boundary surface has the topography $h_s(\theta, \varphi)$, of which pressure is $p_s(\theta, \varphi, t)$, the lower boundary condition of pressure p is

$$p(\theta, \varphi, z, t)|_{z=h_s(\theta,\varphi)} = p_s(\theta, \varphi, t). \tag{1.7.3b}$$

(1.7.3a) and (1.7.3b) are denoted as **dynamic boundary conditions**.

In the pressure coordinate frame, the thermodynamic condition of the lower boundary surface $p = P$ can be briefly written as

$$\frac{\partial T}{\partial p}|_{p=P} = -\alpha_s(T - T_s),$$

where α_s is a parameter about turbulent thermal conductivity, which relies on the properties of the lower surface, and T_s is the reference temperature of the lower surface.

In the pressure coordinate frame, for $\varPhi$ in (1.3.3), the geometrical condition of the lower boundary surface is

$$\varPhi|_{p=P} = \varPhi_s(\theta, \varphi, t).$$

1.7.2 *The Upper Boundary Conditions of the Atmosphere*

First, in the real atmosphere, the upper boundary ($z \to +\infty$) should satisfy

$$\lim_{z \to +\infty} p = 0.$$

Second, since the total energy of vertical air column per unit section is bounded, we get condition

$$\int_0^{+\infty} \left(\frac{v_\theta^2 + v_\varphi^2}{2} + c_v T + gz \right) \rho \mathrm{d}z < +\infty.$$

Thus,

$$\rho v_\theta^2, \rho v_\varphi^2, \rho T, \rho z \to 0, \text{ as } z \to +\infty. \tag{1.7.4}$$

(1.7.4) is generally called a physical boundary condition.

In the practical case, researchers always use homogeneous atmosphere models, such as barotropic models (shallow water models), two-dimensional quasi-geostrophic models and multi-dimensional quasi-geostrophic models. In these models, researchers divide atmosphere into some layers, which are incompressible homogeneous fluids, where the upper surface of each layer is free surface, denoted by $z = h(x, y, t)$ in the local rectangular coordinate frame. The boundary condition on that free surface is given by

$$w|_{z=h} = \frac{\mathrm{d}h}{\mathrm{d}t} = \frac{\partial h}{\partial t} + u\frac{\partial h}{\partial x} + v\frac{\partial h}{\partial y},$$

where w is velocity in the direction of z, and u, v are respectively velocities in the direction of x, y. For the interface of two layers $z = h(x, y, t)$, the boundary condition on the interface is given by

$$w_i|_{z=h} = \frac{\mathrm{d}h}{\mathrm{d}t} = \frac{\partial h}{\partial t} + u_i\frac{\partial h}{\partial x} + v_i\frac{\partial h}{\partial y},$$

where $i = 1, 2$. This condition is called a kinematic condition of interface.

1.7.3 *The Boundary Conditions of the Oceans*

In the local rectangular coordinate frame, boundary conditions of the sea level are given by

$$\frac{\partial \boldsymbol{v}}{\partial z}|_{z=0} = h\tau, \ w|_{z=0} = 0, \tag{1.7.5}$$

$$\frac{\partial T}{\partial z}|_{z=0} = -\alpha(T - T^*), \tag{1.7.6}$$

where $\boldsymbol{v}$ is the horizontal velocity, w denotes velocity in the direction of z, h is the depth of ocean (supposed as constant), τ is the wind stress of the sea level, and T^* is the sea water parameter temperature at the sea level. The first equation of (1.7.5) indicates the effect of wind stress of the ocean surface to the sea water, and (1.7.6) indicates the energy alternation at the sea level. The boundary conditions of the lateral ocean can be written as

$$\boldsymbol{v} \cdot \boldsymbol{n} = 0, \frac{\partial \boldsymbol{v}}{\partial \boldsymbol{n}} \times \boldsymbol{n} = 0, \tag{1.7.7}$$

$$\frac{\partial T}{\partial \boldsymbol{n}} = 0. \tag{1.7.8}$$

Here $\boldsymbol{n}$ denotes the external normal vector of the lateral ocean. (1.7.7) indicates that the normal component of lateral horizontal velocity is zero, and horizontal velocity here is no-ship at the same time. (1.7.8) indicates that there is no lateral heat flux. The boundary conditions of the bottom ocean surface can be written as

$$\frac{\partial \boldsymbol{v}}{\partial z}|_{z=-h} = 0, w|_{z=-h} = 0, \frac{\partial T}{\partial z} = 0.$$

Chapter 2

Some Quasi-Geostrophic Models

As the basic equations and the primitive equations which describe the motions of the atmosphere and oceans are very complicated, researchers should find some models which are simplier but reflect some physical essences. In 1947, Charney established a quasi-geostrophic model in [32], which plays a crucial role in the first 24-hour numerical weather forecast on the ENIAV computer of Institute of Advanced Study in Princeton.

Although with the vigorous development of atmospheric science and the enhancing of data dealing ability and numerical calculation ability of large-scale computer, researchers turn to numerical weather prediction by the primitive equation models from 1960s. But the quasi-geostrophic models still have important effect on the study and application of the atmospheric and oceanic dynamics. Therefore, we introduce various quasi-geostrophic models such as two-dimensional and three-dimensional quasi-geostrophic models, multi-dimensional quasi-geostrophic model and surface quasi-geostrophic model.

The research of two-dimensional and three-dimensional quasi-geostrophic models stresses on some features of the atmosphere and oceans. Some of these models have been the subject of mathematical study, see [220, Chapter 12] and [12,13,66,164,165,209,210,213,214]. Besides the nonlinear stability of the two-dimensional and three-dimensional quasi-geostrophic flows, researchers study the mathematical validity of the two-dimensional and three-dimensional quasi-geostrophic models, see [146,153,164,165,167,218] and references therein.

To describe the potential temperature (or buoyancy) evolution of the lower surface of identical potential vorticity field ($q = 0$), researchers obtain the surface quasi-geostrophic equation, see [100]. Constantin et al., studying the surface quasi-geostrophic equation in [41], unveiled the important similarities between equation (2.4.6) and the three-dimensional incompressible Euler equation, and posed a conjecture of the existence of singular solutions of the surface quasi-geostrophic equation. The well-posedness problem of the surface quasi-geostrophic equation has become a subject of mathematical study afterward. For the related works, see [2,41,42,45,46,117,216,217] and references therein.

2.1 The Barotropic Model and the Two-Dimensional Quasi-Geostrophic Equation

2.1.1 *The Barotropic Model*

The barotropic model, which is also called a shallow water model, is a model describing dynamics of homogeneous incompressible inviscid rotating shallow fluid with a free surface. This model can describe some important features of large-scale atmospheric and oceanic motions. Applying the two-dimensional quasi-geostrophic equation based on the barotropic model, Charney, Fjörtaft and Von Neumann successfully made true numerical weather forecast on the ENIAV computer of the Institute for Advanced Studies in Princeton, which is considered as a great revolution of weather forecast.

Now let's deduce the barotropic model. Suppose that the height of a free surface of homogeneous incompressible inviscid rotating shallow fluid is $h(x, y, t)$ which is relative to the reference plane $z = 0$, and the rigid lower boundary is given by a surface $h_b(x, y)$. Assuming that the vertical characteristic length scale of motion be D, and the horizontal characteristic length scale of motion be L, and we consider the linear flow, we suppose further

$$\frac{D}{L} \ll 1.$$

Assuming that the horizontal velocity characteristic scale is U, and the vertical velocity characteristic scale is W, according to the continuity equation of incompressible fluid, we have

$$\frac{\partial u}{\partial x} + \frac{\partial v}{\partial y} + \frac{\partial w}{\partial z} = 0,$$

where (u, v) is the horizontal velocity, w is the vertical velocity, and we get

$$\frac{W}{U} \ll 1.$$

Making scale analysis of the momentum equations, we know that it's suitable to use the hydrostatic equilibrium equation to substitute the vertical momentum equation. Then we write the barotropic model as

$$\frac{\partial u}{\partial t} + u\frac{\partial u}{\partial x} + v\frac{\partial u}{\partial y} + w\frac{\partial u}{\partial z} - fv = -\frac{1}{\rho}\frac{\partial p}{\partial x}, \tag{2.1.1}$$

$$\frac{\partial v}{\partial t} + u\frac{\partial v}{\partial x} + v\frac{\partial v}{\partial y} + w\frac{\partial v}{\partial z} + fu = -\frac{1}{\rho}\frac{\partial p}{\partial y}, \tag{2.1.2}$$

$$\frac{\partial p}{\partial z} = -\rho g, \tag{2.1.3}$$

$$\frac{\partial u}{\partial x} + \frac{\partial v}{\partial y} + \frac{\partial w}{\partial z} = 0, \tag{2.1.4}$$

where ρ is a constant. As the upper surface of fluid is a free surface, and the lower surface is a rigid surface, the boundary conditions are given by

$$w|_{z=h} = \frac{\mathrm{d}h}{\mathrm{d}t} = \frac{\partial h}{\partial t} + u\frac{\partial h}{\partial x} + v\frac{\partial h}{\partial y},$$

$$w|_{z=h_b} = \frac{\mathrm{d}h_b}{\mathrm{d}t} = u\frac{\partial h_b}{\partial x} + v\frac{\partial h_b}{\partial y}.$$

Next, we shall reduce equations (2.1.1)-(2.1.4). Integrating equation (2.1.3) with respect to z from z to h, we get

$$p(x, y, z, t) = \rho g(h - z) + p_h,$$

where p_h is a constant, denoted the pressure at the free surface. Thus

$$\frac{\partial p}{\partial x} = \rho g\frac{\partial h}{\partial x}, \qquad \frac{\partial p}{\partial y} = \rho g\frac{\partial h}{\partial y}.$$

The above two equations indicate that the horizontal pressure-gradient force of the fluid is independent of z, that is, the horizontal acceleration of the fluid is independent of z. If the initial horizontal velocity field is independent of z, then the following horizontal velocity field is also independent of z. Therefore, we rewrite equations (2.1.1) and (2.1.2) as

$$\frac{\partial u}{\partial t} + u\frac{\partial u}{\partial x} + v\frac{\partial u}{\partial y} - fv = -g\frac{\partial h}{\partial x}, \tag{2.1.5}$$

$$\frac{\partial v}{\partial t} + u\frac{\partial v}{\partial x} + v\frac{\partial v}{\partial y} + fu = -g\frac{\partial h}{\partial y}. \tag{2.1.6}$$

Integrating equation (2.1.4) with respect to z from h_b to h, we get

$$\left(\frac{\partial u}{\partial x} + \frac{\partial v}{\partial y}\right)(h - h_b) + w|_{z=h} - w|_{z=h_b} = 0.$$

Applying the boundary condition of w, we have

$$\frac{\partial(h-h_b)}{\partial t}+u\frac{\partial(h-h_b)}{\partial x}+v\frac{\partial(h-h_b)}{\partial y}+\left(\frac{\partial u}{\partial x}+\frac{\partial v}{\partial y}\right)(h-h_b)=0,$$

that is,

$$\frac{\partial(h-h_b)}{\partial t}+\frac{\partial u(h-h_b)}{\partial x}+\frac{\partial v(h-h_b)}{\partial y}=0, \tag{2.1.7}$$

where $h-h_b$ denotes the total depth. Equations (2.1.5)-(2.1.7) is called the **barotropic model**, also called shallow water equations. According to (2.1.5) and (2.1.6), we have

$$\frac{\mathrm{d}(\zeta+f)}{\mathrm{d}t}=-(\zeta+f)\left(\frac{\partial u}{\partial x}+\frac{\partial v}{\partial y}\right).$$

Combining the above equation with (2.1.7), we get the following potential vorticity conservation law of the barotropic model

$$\frac{\mathrm{d}}{\mathrm{d}t}\left(\frac{\zeta+f}{h-h_b}\right)=0,$$

where $\zeta=\dfrac{\partial v}{\partial x}-\dfrac{\partial u}{\partial y}$ is the vertical vorticity, $\dfrac{\mathrm{d}}{\mathrm{d}t}=\dfrac{\partial}{\partial t}+u\dfrac{\partial}{\partial x}+v\dfrac{\partial}{\partial y}$, and $\dfrac{\zeta+f}{h-h_b}$ is called potential vorticity.

2.1.2 *The Two-Dimensional Quasi-Geostrophic Equation*

The system (2.1.5)-(2.1.7), the inviscid rotating barotropic model, is a hyperbolic system. It consists in various motion forms, not only motions of faster time evolution, but motions of slower time evolution. It is difficult to make precise numerical weather prediction by means of this model, because the numerical solutions of this model have high frequency gravity waves which may seriously distort short-term weather prediction. Therefore, researchers have to simplify the barotropic model rationally. For example, one can use the perturbation method to simplify (2.1.5)-(2.1.7) which is also called a small parameter expansion method, WKB method.

Suppose that $(0,0,D)$ is a static solution of equations (2.1.5)-(2.1.7), where D is a positive constant, and (u,v,η) is a solution of the following equations

$$\frac{\partial u}{\partial t}+u\frac{\partial u}{\partial x}+v\frac{\partial u}{\partial y}-fv=-g\frac{\partial \eta}{\partial x}, \tag{2.1.8}$$

$$\frac{\partial v}{\partial t}+u\frac{\partial v}{\partial x}+v\frac{\partial v}{\partial y}+fu=-g\frac{\partial \eta}{\partial y}, \tag{2.1.9}$$

$$\frac{\partial \eta}{\partial t}+u\frac{\partial(\eta-h_b)}{\partial x}+v\frac{\partial(\eta-h_b)}{\partial y}+(D+\eta-h_b)\left(\frac{\partial u}{\partial x}+\frac{\partial v}{\partial y}\right)=0, \tag{2.1.10}$$

thus $(u,v,D+\eta)$ is a solution of equations (2.1.5)-(2.1.7). Therefore, to get the reductive form of equations (2.1.5)-(2.1.7), one can simplify equations (2.1.8)-(2.1.10) by means of the small parameter method. Firstly, one should make equations (2.1.8)-(2.1.10) dimensionless. Letting

$$(x,y)=L(x_1,y_1),\ t=Tt_1,\ (u,v)=U(u_1,v_1),\ \eta=N_0\eta_1,\ f=f_0f_1,$$

where the quantities with subscript 1 are all dimensionless, we get

$$\begin{aligned}
&\frac{U}{T}\frac{\partial u_1}{\partial t_1}+\frac{U^2}{L}\left(u_1\frac{\partial u_1}{\partial x_1}+v_1\frac{\partial u_1}{\partial y_1}\right)-f_0Uf_1v_1=-g\frac{N_0}{L}\frac{\partial \eta_1}{\partial x_1},\\
&\frac{U}{T}\frac{\partial v_1}{\partial t_1}+\frac{U^2}{L}\left(u_1\frac{\partial v_1}{\partial x_1}+v_1\frac{\partial v_1}{\partial y_1}\right)+f_0Uf_1u_1=-g\frac{N_0}{L}\frac{\partial \eta_1}{\partial y_1},\\
&\frac{N_0}{T}\frac{\partial \eta_1}{\partial t_1}+\frac{U}{L}\left(u_1\frac{\partial\left(N_0\eta_1-h_b\right)}{\partial x_1}+v_1\frac{\partial\left(N_0\eta_1-h_b\right)}{\partial y_1}\right)\\
&+\frac{U}{L}\left(D+N_0\eta_1-h_b\right)\left(\frac{\partial u_1}{\partial x_1}+\frac{\partial v_1}{\partial y_1}\right)=0.
\end{aligned}$$

As we consider large-scale motions, Rossby number $\epsilon=\dfrac{U}{f_0L}\ll 1$, Kibel number $\epsilon_T=\dfrac{1}{f_0T}\ll 1$, in the momentum equations, local acceleration term and advection acceleration term are $O(\epsilon)$ terms relative to the Coriolis acceleration term. To make u_1, v_1 not equal to zero, the pressure gradient term must match the Coriolis acceleration, that is

$$f_0U=g\frac{N_0}{L},\quad N_0=\frac{f_0UL}{g}.$$

Similarly, we also assume that the local acceleration terms match the advection acceleration terms, that is,

$$\frac{U}{T}=\frac{U^2}{L},\ T=\frac{L}{U},\ \epsilon=\epsilon_T,$$

which also indicate that the local changes scale in time is equal to the advection time scale. Therefore, we get the following dimensionless equations of (2.1.8)-(2.1.10)

$$\epsilon\frac{\partial u_1}{\partial t_1}+\epsilon\left(u_1\frac{\partial u_1}{\partial x_1}+v_1\frac{\partial u_1}{\partial y_1}\right)-f_1v_1=-\frac{\partial \eta_1}{\partial x_1}, \tag{2.1.11}$$

$$\epsilon\frac{\partial v_1}{\partial t_1}+\epsilon\left(u_1\frac{\partial v_1}{\partial x_1}+v_1\frac{\partial v_1}{\partial y_1}\right)+f_1u_1=-\frac{\partial \eta_1}{\partial y_1}, \tag{2.1.12}$$

$$\begin{aligned}&\epsilon F\frac{\partial \eta_1}{\partial t_1}+\epsilon F\left(u_1\frac{\partial \eta_1}{\partial x_1}+v_1\frac{\partial \eta_1}{\partial y_1}\right)-\left(u_1\frac{\partial \frac{h_b}{D}}{\partial x_1}+v_1\frac{\partial \frac{h_b}{D}}{\partial y_1}\right)\\&+\left(1+\epsilon F\eta_1-\frac{h_b}{D}\right)\left(\frac{\partial u_1}{\partial x_1}+\frac{\partial v_1}{\partial y_1}\right)=0,\end{aligned} \tag{2.1.13}$$

where the planet Froude $F=\dfrac{f_0^2L^2}{gD}$.

If the friction and viscosity are included, the upper surface is also effected by wind stress, and $h_b=0$, then the dimensionless barotropic model is written by

$$\epsilon\frac{\partial u_1}{\partial t_1}+\epsilon\left(u_1\frac{\partial u_1}{\partial x_1}+v_1\frac{\partial u_1}{\partial y_1}\right)-f_1v_1=-\frac{\partial \eta_1}{\partial x_1}+E_H\left(\frac{\partial^2 u_1}{\partial x_1^2}+\frac{\partial^2 u_1}{\partial y_1^2}\right)+\alpha_{SW}\frac{\tau_{x_1}}{\eta_1}, \tag{2.1.11$'$}$$

$$\epsilon\frac{\partial v_1}{\partial t_1}+\epsilon\left(u_1\frac{\partial v_1}{\partial x_1}+v_1\frac{\partial v_1}{\partial y_1}\right)+f_1u_1=-\frac{\partial \eta_1}{\partial y_1}+E_H\left(\frac{\partial^2 v_1}{\partial x_1^2}+\frac{\partial^2 v_1}{\partial y_1^2}\right)+\alpha_{SW}\frac{\tau_{y_1}}{\eta_1}, \tag{2.1.12$'$}$$

$$\epsilon F\left(\frac{\partial \eta_1}{\partial t_1}+u_1\frac{\partial \eta_1}{\partial x_1}+v_1\frac{\partial \eta_1}{\partial y_1}\right)+(1+\epsilon F\eta_1)\left(\frac{\partial u_1}{\partial x_1}+\frac{\partial v_1}{\partial y_1}\right)=0, \tag{2.1.13$'$}$$

where

$$E_H=\frac{A_H}{f_0L^2},\qquad \alpha_{SW}=\frac{\tau_0}{f_0\rho DU},$$

A_H is the horizontal turbulent viscosity coefficient, and τ_0 is the wind stress scale of size. The models (2.1.11$'$)–(2.1.13$'$) can be referred to p.183 of [60].

In the following, we use the WKB method to deduce the system satisfied by the approximate solutions of (2.1.11)-(2.1.13). For a small parameter ϵ, we denote the unknown functions as power series of ϵ. Let

$$u_1=u_1^{(0)}+u_1^{(1)}\epsilon+u_1^{(2)}\epsilon^2+\cdots,$$

$$v_1=v_1^{(0)}+v_1^{(1)}\epsilon+v_1^{(2)}\epsilon^2+\cdots,$$

$$\eta_1=\eta_1^{(0)}+\eta_1^{(1)}\epsilon+\eta_1^{(2)}\epsilon^2+\cdots.$$

Due to the β-plane approximation in section 1.5, $f = f_0 + \beta_0 y$, and by the potential vorticity conservation law in the above section, we know that $\frac{\partial \zeta}{\partial y} \sim \frac{U}{L^2}$, and $\frac{\partial f}{\partial y} = \beta_0$. Thus, $\beta_1 = \frac{\beta_0}{\frac{U}{L^2}} = 1$, and

$$f = f_0 + \beta_0 y = f_0 \left(1 + \frac{\beta_0}{f_0} y\right) = f_0 \left(1 + \frac{\beta_1 \frac{U}{L^2}}{f_0} y\right) = f_0 \left(1 + \epsilon \beta_1 y_1\right),$$

therefore

$$f_1 = 1 + \epsilon \beta_1 y_1.$$

Substituting the ϵ power series of unknown functions and the expression of f_1 into equations (2.1.11)-(2.1.13), we get the zeroth order approximation equations

$$v_1^{(0)} = \frac{\partial \eta_1^{(0)}}{\partial x_1}, u_1^{(0)} = -\frac{\partial \eta_1^{(0)}}{\partial y_1}. \tag{2.1.14}$$

Recover the above equations to the dimensional equations

$$f_0 v^{(0)} = g\frac{\partial \eta^{(0)}}{\partial x}, f_0 u^{(0)} = -g\frac{\partial \eta^{(0)}}{\partial y},$$

which reflect the important features of large-scale atmospheric or oceanic motions: geostrophic equilibrium (equilibrium between pressure gradient force and Coriolis force) and horizontal non-divergence (because $\frac{\partial u^{(0)}}{\partial x} + \frac{\partial v^{(0)}}{\partial y} = 0$). To obtain the closed system of $(u_1^{(0)}, v_1^{(0)}, \eta_1^{(0)})$, we have to find another relationship. If $\frac{h_b}{D}$ is a $O(1)$ term in (2.1.13), then

$$u_1^{(0)} \frac{\partial \frac{h_b}{D}}{\partial x_1} + v_1^{(0)} \frac{\partial \frac{h_b}{D}}{\partial y_1} = 0.$$

The above relationship and (2.1.14) are the closed equations of $(u_1^{(0)}, v_1^{(0)}, \eta_1^{(0)})$, which describe the geostrophic motions.

Of course other conditions are also concerned on motions in other situations, for example, $\frac{h_b}{D}$ is $O(\varepsilon)$ term, and there is $O(1)$ term $\eta_b(x, y)$ satisfying

$$\frac{h_b}{D} = \epsilon \eta_b.$$

At this time, one can't find the closed equations of $(u_1^{(0)}, v_1^{(0)}, \eta_1^{(0)})$ only from the zeroth order approximation, but to make use of the higher order approximation relationships. $O(\epsilon)$ terms give the first-order approximation relationships

$$\frac{\partial u_1^{(0)}}{\partial t_1}+u_1^{(0)}\frac{\partial u_1^{(0)}}{\partial x_1}+v_1^{(0)}\frac{\partial u_1^{(0)}}{\partial y_1}-\beta_1 y_1 v_1^{(0)}-v_1^{(1)}=-\frac{\partial \eta_1^{(1)}}{\partial x_1}, \tag{2.1.15}$$

$$\frac{\partial v_1^{(0)}}{\partial t_1}+u_1^{(0)}\frac{\partial v_1^{(0)}}{\partial x_1}+v_1^{(0)}\frac{\partial v_1^{(0)}}{\partial y_1}+\beta_1 y_1 u_1^{(0)}+u_1^{(1)}=-\frac{\partial \eta_1^{(1)}}{\partial y_1}, \tag{2.1.16}$$

$$F\left(\frac{\partial \eta_1^{(0)}}{\partial t_1}+u_1^{(0)}\frac{\partial \eta_1^{(0)}}{\partial x_1}+v_1^{(0)}\frac{\partial \eta_1^{(0)}}{\partial y_1}\right)-\left(u_1^{(0)}\frac{\partial \eta_b}{\partial x_1}+v_1^{(0)}\frac{\partial \eta_b}{\partial y_1}\right)$$
$$+\left(\frac{\partial u_1^{(1)}}{\partial x_1}+\frac{\partial v_1^{(1)}}{\partial y_1}\right)=0, \tag{2.1.17}$$

which establish the relationships between the zeroth order approximation and the first-order approximation: in the localized variation, the advection variation and terms with β_1, the geostrophic relationships can describe all the horizontal motions, but the first-order approximation is horizontally divergent. Researchers usually call the motions described by the above first-order approximation equations as the barotropic quasi-geostrophic motions. According to (2.1.15) and (2.1.16), we get

$$\frac{\partial\left(\zeta_1^{(0)}+\beta_1 y_1\right)}{\partial t_1}+u_1^{(0)}\frac{\partial\left(\zeta_1^{(0)}+\beta_1 y_1\right)}{\partial x_1}+v_1^{(0)}\frac{\partial\left(\zeta_1^{(0)}+\beta_1 y_1\right)}{\partial y_1}$$
$$=-\left(\frac{\partial u_1^{(1)}}{\partial x_1}+\frac{\partial v_1^{(1)}}{\partial y_1}\right),$$

where $\zeta_1^{(0)}=\frac{\partial v_1^{(0)}}{\partial x_1}-\frac{\partial u_1^{(0)}}{\partial y_1}=\left(\frac{\partial^2}{\partial x_1^2}+\frac{\partial^2}{\partial y_1^2}\right)\eta_1^{(0)}$. Subtracting the above equation by (2.1.17), we get

$$\left(\frac{\partial}{\partial t_1}+\frac{\partial \eta_1^{(0)}}{\partial x_1}\frac{\partial}{\partial y_1}-\frac{\partial \eta_1^{(0)}}{\partial y_1}\frac{\partial}{\partial x_1}\right)\left(\zeta_1^{(0)}+\beta_1 y_1-F\eta_1^{(0)}+\eta_b\right)=0. \tag{2.1.18}$$

This equation is called the two-dimensional quasi-geostrophic equation. References [19,69,190] proved that the two-dimensional non-viscous quasi-geostrophic equation is a valid approximation of the rotating inviscid shallow water model without friction and external force terms with some initial data as Rossby number tends to zero.

In the same way, we get the two-dimensional quasi-geostrophic equation with friction and viscosity from (2.1.11′)-(2.1.13′)

$$\left(\frac{\partial}{\partial t_1}+\frac{\partial \eta_1^{(0)}}{\partial x_1}\frac{\partial}{\partial y_1}-\frac{\partial \eta_1^{(0)}}{\partial y_1}\frac{\partial}{\partial x_1}\right)\left(\zeta_1^{(0)}+\beta_1 y_1-F\eta_1^{(0)}+\eta_b\right)$$

$$= \frac{1}{Re}\left(\frac{\partial^2 \zeta_1^{(0)}}{\partial x_1^2} + \frac{\partial^2 \zeta_1^{(0)}}{\partial y_1^2}\right) + \alpha_{QG}\left(\frac{\partial \tau_{y_1}}{\partial x_1} - \frac{\partial \tau_{x_1}}{\partial y_1}\right) - r_{b_1}\zeta_1^{(0)}, \qquad (2.1.19)$$

where $Re = \frac{E_H}{\epsilon}$, $\alpha_{QG} = \frac{\tau_0 L}{\rho D U^2}$, $r_{b_1} = \frac{\sqrt{E_V}}{\epsilon}$, $E_H = \frac{A_H}{f_0 L^2}$, $E_V = \frac{A_V}{f_0 D^2}$, A_H is the horizontal turbulent viscosity coefficient, A_V is the vertical turbulent viscosity coefficient, and $\tau = (\tau_{x_1}, \tau_{y_1})$ is the wind stress of the upper surface.

2.2 Three-Dimensional Quasi-Geostrophic Equation

In this section, we present a three-dimensional quasi-geostrophic equation reduced from the equations of the atmosphere. The equations of the adiabatic frictionless dry atmosphere in the local rectangular coordinate frame are

$$\frac{\mathrm{d}u}{\mathrm{d}t} = -\frac{1}{\rho}\frac{\partial p}{\partial x} + fv, \qquad (2.2.1)$$

$$\frac{\mathrm{d}v}{\mathrm{d}t} = -\frac{1}{\rho}\frac{\partial p}{\partial y} - fu, \qquad (2.2.2)$$

$$\frac{\mathrm{d}w}{\mathrm{d}t} = -\frac{1}{\rho}\frac{\partial p}{\partial z} - g, \qquad (2.2.3)$$

$$\frac{\mathrm{d}\ln\rho}{\mathrm{d}t} + \left(\frac{\partial u}{\partial x} + \frac{\partial v}{\partial y} + \frac{\partial w}{\partial z}\right) = 0, \qquad (2.2.4)$$

$$p = R\rho T, \qquad (2.2.5)$$

$$\frac{\mathrm{d}\ln\vartheta}{\mathrm{d}t} = 0, \qquad (2.2.6)$$

where $f = f_0 + \beta_0 y$, potential temperature $\vartheta = T(\frac{P_0}{p})^{\frac{R}{c_p}}$, and equation $\frac{\mathrm{d}\ln\vartheta}{\mathrm{d}t} = 0$ is derived from equation (1.5.13) when $Q = 0$ and the potential temperature formula.

Equations (2.2.1)-(2.2.6) are far more complicated than the barotropic model (2.1.5)-(2.1.7). So far researchers can't solve them theoretically and numerically, and thus can't improve numerical weather forecast by making use of them directly. Therefore, researchers may omit some small and medium scale factors, and reduce the adiabatic frictionless atmospheric fundermental equations (2.2.1)-(2.2.6) rationally, then improve numerical weather forecast. Firstly, we consider the reduced equations satisfied by the perturbation solution $(u, v, w, p+p_0(z), \rho+\rho_0(z), \vartheta+\vartheta_0(z))$ of (2.2.1)-(2.2.6) near a given static solution $(0, 0, 0, p_0(z), \rho_0(z), \vartheta_0(z))$, that is,

$$\frac{\mathrm{d}u}{\mathrm{d}t} - fv = -\frac{1}{\rho_0}\frac{\partial p}{\partial x}, \tag{2.2.7}$$

$$\frac{\mathrm{d}v}{\mathrm{d}t} + fu = -\frac{1}{\rho_0}\frac{\partial p}{\partial y}, \tag{2.2.8}$$

$$\frac{\mathrm{d}w}{\mathrm{d}t} = -\frac{1}{\rho_0}\frac{\partial p}{\partial z} - g\frac{\rho}{\rho_0}, \tag{2.2.9}$$

$$\frac{\mathrm{d}\frac{\rho}{\rho_0}}{\mathrm{d}t} + \frac{\partial u}{\partial x} + \frac{\partial v}{\partial y} + \frac{1}{\rho_0}\frac{\partial \rho_0 w}{\partial z} = 0, \tag{2.2.10}$$

$$\frac{\vartheta}{\vartheta_0} = \frac{1}{\gamma}\frac{p}{p_0} - \frac{\rho}{\rho_0}, \tag{2.2.11}$$

$$\frac{\mathrm{d}\frac{\vartheta}{\vartheta_0}}{\mathrm{d}t} + \frac{N^2}{g}w = 0, \tag{2.2.12}$$

where $\gamma = \frac{c_p}{c_v}$ (c_p, c_v appear in section 1.1), and $N^2 = g\frac{\partial \ln \vartheta_0}{\partial z}$, for details, see section 4.4 of [145]. Then one can reduce equations (2.2.7)-(2.2.12) by means of the small parameter method. For this purpose, we make equations (2.2.7)-(2.2.12) dimensionless first. Letting

$$(x, y) = L(x_1, y_1),\ z = Dz_1,\ t = \frac{L}{U}t_1,\ (u, v) = U(u_1, v_1),$$

$$w = N_1 w_1,\ p = N_2 p_1,\ \rho = N_3 \rho_1,\ \vartheta = N_4 \vartheta_1,\ f = f_0 f_1,$$

where quantities with subscript 1 are all dimensionless, and N_i is undetermined, we get

$$\frac{U^2}{L}\frac{\partial u_1}{\partial t_1} + \frac{U^2}{L}\left(u_1\frac{\partial u_1}{\partial x_1} + v_1\frac{\partial u_1}{\partial y_1}\right) + \frac{UN_1}{D}w_1\frac{\partial u_1}{\partial z_1} - f_0 U f_1 v_1 = -\frac{N_2}{L\rho_0}\frac{\partial p_1}{\partial x_1},$$

$$\frac{U^2}{L}\frac{\partial v_1}{\partial t_1} + \frac{U^2}{L}\left(u_1\frac{\partial v_1}{\partial x_1} + v_1\frac{\partial v_1}{\partial y_1}\right) + \frac{UN_1}{D}w_1\frac{\partial v_1}{\partial z_1} + f_0 U f_1 u_1 = -\frac{N_2}{L\rho_0}\frac{\partial p_1}{\partial y_1},$$

$$\frac{UN_1}{L}\frac{\partial w_1}{\partial t_1} + \frac{UN_1}{L}\left(u_1\frac{\partial w_1}{\partial x_1} + v_1\frac{\partial w_1}{\partial y_1}\right) + \frac{N_1^2}{D}w_1\frac{\partial w_1}{\partial z_1} = -\frac{N_2}{D\rho_0}\frac{\partial p_1}{\partial z_1} - g\frac{N_3}{\rho_0}\rho_1,$$

$$\frac{UN_3}{L\rho_0}\frac{\partial \rho_1}{\partial t_1} + \frac{UN_3}{L\rho_0}\left(u_1\frac{\partial \rho_1}{\partial x_1} + v_1\frac{\partial \rho_1}{\partial y_1}\right) + \frac{N_1}{D}w_1\frac{\partial\frac{N_3\rho_1}{\rho_0}}{\partial z_1} + \frac{U}{L}\left(\frac{\partial u_1}{\partial x_1} + \frac{\partial v_1}{\partial y_1}\right)$$
$$+ \frac{N_1}{D\rho_0}\frac{\partial \rho_0 w_1}{\partial z_1} = 0,$$

$$N_4\frac{\vartheta_1}{\vartheta_0} = \frac{N_2}{\gamma}\frac{p_1}{p_0} - N_3\frac{\rho_1}{\rho_0},$$

$$\frac{UN_4}{L\vartheta_0}\frac{\partial\vartheta_1}{\partial t_1}+\frac{UN_4}{L\vartheta_0}\left(u_1\frac{\partial\vartheta_1}{\partial x_1}+v_1\frac{\partial\vartheta_1}{\partial y_1}\right)+\frac{N_1}{D}w_1\frac{\partial\frac{N_4\vartheta_1}{\vartheta_0}}{\partial z_1}+N_1\frac{N^2}{g}w_1=0.$$

As $N_1\le\frac{D}{L}U$ in the horizontal momentum equations, the local acceleration term and the advection acceleration term are $O(\epsilon)$ terms (Rossby number $\epsilon=\frac{U}{f_0L}$) relative to the Coriolis acceleration term. For making u_1, v_1 not equal to zero, the horizontal pressure gradient force must be equivalent to Coriolis force, that is,

$$f_0U=\frac{N_2}{L\rho_0},\ N_2=\rho_0f_0UL.$$

As $p_0=\rho_0gH$, where H is the atmospheric elevation ($H=D$), we know $\frac{N_3}{\rho_0}=\frac{N_2}{p_0}=\frac{N_4}{\vartheta_0}$ from equation $N_4\frac{\vartheta_1}{\vartheta_0}=\frac{N_2}{\gamma}\frac{p_1}{p_0}-N_3\frac{\rho_1}{\rho_0}$, that is,

$$N_3=\rho_0\frac{f_0UL}{gH}=\rho_0\mu_0^2\epsilon,\quad N_4=\vartheta_0\frac{f_0UL}{gH}=\vartheta_0\mu_0^2\epsilon,$$

where $\mu_0=\frac{L}{L_0}$ is the Obukhov parameter, and the barotrophic atmospheric Rossby deformed radius is $L_0=\frac{\sqrt{gH}}{f_0}$.

Next, let's determine the scale N_1. Combining the scale analysis of the last equation of the above system with $\frac{N_1}{D}\le\frac{U}{L}$, we know $N_1\frac{N^2}{g}=\frac{UN_4}{L\vartheta_0}$, that is, $N_1=\frac{f_0U^2}{N^2H}=\frac{f_0^2L^2}{N^2DH}\cdot\frac{DU}{L}\epsilon$. For the large-scale atmosphere, we assume that $\frac{f_0U^2}{N^2H}=\frac{f_0^2L^2}{N^2DH}$ is equal to 1. Let

$$N_1=\frac{DU}{L}\epsilon.$$

Thus, we derive from the above system that

$$\begin{aligned}
&\epsilon\left(\frac{\partial u_1}{\partial t_1}+u_1\frac{\partial u_1}{\partial x_1}+v_1\frac{\partial u_1}{\partial y_1}\right)+\epsilon^2w_1\frac{\partial u_1}{\partial z_1}-f_1v_1=-\frac{\partial p_1}{\partial x_1},\\
&\epsilon\left(\frac{\partial v_1}{\partial t_1}+u_1\frac{\partial v_1}{\partial x_1}+v_1\frac{\partial v_1}{\partial y_1}\right)+\epsilon^2w_1\frac{\partial v_1}{\partial z_1}+f_1u_1=-\frac{\partial p_1}{\partial y_1},\\
&\delta^2\epsilon^2\left(\frac{\partial w_1}{\partial t_1}+u_1\frac{\partial w_1}{\partial x_1}+v_1\frac{\partial w_1}{\partial y_1}\right)+\delta^2\epsilon^3w_1\frac{\partial w_1}{\partial z_1}=-\frac{\partial p_1}{\partial z_1}+\sigma_1p_1-\rho_1,
\end{aligned}$$

$$\mu_0^2\epsilon\left(\frac{\partial\rho_1}{\partial t_1}+u_1\frac{\partial\rho_1}{\partial x_1}+v_1\frac{\partial\rho_1}{\partial y_1}\right)+\mu_0^2\epsilon^2 w_1\frac{\partial\rho_1}{\partial z_1}+\left(\frac{\partial u_1}{\partial x_1}+\frac{\partial v_1}{\partial y_1}\right)+\frac{\epsilon}{\rho_0}\frac{\partial\rho_0 w_1}{\partial z_1}$$
$$=0,$$
$$\vartheta_1=\frac{1}{\gamma}p_1-\rho_1,$$
$$\epsilon\left(\frac{\partial\vartheta_1}{\partial t_1}+u_1\frac{\partial\vartheta_1}{\partial x_1}+v_1\frac{\partial\vartheta_1}{\partial y_1}\right)+\epsilon^2 w_1\frac{\partial\vartheta_1}{\partial z_1}+\epsilon\frac{a_0}{\mu_0^2}w_1=0,$$

where $\delta=\frac{D}{L}$, $a_0=\frac{N^2H}{g}$, $\sigma_1=-\frac{\partial\ln\rho_0}{\partial z_1}=\frac{\partial\ln\vartheta_0}{\partial z_1}-\frac{1}{\gamma}\frac{\partial\ln p_0}{\partial z_1}=\frac{N^2D}{g}+\frac{1}{\gamma}=a_0+\frac{1}{\gamma}$, which is derived from the continuity equation, potential temperature formula and $D=H$. For the large-scale atmosphere, $\delta^2\ll 1$, $\mu_0^2\ll 1$, one can omit $\delta^2\epsilon^2(\frac{\partial w_1}{\partial t_1}+u_1\frac{\partial w_1}{\partial x_1}+v_1\frac{\partial w_1}{\partial y_1})+\delta^2\epsilon^3 w_1\frac{\partial w_1}{\partial z_1}$ and $\mu_0^2\epsilon(\frac{\partial\rho_1}{\partial t_1}+u_1\frac{\partial\rho_1}{\partial x_1}+v_1\frac{\partial\rho_1}{\partial y_1})+\mu_0^2\epsilon^2 w_1\frac{\partial\rho_1}{\partial z_1}$. Thus, we have

$$\epsilon\left(\frac{\partial u_1}{\partial t_1}+u_1\frac{\partial u_1}{\partial x_1}+v_1\frac{\partial u_1}{\partial y_1}\right)+\epsilon^2 w_1\frac{\partial u_1}{\partial z_1}-f_1v_1=-\frac{\partial p_1}{\partial x_1},\tag{2.2.13}$$
$$\epsilon\left(\frac{\partial v_1}{\partial t_1}+u_1\frac{\partial v_1}{\partial x_1}+v_1\frac{\partial v_1}{\partial y_1}\right)+\epsilon^2 w_1\frac{\partial v_1}{\partial z_1}+f_1u_1=-\frac{\partial p_1}{\partial y_1},\tag{2.2.14}$$
$$\frac{\partial p_1}{\partial z_1}-a_0p_1-\vartheta_1=0,\tag{2.2.15}$$
$$\frac{\partial u_1}{\partial x_1}+\frac{\partial v_1}{\partial y_1}+\frac{\epsilon}{\rho_0}\frac{\partial\rho_0 w_1}{\partial z_1}=0,\tag{2.2.16}$$
$$\epsilon\left(\frac{\partial\vartheta_1}{\partial t_1}+u_1\frac{\partial\vartheta_1}{\partial x_1}+v_1\frac{\partial\vartheta_1}{\partial y_1}\right)+\epsilon^2 w_1\frac{\partial\vartheta_1}{\partial z_1}+\epsilon\frac{1}{\mu_1^2}w_1=0,\tag{2.2.17}$$

where the baroclinic Obukhov parameter $\mu_1=\frac{L}{L_1}$, and the baroclinic atmospheric Rossby deformed radius $L_1=\frac{NH}{f_0}=\frac{ND}{f_0}$. In the following we assume $a_0\approx\epsilon$, $\mu_1\approx 1$, which is suitable for the large-scale atmosphere. For the small parameter ϵ, the unknown functions are denoted as ϵ power series, that is

$$u_1=u_1^{(0)}+u_1^{(1)}\epsilon+u_1^{(2)}\epsilon^2+\cdots,$$
$$v_1=v_1^{(0)}+v_1^{(1)}\epsilon+v_1^{(2)}\epsilon^2+\cdots,$$

$$\epsilon w_1 = w_1^{(1)}\epsilon + w_1^{(2)}\epsilon^2 + \cdots,$$

$$p_1 = p_1^{(0)} + p_1^{(1)}\epsilon + p_1^{(2)}\epsilon^2 + \cdots,$$

$$\vartheta_1 = \vartheta_1^{(0)} + \vartheta_1^{(1)}\epsilon + \vartheta_1^{(2)}\epsilon^2 + \cdots.$$

Taking the zeroth order approximation of (2.2.13) and (2.2.14), we get $\dfrac{\partial u_1^{(0)}}{\partial x_1} + \dfrac{\partial v_1^{(0)}}{\partial y_1} = 0$. Combining this equation with (2.2.16), we obtain $\dfrac{1}{\rho_0}\dfrac{\partial \rho_0(\epsilon w_1)^{(0)}}{\partial z_1} = 0$, which means that if there is a z_1 such that $(\epsilon w_1)^{(0)}(z_1) = 0$, then $(\epsilon w_1)^{(0)} \equiv 0$. So we let $\epsilon w_1 = w_1^{(1)}\epsilon + w_1^{(2)}\epsilon^2 + \cdots$. Substituting the ϵ power series of unknown functions and the expression of $f_1 = 1 + \epsilon\beta_1 y_1$ into equations (2.2.13)-(2.2.17), we get the zeroth order approximation relationships

$$v_1^{(0)} = \frac{\partial p_1^{(0)}}{\partial x_1}, \quad u_1^{(0)} = -\frac{\partial p_1^{(0)}}{\partial y_1}, \quad \frac{\partial p_1^{(0)}}{\partial z_1} = \vartheta_1^{(0)}. \tag{2.2.18}$$

Recover the above equations to the dimensional equations

$$f_0 v^{(0)} = \frac{1}{\rho_0}\frac{\partial p^{(0)}}{\partial x}, \quad f_0 u^{(0)} = -\frac{1}{\rho_0}\frac{\partial \eta^{(0)}}{\partial y}, \quad \frac{\partial \frac{p^{(0)}}{\rho_0}}{\partial z} = g\frac{\vartheta^{(0)}}{\vartheta_0},$$

which reflect the important features of the large-scale baroclinic atmospheric motions: geostrophic equilibrium (equilibrium between horizontal pressure-gradient force and Coriolis force), hydrostatic equilibrium and horizontal divergence (because $\dfrac{\partial u^{(0)}}{\partial x} + \dfrac{\partial v^{(0)}}{\partial y} = 0$). To obtain the closed equations of $(u_1^{(0)}, v_1^{(0)}, p_1^{(0)}, \vartheta_1^{(0)})$, we must find other relationships. Therefore, we make use of high-order approximation relationships. $O(\epsilon)$ terms give the first-order approximation relationships

$$\frac{\partial u_1^{(0)}}{\partial t_1} + u_1^{(0)}\frac{\partial u_1^{(0)}}{\partial x_1} + v_1^{(0)}\frac{\partial u_1^{(0)}}{\partial y_1} - \beta_1 y_1 v_1^{(0)} - v_1^{(1)} = -\frac{\partial p_1^{(1)}}{\partial x_1},$$

$$\frac{\partial v_1^{(0)}}{\partial t_1} + u_1^{(0)}\frac{\partial v_1^{(0)}}{\partial x_1} + v_1^{(0)}\frac{\partial v_1^{(0)}}{\partial y_1} + \beta_1 y_1 u_1^{(0)} + u_1^{(1)} = -\frac{\partial p_1^{(1)}}{\partial y_1},$$

$$\frac{\partial p_1^{(1)}}{\partial z_1} - \frac{a_0}{\epsilon}p_1^{(0)} - \vartheta_1^{(1)} = 0,$$

$$\frac{\partial u_1^{(1)}}{\partial x_1} + \frac{\partial v_1^{(1)}}{\partial y_1} + \frac{1}{\rho_0}\frac{\partial \rho_0 w_1^{(1)}}{\partial z_1} = 0,$$

$$\frac{\partial \vartheta_1^{(0)}}{\partial t_1} + u_1^{(0)}\frac{\partial \vartheta_1^{(0)}}{\partial x_1} + v_1^{(0)}\frac{\partial \vartheta_1^{(0)}}{\partial y_1} + \frac{1}{\mu_1^2}w_1^{(1)} = 0.$$

Researchers usually call the flows, which the above first order approximation equations describe, as baroclinic quasi-geostrophic flows (also called continuously stratified quasi-qeostrophic flow). Using the fourth equation of the above equations, taking the vorticity equation of the first two equations, we get

$$\frac{\partial \zeta_1^{(0)}}{\partial t_1} + u_1^{(0)} \frac{\partial \zeta_1^{(0)}}{\partial x_1} + v_1^{(0)} \frac{\partial \zeta_1^{(0)}}{\partial y_1} + \beta_1 v_1^{(0)} = \frac{1}{\rho_0} \frac{\partial \rho_0 w_1^{(1)}}{\partial z_1},$$

$$\frac{\partial \vartheta_1^{(0)}}{\partial t_1} + u_1^{(0)} \frac{\partial \vartheta_1^{(0)}}{\partial x_1} + v_1^{(0)} \frac{\partial \vartheta_1^{(0)}}{\partial y_1} + \frac{1}{\mu_1^2} w_1^{(1)} = 0,$$

where $\zeta_1^{(0)} = \frac{\partial v_1^{(0)}}{\partial x} - \frac{\partial u_1^{(0)}}{\partial y} = (\frac{\partial^2}{\partial x_1^2} + \frac{\partial^2}{\partial y_1^2}) p_1^{(0)}$. According to (2.2.18) and the above equations, we have

$$\frac{\partial q}{\partial t_1} + \frac{\partial p_1^{(0)}}{\partial x_1} \frac{\partial q}{\partial y_1} - \frac{\partial p_1^{(0)}}{\partial y_1} \frac{\partial q}{\partial x_1} = 0, \tag{2.2.19}$$

where $q = (\frac{\partial^2}{\partial x_1^2} + \frac{\partial^2}{\partial y_1^2}) p_1^{(0)} + \frac{1}{\rho_0} \frac{\partial}{\partial z_1} (\mu_1^2 \rho_0 \frac{\partial p_1^{(0)}}{\partial z_1}) + \beta_1 y_1$ is called the baroclinic quasi-geostrophic potential vorticity. (2.2.19) is called the baroclinic atmospheric quasi-geostrophic equation in adiabatic frictionless situation (or three-dimensional quasi-geostrophic equation).

Like the deduction of (2.2.19), we obtain the baroclinic quasi-geostrophic equation from inviscous equation (1.6.10)-(1.6.14) of the oceans (take $\nu = 0$, $\kappa = 0$) under hydrostatic approximation, Boussinesq approximation and stratify approximation

$$\frac{\partial q}{\partial t_1} + \frac{\partial p_1^{(0)}}{\partial x_1} \frac{\partial q}{\partial y_1} - \frac{\partial p_1^{(0)}}{\partial y_1} \frac{\partial q}{\partial x_1} = 0, \tag{2.2.20}$$

where $q = (\frac{\partial^2}{\partial x_1^2} + \frac{\partial^2}{\partial y_1^2}) p_1^{(0)} + \frac{\partial}{\partial z_1} (\frac{1}{S} \frac{\partial p_1^{(0)}}{\partial z_1}) + \beta_1 y_1, S = (\frac{L_D}{L})^2$, $L_D = \frac{N_S D}{f_0}$, $N_S = (-\frac{g}{\bar{\rho}} \frac{\partial \bar{\rho}}{\partial z})^{\frac{1}{2}}$.

The problems of the global well-posedness of the two-dimensional and the three-dimensional quasi-geostrophic models are well studied, for more detail, see [217, chapter 12] and [12, 13, 66, 146, 164, 165, 209, 210].

2.3 The Multi-Layer Quasi-Geostrophic Model

The multi-layer quasi-geostrophic model is a model between the barotrophic and baroclinic quasi-geostrophic model, which approximately describes the large-scale motions between finitely homogeneous flow layers with uniform density which is different from each other. It's two-dimensional as the barotrophic quasi-geostrophic model, but reserves some baroclinic features. Without loss of gernerality, we here give the derivation of the two-layer quasi-geostrophic model.

Suppose that the fluid consists of two layers, the densities of the upper and lower layer are constants $\rho_1,\ \rho_2\ (\rho_1 < \rho_2)$ respectively, D is the characteristic scale of the fluid total vertical height, and $U,\ L$ are respectively the characteristic scale of the horizontal velocity and length of motions. And suppose further that $\epsilon = \dfrac{U}{f_0 L} \ll 1$, and $\dfrac{D}{L} \ll 1$. Since the hydrostatic approximation is valid between two layers, the pressure for the upper and lower layers can be denoted as

$$P_1 = \rho_1 g(H_1 - z) + p_1(t,x,y), \quad P_2 = \rho_1 g D_1 + \rho_2 g(H_2 - z) + p_2(t,x,y),$$

where $H_i\ (i = 1, 2)$ is the fixed dimension height of the upper surface of the i-th layer, D_i is the dimension thickness of the i-th layer, $D_1 = H_1 - H_2$, and p_i denotes the deviation between pressure and the value of the static pressure of the fluid. Like N_0, $p_i = \rho_i f_0 U L p_{i1}$ in section 2.1.2, here p_{i1} is dimensionless. Because the fluid is homogeneous incompressible, its motions are described by equations (2.1.1)-(2.1.4). Thus, we use the method in section 2.1, and assume

$$u_{i1} = u_{i1}^{(0)} + u_{i1}^{(1)}\epsilon + u_{i1}^{(2)}\epsilon^2 + \cdots,$$

$$v_{i1} = v_{i1}^{(0)} + v_{i1}^{(1)}\epsilon + v_{i1}^{(2)}\epsilon^2 + \cdots,$$

$$p_{i1} = p_{i1}^{(0)} + p_{i1}^{(1)}\epsilon + p_{i1}^{(2)}\epsilon^2 + \cdots,$$

where u_{i1} and v_{i1} are independence of vertical axis z_1. Taking the zeroth order approximation, we have

$$v_{i1}^{(0)} = \frac{\partial p_{i1}^{(0)}}{\partial x_1}, \quad u_{i1}^{(0)} = -\frac{\partial p_{i1}^{(0)}}{\partial y_1}. \tag{2.3.1}$$

And taking the first order approximation, we obtain

$$\frac{\partial(\zeta_{i1}^{(0)} + \beta_1 y_1)}{\partial t_1} + u_{i1}^{(0)}\frac{\partial(\zeta_{i1}^{(0)} + \beta_1 y_1)}{\partial x_1} + v_{i1}^{(0)}\frac{\partial(\zeta_{i1}^{(0)} + \beta_1 y_1)}{\partial y_1} = \frac{\partial w_{i1}^{(1)}}{\partial z_1}, \tag{2.3.2}$$

where $\zeta_{i1}^{(0)} = \frac{\partial v_{i1}^{(0)}}{\partial x_1} - \frac{\partial u_{i1}^{(0)}}{\partial y_1} = (\frac{\partial^2}{\partial x_1^2} + \frac{\partial^2}{\partial y_1^2})p_{i1}^{(0)}$.

Next, we reduce (2.3.2) by applying boundary conditions. For this purpose, we give boundary and interface conditions. Let η_i be the deviation between upper surface of the i-th layer and H_i, we know that the height h_i of the i-th layer is

$$h_i = H_i + \eta_i = H_i + N_{0i}\eta_{i1},$$

which is dimensionless. Firstly, according to the deduction of N_0 in section 2.1.1, we get

$$N_{01} = \frac{f_0 UL}{g} = \epsilon FD, \quad \eta_{11} = p_{11},$$

where the definition of F appears in section 2.1.2. Since $P_1 = P_2$ on $z = h_2$, that is,

$$\begin{aligned}&\rho_1 g(H_1 - H_2 - N_{02}\eta_{21}) + \rho_1 f_0 UL p_{11}\\ = &\rho_1 g(H_1 - H_2) + \rho_2 g(-N_{02}\eta_{21}) + \rho_2 f_0 UL p_{21},\end{aligned}$$

there is

$$N_{02} g(\rho_2 - \rho_1)\eta_{21} = \rho_2 f_0 UL p_{21} - \rho_1 f_0 UL p_{11}. \tag{2.3.3}$$

Thus, as η_{21}, p_{21} and p_{11} are dimensionless, we get

$$N_{02} = \frac{\rho_0 f_0 UL}{g(\rho_2 - \rho_1)} = \epsilon FD \frac{\rho_0}{\rho_2 - \rho_1},$$

where ρ_0 is a constant denoting the characteristic value of the fluid density. Supposing

$$\frac{\rho_2 - \rho_1}{\rho_0} \ll 1,$$

we derive from (2.3.3)

$$\eta_{21}^{(0)} = p_{21}^{(0)} - p_{11}^{(0)}.$$

Since

$$w_1|_{z=h_1} = \frac{\mathrm{d}\eta_1}{\mathrm{d}t} = \frac{\partial \eta_1}{\partial t} + u_1 \frac{\partial \eta_1}{\partial x} + v_1 \frac{\partial \eta_1}{\partial y},$$

$$w_2|_{z=h_2} = \frac{\mathrm{d}\eta_2}{\mathrm{d}t} = \frac{\partial \eta_2}{\partial t} + u_2 \frac{\partial \eta_2}{\partial x} + v_2 \frac{\partial \eta_2}{\partial y},$$

$$w_2|_{z=h_3} = \frac{\mathrm{d}h_b}{\mathrm{d}t} = u_2\frac{\partial h_b}{\partial x} + v_2\frac{\partial h_b}{\partial y},$$

where $h_3 = h_b(x, y)$ is the rigid lower boundary surface of the lower layer and $\frac{h_b}{D_2} = \epsilon\eta_b$. Making the boundary and interface conditions dimensionless, we get

$$\frac{DU}{L}w_{11}|_{z_1=\frac{h_1}{D}} = \frac{UN_{01}}{L}\frac{\mathrm{d}\eta_{11}}{\mathrm{d}t_1} = \frac{UN_{01}}{L}\left(\frac{\partial\eta_{11}}{\partial t_1} + u_{11}\frac{\partial\eta_{11}}{\partial x_1} + v_{11}\frac{\partial\eta_{11}}{\partial y_1}\right),$$

$$\frac{DU}{L}w_{21}|_{z_1=\frac{h_2}{D}} = \frac{UN_{02}}{L}\frac{\mathrm{d}\eta_{21}}{\mathrm{d}t_1} = \frac{UN_{02}}{L}\left(\frac{\partial\eta_{21}}{\partial t_1} + u_{21}\frac{\partial\eta_{21}}{\partial x_1} + v_{21}\frac{\partial\eta_{21}}{\partial y_1}\right),$$

$$\frac{DU}{L}w_{21}|_{z_1=\frac{h_b}{D}} = \frac{UD_2\epsilon}{L}\frac{\mathrm{d}\eta_b}{\mathrm{d}t_1} = \frac{UD_2\epsilon}{L}\left(u_{21}\frac{\partial\eta_b}{\partial x_1} + v_{21}\frac{\partial\eta_b}{\partial y_1}\right).$$

The zero approximation of horizontal velocity field (u_{i1}, v_{i1}) is horizontal non-divergence, and the fluid is incompressible. So

$$w_{i1} = w_{i1}^{(1)}\epsilon + w_{i1}^{(2)}\epsilon^2 + \cdots.$$

Therefore, we have

$$w_{11}^{(1)}|_{z_1=\frac{h_1}{D}} = F\frac{\mathrm{d}\eta_{11}^{(0)}}{\mathrm{d}t_1},$$

$$w_{21}^{(1)}|_{z_1=\frac{h_2}{D}} = F\frac{\rho_0}{\rho_2-\rho_1}\frac{\mathrm{d}\eta_{21}^{(0)}}{\mathrm{d}t_1},$$

$$w_{21}^{(1)}|_{z_1=\frac{h_b}{D}} = \frac{D_2}{D}\frac{\mathrm{d}\eta_b}{\mathrm{d}t_1}.$$

Integrating (2.3.2) with respect to z_1 from h_{i+1} to h_i, we have

$$\left(\frac{D_1}{D} + O(\epsilon F)\right)\left(\frac{\partial}{\partial t_1} + u_{11}^{(0)}\frac{\partial}{\partial x_1} + v_{11}^{(0)}\frac{\partial}{\partial y_1}\right)\left(\zeta_{11}^{(0)} + \beta_1 y_1\right)$$
$$= F\frac{\mathrm{d}\eta_{11}^{(0)}}{\mathrm{d}t_1} - F\frac{\rho_0}{\rho_2-\rho_1}\frac{\mathrm{d}\eta_{21}^{(0)}}{\mathrm{d}t_1},$$

$$\left(\frac{D_2}{D} + \frac{O(\epsilon F)\rho_0}{\rho_2-\rho_1}\right)\left(\frac{\partial}{\partial t_1} + u_{21}^{(0)}\frac{\partial}{\partial x_1} + v_{21}^{(0)}\frac{\partial}{\partial y_1}\right)\left(\zeta_{21}^{(0)} + \beta_1 y_1\right)$$
$$= F\frac{\rho_0}{\rho_2-\rho_1}\frac{\mathrm{d}\eta_{21}^{(0)}}{\mathrm{d}t_1} - \frac{D_2}{D}\frac{\mathrm{d}\eta_b}{\mathrm{d}t_1}.$$

If $\dfrac{F\rho_0}{\rho_2-\rho_1}=O(1)$, then the above equations are rewritten as

$$\left(\frac{\partial}{\partial t_1}+u_{11}^{(0)}\frac{\partial}{\partial x_1}+v_{11}^{(0)}\frac{\partial}{\partial y_1}\right)\left(\zeta_{11}^{(0)}-F_1\left(p_{11}^{(0)}-p_{21}^{(0)}\right)+\beta_1 y_1\right)=0,$$

$$\left(\frac{\partial}{\partial t_1}+u_{21}^{(0)}\frac{\partial}{\partial x_1}+v_{21}^{(0)}\frac{\partial}{\partial y_1}\right)\left(\zeta_{21}^{(0)}+F_2\left(p_{11}^{(0)}-p_{21}^{(0)}\right)+\beta_1 y_1+\eta_b\right)=0,$$

where $F_1=\dfrac{DF}{D_1}\dfrac{\rho_0}{\rho_2-\rho_1}$, $F_2=\dfrac{DF}{D_2}\dfrac{\rho_0}{\rho_2-\rho_1}$. If

$$\psi_1=p_{11}^{(0)},\quad \psi_2=p_{21}^{(0)},$$

then the above equations are rewritten as

$$\left(\frac{\partial}{\partial t_1}+\frac{\partial\psi_1}{\partial x_1}\frac{\partial}{\partial y_1}-\frac{\partial\psi_1}{\partial y_1}\frac{\partial}{\partial x_1}\right)\left(\frac{\partial^2\psi_1}{\partial x_1^2}+\frac{\partial^2\psi_1}{\partial y_1^2}-F_1(\psi_1-\psi_2)+\beta_1 y_1\right)=0, \tag{2.3.4}$$

$$\left(\frac{\partial}{\partial t_1}+\frac{\partial\psi_2}{\partial x_1}\frac{\partial}{\partial y_1}-\frac{\partial\psi_2}{\partial y_1}\frac{\partial}{\partial x_1}\right)\left(\frac{\partial^2\psi_2}{\partial x_1^2}+\frac{\partial^2\psi_2}{\partial y_1^2}+F_2(\psi_1-\psi_2)+\beta_1 y_1+\eta_b\right)=0. \tag{2.3.5}$$

(2.3.4) and (2.3.5) are called the two-layer quasi-geostrophic model without viscous and friction, which is the simplest model reserving the baroclinic features. If the wind stress of the upper surface, the viscosity and friction are considered, then the two-layer quasi-geostrophic model is given as follows

$$\begin{aligned}&\left(\frac{\partial}{\partial t_1}+\frac{\partial\psi_1}{\partial x_1}\frac{\partial}{\partial y_1}-\frac{\partial\psi_1}{\partial y_1}\frac{\partial}{\partial x_1}\right)\left(\zeta_1-F_1\left(\psi_1-\psi_2\right)+\beta_1 y_1\right)\\&=\frac{1}{Re}\left(\frac{\partial^2\zeta_1}{\partial x_1^2}+\frac{\partial^2\zeta_1}{\partial y_1^2}\right)+\alpha_{QG}\left(\frac{\partial\tau_{y_1}}{\partial x_1}-\frac{\partial\tau_{x_1}}{\partial y_1}\right),\end{aligned} \tag{2.3.4$'$}$$

$$\begin{aligned}&\left(\frac{\partial}{\partial t_1}+\frac{\partial\psi_2}{\partial x_1}\frac{\partial}{\partial y_1}-\frac{\partial\psi_2}{\partial y_1}\frac{\partial}{\partial x_1}\right)\left(\zeta_2+F_2\left(\psi_1-\psi_2\right)+\beta_1 y_1+\eta_b\right)\\&=\frac{1}{Re}\left(\frac{\partial^2\zeta_2}{\partial x_1^2}+\frac{\partial^2\zeta_2}{\partial y_1^2}\right)-r_{b_2}\zeta_2,\end{aligned} \tag{2.3.5$'$}$$

where $\zeta_1=\dfrac{\partial^2\psi_1}{\partial x_1^2}+\dfrac{\partial^2\psi_1}{\partial y_1^2}$, $\zeta_2=\dfrac{\partial^2\psi_2}{\partial x_1^2}+\dfrac{\partial^2\psi_2}{\partial y_1^2}$, $Re=\dfrac{E_H}{\epsilon}$, $\alpha_{QG}=\dfrac{\tau_0 L}{\rho_0 DU^2}$, $r_{b_2}=\dfrac{D}{D_2}\dfrac{r_{b_1}}{2}$, $r_{b_1}=\dfrac{\sqrt{E_V}}{\epsilon}$, $E_H=\dfrac{A_H}{f_0L^2}$, $E_V=\dfrac{A_V}{f_0D^2}$, A_H is the horizontal turbulent viscosity coefficient, and A_V is the vertical turbulent viscosity coefficient.

Like the deduction of (2.3.4) and (2.3.5), we can get the N-layer quasi-geostrophic model ($N \geq 3$).
If $i = 1$,

$$\left(\frac{\partial}{\partial t_1} + \frac{\partial \psi_1}{\partial x_1}\frac{\partial}{\partial y_1} - \frac{\partial \psi_1}{\partial y_1}\frac{\partial}{\partial x_1}\right)\left(\frac{\partial^2 \psi_1}{\partial x_1^2} + \frac{\partial^2 \psi_1}{\partial y_1^2}\right.$$
$$\left. - F_1(\psi_1 - \psi_2) + \beta_1 y_1\right) = 0.$$

If $i \neq 1, N$,

$$\left(\frac{\partial}{\partial t_1} + \frac{\partial \psi_i}{\partial x_1}\frac{\partial}{\partial y_1} - \frac{\partial \psi_i}{\partial y_1}\frac{\partial}{\partial x_1}\right)\left(\frac{\partial^2 \psi_i}{\partial x_1^2} + \frac{\partial^2 \psi_i}{\partial y_1^2}\right.$$
$$\left. - F_i(2\psi_i - \psi_{i-1} - \psi_{i+1}) + \beta_1 y_1\right) = 0.$$

If $i = N$,

$$\left(\frac{\partial}{\partial t_1} + \frac{\partial \psi_N}{\partial x_1}\frac{\partial}{\partial y_1} - \frac{\partial \psi_N}{\partial y_1}\frac{\partial}{\partial x_1}\right)\left(\frac{\partial^2 \psi_N}{\partial x_1^2} + \frac{\partial^2 \psi_N}{\partial y_1^2}\right.$$
$$\left. + F_N(\psi_{N-1} - \psi_N) + \beta_1 y_1 + \eta_b\right) = 0,$$

where $F_i = \dfrac{DF}{D_i}\dfrac{\rho_0}{\rho_{i+1} - \rho_i}$ for $i \leq N-1$ (here we suppose $\dfrac{\rho_0}{\rho_{i+1} - \rho_i} = \dfrac{\rho_0}{\rho_i - \rho_{i-1}}$), and $F_N = \dfrac{DF}{D_N}\dfrac{\rho_0}{\rho_N - \rho_{N-1}}$ when $i = N$.

In [159], Majda and Wang proved that the solutions of equations (2.3.4) and (2.3.5) converge to solutions of the equations as follows when $\eta_b = 0$ and $\dfrac{F_2}{F_1}$ tend to zero,

$$\left(\frac{\partial}{\partial t_1} + \frac{\partial \phi_1}{\partial x_1}\frac{\partial}{\partial y_1} - \frac{\partial \phi_1}{\partial y_1}\frac{\partial}{\partial x_1}\right)\left(\frac{\partial^2 \phi_1}{\partial x_1^2} + \frac{\partial^2 \phi_1}{\partial y_1^2} - F_1\phi_1 + \beta_1 y_1 + F_1\phi_2\right) = 0,$$
$$\left(\frac{\partial}{\partial t_1} + \frac{\partial \phi_2}{\partial x_1}\frac{\partial}{\partial y_1} - \frac{\partial \phi_2}{\partial y_1}\frac{\partial}{\partial x_1}\right)\left(\frac{\partial^2 \phi_2}{\partial x_1^2} + \frac{\partial^2 \phi_2}{\partial y_1^2} + \beta_1 y_1\right) = 0.$$

Colin proved the global well-posedness of Cauchy problem of the multi-layer quasi-geostrophic model in [39], and proved that when the layer of multi-layer quasi-geostrophic model tends to infinity and the thickness of each layer tends to zero, the continuity limit of the multi-layer quasi-geostrophic model is the three-dimensional quasi-geostrophic model.

2.4 The Surface Quasi-Geostrophic Equation

2.4.1 *Introduction of Surface Quasi-Geostrophic Equation*

As is in the deduction in section 2.2, expanding the adiabatic frictionless atmospheric or oceanic equations under hydrostatic approximation and Boussinesq approximation with respect to small Rossby number, we get the first order approximation

$$\partial_t \xi + J(\psi, \xi) = f \partial_z w,$$

$$\partial_t \theta + J(\psi, \theta) = -\frac{N^2}{f} w,$$

where $\xi = \psi_{xx} + \psi_{yy}$, ψ is a stream function, $J(\psi, \xi) = \psi_x \xi_y - \psi_y \xi_x$, w is the vertical velocity, $\theta = \psi_z$, θ is the potential temperature for the atmosphere, θ is buoyancy for the oceans, f is Coriolis parameter, and N is buoyancy frequency. For simplicity, we suppose that f and N are both constants. Eliminate w in the above two equations, we obtain

$$\partial_t q + J(\psi, q) = 0, \tag{2.4.1}$$

where $q = (\partial_{xx} + \partial_{yy} + \partial_{zz})\psi$, and q denotes the potential vorticity. If there are no heat exchange on the boundary, and no friction on the boundary, and the bottom surface is a plane, then the lower boundary ($z = 0$) condition of equation (2.4.1) can be written as

$$\frac{\mathrm{d}}{\mathrm{d}t}\left(\frac{\partial \psi}{\partial z}\right) = 0,$$

that is,

$$\partial_t \theta + J(\psi, \theta) = 0, \tag{2.4.2}$$

where $\theta(x, y) = \partial_z \psi|_{z=0}$.

One special example of (2.4.1) and (2.4.2) is the following system

$$(\partial_{xx} + \partial_{yy} + \partial_{zz})\psi = 0, \quad z > 0, \tag{2.4.3}$$

$$\partial_t \theta + J(\psi, \theta) = 0, \quad z = 0, \tag{2.4.4}$$

$$\psi \to 0, \quad z \to \infty. \tag{2.4.5}$$

System (2.4.3)-(2.4.5) describes the evolution of the lower surface potential temperature (or buoyancy) in the identical potential vorticity field($q = 0$), which can be used to simulate the frontogenesis of the edge between warm and cold air mass. For the detail of application (2.4.3)-(2.4.5), see [100].

If $\theta(x, y)$ is given, by solving the boundary value of elliptic equation

$$(\partial_{xx} + \partial_{yy} + \partial_{zz})\psi = 0, \quad z > 0,$$

$$\partial_z \psi|_{z=0} = \theta(x, y),$$

$$\psi \to 0, \quad z \to \infty,$$

we get the distribution of ψ in $z > 0$. Therefore, for studying system (2.4.3)-(2.4.5), we consider the Cauchy problem of the following equation

$$\partial_t \theta + J(\psi, \theta) = 0,$$

that is,

$$\partial_t \theta + \boldsymbol{u} \cdot \boldsymbol{\nabla} \theta = 0, \quad \text{in } \mathbb{R}^2, \tag{2.4.6}$$

where

$$-\theta = (-\Delta)^{\frac{1}{2}} \psi, \tag{2.4.7}$$

$$\boldsymbol{u} = (u_1, u_2) = (-\psi_y, \psi_x), \tag{2.4.8}$$

where $\mathbb{R}^2$ can be replaced by two-dimensional torus $\mathbb{T}^2$, and the non-local operator $(-\Delta)^{\frac{1}{2}}$ is defined by Fourier transformation as follows,

$$\widehat{(-\Delta)^{\frac{1}{2}} f}(\xi) = |\xi| \hat{f}(\xi), \tag{2.4.9}$$

here

$$\hat{f}(\xi) = \frac{1}{2\pi} \int_{\mathbb{R}^2} \mathrm{e}^{-\mathrm{i}\eta \cdot \xi} f(\eta) \mathrm{d}\eta.$$

(2.4.8) is rewritten as

$$\boldsymbol{u} = (\partial_y (-\Delta)^{-\frac{1}{2}} \theta, -\partial_x (-\Delta)^{-\frac{1}{2}} \theta) = (-R_2 \theta, R_1 \theta), \tag{2.4.10}$$

where R_1, R_2 denote Riesz transformation,

$$\widehat{R_j f}(\xi) = -\frac{\mathrm{i}\xi_j}{|\xi|} \hat{f}(\xi).$$

Equation (2.4.6) is called the **surface quasi-geostrophic equation**.

Another important reason for studying equation (2.4.6) is that (2.4.6) has some important similarities with the three-dimensional incompressible Euler equation. In fact, taking derivative of equation (2.4.6) with respect to y and x, we get

$$\frac{\partial \boldsymbol{\nabla}^{\perp} \theta}{\partial t} + (\boldsymbol{u} \cdot \boldsymbol{\nabla}) \boldsymbol{\nabla}^{\perp} \theta = (\boldsymbol{\nabla} \boldsymbol{u}) \boldsymbol{\nabla}^{\perp} \theta, \tag{2.4.11}$$

where $\nabla^{\perp}\theta = \begin{pmatrix} -\theta_y \\ \theta_x \end{pmatrix}, (\nabla \boldsymbol{u}) = \begin{pmatrix} u_{1x} & u_{1y} \\ u_{2x} & u_{2y} \end{pmatrix}$, and the vorticity form for the three-dimensional incompressible Euler equation is

$$\frac{\partial \omega}{\partial t} + (\boldsymbol{v} \cdot \nabla)\omega = (\nabla \boldsymbol{v})\omega, \tag{2.4.12}$$

where $\omega = \mathbf{curl}\boldsymbol{v}, \boldsymbol{v} = (v_1, v_2, v_3), (\nabla \boldsymbol{v}) = (\partial_j v_i), \partial_j = \partial_x, \partial_y, \partial_z, i = 1, 2, 3.$

(2.4.11) and (2.4.12) have a similar construction: vortex stretching. In fact, the stream function in (2.4.7) is given by the equation as follows

$$\psi(\cdot) = -\int_{\mathbb{R}^2} \frac{\theta(\cdot + \eta)}{|\eta|} \mathrm{d}\eta. \tag{2.4.13}$$

Thus

$$\boldsymbol{u}(\cdot) = -\int_{\mathbb{R}^2} \frac{1}{|\eta|} \nabla^{\perp}\theta(\cdot + \eta) \mathrm{d}\eta. \tag{2.4.14}$$

The symmetric part of $(\nabla \boldsymbol{u})$ is

$$S(\cdot) = \frac{1}{2}\left[(\nabla \boldsymbol{u}) + (\nabla \boldsymbol{u})^*\right],$$

with (2.4.14), $S(x)$ is denoted as a singular integral

$$S(\cdot) = P.V. \int_{\mathbb{R}^2} N(\hat{\eta}, (\nabla^{\perp}\theta)(\cdot + \eta)) \frac{\mathrm{d}\eta}{|\eta|^2}, \tag{2.4.15}$$

where $P.V.$ denotes the integral principal value, $\hat{\eta} = \frac{\eta}{|\eta|}$, N is the function of two variables, and

$$N(\hat{\eta}, (\nabla^{\perp}\theta)) = \frac{1}{2}\left[\hat{\eta}^{\perp} \otimes (\nabla^{\perp}\theta)^{\perp} + (\nabla^{\perp}\theta)^{\perp} \otimes \hat{\eta}^{\perp}\right].$$

Here $\otimes$ denotes tensor product. According to Biot-Savart principle, and $\boldsymbol{v}$ in (2.4.12) can be denoted by ω, that is,

$$\boldsymbol{v}(\cdot) = -\frac{1}{4\pi}\int_{\mathbb{R}^3} \left(\nabla^{\perp}\frac{1}{|\eta|}\right) \times \omega(\cdot + \eta)\mathrm{d}\eta,$$

thus the symmetric part of $(\nabla \boldsymbol{v})$ is

$$S(\cdot) = \frac{1}{2}((\nabla \boldsymbol{v}) + (\nabla \boldsymbol{v})^*),$$

which is denoted as

$$S(\cdot) = \frac{3}{4\pi} P.V. \int M(\hat{\eta}, \omega(\cdot + \eta)) \frac{\mathrm{d}\eta}{|\eta|^3}, \tag{2.4.16}$$

where M is the function of two variables, and

$$M(\hat{\eta}, \omega) = \frac{1}{2}[\hat{\eta} \otimes (\hat{\eta} \times \omega) + (\hat{\eta} \times \omega) \otimes \hat{\eta}].$$

(2.4.15) and (2.4.16) respectively are the important quantity describing the vortex stretching of (2.4.11) and (2.4.12), which respectively are the only reason of the singularity occurrence of (2.4.11) and (2.4.12), for details, see [41].

2.4.2 *Some Results of Surface Quasi-Geostrophic Equation*

2.4.2.1 *The Surface Quasi-Geostrophic Equation*

The research of statistical turbulence theory of the quasi-geostrophic flows described by system (2.4.3)-(2.4.5) appears in [18] and [179]. Afterward, Held et al. studied some qualitative features of solutions for (2.4.3)-(2.4.5) through numeric calculation in [100], and that of solutions to equation (2.4.6) with some special initial data.

In [41], Constantin et al. studied the Cauchy problem of the surface quasi-geostrophic equation through combining mathematical theory with numeric analysis, (2.4.6) with the initial condition

$$\theta(t,x,y)|_{t=0} = \theta_0(x,y). \tag{2.4.17}$$

Firstly, they noticed that one can get the local existence of smooth solutions of (2.4.6) and (2.4.17) by applying the method for the proof of the local existence of smooth solutions in the conservation laws system, that is,

Theorem 2.4.1. If $\theta_0 \in H^k(\mathbb{R}^2)$, $k \geq 3$, then $T^* > 0$ exists, such that $0 \leq t < T^*$, $\theta(t,x,y)$ is a solution of equation (2.4.6), $\theta(t,x,y) \in H^k(\mathbb{R}^2)$, and if $[0,T^*)$ is the maximum interval of life-time of solutions to problem (2.4.6) and (2.4.17), then as $t \to T^*$,

$$\|\theta(\cdot,t)\|_{H^k} \to +\infty.$$

Moreover, they got a blow-up criterion of the Cauchy problem of the surface quasi-geostrophic equation.

Theorem 2.4.2. Suppose $\theta_0 \in H^k(\mathbb{R}^2)$, $k \geq 3, \theta(t,x,y)$ be the unique solution of equation (2.4.6) with (2.4.17) in interval $[0,T^*)$. The following three events are equivalent:

(i) $[0,T^*)$ is the maximum existence interval of life-time of solutions to (2.4.6) and (2.4.17);

(ii) when $T \to T^*$, $\displaystyle\int_0^T |\nabla\theta(s)|_{L^\infty} \mathrm{d}s \to +\infty$;

(iii) when $T \to T^*$, $\displaystyle\int_0^T \alpha^*(s)\mathrm{d}s \to +\infty$, where $\alpha^*(s) = \max\limits_{x\in\mathbb{R}^2}(S(x)\xi\cdot\xi) = \max\limits_{x\in\mathbb{R}^2}\alpha(x,s)$, $S(x)$ is given in (2.4.15), and $\xi = \dfrac{\nabla^\perp\theta}{|\nabla^\perp\theta|}$.

The proof of the equivalence of (i) and (ii) resembles the famous Beale-Kato-Majda criterion (see [11]). The results of Theorem 2.4.1 and Theorem 2.4.2 are also suitable for the periodic boundary value problem of the surface quasi-geostrophic equation.

Although the local smooth solutions of both the Cauchy problem and the periodic boundary value problem of the non-dissipative surface quasi-geostrophic equation exist, but whether they have blow-up solutions in a limited time or not is still an open problem. To solve this problem, researchers try to find the limited time blow-up solutions by numerical simulation. In [41], Constantin et al. chose

$$\theta_0 = \sin x \sin y + \cos y, \tag{2.4.18}$$

and obtained a strong frontogenesis solution of the periodic boundary value problem of the non-dissipative surface quasi-geostrophic equation by numerical simulation, which generates blow-up in a limited time. And inspired by the results of the numerical simulation, they studied the topology features of the potential temperature tensor level set which is blow-up. References [43,45,169] and reference [47] respectively give the numerical and theoretical evidences that the solution to the surface quasi-geostrophic equation with the initial data (2.4.18) is not blow-up in a limited time.

As far as we know about the theoretical research of Cauchy problem of equation (2.4.6), there are still three results:

(1) Resnick proved the global existence of weak solutions, the Cauchy problem, to the equation (2.4.6) in his doctoral dissertation. But the uniqueness of the weak solutions is still an open problem;

(2) D. Cordoba and Fefferman studied the growth of solutions to the Cauchy problem of equation (2.4.6) in reference [49];

(3) Khouiner and Titi got a blow-up criterion of the periodic boundary value problem of the surface quasi-geostrophic equation in [121], also see the following theorems.

Theorem 2.4.3. Let θ be the solution of the following problem in $[0, T^*)$,

$$\frac{\partial \theta}{\partial t} + \boldsymbol{u} \cdot \boldsymbol{\nabla} \theta = 0, \text{ in } \Omega, \tag{2.4.19a}$$

$$(-\Delta)^{\frac{1}{2}} \psi = -\theta, \boldsymbol{\nabla}^{\perp} \psi = \boldsymbol{u}, \text{ in } \Omega, \tag{2.4.19b}$$

$$\theta(x, y, 0) = \theta_0(x, y), \text{ in } \Omega, \tag{2.4.19c}$$

$$\int_\Omega \theta = \int_\Omega \psi = \int_\Omega \boldsymbol{v} = 0, \tag{2.4.19d}$$

where $\Omega = [0, 1] \times [0, 1]$, and ψ is periodical in the directions of x, y. Then $\limsup_{t \to T^*} \|\boldsymbol{\nabla}\theta\|_{L^2} = +\infty$ if and only if $\sup_{[0,T^*)} \liminf_{\alpha \to 0^+} \alpha^2 \|\boldsymbol{\nabla}\theta^\alpha\|_{L^2}^2 = \varepsilon > 0$, where θ^α is the solution of inviscous regularization problem of (2.4.19), that is, θ^α is the solution of the following system,

$$\begin{aligned}
&(1-\alpha^2)\theta^\alpha = \tilde{\theta}^\alpha,\\
&\frac{\partial\tilde{\theta}^\alpha}{\partial t} + \operatorname{div}(\boldsymbol{u}^\alpha\theta^\alpha) = 0,\\
&(-\Delta)^{-\frac{1}{2}}\psi^\alpha = \theta^\alpha, \boldsymbol{\nabla}^\perp\psi^\alpha = \boldsymbol{u}^\alpha,\\
&\int_\Omega \theta^\alpha = \int_\Omega \psi^\alpha = 0,\\
&\theta^\alpha(x,0) = \theta_0(x),
\end{aligned}$$

where ψ^α is periodical in the directions of x, y.

2.4.2.2 *Dissipative Surface Quasi-Geostrophic Equation*

The dissipative surface quasi-geostrophic equation is

$$\partial_t\theta + \boldsymbol{u}\cdot\boldsymbol{\nabla}\theta + \kappa(-\Delta)^{-\alpha}\theta = 0, \text{ in } \Omega, \tag{2.4.20}$$

where

$$-\theta = (-\Delta)^{\frac{1}{2}}\psi, \tag{2.4.21}$$

$$\boldsymbol{u} = (-\psi_y, \psi_x) = (-R_2\theta, R_1\theta), \tag{2.4.22}$$

where $\kappa > 0$, the non-local norm $(-\Delta)^\alpha$ $(1 \geq \alpha \geq 0)$ is defined by the Fourier transform:

$$\widehat{(-\Delta)^\alpha\theta}(\xi) = |\xi|^{2\alpha}\hat{\theta}(\xi),$$

R_j is the Riesz transform, and Ω can be $\mathbb{R}^2$ or two-dimensional torus $\mathbb{T}^2$.

When $\alpha = 1/2$, equation (2.4.20) is deduced by the following system,

$$(\partial_{xx} + \partial_{yy} + \partial_{zz})\psi = 0, z > 0, \tag{2.4.23a}$$

$$\frac{\mathrm{d}\dfrac{\partial\psi}{\partial z}}{\mathrm{d}t} = \frac{\mathrm{d}\theta}{\mathrm{d}t} = \frac{\partial\theta}{\partial t} + J(\psi,\theta) = -\frac{E_v^{\frac{1}{2}}}{2\varepsilon}(\partial_{xx} + \partial_{yy})\psi, z = 0, \tag{2.4.23b}$$

$$\psi \to 0, z \to \infty, \tag{2.4.23c}$$

where (2.4.23b) is just the equation (6.6.10) in [171] (when $\eta_B = 0, \mathscr{H} = 0$, that is, the lower boundary is a plane, and there is no heat exchange on the boundary), E_v is Ekman number, ε is Rossby number, and $-\frac{E_v^{\frac{1}{2}}}{2\varepsilon}(\partial_{xx} + \partial_{yy})\psi$ denotes friction.

In the following, we recall some results of the well-posedness of the dissipative surface quasi-geostrophic equation. Resnick obtained the global existence of weak solutions of the dissipative surface quasi-geostrophic equation with external force.

Theorem 2.4.4. Assume $\theta_0 \in L^2(\Omega)$, $f \in L^2([0,T]; H^{-\alpha}(\Omega))$, $\Omega = \mathbb{R}^2$ or two-dimensional torus $\mathbb{T}^2$. Then for any $T > 0$, the initial value problem

$$\partial_t \theta + \boldsymbol{u} \cdot \boldsymbol{\nabla} \theta + \kappa(-\Delta)^\alpha \theta = f, \tag{2.4.24}$$

$$\theta(x, y, 0) = \theta_0(x, y) \tag{2.4.25}$$

has a weak solution θ such that $\theta \in L^\infty([0,T]; L^2(\Omega)) \cap L^2([0,T]; H^\alpha(\Omega))$.

Constantin and Wu obtained, in [45] the global existence of strong solutions of (2.4.24) and (2.4.25) when $1/2 < \alpha \leq 1$.

Theorem 2.4.5. Assume $\dfrac{1}{2} < \alpha \leq 1$, $\beta > 2 - 2\alpha$. If $\theta_0 \in H^\beta(\mathbb{T}^2)$, for any $T > 0$, $f \in L^2([0,T]; H^{\beta-\alpha}(\mathbb{T}^2))$, $\displaystyle\int_0^T \|f(s)\|_{L^q} \mathrm{d}s < +\infty$, $q = \infty$ when $\beta \geq 1$, $q = \dfrac{2}{1-\beta}$ when $\beta < 1$, then the solutions of (2.4.24) and (2.4.25) satisfy

$$\|(-\Delta)^{\frac{\beta}{2}} \theta(t)\|_{L^2} \leq C.$$

$\forall t \leq T$, where C depends on T, $\|\theta_0\|_{H^\beta}, \|f\|_{L^2([0,T];H^{\beta-\alpha})}$, and $\displaystyle\int_0^T \|f(s)\|_{L^q}$.

Now we give the proof of Theorem 2.4.5 in [45].

Proof. Taking L^2 inner product of equation (2.4.24) with $(-\Delta)^\beta \theta$, we get

$$\frac{1}{2}\frac{\mathrm{d}}{\mathrm{d}t} \int |(-\Delta)^{\frac{\beta}{2}} \theta|^2 + \kappa \int |(-\Delta)^{\frac{\alpha+\beta}{2}} \theta|^2 = -\int (\boldsymbol{u} \cdot \boldsymbol{\nabla} \theta)(-\Delta)^\beta \theta + \int (-\Delta)^\beta \theta f. \tag{2.4.26}$$

Let $\Lambda = (-\Delta)^{\frac{1}{2}}, \dfrac{1}{p} + \dfrac{1}{q} = \dfrac{1}{2}$, p, $q > 2$. Applying

$$\|\Lambda^{\beta-\alpha+1}(gh)\|_{L^2} \leq C(\|g\|_{L^q} \|\Lambda^{\beta-\alpha+1} h\|_{L^p} + \|h\|_{L^q} \|\Lambda^{\beta-\alpha+1} g\|_{L^p}),$$

with the boundedness of Riesz operator (see [197]) and Gagliardo-Nirenberg Inequality, we get

$$-\int (\boldsymbol{u} \cdot \boldsymbol{\nabla} \theta) \Lambda^{2\beta\theta}$$

$$
\begin{aligned}
&\le \|\Lambda^{\beta-\alpha}(\boldsymbol{u}\cdot\boldsymbol{\nabla}\theta)\|_{L^2}\|\Lambda^{\beta+\alpha}\theta\|_{L^2}\\
&\le \|\Lambda^{\beta-\alpha}\mathrm{div}(\boldsymbol{u}\cdot\theta)\|_{L^2}\|\Lambda^{\beta+\alpha}\theta\|_{L^2}\\
&\le \|\Lambda^{\beta-\alpha+1}(\boldsymbol{u}\cdot\theta)\|_{L^2}\|\Lambda^{\beta+\alpha}\theta\|_{L^2}\\
&\le C(\|\Lambda^{\beta-\alpha+1}\theta\|_{L^{\frac{2}{\beta}}}\|u\|_{L^q}+\|\Lambda^{\beta-\alpha+1}u\|_{L^{\frac{2}{\beta}}}\|\theta\|_{L^q})\|\Lambda^{\beta+\alpha}\theta\|_{L^2}\\
&\le C\|\Lambda^{\beta-\alpha+1}\theta\|_{L^{\frac{2}{\beta}}}\|\theta\|_{L^q}\|\Lambda^{\beta+\alpha}\theta\|_{L^2}\\
&\le C\|\Lambda^{2-\alpha}\theta\|_{L^2}\|\theta\|_{L^q}\|\Lambda^{\beta+\alpha}\theta\|_{L^2}\\
&\le C\|\Lambda^{\beta+\alpha}\theta\|_{L^2}^{\frac{2-\alpha-\beta}{\alpha}}\|\Lambda^{\beta}\theta\|_{L^2}^{1-\frac{2-\alpha-\beta}{\alpha}}\|\theta\|_{L^q}\|\Lambda^{\beta+\alpha}\theta\|_{L^2}\\
&= C\|\Lambda^{\beta+\alpha}\theta\|_{L^2}^{\frac{2-\beta}{\alpha}}\|\Lambda^{\beta}\theta\|_{L^2}^{2-\frac{2-\beta}{\alpha}}\|\theta\|_{L^q}\\
&\le \frac{\kappa}{4}\|\theta\|_{H^{\alpha+\beta}}^2+C(\kappa,\theta_0,f)\|\theta\|_{H^\beta}^2,
\end{aligned}\tag{2.4.27}
$$

where we use $\beta+\alpha>2-\alpha$, that is, $\beta>2-2\alpha$, and $1-\beta\ge 0$, that is, $\beta\le 1$.

By (2.4.26), (2.4.27) and $|\int\Lambda^{2\beta}\theta f|\le\frac{\kappa}{4}\|\Lambda^{\alpha+\beta}\theta\|_{L^2}^2+\frac{1}{\kappa}\|\Lambda^{\beta-\alpha}f\|_{L^2}^2$, we have

$$
\frac{\mathrm{d}}{\mathrm{d}t}\int|\Lambda^\beta\theta|^2+\kappa\|\Lambda^{\alpha+\beta}\theta\|_{L^2}^2\le C(\kappa,\theta_0,f)\|\theta\|_{H^\beta}^2+C\|\Lambda^{\beta-\alpha}f\|_{L^2}^2. \tag{2.4.28}
$$

Applying Gronwall Inequality, we derive from (2.4.28) that $\forall t\le T, 2-2\alpha<\beta\le 1$,

$$
\|\Lambda^\beta\theta(t)\|_{L^2}^2\le C. \tag{2.4.29}
$$

It's easy to prove that $\forall\beta>1,\theta_0\in H^\beta(\mathbb{T}^2)$, $\|\Lambda^\beta\theta(t)\|_{L^2}^2\le C$.

Remark 2.4.6. In the proof of (2.4.27), we use the maximum principle, $\|\theta(t)\|_{L^q}\le\|\theta_0\|_{L^q}+\int_0^t\|f(s)\|_{L^q}\mathrm{d}s,\ 0\le t\le T$. This maximum principle is proved by Resnick in [182]. Afterward, for the situation of $f=0$, A. Cordoba and D. Cordoba proposed a different method to prove the above maximum principle when $q=2^n$(n is a positive integer) in [48], for $q=2$, Ju gave another proof in [117].

Remark 2.4.7. When $1\ge\alpha>1/2$, the smooth solutions of the dissipative surface quasi-geostrophic equation (2.4.24) globally exist. Researchers denote (2.4.24) when $\alpha>1/2$ as **the subcritical dissipative surface quasi-geostrophic equation**, (2.4.24) when $\alpha=1/2$ as **the critical dissipative surface quasi-geostrophic equation**, and the (2.4.24) when $0\le\alpha<1/2$ as **the supercritical dissipative surface quasi-geostrophic equation**.

Here we introduce several results about the global well-posedness of the critical dissipative surface quasi-geostrophic equation:

(1) Constantin and others obtained the global existence of smooth solutions in [40] for small enough $\|\theta_0\|_{L^\infty}$. To obtain this conclusion, they mainly proved the following theorem:

Theorem 2.4.8. If $\theta_0 \in H^2(\Omega)$, $\Omega = [0, 2\pi]^2$, θ_0 satisfies the periodical boundary conditions, and there exist a positive constant C_∞ such that $\|\theta_0\|_{L^\infty} \leq c_\infty \kappa$, then, the initial value problem of the critical dissipative surface quasi-geostrophic equation

$$\begin{cases} \theta_t + \boldsymbol{u} \cdot \nabla\theta + \kappa(-\Delta)^{\frac{1}{2}}\theta = 0, \text{ in } \Omega, & (2.4.30a) \\ \boldsymbol{u} = (u_1, u_2) = (-R_2\theta, R_1\theta), \text{ in } \Omega, & (2.4.30b) \\ \theta(x, 0) = \theta_0(x) & (2.4.30c) \end{cases}$$

has a unique global solution, and $\forall t \geq 0$,

$$\|\theta(\cdot, t)\|_{H^2} \leq \|\theta_0\|_{H^2}.$$

Proof. It's easy to prove that the initial value problem (2.4.30) has a unique local solution $\theta(t) \in H^2(\Omega)$, $t \in [0, T^*]$, where T^* is dependent on $\|\theta_0\|_{H^2(\Omega)}$. In order to prove the global existence in the situation of $\|\theta_0\|_{L^\infty} \leq c_\infty \kappa$, we make uniform *a priori* estimates of local solutions.

Taking $L^2(\Omega)$ inner product of equation (2.4.30a) with $\Delta^2\theta$, we have

$$\frac{1}{2}\frac{\mathrm{d}}{\mathrm{d}t}\int |\Delta\theta|^2 \mathrm{d}x + \kappa \int |(-\Delta)^{5/4}\theta|^2 \mathrm{d}x = -\int \Delta^2\theta(\boldsymbol{u} \cdot \nabla\theta)\mathrm{d}x.$$

By integration by parts and Hölder Inequality, we get

$$\int \Delta^2\theta(\boldsymbol{u} \cdot \nabla\theta)\mathrm{d}x = 2\int \nabla\boldsymbol{u} \cdot (\nabla(\nabla\theta))\Delta\theta \mathrm{d}x + \int (\Delta\boldsymbol{u} \cdot \nabla\theta)\Delta\theta \mathrm{d}x$$

$$\leq C[\|\nabla\boldsymbol{u}\|_{L^3}\|\Delta\theta\|_{L^3}^2 + \|\Delta\boldsymbol{u}\|_{L^3}\|\nabla\theta\|_{L^3}\|\Delta\theta\|_{L^3}].$$

According to the boundedness of the Riesz operator,

$$\|\Delta\boldsymbol{u}\|_{L^3} \leq C\|\Delta\theta\|_{L^3}, \|\nabla\boldsymbol{u}\|_{L^3} \leq C\|\nabla\theta\|_{L^3},$$

and the Gagliardol-Nirenberg Inequality

$$\|\nabla\theta\|_{L^3} \leq C\|\Delta\theta\|_{L^3}^{7/9}\|(-\Delta)^{5/4}\theta\|_{L^2}^{2/9}, \|\Delta\theta\|_{L^3} \leq C\|\theta\|_{L^\infty}^{1/9}\|(-\Delta)^{5/4}\theta\|_{L^2}^{8/9},$$

we obtain

$$\left|\int (-\Delta)^2\theta(\boldsymbol{u} \cdot \nabla\theta)\mathrm{d}x\right| \leq C\|\theta\|_{L^\infty}\|(-\Delta)^{5/4}\theta\|_{L^2}^2.$$

Thus, we get

$$\frac{1}{2}\frac{\mathrm{d}}{\mathrm{d}t}\int |\Delta\theta|^2 \mathrm{d}x + \kappa \int |(-\Delta)^{5/4}\theta|^2 \mathrm{d}x \leq C_\infty\|\theta\|_{L^\infty}\|(-\Delta)^{5/4}\theta\|_{L^2}^2.$$

According to the results of [172], θ satisfies the maximum principle

$$\|\theta(\cdot,t)\|_{L^\infty} \le \|\theta_0\|_{L^\infty}, \quad \forall\, t \ge 0,$$

letting $c_\infty = (C_\infty)^{-1}$, we have

$$\|\theta(\cdot,t)\|_{H^2} \le \|\theta_0\|_{H^2}.$$

The proof of Theorem 2.4.8 is complete.

(2) A. Cordoba and D. Cordoba proved the following theorem in [48].

Theorem 2.4.9. Let θ be a viscosity solution of (2.4.30) with initial data $\theta_0 \in H^s(\Omega)$ $(s > \frac{3}{2})$, $\Omega = \mathbb{R}^2$ or $\mathbb{T}^2$, that is, θ is the weak limit of a sequence of solution to problem $\theta_t^\varepsilon + \boldsymbol{u}^\varepsilon \cdot \boldsymbol{\nabla}\theta^\varepsilon + \kappa(-\Delta)^{\frac{1}{2}}\theta^\varepsilon = \varepsilon\Delta\theta^\varepsilon, \boldsymbol{u}^\varepsilon = (-R_2\theta^\varepsilon, R_1\theta^\varepsilon), \theta^\varepsilon(x,0) = \theta_0$, as $\varepsilon \to 0$. Then exist T_1, T_2 depending only on κ and θ_0, $0 < T_1 < T_2$, such that if $t \le T_1$, then $\theta(\cdot,t) \in C^1([0,T_1); H^s(\Omega))$ is the solution of (2.4.30), and

$$\|\theta(\cdot,t)\|_{H^s} \ll \|\theta_0\|_{H^s};$$

and if $t \ge T_2$, then $\theta(\cdot,t) \in C^1([T_2,\infty); H^s)$ is the solution of (2.4.30), and

$$\|\theta(\cdot,t)\|_{H^s} \le C\|\theta_0\|_{H^s},$$

with $\|\theta(\cdot,t)\|_{H^s}$ monotonically decreasing with respect to t, and satisfying

$$\int_{T_2}^\infty \|\theta(t)\|_{H^s}^2 \mathrm{d}t < +\infty;$$

in particular, as $t \to \infty$,

$$\|\theta(\cdot,t)\|_{H^s} = O(t^{-\frac{1}{2}}).$$

Recently, there are two important results about the global well-posedness of the critical dissipative surface quasi-geostrophic equation.

(1) Kiselev et al. in [122] constructed a proper continuous norm with nonlocal maximum principle, thus proved the global well-posedness of the periodic problem of the critical dissipative surface quasi-geostrophic equation, that is,

Theorem 2.4.10. The periodic problem of the critical dissipative surface quasi-geostrophic equation $\theta_t + \boldsymbol{u} \cdot \boldsymbol{\nabla}\theta + \kappa(-\Delta)^{\frac{1}{2}}\theta = 0$, $\theta(x,y,0) = \theta_0$, where θ_0 is smooth and periodic, has a unique and smooth solution globally, moreover,

$$\|\boldsymbol{\nabla}\theta\|_\infty \le C\|\boldsymbol{\nabla}\theta_0\|_\infty \exp\exp c\|\theta_0\|_\infty.$$

(2) In [23], Caffarelli and Vasseur proved the regularity of the weak solutions with the De Giorgi Nash Moser method and the harmonic continuation, thus obtained the global well-posedness of the Cauchy problem of the critical dissipative surface quasi-geostrophic equation.

The global well-posedness of the supercritical dissipative surface quasi-geostrophic equation is still open now. Researchers have proved the global existence of small solutions or the local existence of the solutions in Sobolev and Besov spaces of the Cauchy problem or the periodic problem of this equation, which can be seen in [30,36,48,103,116,118]. Now, we recall several results.

Chae and Lee obtained the global existence of the small solutions of the critical and supercritical dissipative surface quasi-geostrophic equations in [30].

Theorem 2.4.11. Suppose $0 \le \alpha \le 1/2$. There exists $\varepsilon > 0$ such that for any $\theta_0 \in B_{2,1}^{2-2\alpha}$, $\|\theta_0\|_{\dot{B}_{2,1}^{2-2\alpha}} < \varepsilon$, the initial value problem (2.2.24)-(2.2.25) in $\mathbb{R}^2$ has a unique global solution

$$\theta \in L^\infty(0,\infty; B_{2,1}^{2-2\alpha}) \cap L^1(0,\infty; \dot{B}_{2,1}^{2-2\alpha}) \cap C(0,\infty; B_{2,1}^{\beta}),$$

where the definitions of Besov spaces $B_{2,1}^{2-2\alpha}$, $\dot{B}_{2,1}^{2-2\alpha}$ will be given below, $\beta = \max\{2-2\alpha-\delta_1, 1\}$, and $\delta_1 > 0$.

Before the proof of Theorem 2.4.11, we give some symbols and the definition of Besov space first, for the detail, see [205]. Assume that S is the Schwarz class of rapidly decreasing functions, and the Fourier transform $\mathcal{F}(f) = \hat{f}$ of any $f \in S$ is defined by

$$\hat{f}(\xi) = \frac{1}{(2\pi)^{n/2}} \int \mathrm{e}^{-\mathrm{i}x\cdot\xi} f(x)\mathrm{d}x.$$

Let $\varphi \in S$ satisfy

$$\mathrm{Supp}\hat{\varphi} \subset \left\{\xi \in \mathbb{R}^n \Big| \frac{1}{2} \le |\xi| \le 2\right\}, \text{ and } \hat{\varphi}(\xi) \ge 0 \text{ if } \frac{1}{2} < |\xi| < 2\ ,$$

and let $\hat{\varphi}_j = \hat{\varphi}(2^{-j}\xi)$, that is, $\varphi_j(x) = 2^{jn}\varphi(2^j x)$. We adjust the normalization constant of $\hat{\varphi}$ such that

$$\sum_{j\in\mathbb{Z}} \hat{\varphi}_j(\xi) = 1\ \forall \xi \in \mathbb{R}^n \backslash \{0\}.$$

For any $k \in \mathbb{Z}$, define $S_k \in S$ by

$$\hat{S}_k(\xi) = 1 - \sum_{j \ge k+1} \hat{\varphi}_j(\xi).$$

It is notable that

$$\mathrm{Supp}\hat{\varphi}_j \cap \mathrm{Supp}\hat{\varphi}_{j'} = \phi \text{ if } |j - j'| \geq 2.$$

Let $s \in \mathbb{R}$, p, $q \in [0, \infty]$. For any $f \in S'$ given, we denote $\Delta_j f = \varphi_j * f$. Then we define the homogeneous Besov semi-norm $\|f\|_{\dot{B}^s_{p,q}}$ by

$$\|f\|_{\dot{B}^s_{p,q}} = \begin{cases} \left[\sum_{-\infty}^{\infty} 2^{jqs}\|\varphi_j * f\|_{L^p}^q\right]^{\frac{1}{q}}, & \text{if } q \in [1, \infty), \\ \sup_j [2^{js}\|\varphi_j * f\|_{L^p}], & \text{if } q = \infty. \end{cases}$$

Homogeneous Besov space $\dot{B}^s_{p,q}$ is a semi-normed space with semi-norm $\|\cdot\|_{\dot{B}^s_{p,q}}$. For any $s > 0$, define the norm of non-homogeneous Besov space by $\|f\|_{B^s_{p,q}}$, where $f \in S'$, $\|f\|_{B^s_{p,q}} = \|f\|_{L_p} + \|f\|_{\dot{B}^s_{p,q}}$.

After defining Besov space, we give the following proposition of the basic properties of Besov space used in the proof of Theorem 2.4.11.

Proposition 2.4.12.

(i) **Bernstein's Lemma**. Assume that $f \in L^p, 1 \leq p \leq \infty$, and $\mathrm{supp}\hat{f} \subset \{2^{j-2} \leq |\xi| < 2^j\}$. Then, there exists a positive constant C_k such that the following inequality holds,

$$C_k^{-1} 2^{jk}\|f\|_{L_p} \leq \|D^k f\|_{L_p} \leq C_k 2^{jk}\|f\|_{L_p}.$$

(ii) The following norms are equivalent,

$$\|D^k f\|_{\dot{B}^s_{p,q}} \sim \|f\|_{\dot{B}^{s+k}_{p,q}}.$$

(iii) Assume $s > 0$, $q \in [1, \infty]$. Then exists a positive constant C such that

$$\|fg\|_{\dot{B}^s_{p,q}} \leq C(\|f\|_{L^{p_1}}\|g\|_{\dot{B}^s_{p_2,q}} + \|g\|_{L_{r_1}}\|f\|_{\dot{B}^s_{r_2,q}}),$$

where $p_1, r_1 \in [1, \infty], \dfrac{1}{p} = \dfrac{1}{p_1} + \dfrac{1}{p_2} = \dfrac{1}{r_1} + \dfrac{1}{r_2}$.

Let $s_1, s_2 \leq \dfrac{N}{p}, s_1 + s_2 > 0, f \in \dot{B}^{s_1}_{p,1}, g \in \dot{B}^{s_2}_{p,1}$. Then $fg \in \dot{B}^{s_1+s_2-\frac{N}{p}}_{p,1}$, and

$$\|fg\|_{\dot{B}^{s_1+s_2-\frac{N}{p}}_{p,1}} \leq C\|f\|_{\dot{B}^{s_1}_{p,1}}\|g\|_{\dot{B}^{s_2}_{p,1}}.$$

(iv) If $s \in \left(-\dfrac{N}{p} - 1, \dfrac{N}{p}\right]$, then

$$\|[\boldsymbol{u}, \Delta_q]w\|_{L^p} \leq C_q 2^{-q(s+1)}\|\boldsymbol{u}\|_{\dot{B}^{\frac{N}{p}+1}_{p,1}}\|\boldsymbol{w}\|_{\dot{B}^s_{p,1}},$$

where $\sum_{q\in\mathbb{Z}} C_q \leq 1$,

$$[\boldsymbol{u}, \Delta_q]\boldsymbol{w} = \boldsymbol{u}\Delta_q \boldsymbol{w} - \Delta_q(\boldsymbol{u}\boldsymbol{w}).$$

(v) (Embedding Theorem) $\dot{B}^{\frac{N}{p}}_{p,1}(\mathbb{R}^N)$ is an algebra included in $C_0(\mathbb{R}^N)$. Assume $s \in \mathbb{R}, \varepsilon > 0$, $p, q \in [1, \infty]$. Then,

$$\dot{B}^s_{p,1} \hookrightarrow \dot{H}^s_p \hookrightarrow \dot{B}^s_{p,\infty},$$

and

$$B^{s+\varepsilon}_{p,\infty} \hookrightarrow B^s_{p,q}.$$

(vi) (Two Interpolation Inequalities) Assume $s_1, s_2 \in \mathbb{R}, \theta \in [0,1]$. Then,

$$\|\boldsymbol{u}\|_{\dot{B}^{\theta s_1+(1-\theta)s_2}_{p,1}} \leq C\|\boldsymbol{u}\|^{\theta}_{\dot{B}^{s_1}_{p,1}} \|\boldsymbol{u}\|^{1-\theta}_{\dot{B}^{s_2}_{p,1}},$$

$$\|\boldsymbol{u}\|_{B^{\theta s_1+(1-\theta)s_2}_{p,1}} \leq C\|\boldsymbol{u}\|^{\theta}_{B^{\theta s_1}_{p,1}} \|\boldsymbol{u}\|^{1-\theta}_{B^{s_2}_{p,1}}.$$

Proof of Theorem 2.4.11.

(1) *A priori* estimates. Applying the operator Δ_q to equation (2.2.24), we obtain

$$\partial_t \Delta_q \theta + (\boldsymbol{u}\cdot\boldsymbol{\nabla})\Delta_q\theta + \kappa\Lambda^{2\alpha}\Delta_q\theta = -[\Delta_q, \boldsymbol{u}]\cdot\boldsymbol{\nabla}\theta. \tag{2.4.31}$$

Taking L^2 inner product of equation (2.4.31) with $\Delta_q\theta$, and then applying (iv) in Proposition 2.4.12, we have

$$\begin{aligned}\frac{1}{2}\frac{\mathrm{d}}{\mathrm{d}t}\|\Delta_q\theta\|^2_{L^2} + C\kappa 2^{2\alpha q}\|\Delta_q\theta\|^2_{L^2} \leq &\|[\Delta_q, \boldsymbol{u}]\cdot\boldsymbol{\nabla}\theta\|_{L^2}\|\Delta_q\theta\|_{L^2}\\ \leq &C_q 2^{-(2-2\alpha)q}\|\boldsymbol{u}\|_{\dot{B}^2_{2,1}}\|\boldsymbol{\nabla}\theta\|_{\dot{B}^{1-2\alpha}_{2,1}}\|\Delta_q\theta\|_{L^2}.\end{aligned}$$

Dividing both sides of the above inequality by $\|\Delta_q\theta\|_{L^2}$, then multiplying them by $2^{(2-2\alpha)q}$, and summing up in $\mathbb{Z}$ with respect to q, we have

$$\begin{aligned}\frac{\mathrm{d}}{\mathrm{d}t}\|\theta(t)\|_{\dot{B}^{2-2\alpha}_{2,1}} + C_1\kappa\|\theta(t)\|_{\dot{B}^2_{2,1}} &\leq C\|\boldsymbol{u}(t)\|_{\dot{B}^2_{2,1}}\|\theta(t)\|_{\dot{B}^{2-2\alpha}_{2,1}}\\ &\leq C_2\|\theta(t)\|_{\dot{B}^2_{2,1}}\|\theta(t)\|_{\dot{B}^{2-2\alpha}_{2,1}}.\end{aligned} \tag{2.4.32}$$

Applying Calderon-Zygmund type inequality, we get

$$\|\boldsymbol{u}(t)\|_{\dot{B}^2_{2,1}} \leq C\|\theta(t)\|_{\dot{B}^2_{2,1}}.$$

By Gronwall Inequality, we derive from (2.4.32)

$$\begin{aligned}&\sup_{0\leq t<\infty}\|\theta(t)\|_{\dot{B}^{2-2\alpha}_{2,1}} + C_1\kappa\int_0^\infty \|\theta(t)\|_{\dot{B}^2_{2,1}}\mathrm{d}t\\ \leq &\|\theta_0\|_{\dot{B}^{2-2\alpha}_{2,1}}\exp\left(C_2\int_0^\infty \|\theta(t)\|_{\dot{B}^2_{2,1}}\mathrm{d}t\right).\end{aligned}$$

According to the above inequality and

$$\|\theta(t)\|_{L^2} \leq \|\theta_0\|_{L^2},$$

we get

$$\sup_{0\leq t<\infty} \|\theta(t)\|_{B^{2-2\alpha}_{2,1}} + C_1\kappa \int_0^\infty \|\theta(t)\|_{\dot{B}^2_{2,1}} \mathrm{d}t$$
$$\leq \|\theta_0\|_{B^{2-2\alpha}_{2,1}} \exp\left(C_2 \int_0^\infty \|\theta(t)\|_{\dot{B}^2_{2,1}} \mathrm{d}t \right). \tag{2.4.33}$$

(2) The uniform estimate of the approximate solutions. Define the sequence $\{\theta^n\}$ satisfying

$$\begin{cases} \partial_t \theta^{n+1} + (\boldsymbol{u}^n \cdot \boldsymbol{\nabla})\theta^{n+1} + \kappa \Lambda^{2\alpha}\theta^{n+1} = 0, \mathbb{R}^2 \times \mathbb{R}_+, \\ \boldsymbol{u}^n = \boldsymbol{\nabla}^\perp(-\Delta)^{-\frac{1}{2}}\theta^n, \\ \theta^{n+1}(0,x) = \theta_0^{n+1}(x) = \sum\limits_{q\leq n+1} \Delta_q \theta_0, \end{cases} \tag{I}$$

letting $(\theta^0, u^0) = (0,0)$, we obtain the existence of solutions $(\theta^n, \boldsymbol{u}^n)$ to the above system. Like *a priori* estimates in (1), we have

$$\sup_{0\leq t<\infty} \|\theta^{n+1}(t)\|_{\dot{B}^{2-2\alpha}_{2,1}} + C_1\kappa \int_0^\infty \|\theta^{n+1}(t)\|_{\dot{B}^2_{2,1}} \mathrm{d}t$$
$$\leq \|\theta_0^{n+1}\|_{\dot{B}^{2-2\alpha}_{2,1}} \exp\left(C_2 \int_0^\infty \|\theta^n(t)\|_{\dot{B}^2_{2,1}} \mathrm{d}t \right). \tag{2.4.34}$$

Next we shall prove that if $\|\theta_0\|_{B^{2-2\alpha}_{2,1}} \leq \varepsilon$, then, for any n,

$$\sup_{0\leq t<\infty} \|\theta^{n+1}(t)\|_{\dot{B}^{2-2\alpha}_{2,1}} + C_1\kappa \int_0^\infty \|\theta^{n+1}(t)\|_{\dot{B}^2_{2,1}} \mathrm{d}t \leq M\varepsilon, \tag{2.4.35}$$

where $M > 0$. Assume $M > 1$. Then choose ε small enough such that

$$\exp \frac{C_2 M\varepsilon}{C_1\kappa} \leq M.$$

If $\|\theta^n\|_{L^\infty(0,T;\dot{B}^{2-2\alpha}_{2,1})} + C_1\kappa\|\theta^n\|_{L^1(0,T;\dot{B}^2_{2,1})} \leq M\varepsilon$, then, according to (2.4.34), we get

$$\sup_{0\leq t<\infty} \|\theta^{n+1}(t)\|_{\dot{B}^{2-2\alpha}_{2,1}} + C_1\kappa \int_0^\infty \|\theta^{n+1}(t)\|_{\dot{B}^2_{2,1}} \mathrm{d}t$$
$$\leq \|\theta_0^{n+1}\|_{\dot{B}^{2-2\alpha}_{2,1}} \exp\left(C_2 \int_0^\infty \|\theta^n(t)\|_{\dot{B}^2_{2,1}} \mathrm{d}t \right)$$
$$\leq \|\theta_0^{n+1}\|_{\dot{B}^{2-2\alpha}_{2,1}} \exp\left(\frac{C_2 M\varepsilon}{C_1\kappa} \right) \leq M\varepsilon,$$

which means that (2.4.35) is valid.

(3) The convergence of the approximate solutions. In this part, we prove $\{\theta^n\}$ is a Cauchy convergence sequence in $L^\infty(0,\infty;B^1_{2,1})\cap L^1(0,\infty;\dot{B}^{1+2\alpha}_{2,1})$. Assume $\delta\theta^{n+1}=\theta^{n+1}-\theta^n$, $\delta\boldsymbol{u}^{n+1}=\boldsymbol{u}^{n+1}-\boldsymbol{u}^n$, then the sequence $\{\delta\theta^n\}$ satisfies

$$\begin{cases}\partial_t\delta\theta^{n+1}+(\boldsymbol{u}^n\cdot\boldsymbol{\nabla})\delta\theta^{n+1}+(\delta\boldsymbol{u}^n\cdot\boldsymbol{\nabla})\theta^n+\kappa\Lambda^{2\alpha}\delta\theta^{n+1}=0,\ \mathbb{R}^2\times\mathbb{R}_+,\\ \delta\boldsymbol{u}^n=\boldsymbol{\nabla}^\perp(-\Delta)^{-\frac{1}{2}}\delta\theta^n,\\ \delta\theta^{n+1}(0,x)=\Delta_{n+1}\theta_0,\end{cases}\tag{I'}$$

similarly to the *a priori* estimates of (1), applying (iv) in Proposition 2.4.12, that is

$$\|[\boldsymbol{u}^n,\Delta_q]\boldsymbol{\nabla}\delta\theta^{n+1}\|_{L^2}\le C_q2^{-q}\|\boldsymbol{u}^n\|_{\dot{B}^2_{2,1}}\|\boldsymbol{\nabla}\delta\theta^{n+1}\|_{\dot{B}^0_{2,1}},$$

we obtain

$$\begin{aligned}&\frac{1}{2}\frac{\mathrm{d}}{\mathrm{d}t}\|\Delta_q\delta\theta^{n+1}\|^2_{L^2}+C_3\kappa2^{2\alpha q}\|\Delta_q\delta\theta^{n+1}\|^2_{L^2}\\ \le&C2^{-q}\|\boldsymbol{u}^n\|_{\dot{B}^2_{2,1}}\|\delta\theta^{n+1}\|_{\dot{B}^1_{2,1}}\|\Delta_q\delta\theta^{n+1}\|_{L^2}\\ &+C\|\Delta_q(\delta\boldsymbol{u}^n\cdot\boldsymbol{\nabla}\theta^n)\|_{L^2}\|\Delta_q\delta\theta^{n+1}\|_{L^2}.\end{aligned}$$

Dividing both sides of the above inequality by $\|\Delta_q\delta\theta^{n+1}\|_{L^2}$, then multiplying it by 2^q, at last summing up in $\mathbb{Z}$ with respect to q, we have

$$\begin{aligned}&\frac{\mathrm{d}}{\mathrm{d}t}\|\delta\theta^{n+1}\|_{\dot{B}^1_{2,1}}+C_3\kappa\|\delta\theta^{n+1}\|_{\dot{B}^{1+2\alpha}_{2,1}}\\ \le&C\|\boldsymbol{u}^n\|_{\dot{B}^1_{2,1}}\|\delta\theta^{n+1}\|_{\dot{B}^1_{2,1}}+C\|(\delta\boldsymbol{u}^n\cdot\boldsymbol{\nabla})\theta^n\|_{\dot{B}^1_{2,1}}.\end{aligned}$$

Applying (iii) of Proposition 2.4.12, that is,

$$\|(\delta\boldsymbol{u}^n\cdot\boldsymbol{\nabla})\theta^n\|_{\dot{B}^1_{2,1}}\le C\|\delta\boldsymbol{u}^n\|_{\dot{B}^1_{2,1}}\|\boldsymbol{\nabla}\theta^n\|_{\dot{B}^1_{2,1}},$$

and the Calderon-Zygmund type inequality

$$\|\delta\boldsymbol{u}^n\|_{\dot{B}^1_{2,1}}\le C\|\delta\theta^n\|_{\dot{B}^1_{2,1}},$$

we get

$$\begin{aligned}&\frac{\mathrm{d}}{\mathrm{d}t}\|\delta\theta^{n+1}\|_{\dot{B}^1_{2,1}}+C_3\kappa\|\delta\theta^{n+1}\|_{\dot{B}^{1+2\alpha}_{2,1}}\\ \le&C_4\|\theta^n\|_{\dot{B}^2_{2,1}}\|\delta\theta^{n+1}\|_{\dot{B}^1_{2,1}}+C_5\|\delta\theta^n\|_{\dot{B}^1_{2,1}}\|\boldsymbol{\nabla}\theta^n\|_{\dot{B}^1_{2,1}}.\end{aligned}$$

Applying Gronwall's Inequality, we derive from the above inequality

$$\begin{aligned}&\sup_{0\le t<\infty}\|\delta\theta^{n+1}(t)\|_{\dot{B}^1_{2,1}}+C_3\kappa\int_0^\infty\|\delta\theta^{n+1}\|_{\dot{B}^{1+2\alpha}_{2,1}}\mathrm{d}t\\ \le&\|\delta\theta_0^{n+1}\|_{\dot{B}^1_{2,1}}\exp\left(C_5\int_0^\infty\|\theta^n(t)\|_{\dot{B}^2_{2,1}}\mathrm{d}t\right)\\ &+C_4\sup_{0\le t<\infty}\|\delta\theta^n(t)\|_{\dot{B}^1_{2,1}}\int_0^\infty\|\theta^n(t)\|_{\dot{B}^2_{2,1}}\mathrm{d}t\exp\left(C_5\int_0^\infty\|\theta^n(\tau)\|_{\dot{B}^2_{2,1}}\mathrm{d}\tau\right).\end{aligned}$$

According to (2.4.35), taking ε small enough such that

$$\exp\left(C_5\int_0^\infty \|\theta^n(t)\|_{\dot{B}^2_{2,1}}\mathrm{d}t\right)\le 2,$$

$$C_4\int_0^\infty \|\theta^n(t)\|_{\dot{B}^2_{2,1}}\mathrm{d}t<\frac{1}{8},$$

where $n\in\mathbb{N}$, then

$$\begin{aligned}&\sup_{0\le t<\infty}\|\delta\theta^{n+1}(t)\|_{\dot{B}^1_{2,1}}+C_3\kappa\int_0^\infty\|\delta\theta^{n+1}(t)\|_{\dot{B}^{1+2\alpha}_{2,1}}\mathrm{d}t\\ \le&2\|\delta\theta_0^{n+1}\|_{\dot{B}^1_{2,1}}+\frac{1}{4}\sup_{0\le t<\infty}\|\delta\theta^n(t)\|_{\dot{B}^1_{2,1}}.\end{aligned}$$

Therefore, applying the above inequality repeatedly, for any positive integer N, we obtain

$$\begin{aligned}&\sum_{n=1}^N\sup_{0\le t<\infty}\|\delta\theta^{n+1}(t)\|_{\dot{B}^1_{2,1}}+C_3\kappa\sum_{n=1}^N\int_0^\infty\|\delta\theta^{n+1}(t)\|_{\dot{B}^{1+2\alpha}_{2,1}}\mathrm{d}t\\ \le&2\sum_{n=1}^N\|\delta\theta_0^{n+1}\|_{\dot{B}^1_{2,1}}+\frac{1}{4}\sum_{n=2}^N\sup_{0\le t<\infty}\|\delta\theta^n(t)\|_{\dot{B}^1_{2,1}}\\ \le&2\sum_{n=1}^N\|\Delta_{n+1}\theta_0\|_{\dot{B}^1_{2,1}}+\frac{1}{4}\sum_{n=2}^N\sup_{0\le t<\infty}\|\delta\theta^n(t)\|_{\dot{B}^1_{2,1}}\\ \le&2C\|\theta_0\|_{\dot{B}^1_{2,1}}+\frac{1}{4}\sum_{n=2}^N\sup_{0\le t<\infty}\|\delta\theta^n(t)\|_{\dot{B}^1_{2,1}}\\ \le&2C\|\theta_0\|_{\dot{B}^1_{2,1}}+\cdots+\frac{2C}{4^{N-1}}\|\theta_0\|_{\dot{B}^1_{2,1}}\le\frac{8C}{3}\|\theta_0\|_{\dot{B}^1_{2,1}}.\end{aligned}\tag{2.4.36}$$

Taking L^2 inner product of the first equation of (I) with $\delta\theta^{n+1}$, we get

$$\begin{aligned}\frac{1}{2}\frac{\mathrm{d}}{\mathrm{d}t}\|\delta\theta^{n+1}\|^2_{L^2}+\kappa\|\delta\theta^{n+1}\|^2_{\dot{H}^\alpha}&\le C\|(\delta\boldsymbol{u}^n\cdot\boldsymbol{\nabla})\theta^n\|_{L^2}\|\delta\theta^{n+1}\|_{L^2}\\ &\le C\|\delta\boldsymbol{u}^n\|_{L^2}\|\boldsymbol{\nabla}\theta^n\|_{L^\infty}\|\delta\theta^{n+1}\|_{L^2}.\end{aligned}$$

Applying (v) in Proposition 2.4.12 and Calderon-Zygmund Inequality, we know that there exists a $C_6>0$ such that

$$\frac{\mathrm{d}}{\mathrm{d}t}\|\delta\theta^{n+1}\|_{L^2}\le C\|\delta\boldsymbol{u}^n\|_{L^2}\|\theta^n\|_{\dot{B}^2_{2,1}}\le C_6\|\delta\theta^n\|_{L^2}\|\theta^n\|_{\dot{B}^2_{2,1}}.$$

In (2.4.35), taking ε small enough such that $C_6\int_0^\infty\|\theta^n(t)\|_{\dot{B}^2_{2,1}}\mathrm{d}t<\frac{1}{4}$, applying Gronwall Inequality, we get

$$\sup_{0\le t<\infty}\|\delta\theta^{n+1}(t)\|_{L^2}\le\|\delta\theta_0^{n+1}\|_{L^2}+C_6\sup_{0\le\tau<\infty}\|\delta\theta^n(\tau)\|_{L^2}\int_0^\infty\|\theta^n(t)\|_{\dot{B}^2_{2,1}}\mathrm{d}t$$

$$\leq \|\delta\theta_0^{n+1}\|_{L^2} + \frac{1}{4} \sup_{0\leq\tau<\infty} \|\delta\theta^n(\tau)\|_{L^2}.$$

Therefore, for any positive integer N,

$$\sum_{n=1}^{N} \sup_{0\leq t<\infty} \|\delta\theta^{n+1}(t)\|_{L^2} \leq \sum_{n=1}^{N} \|\delta\theta_0^{n+1}\|_{L^2} + \frac{1}{4}\sum_{n=2}^{N} \sup_{0\leq t<\infty} \|\delta\theta^n(t)\|_{L^2}$$

$$\leq \sum_{n=1}^{N} \|\Delta_{n+1}\theta_0\|_{L^2} + \frac{1}{4}\sum_{n=2}^{N} \sup_{0\leq t<\infty} \|\delta\theta^n(t)\|_{L^2}$$

$$\leq \|\theta_0\|_{B^1_{2,1}} + \frac{1}{4}\|\theta_0\|_{B^1_{2,1}} + \cdots + \frac{1}{4^{N-1}}\|\theta_0\|_{B^1_{2,1}} \leq \frac{4}{3}\|\theta_0\|_{B^1_{2,1}}. \qquad (2.4.37)$$

According to (2.4.36) and (2.4.37), with iteration, we prove that $\{\theta^n\}$ is a Cauchy convergence sequence of $L^\infty(0,\infty;B^1_{2,1}) \cap L^1(0,\infty;\dot{B}^{1+2\alpha}_{2,1})$. Thus there exists $\theta \in L^\infty(0,\infty;B^1_{2,1}) \cap L^1(0,\infty;\dot{B}^{1+2\alpha}_{2,1})$ such that θ^n converges to θ in $L^\infty(0,\infty;B^1_{2,1}) \cap L^1(0,\infty;\dot{B}^{1+2\alpha}_{2,1})$, and θ is a global solution of the initial value problem (2.2.24)-(2.2.25). According to the *a priori* estimates of (1), we know $\theta \in L^\infty(0,\infty;B^{2-2\alpha}_{2,1}) \cap L^1(0,\infty;\dot{B}^2_{2,1})$. We prove in a similar way that the global solution of the initial value problem (2.2.24)-(2.2.25) is unique.

Next, we prove that $\theta(t)$ is continuous in $B^\beta_{2,1}$. For any $\delta_1 > 0, \beta = \max\{2-2\alpha-\delta_1, 1\}, \theta^{n+1}$ satisfies

$$\partial_t\theta^{n+1} = -(\boldsymbol{u}^n \cdot \boldsymbol{\nabla})\theta^{n+1} - \Lambda^{2\alpha}\theta^{n+1}.$$

The right-hand side of the above equality belongs to $L^1(0,\infty;B^1_{2,1})$. It's easy to prove that $\theta^{n+1} \in C([0,\infty);B^1_{2,1})$. Since

$$\|\theta(t)-\theta(s)\|_{B^1_{2,1}} \leq \|\theta(t)-\theta^{n+1}(t)\|_{B^1_{2,1}} + \|\theta^{n+1}(t)-\theta^{n+1}(s)\|_{B^1_{2,1}} + \|\theta^{n+1}(s)-\theta(s)\|_{B^1_{2,1}},$$

$\theta \in C([0,\infty);B^1_{2,1})$. Applying the interpolation inequality, we get $\theta \in C([0,\infty);B^\beta_{2,1})$, for any $\delta_1 > 0$, $\beta = \max\{2-2\alpha-\delta_1, 1\}$.

In [116], Ju proved the existence and uniqueness of small solutions and local solutions in Sobolev space of the initial value problem (2.4.24)-(2.4.25) of the surface quasi-geostrophic equation.

Theorem 2.4.13. Assume $\alpha \in (0,1)$, $\kappa > 0$, $\Omega = \mathbb{R}^2, \theta_0 \in H^s$.

(i) If $s = 2(1-\alpha)$, then there exists a constant $C_0 > 0$ such that for any weak solution θ of the initial value problem (2.4.24)-(2.4.25), if $\|\Lambda^s\theta_0\|_{L^2} \leq \dfrac{\kappa}{C_0}$, then

$$\|\Lambda^s\theta(t)\|_{L^2} \leq \|\Lambda^s\theta_0\|_{L^2}, \quad \forall t > 0,$$

if $\|\Lambda^s\theta_0\|_{L^2} < \dfrac{\kappa}{C_0}$, then $\theta \in L^2(0, +\infty; H^{s+\alpha})$, and the solution θ is unique.

(ii) If $s \in (2(1-\alpha), 2-\alpha]$, then there exists a $T = T(\kappa, \|\Lambda^s\theta_0\|_{L^2}) > 0$ such that for any weak solution θ of the initial value problem (2.4.24)-(2.4.25),

$$\theta \in L^\infty(0, T; H^s) \cap L^2(0, T; H^{s+\alpha}).$$

Moreover, if $\theta_0 \in L^2$, then θ is unique.

(iii) If $s > 2-\alpha$, then there exists a $T = T(\kappa, \|\theta_0\|_{L^2}, \|\Lambda^s\theta_0\|_{L^2}) > 0$ such that the weak solution θ of the initial value problem (2.4.24)-(2.4.25) satisfies

$$\theta \in L^\infty(0, T; H^s \cap L^2) \cap L^2(0, T; H^{s+\alpha}), \text{ if } \theta_0 \in H^s \cap L^2$$

and the solution θ is unique.

(iv) If $s > 2(1-\alpha)$, then there exists a $C_0 > 0$ such that if a weak solution θ of the initial value problem (2.4.24)-(2.4.25) satisfies

$$\|\theta_0\|_{L^2}^{\frac{s-2(1-\alpha)}{s}} \|\Lambda^s\theta_0\|_{L^2}^{\frac{2(1-\alpha)}{s}} \le \frac{\kappa}{C_0}, \tag{2.4.38}$$

then, for any $t > 0$,

$$\|\Lambda^s\theta(t)\|_{L^2} \le \|\Lambda^s\theta_0\|_{L^2}$$

and the solution θ is unique. Moreover, if inequality (2.4.38) strictly holds, then $\theta \in L^2(0, +\infty; H^{s+\alpha})$.

In [217], Wu obtained the global existence of small solutions in Besov space of the critical and supercritical dissipative surface quasi-geostrophic equations.

Theorem 2.4.14. Assume $\kappa > 0$, $0 \le \alpha \le \dfrac{1}{2}$, $\Omega = \mathbb{R}^2$. If the initial value θ_0 is in Besov space $B^r_{2,\infty}$, $r > 2 - 2\alpha$, then there exists a constant C_0 relying on α and r, such that. If

$$\|\theta_0\|_{B^r_{2,\infty}} \le C_0\kappa,$$

then the initial value problem (2.4.24)-(2.4.25) has a unique global solution θ,

$$\begin{aligned}\theta \in\ & L^\infty([0,\infty); B^r_{2,\infty}) \cap L^1([0,\infty); B^{r+2\alpha}_{2,\infty}) \cap \mathrm{Lip}([0,\infty); B^{r-1}_{2,\infty}) \\ & \cap C([0,\infty); B^\delta_{2,\infty}),\end{aligned}$$

where $\delta \in [r-1, r)$, and

$$\|\theta(\cdot, t)\|_{B^r_{2,\infty}} \le C_0\kappa, \text{ for any } t \ge 0.$$

Remark 2.4.15. Because

$$B^s_{2,1} \hookrightarrow H^s_2 \hookrightarrow B^s_{2,\infty},$$

Theorem 2.4.14 means the global existence of the small initial value solution in $B^s_{2,1}$ or H^s $(s > 2-2\alpha)$ of the initial value problem (2.4.24)-(2.4.25).

Chen, Miao and Zhang in [36] proved the global existence of the small initial value solution and the local existence in critical Besov space of the critical and supercritical dissipative surface quasi-geostrophic equation with Bernstein Inequality. Their results improved Theorem 2.4.11 and the first part of Theorem 2.4.13.

Theorem 2.4.16. Suppose $\kappa > 0, 0 < \alpha \leq 1/2$, $\Omega = \mathbb{R}^2, 2 \leq p < +\infty$, $1 \leq q < +\infty$. If the initial value θ_0 is in the Besov space $B^\sigma_{p,q}$, $\sigma = \dfrac{2}{p} + 1 - 2\alpha$, then, there exists a $T > 0$, such that the initial value problem (2.4.24)-(2.4.25) possesses a unique local solution θ, which satisfies

$$\theta \in C([0,T]; B^\sigma_{p,q}) \cap \tilde{L}^1\left([0,T]; \dot{B}^{\frac{2}{p}+1}_{p,q}\right),$$

where the lower bound of T is

$$\sup\{T' > 0;\ \|(1-\exp(-\kappa c_p 2^{2\alpha j} T'))^{\frac{1}{2}} 2^{j\sigma} \|\Delta_j \theta_0\|_{L^p}\|_{l^p(\mathbb{Z})} \leq c\kappa\},$$

and the constant c_p is in Bernstein Inequality. Moreover, there exists a positive constant ε, such that if $\|\theta_0\|_{\dot{B}^\sigma_{p,q}} \leq \varepsilon\kappa$, then the initial value problem (2.4.24)-(2.4.25) has a unique global solution θ. For the definition of space $\tilde{L}^1([0,T];\ \dot{B}^{\frac{2}{p}+1}_{p,q})$, see [36, Definition 2.3].

Remark 2.4.17.

(i) By the embedding

$$B^s_{2,1} \hookrightarrow H^s_2 \hookrightarrow B^s_{2,q},$$

where $q > 2$, we know that Theorem 2.4.16 improves Theorem 2.4.11 and the first part of Theorem 2.4.13.

(ii) Suppose $\kappa > 0, 0 \leq \alpha \leq 1$, $\Omega = \mathbb{R}^2$. If $\theta(x,t)$ is the solution of (2.4.24)-(2.4.25), then $\theta^\lambda(x,t) = \lambda^{2\alpha-1}\theta(\lambda x, \lambda^{2\alpha} t)$ is a solution of the initial value problem (2.4.24)-(2.4.25). Based on this scale-invariant property, researchers denote $B^\sigma_{p,q}$ (here $\sigma = \dfrac{2}{p} + 1 - 2\alpha$) as the critical Besov space of the initial value problem (2.4.24)-(2.4.25), $H^{2-2\alpha}$ as the critical Sobolev space of the initial value problem (2.4.24)-(2.4.25). The initial value problem (2.4.24)-(2.4.25) is ill-posed in $B^\sigma_{p,q}$ (here $\sigma < \dfrac{2}{p} + 1 - 2\alpha$) and $H^s(s < 2-2\alpha)$.

About the critical or supercritical dissipative surface quasi-geostrophic equation, researchers achieve some other results, see [2,46,117,216].

Chapter 3

Well-Posedness and Global Attractors of the Primitive Equations

In 1979, Zeng discussed the well-posedness problem of the atmospheric primitive equations without viscosity with Galerkin method in reference [209], and obtained the existence of weak solutions. In the early 1990s, many mathematicians (such as Lions, Temam and Wang) started the mathematical research of the primitive equations of large-scale atmosphere, oceans and the coupled atmospheric and oceanic primitive equations (see [140,141,142,143,144,203] and the references therein). In [140], through the introduction of viscosity and some technical processes, Lions, Temam and Wang obtained the new formulation of the large-scale dry atmospheric primitive equations suitable for mathematical treatments. In phase space H, the initial-boundary value problem of the new formulation of the primitive equations of the large-scale dry atmosphere is abbreviated as

$$\frac{\mathrm{d}\boldsymbol{U}}{\mathrm{d}t} + A\boldsymbol{U} + B(\boldsymbol{U},\boldsymbol{U}) + E(\boldsymbol{U}) = f,$$
$$\boldsymbol{U}(0) = \boldsymbol{U}_0,$$

where $\boldsymbol{U} = (\boldsymbol{v}, T)$, for more detail, see section 3.1. In the pressure coordinate frame, this new formulation of the primitive equations resembles the Navier-Stokes equation of incompressible fluid (of course there are some differences, such as, the nonlinear term of Navier-Stokes equation is $(\boldsymbol{u}\cdot\boldsymbol{\nabla})\boldsymbol{u}$, but the nonlinear term of the new primitive equations expression consists of $(\int_\xi^1 \mathrm{div}\boldsymbol{v}\mathrm{d}\xi')\frac{\partial \boldsymbol{v}}{\partial \xi}$, where $\boldsymbol{u}$ is the three-dimensional velocity field of Navier-Stokes equation, and $\boldsymbol{v}$ is the horizontal velocity field of the atmosphere). By means of the methods used in studying Navier-Stokes equation in [137], they proved the global existence of weak solutions to the initial boundary value problem of the primitive equations (but they did not study the global well-posedness of the strong solutions). Under the assumption of global existence of strong solutions to the initial boundary value problem of the primitive equations of the atmosphere with vertical viscosity where the strong solutions satisfy the uniform boundedness of H^1 norm in time, they obtained Hausdorff and fractal dimension of the global attractors. With the same method, in references [141,142], they established the mathematical theories respectively about the primitive equations of the oceans and the coupled atmosphere-ocean model introduced in [142], where they mainly proved the global existence of weak solutions, and studied the estimates for the Hausdorff and fractal dimension estimations of the global attractors under the hypothesis of the global existence of strong solutions. In [130,131], in the hypothesis of global existence of strong solutions to the dry and moist atmospheric primitive equations, Li and Chou studied the asymptotic behavior of the corresponding solutions.

In recent years, some mathematicians have again considered the existence, uniqueness and stability of the three-dimensional viscous primitive equations of the large-scale atmosphere and oceans (see [24, 25, 26, 50, 87, 107, 188, 193] and references therein). In [87], Guillén-González et al. cleverly treated the nonlinear term $(\int_\xi^1 \mathrm{div}\boldsymbol{v}d\xi')\frac{\partial \boldsymbol{v}}{\partial \xi}$ with anisotropic estimaties, thus obtained the global existence of strong solutions of the primitive equations of the large-scale oceans under the assumption of the initial data small enough, and proved the local existence of strong solutions for all the initial data. In [203], Temam and Ziane studies the local existence of strong solutions for the primitive equations of the atmosphere, the oceans and the coupling atmosphere-ocean primitive equations. References [24, 25, 26] are available to considering the dimensionless Boussinesq equations and the modified models, which are seen in [172,188]. In [24], Cao and Titi considered the global well-posedness and the finite-dimensional attractors of the three-dimensional planetary geostrophic model. The authors of [26] studied the global well-posedness of the three-dimensional viscous primitive equations of the large-scale oceans. In [26], Cao and Titi introduced a beautiful method: with hydrostatic approximation, decompose the horizontal velocity field $\boldsymbol{v}$ into the barotropic part $\bar{\boldsymbol{v}}$ and baroclinic part $\tilde{\boldsymbol{v}}$, and obtained the uniform boundness of $\tilde{\boldsymbol{v}}$ L^6-norm about time through complex calculations. With the uniform boundness of $\tilde{\boldsymbol{v}}$ L^6-norm about time, Cao and Titi completely proved the existence, uniqueness and stability of the strong solutions of three-dimensional viscous primitive equations of the large-scale ocean. Because of this result, researchers can say that the three-dimensional viscous primitive equations of the large-scale oceans are easier than the impressible Navier-Stokes equation, which is coincident with physical view, because the primitive equations is obtained by taking hydrostatic approximation.

In this chapter, we introduce some results of the primitive equations of the large-scale atmosphere and oceans and the infinite-dimensional dynamical system in geophysical fluid dynamics. In section 3.1, we consider the existence of atmospheric global attractors existence in weak meaning of the primitive equations of the moist atmosphere, see [88]. In section 3.2, we study the existence of global attractors of the primitive, see [93]. In section 3.3, we give the proof of the global existence of smooth solutions of viscous primitive equations, see [90]. In the last section, we give the well-posed results of the large-scale ocean primitive equations of the large-scale oceans, see [26].

3.1 Existence of Weak Solutions and Trajectory Attractors of the Moist Atmospheric Equations

In this section, we study the initial boundary value problem of the primitive equations of the large-scale moist atmosphere. We obtain the globally existence of weak solutions and trajectory attractors of the primitive equations of the moist atmosphere. And these results will appear in Theorem 3.1.10 and Theorem 3.1.25. Without the assumption of the global existence of strong solutions, and without the assumption of uniform boundedness for strong solutions in a space in time, we obtain the existence of global attractors of the primitive equations of large-scale moist atmosphere in the weak sense. Under some technical treatment, the initial boundary value problem of the primitive equations of the moist atmosphere corresponds to problem 3.1.5, which is given in section 3.1.2. With the methods in [140], we make full use of the hydrostatic equilibrium equation which reflects the relationship between pressure and density, thus obtain the global existence of weak solutions to problem 3.1.5. As the uniqueness of weak solutions of problem 3.1.5 is still not proved, we can't use the common method of studying global attractors which is based on the corresponding semigroup. However, we consider the trajectory and global attractors related to the time translation semigroup. Our methods to proving the existence of trajectory and global attractors to problem 3.1.5 is inspired by [37, 206].

The arrangement of this section is as follows. In subsection 3.1.1, we give the new formulation of the primitive equations of the large-scale moist atmosphere. In subsection 3.1.2, we give the mathematical setting of the initial boundary value problem of the primitive equations of the large-scale moist atmosphere, and obtain the global existence of weak solutions to problem 3.1.5. We shall prove the existence of the trajectory attractors and global attractors of the primitive equations of the large-scale moist atmosphere in section 3.1.3.

3.1.1 *Primitive Equations of the Large-Scale Moist Atmosphere*

Inspired by the method used by Lions, Temam and Wang in [140], we shall give the new formulation of the primitive equations of the large-scale moist atmosphere in this section.

Like the deduction of equations (1.6.6)-(1.6.9), we obtain the primitive equations of the large-scale moist atmosphere as follows,

$$\frac{\partial \boldsymbol{v}}{\partial t}+\nabla_{\boldsymbol{v}}\boldsymbol{v}+\omega\frac{\partial \boldsymbol{v}}{\partial p}+f\boldsymbol{k}\times\boldsymbol{v}+\operatorname{grad}\Phi-\mu_1\Delta\boldsymbol{v}-\nu_1\frac{\partial}{\partial p}\left[\left(\frac{gp}{R\bar{T}}\right)^2\frac{\partial \boldsymbol{v}}{\partial p}\right]=0, \tag{3.1.1}$$

$$\operatorname{div}\boldsymbol{v}+\frac{\partial\omega}{\partial p}=0, \tag{3.1.2}$$

$$\frac{\partial\Phi}{\partial p}+\frac{R}{p}(1+cq)T=0, \tag{3.1.3}$$

$$\frac{R^2}{C^2}\left(\frac{\partial T}{\partial t}+\nabla_{\boldsymbol{v}}T+\omega\frac{\partial T}{\partial p}\right)-\frac{R}{p}(1+cq)\omega-\mu_2\Delta T-\nu_2\frac{\partial}{\partial p}\left[\left(\frac{gp}{R\bar{T}}\right)^2\frac{\partial T}{\partial p}\right]$$
$$=Q_1, \tag{3.1.4}$$

$$\frac{\partial q}{\partial t}+\nabla_{\boldsymbol{v}}q+\omega\frac{\partial q}{\partial p}-\mu_3\Delta q-\nu_3\frac{\partial}{\partial p}\left[\left(\frac{gp}{R\bar{T}}\right)^2\frac{\partial q}{\partial p}\right]=Q_2, \tag{3.1.5}$$

where the unknown functions are $\boldsymbol{v},\omega,\Phi,T,q,\boldsymbol{v}=(v_\theta,v_\varphi)$ the horizontal velocity, $\omega=\dfrac{\mathrm{d}p}{\mathrm{d}t}$ the vertical velocity in pressure coordinates, $\Phi=gz$ the terrain, T temperature, $q=\dfrac{\rho_1}{\rho}$ the mixing ratio of vapor in the air, ρ_1 the density of vapor in the air, $f=2\Omega\cos\theta$ Coriolis parameter, $\boldsymbol{k}$ the vertical unit vector, $\bar{T}$ the referenced air temperature, μ_i,ν_i,c positive constants $(i=1,2,3,c\approx 0.618)$, Q_1 the heat source, and Q_2 the source of salinity. The definitions of $\nabla_{\boldsymbol{v}}\boldsymbol{v},\Delta\boldsymbol{v},\Delta T,\Delta q,\nabla_{\boldsymbol{v}}q,\nabla_{\boldsymbol{v}}T,\operatorname{div}\boldsymbol{v},\operatorname{grad}\Phi$ will be given in section 3.1.2.

Remark 3.1.1. Equations (3.1.1)-(3.1.5) are different from atmospheric equations in [142]. Equations in [142] are (3.1.1),(3.1.2),(3.1.5) with (3.1.3) and (3.1.4) when $c = 0$.

We consider the primitive equations (3.1.1)–(3.1.5) with boundary conditions as follows

$$p = P: \ (\boldsymbol{v}, \omega) = 0, \ \frac{\partial T}{\partial p} = \tilde{\alpha}_s(T_s - T), \ \frac{\partial q}{\partial p} = \tilde{\beta}_s(q_s - q), \tag{3.1.6}$$

$$p = p_0: \ (\boldsymbol{v}, \omega) = 0, \ \frac{\partial T}{\partial p} = 0, \ \frac{\partial q}{\partial p} = 0, \tag{3.1.7}$$

where P is an approximate value of pressure on the earth surface, p_0 is the top atmospheric pressure, assume $p_0 > 0$, $\tilde{\alpha}_s$, $\tilde{\beta}_s$ are positive constants, T_s is the given temperature on the earth surface, and q_s is the given mixing ratio of water vapor on the earth surface.

Next we nondimensionalize the boundary value problem (3.1.1)-(3.1.7) of the primitive equations of the large-scale moist atmosphere. Suppose that the characteristic scale of horizontal velocity is U, the characteristic length scale of horizontal motion is L, the characteristic scale of time is $\frac{L}{U}$, and the characteristic scale of pressure is P. Then,

$$\boldsymbol{v} = U\boldsymbol{v}', \ \omega = \frac{P - p_0}{L}U\omega', \ \Phi = U^2\Phi', \ T = \bar{T}_0 T', \ q = \bar{q_0} q', \ t = \frac{L}{U}t',$$

$$p = (P - p_0)\xi + p_0, \ R_0 = \frac{U}{L\Omega}, \ Re_1 = \frac{LU}{\mu_1}, \ Re_2 = \frac{LR^2\bar{T_0}^2}{\nu_1 L g^2}\frac{(P - p_0)^2}{P^2},$$

$$Rt_1 = \frac{LU^3}{\mu_2 \bar{T_0}^2}, \ Rt_2 = \frac{U^3R^2}{\nu_2 L g^2}\frac{(P - p_0)^2}{P^2}, \ Rt_3 = \frac{LU}{\mu_3},$$

$$Rt_4 = \frac{LR^2\bar{T_0}^2}{\nu_3 L g^2}\frac{(P - p_0)^2}{P^2}, a_1 = \frac{R^2\bar{T_0}^2}{C^2U^2}, \ b = \frac{R\bar{T_0}(P - p_0)}{U^2P}, \ \tilde{\alpha}_s = \frac{\alpha_s}{P - p_0},$$

$$\tilde{\beta}_s = \frac{\beta_s}{P - p_0}, \ T_s = \bar{T_0}\overline{T_s}, \ q_s = \bar{q_0}\overline{q_s}, \ f' = 2\cos\theta.$$

Thus, equations (3.1.1)-(3.1.5) are written as the dimensionless forms as follows (the symbol $'$ is omitted)

$$\frac{\partial \boldsymbol{v}}{\partial t} + \boldsymbol{\nabla}_{\boldsymbol{v}}\boldsymbol{v} + \omega\frac{\partial \boldsymbol{v}}{\partial \xi} + \frac{f}{R_0}\boldsymbol{k} \times \boldsymbol{v} + \operatorname{grad}\Phi - \frac{1}{Re_1}\Delta\boldsymbol{v} - \frac{1}{Re_2}\frac{\partial}{\partial \xi}\left[\left(\frac{p\bar{T_0}}{P\bar{T}}\right)^2\frac{\partial \boldsymbol{v}}{\partial \xi}\right]$$

$$= 0, \tag{3.1.8}$$

$$\operatorname{div}\boldsymbol{v} + \frac{\partial \omega}{\partial \xi} = 0, \tag{3.1.9}$$

$$\frac{\partial \Phi}{\partial \xi} + \frac{bP}{p}(1 + cq)T = 0, \tag{3.1.10}$$

$$a_1\left(\frac{\partial T}{\partial t}+\nabla_{\boldsymbol{v}}T+\omega\frac{\partial T}{\partial\xi}\right)-\frac{bP}{p}(1+cq)\omega-\frac{1}{Rt_1}\Delta T$$
$$-\frac{1}{Rt_2}\frac{\partial}{\partial\xi}\left[\left(\frac{p\overline{T_0}}{P\overline{T}}\right)^2\frac{\partial T}{\partial\xi}\right]=f_1, \tag{3.1.11}$$

$$\frac{\partial q}{\partial t}+\nabla_{\boldsymbol{v}}q+\omega\frac{\partial q}{\partial\xi}-\frac{1}{Rt_3}\Delta q-\frac{1}{Rt_4}\frac{\partial}{\partial\xi}\left[\left(\frac{p\overline{T_0}}{P\overline{T}}\right)^2\frac{\partial q}{\partial\xi}\right]=f_2, \tag{3.1.12}$$

where R_0 is Rossby number, Re_1, Re_2 denote horizontal and vertical Reynolds numbers respectively, Rt_1, Rt_2 denote horizontal and vertical heat dissipative coefficents respectively, Rt_3, Rt_4 are the horizontal and vertical diffusion coefficients of vapor respectively, and f_1, f_2 are functions on $S^2\times(0,1)$.

The space domain of the above equations (3.1.8)-(3.1.12) is

$$\Omega=S^2\times(0,1),$$

where S^2 is the two-dimensional unit sphere. The boundary conditions of the above equations are

$$\xi=1(p=P):\ (\boldsymbol{v},\omega)=0,\ \frac{\partial T}{\partial\xi}=\alpha_s(\overline{T_s}-T),\ \frac{\partial q}{\partial\xi}=\beta_s(\overline{q_s}-q), \tag{3.1.13}$$

$$\xi=0(p=p_0):\ (\boldsymbol{v},\omega)=0,\ \frac{\partial T}{\partial\xi}=0,\ \frac{\partial q}{\partial\xi}=0. \tag{3.1.14}$$

Integrating (3.1.9), and using of boundary conditions (3.1.13) and (3.1.14), we have

$$\omega(t;\theta,\varphi,\xi)=W(\boldsymbol{v})(t;\theta,\varphi,\xi)=\int_\xi^1\operatorname{div}\boldsymbol{v}(t;\theta,\varphi,\xi')\,\mathrm{d}\xi', \tag{3.1.15}$$

$$\int_0^1\operatorname{div}\boldsymbol{v}\,\mathrm{d}\xi=0. \tag{3.1.16}$$

Assume that Φ_s is an unknown function on isobaric surface $p=P$. Integrating (3.1.10), we obtain

$$\Phi(t;\theta,\varphi,\xi)=\Phi_s(t;\theta,\varphi,\xi)+\int_\xi^1\frac{bP}{p}(1+cq)T\,\mathrm{d}\xi'.$$

Then, equations (3.1.8)-(3.1.12) can be written as

$$\frac{\partial\boldsymbol{v}}{\partial t}+\nabla_{\boldsymbol{v}}\boldsymbol{v}+W(\boldsymbol{v})\frac{\partial\boldsymbol{v}}{\partial\xi}+\frac{f}{R_0}\boldsymbol{k}\times\boldsymbol{v}+\operatorname{grad}\Phi_s+\int_\xi^1\frac{bP}{p}\operatorname{grad}(1+cq)T\,\mathrm{d}\xi'$$
$$-\frac{1}{Re_1}\Delta\boldsymbol{v}-\frac{1}{Re_2}\frac{\partial}{\partial\xi}\left[\left(\frac{p\overline{T_0}}{P\overline{T}}\right)^2\frac{\partial\boldsymbol{v}}{\partial\xi}\right]=0, \tag{3.1.17}$$

$$a_1(\frac{\partial T}{\partial t} + \nabla_v T + W(v)\frac{\partial T}{\partial \xi}) - \frac{bP}{p}(1+cq)W(v) - \frac{1}{Rt_1}\Delta T$$
$$- \frac{1}{Rt_2}\frac{\partial}{\partial \xi}\left[\left(\frac{p\overline{T_0}}{P\overline{T}}\right)^2 \frac{\partial T}{\partial \xi}\right] = f_1, \tag{3.1.18}$$
$$\frac{\partial q}{\partial t} + \nabla_v q + W(v)\frac{\partial q}{\partial \xi} - \frac{1}{Rt_3}\Delta q - \frac{1}{Rt_4}\frac{\partial}{\partial \xi}\left[\left(\frac{p\overline{T_0}}{P\overline{T}}\right)^2 \frac{\partial q}{\partial \xi}\right] = f_2, \tag{3.1.19}$$
$$\int_0^1 \mathrm{div} v \, \mathrm{d}\xi = 0, \tag{3.1.20}$$

where the definitions of grad$T$, gradΦ_s will be given in section 3.1.2. The boundary conditions of equations (3.1.17)-(3.1.20) are

$$\xi = 1 (p = P): \; v = 0, \; \frac{\partial T}{\partial \xi} = \alpha_s(\overline{T_s} - T), \; \frac{\partial q}{\partial \xi} = \beta_s(\overline{q_s} - q), \tag{3.1.21}$$
$$\xi = 0 (p = p_0): \; v = 0, \; \frac{\partial T}{\partial \xi} = 0, \; \frac{\partial q}{\partial \xi} = 0, \tag{3.1.22}$$

the initial condition is given as

$$U_0 = (v_0, T_0, q_0). \tag{3.1.23}$$

(3.1.17)-(3.1.23) is called the initial boundary value problem of the new formulation of the primitive equations of the large-scale moist atmosphere, which is denoted as IBVP.

3.1.2 *The Global Existence of Global Weak Solutions of Problem IBVP*

This subsection is divided into four parts. First, we define some function spaces of problem IBVP. Next, we shall give some functional properties corresponding to equations (3.1.17)-(3.1.20). In the last two parts, we introduce the weak formulation of problem IBVP and prove the global existence of weak solutions to problem IBVP.

3.1.2.1 *Some Function Spaces*

Suppose that $e_\theta, e_\varphi, e_\xi$ are unit vectors in θ, φ and ξ directions of the space domain Ω, respectively,

$$e_\theta = \frac{\partial}{\partial \theta}, \quad e_\varphi = \frac{1}{\sin\theta}\frac{\partial}{\partial \varphi}, \quad e_\xi = \frac{\partial}{\partial \xi}.$$

The inner product and norm on $T_{(\theta,\varphi,\xi)}\Omega$, which is the tangent space of Ω at point (θ,φ,ξ), are given by

$$(X,Y)_T = X\cdot Y = X_1Y_1 + X_2Y_2 + X_3Y_3, \quad |X|_T = (X,X)^{\frac{1}{2}},$$

for $X = X_1\boldsymbol{e}_\theta + X_2\boldsymbol{e}_\varphi + X_3\boldsymbol{e}_\xi$, $Y = Y_1\boldsymbol{e}_\theta + Y_2\boldsymbol{e}_\varphi + Y_3\boldsymbol{e}_\xi \in T_{(\theta,\varphi,\xi)}\Omega$.

The norm of space $L^p(\Omega) := \{h; h:\Omega\to\mathbb{R}, \int_\Omega |h|^p < +\infty\}$ is $|h|_p = (\int_\Omega |h|^p)^{\frac{1}{p}}$, $1\le p<+\infty$. $L^2(T\Omega|TS^2) = \{v; v:\Omega\to TS^2\}$ is a space consisted of the former two components of the L^2 vector field in Ω, with the norm $|\boldsymbol{v}|_2 = (\int_\Omega(|v_\theta|^2+|v_\varphi|^2))^{\frac{1}{2}}$, where $T\Omega, TS^2$ are tangent spaces of Ω and S^2, respectively, $\boldsymbol{v}=(v_\theta,v_\varphi)$. $C^\infty(S^2)$ is the space consisted of all the smooth functions from S^2 to R. $C^\infty(\Omega)$ is the space consisted of all the smooth functions from Ω to R. $C^\infty(T\Omega|TS^2)$ is the space consisted of the former two components of the smooth vector field in Ω. $C_0^\infty(\Omega) := \{h;\ h\in C^\infty(\Omega),\ \text{supp}h \text{ is the compact subset of } \Omega\}$. $C_0^\infty(T\Omega|TS^2) := \{\boldsymbol{v};\ \boldsymbol{v}\in C^\infty(T\Omega|TS^2),\ \text{supp}\boldsymbol{v} \text{ is the compact subset of } \Omega\}$. $H^m(\Omega)$ is the Sobolev space consisted of functions which are in L^2 together with all their covariant derivatives with respect to $\boldsymbol{e}_\theta$, $\boldsymbol{e}_\varphi$, $\boldsymbol{e}_\xi$ of order less than m, with the norm

$$\|h\|_m = \left(\int_\Omega\left(\sum_{1\le k\le m}\ \sum_{i_j=1,2,3;j=1,\cdots,k} |\nabla_{i_1}\cdots\nabla_{i_k}h|^2 + |h|^2\right)\right)^{\frac{1}{2}},$$

where $\nabla_1 = \nabla_{\boldsymbol{e}_\theta}$, $\nabla_2 = \nabla_{\boldsymbol{e}_\varphi}$, $\nabla_3 = \nabla_{\boldsymbol{e}_\xi} = \dfrac{\partial}{\partial\xi}$ (the definitions of $\nabla_{\boldsymbol{e}_\theta}, \nabla_{\boldsymbol{e}_\xi}$ will be given later). $H^m(T\Omega|TS^2) = \{\boldsymbol{v};\ \boldsymbol{v}=(v_\theta,v_\varphi):\Omega\to TS^2,\ v_\theta,v_\varphi\in H^m(\Omega)\}$, the norm of which is similar to $H^m(\Omega)$, that is, we let $h=(v_\theta,v_\varphi)=v_\theta\boldsymbol{e}_\theta+v_\varphi\boldsymbol{e}_\varphi$ in the above norm formula of $H^m(\Omega)$.

The definitions of the horizontal divergence div, the horizontal gradient ∇ = grad, the horizontal covariant derivative $\nabla_{\boldsymbol{v}}$ and horizontal Laplace-Beltrami operator Δ of scalar and vector functions are defined by

$$\text{div}\boldsymbol{v} = \text{div}\,(v_\theta\boldsymbol{e}_\theta + v_\varphi\boldsymbol{e}_\varphi) = \frac{1}{\sin\theta}\left(\frac{\partial v_\theta\sin\theta}{\partial\theta} + \frac{\partial v_\varphi}{\partial\varphi}\right),$$

$$\nabla T = \text{grad}T = \frac{\partial T}{\partial\theta}\boldsymbol{e}_\theta + \frac{1}{\sin\theta}\frac{\partial T}{\partial\varphi}\boldsymbol{e}_\varphi, \quad \text{grad}\,\Phi_s = \frac{\partial\Phi_s}{\partial\theta}\boldsymbol{e}_\theta + \frac{1}{\sin\theta}\frac{\partial\Phi_s}{\partial\varphi}\boldsymbol{e}_\varphi,$$

$$\nabla_{\boldsymbol{v}}\tilde{\boldsymbol{v}} = \left(v_\theta\frac{\partial\tilde{v}_\theta}{\partial\theta} + \frac{v_\varphi}{\sin\theta}\frac{\partial\tilde{v}_\theta}{\partial\varphi} - v_\varphi\tilde{v}_\varphi\cot\theta\right)\boldsymbol{e}_\theta + \left(v_\theta\frac{\partial\tilde{v}_\varphi}{\partial\theta} + \frac{v_\varphi}{\sin\theta}\frac{\partial\tilde{v}_\varphi}{\partial\varphi}\right.$$
$$\left. + v_\varphi\tilde{v}_\theta\cot\theta\right)\boldsymbol{e}_\varphi,$$

$$\nabla_{\boldsymbol{v}}T = v_\theta\frac{\partial T}{\partial\theta} + \frac{v_\varphi}{\sin\theta}\frac{\partial T}{\partial\varphi}, \quad \nabla_{\boldsymbol{v}}q = v_\theta\frac{\partial q}{\partial\theta} + \frac{v_\varphi}{\sin\theta}\frac{\partial q}{\partial\varphi},$$

$$\Delta T = \frac{1}{\sin\theta}\left(\frac{\partial}{\partial\theta}\left(\sin\theta\frac{\partial T}{\partial\theta}\right) + \frac{1}{\sin\theta}\frac{\partial^2 T}{\partial\varphi^2}\right),$$
$$\Delta q = \frac{1}{\sin\theta}\left(\frac{\partial}{\partial\theta}\left(\sin\theta\frac{\partial q}{\partial\theta}\right) + \frac{1}{\sin\theta}\frac{\partial^2 q}{\partial\varphi^2}\right),$$
$$\Delta v = \left(\Delta v_\theta - \frac{2\cos\theta}{\sin^2\theta}\frac{\partial v_\varphi}{\partial\varphi} - \frac{v_\theta}{\sin^2\theta}\right)\boldsymbol{e}_\theta + \left(\Delta v_\varphi + \frac{2\cos\theta}{\sin^2\theta}\frac{\partial v_\theta}{\partial\varphi} - \frac{v_\varphi}{\sin^2\theta}\right)\boldsymbol{e}_\varphi,$$

where $\boldsymbol{v} = v_\theta\boldsymbol{e}_\theta + v_\varphi\boldsymbol{e}_\varphi$, $\widetilde{\boldsymbol{v}} = \widetilde{v}_\theta\boldsymbol{e}_\theta + \widetilde{v}_\varphi\boldsymbol{e}_\varphi \in C^\infty(T\Omega|TS^2)$, $T,\ q \in C^\infty(\Omega)$, and $\Phi_s \in C^\infty(S^2)$.

Next, we define some working spaces for the problem IBVP. Assume
$\widetilde{V} := \left\{\boldsymbol{v};\ \boldsymbol{v} \in C_0^\infty(T\Omega|TS^2),\ \int_0^1 \mathrm{div}\boldsymbol{v}\,\mathrm{d}\xi = 0\right\}$,
$V_1 =$ the closure of $\widetilde{V}$ with respect to the norm $\|\cdot\|_1$, $V_2 = H^1(\Omega)$,
$H_1 =$ the closure of $\widetilde{V}$ with respect to the norm $|\cdot|_2$, $H_2 = L^2(\Omega)$,
$V = V_1 \times V_2 \times V_2$, $\quad H = H_1 \times H_2 \times H_2$,
$V_1^{(3)} =$ the closure of $\widetilde{V}$ with respect to the norm $\|\cdot\|_3$, $V_2^{(3)} = H^3(\Omega)$,
$V^{(3)} = V_1^{(3)} \times V_2^{(3)} \times V_2^{(3)}$, $\quad V^{(-3)} = (V^{(3)})'$,
where $(V^{(3)})'$ is the dual space of $V^{(3)}$. To study the nonlinear terms of equations (3.1.17)-(3.1.19), we have to introduce spaces $V^{(3)}$ and $V^{(-3)}$. The inner product and norm of $V,\ V_1,\ V_2$ are given by

$$(\boldsymbol{v},\boldsymbol{v}_1)_{V_1} = \int_\Omega \left(\nabla_{\boldsymbol{e}_\theta}\boldsymbol{v}\cdot\nabla_{\boldsymbol{e}_\theta}\boldsymbol{v}_1 + \nabla_{\boldsymbol{e}_\varphi}\boldsymbol{v}\cdot\nabla_{\boldsymbol{e}_\varphi}\boldsymbol{v}_1 + \frac{\partial\boldsymbol{v}}{\partial\xi}\frac{\partial\boldsymbol{v}_1}{\partial\xi} + \boldsymbol{v}\cdot\boldsymbol{v}_1\right),$$
$$\|\boldsymbol{v}\|_{V_1} = (\boldsymbol{v},\boldsymbol{v})_{V_1}^{\frac{1}{2}},\quad \forall\boldsymbol{v},\ \boldsymbol{v}_1 \in V_1,$$
$$(T,T_1)_{V_2} = \int_\Omega \left(\mathrm{grad}T\cdot\mathrm{grad}T_1 + \frac{\partial T}{\partial\xi}\frac{\partial T_1}{\partial\xi} + TT_1\right),$$
$$\|T\|_{V_2} = (T,T)_{V_2}^{\frac{1}{2}},\quad \forall T,\ T_1 \in V_2,$$
$$(q,q_1)_{V_2} = \int_\Omega \left(\mathrm{grad}q\cdot\mathrm{grad}q_1 + \frac{\partial q}{\partial\xi}\frac{\partial q_1}{\partial\xi} + qq_1\right),$$
$$\|q\|_{V_2} = (q,q)_{V_2}^{\frac{1}{2}},\quad \forall q,\ q_1 \in V_2,$$
$$(\boldsymbol{U},\boldsymbol{U}_1)_H = (\boldsymbol{v},\boldsymbol{v}_1) + (a_1T,T_1) + (q,q_1),$$
$$(\boldsymbol{U},\boldsymbol{U}_1)_V = (\boldsymbol{v},\boldsymbol{v}_1)_{V_1} + (T,T_1)_{V_2} + (q,q_1)_{V_2},$$
$$\|\boldsymbol{U}\| = (\boldsymbol{U},\boldsymbol{U})_V^{\frac{1}{2}},\ |\boldsymbol{U}|_2 = (\boldsymbol{U},\boldsymbol{U})_H^{\frac{1}{2}},\ \forall\boldsymbol{U} = (\boldsymbol{v},T,q),\ \boldsymbol{U}_1 = (\boldsymbol{v}_1,T_1,q_1) \in V,$$
$$\|\boldsymbol{U}\|_{(3)} = (\|\boldsymbol{v}\|_3^2 + \|T\|_3^2 + \|q\|_3^2)^{\frac{1}{2}}, \forall\boldsymbol{U} \in V^{(3)},$$
$$(T,T_1)_{H_2} = (a_1T,T_1),\ |T|_2 = (T,T)_{H_2}^{\frac{1}{2}},\ |\boldsymbol{v}|_2 = (\boldsymbol{v},\boldsymbol{v})^{\frac{1}{2}},$$

where $(\cdot,\cdot)$ is the L^2 inner product in $H_1,\ H_2$. According to the definitions of $V,\ H,\ V^{(-3)}$, we know

$$V \subset H = H' \subset V' \subset V^{(-3)},$$

where V' is the dual space of space V.

3.1.2.2 *Some Functionals and the Properties of Their Corresponding Operators*

In this section, we shall define some functionals and their corresponding operators which are related to equations (3.1.17)-(3.1.20), and give some estimates about these functionals.

Define functionals $\widetilde{a}_1 : V_1 \times V_1 \to \mathbb{R},\ \widetilde{a}_2 : V_2 \times V_2 \to \mathbb{R},\ \widetilde{a}_3 : V_2 \times V_2 \to \mathbb{R},\ \widetilde{a} : V \times V \to \mathbb{R}$ and their corresponding linear operators $A_1 : V_1 \to V_1',\ A_2 : V_2 \to V_2',\ A_3 : V_2 \to V_2',\ A : V \to V'$ by

$$\begin{aligned}
&\widetilde{a}_1(\boldsymbol{v}, \boldsymbol{v}_1) = (A_1\boldsymbol{v}, \boldsymbol{v}_1)\\
&= \int_\Omega \left(\frac{1}{Re_1}(\boldsymbol{\nabla}_{\boldsymbol{e}_\theta}\boldsymbol{v} \cdot \boldsymbol{\nabla}_{\boldsymbol{e}_\theta}\boldsymbol{v}_1 + \boldsymbol{\nabla}_{\boldsymbol{e}_\varphi}\boldsymbol{v} \cdot \boldsymbol{\nabla}_{\boldsymbol{e}_\varphi}\boldsymbol{v}_1 + \boldsymbol{v} \cdot \boldsymbol{v}_1) + \frac{1}{Re_2}\left(\frac{p\overline{T}_0}{P\overline{T}}\right)^2 \frac{\partial \boldsymbol{v}}{\partial \xi}\frac{\partial \boldsymbol{v}_1}{\partial \xi} \right),\\
&\widetilde{a}_2(T, T_1) = (A_2T, T_1)\\
&= \int_\Omega \left(\frac{1}{Rt_1}\mathrm{grad}T \cdot \mathrm{grad}T_1 + \frac{1}{Rt_2}\left(\frac{p\overline{T}_0}{P\overline{T}}\right)^2 \frac{\partial T}{\partial \xi}\frac{\partial T_1}{\partial \xi} \right) + \int_{\Gamma_1} \frac{\alpha_s}{Rt_2}\left(\frac{p\overline{T}_0}{P\overline{T}}\right)^2 TT_1,\\
&\widetilde{a}_3(q, q_1) = (A_3q, q_1)\\
&= \int_\Omega \left(\frac{1}{Rt_3}\mathrm{grad}q \cdot \mathrm{grad}q_1 + \frac{1}{Rt_4}\left(\frac{p\overline{q}_0}{P\overline{q}}\right)^2 \frac{\partial q}{\partial \xi}\frac{\partial q_1}{\partial \xi} \right) + \int_{\Gamma_1} \frac{\beta_s}{Rt_4}\left(\frac{p\overline{q}_0}{P\overline{q}}\right)^2 qq_1,\\
&\widetilde{a}(\boldsymbol{U}, \boldsymbol{U}_1) = (A\boldsymbol{U}, \boldsymbol{U}_1) = \widetilde{a}_1(\boldsymbol{v}, \boldsymbol{v}_1) + \widetilde{a}_2(T, T_1) + \widetilde{a}_3(q, q_1),
\end{aligned}$$

where $\Gamma_1 = S^2 \times 1$.

Lemma 3.1.2.

(i) a is coercive and continuous. $A : V \to V'$ is isomorphism. Moreover,

$$\begin{aligned}
a(\boldsymbol{U}, \boldsymbol{U}_1) \le & c_1 \max\left\{\frac{1}{Re_1}, \frac{1}{Re_2}\right\} \|\boldsymbol{v}\|_{V_1}\|\boldsymbol{v}_1\|_{V_1}\\
&+ c_2 \max\left\{\frac{1}{Rt_1}, \frac{1}{Rt_2}, \frac{\alpha_s}{Rt_2}\right\} \|T\|_{V_2}\|T_1\|_{V_2}\\
&+ c_3 \max\left\{\frac{1}{Rt_3}, \frac{1}{Rt_4}, \frac{\beta_s}{Rt_4}\right\} \|q\|_{V_2}\|q_1\|_{V_2}\\
\le & \frac{1}{R_{\min}}\|\boldsymbol{U}\|\|\boldsymbol{U}_1\|,
\end{aligned} \tag{3.1.24}$$

$$\begin{aligned} a(\boldsymbol{U},\boldsymbol{U}) \geq & c_4 \min\left\{\frac{1}{Re_1},\frac{1}{Re_2}\right\}\|\boldsymbol{v}\|_{V_1}^2 + c_5\min\left\{\frac{1}{Rt_1},\frac{1}{Rt_2},\frac{\alpha_s}{Rt_2}\right\}\|T\|_{V_2}^2 \\ & + c_6\min\left\{\frac{1}{Rt_3},\frac{1}{Rt_4},\frac{\beta_s}{Rt_4}\right\}\|q\|_{V_2}^2 \\ \geq & \frac{1}{R_{\max}}\|\boldsymbol{U}\|^2, \end{aligned} \tag{3.1.25}$$

where

$$R_{\min} = \frac{1}{\min\{c_1,c_2,c_3\}}\min\left\{Re_1, Re_2, Rt_1, Rt_2, Rt_3, Rt_4, \frac{Rt_2}{\alpha_s}, \frac{Rt_4}{\beta_s}\right\},$$

$$R_{\max} = \frac{1}{\max\{c_4,c_5,c_6\}}\min\left\{Re_1, Re_2, Rt_1, Rt_2, Rt_3, Rt_4, \frac{Rt_2}{\alpha_s}, \frac{Rt_4}{\beta_s}\right\}.$$

In this section, c_i will denote positive constants and can be determined in concrete conditions.

(ii) The isomorphism $A : V \to V'$ can be extended to a self-adjoint unbounded operator on H with a compact inverse operator $A^{-1} : H \to H$, and with the domain of definition of the operator A is $D(A) = V \cap (H^2(T\Omega|TS^2) \times H^2(\Omega) \times H^2(\Omega))$.

Proof. The operator A is similar to the usual positive symmetric operator $-\Delta$ on H_0^1. Therefore we omit the details of the proof. For details, the readers can refer to [140, Lemma 2.3].

Concerning the nonlinear terms of equations (3.1.17)-(3.1.19), we define the functional $\widetilde{b} : V \times V \times V \to \mathbb{R}$ and its corresponding operator $B : H \times H \to H$ by

$$\widetilde{b}(\boldsymbol{U},\boldsymbol{U}_1,\boldsymbol{U}_2) = (B(\boldsymbol{U},\boldsymbol{U}_1),\boldsymbol{U}_2)_H = b_1(\boldsymbol{v},\boldsymbol{v}_1,\boldsymbol{v}_2)+b_2(\boldsymbol{v},T_1,T_2)+b_3(\boldsymbol{v},q_1,q_2),$$

where

$$\begin{aligned} b_1(\boldsymbol{v},\boldsymbol{v}_1,\boldsymbol{v}_2) &= \int_\Omega \left(\boldsymbol{\nabla}_{\boldsymbol{v}}\boldsymbol{v}_1 + \left(\int_\xi^1 \operatorname{div}\boldsymbol{v}\, \mathrm{d}\xi'\right)\frac{\partial \boldsymbol{v}_1}{\partial\xi}\right)\cdot \boldsymbol{v}_2, \\ b_2(\boldsymbol{v},T_1,T_2) &= \int_\Omega \left(\boldsymbol{\nabla}_{\boldsymbol{v}}T_1 + \left(\int_\xi^1 \operatorname{div}\boldsymbol{v}\, \mathrm{d}\xi'\right)\frac{\partial T_1}{\partial\xi}\right)T_2, \\ b_3(\boldsymbol{v},q_1,q_2) &= \int_\Omega \left(\boldsymbol{\nabla}_{\boldsymbol{v}}q_1 + \left(\int_\xi^1 \operatorname{div}\boldsymbol{v}\, \mathrm{d}\xi'\right)\frac{\partial q_1}{\partial\xi}\right)q_2. \end{aligned}$$

Let

$$b_4(\boldsymbol{U},\boldsymbol{U},\boldsymbol{U}_2) = \int_\Omega \left\{\left[\int_\xi^1 \frac{bP}{p}\operatorname{grad}(Tcq)\,\mathrm{d}\xi'\right]\cdot \boldsymbol{v}_2 - \frac{bP}{p}cqW(\boldsymbol{v})\,T_2\right\},$$

and

$$b\,(\boldsymbol{U},\boldsymbol{U}_1,\boldsymbol{U}_2)=\widetilde{b}\,(\boldsymbol{U},\boldsymbol{U}_1,\boldsymbol{U}_2)+b_4\,(\boldsymbol{U},\boldsymbol{U},\boldsymbol{U}_2)\,.$$

Lemma 3.1.3.

(i) For any $\boldsymbol{U},\ \boldsymbol{U}_1\in D(A)$,

$$b_1(\boldsymbol{v},\boldsymbol{v}_1,\boldsymbol{v}_1)=b_2(\boldsymbol{v},T_1,T_1)=b_3(\boldsymbol{v},q_1,q_1)=0,\ b_4(\boldsymbol{U},\boldsymbol{U},\boldsymbol{U})=0.$$

(ii) For any $\boldsymbol{U}\in D(A),\ \boldsymbol{U}_2\in D(A)\cap V^{(3)}$, we have

$$|b(\boldsymbol{U},\boldsymbol{U},\boldsymbol{U}_2)|\leq c_7\|\boldsymbol{U}\||\boldsymbol{U}|_2\|U_2\|_{(3)}. \tag{3.1.26}$$

Proof. (i) For any $\boldsymbol{U}=(\boldsymbol{v},T,q),\ \boldsymbol{U}_1=(\boldsymbol{v}_1,T_1,q_1)\in D(A)$, we have

$$\boldsymbol{\nabla}_{\boldsymbol{v}}|\boldsymbol{v}_1|^2=\boldsymbol{\nabla}_{\boldsymbol{v}}\boldsymbol{v}_1\cdot\boldsymbol{v}_1+\boldsymbol{v}_1\cdot\boldsymbol{\nabla}_{\boldsymbol{v}}\boldsymbol{v}_1=2\boldsymbol{\nabla}_{\boldsymbol{v}}\boldsymbol{v}_1\cdot\boldsymbol{v}_1.$$

Then

$$\begin{aligned}
b_1\,(\boldsymbol{v},\boldsymbol{v}_1,\boldsymbol{v}_1)=&\int_\Omega\left(\boldsymbol{\nabla}_{\boldsymbol{v}}\boldsymbol{v}_1\cdot\boldsymbol{v}_1+\frac{1}{2}\left(\int_\xi^1\mathrm{div}\boldsymbol{v}\ \mathrm{d}\xi'\right)\frac{\partial|\boldsymbol{v}_1|^2}{\partial\xi}\right)\\
=&\int_\Omega\left(\frac{1}{2}\boldsymbol{\nabla}_{\boldsymbol{v}}|\boldsymbol{v}_1|^2+\frac{1}{2}\left(\int_\xi^1\mathrm{div}\boldsymbol{v}\ \mathrm{d}\xi'\right)\frac{\partial|\boldsymbol{v}_1|^2}{\partial\xi}\right)\\
=&\frac{1}{2}\int_\Omega\left[\mathrm{div}\,(\boldsymbol{v}|\boldsymbol{v}_1|^2)-|\boldsymbol{v}_1|^2\mathrm{div}\boldsymbol{v}+\left(\int_\xi^1\mathrm{div}\boldsymbol{v}\ \mathrm{d}\xi'\right)\frac{\partial|\boldsymbol{v}_1|^2}{\partial\xi}\right]\\
=&\frac{1}{2}\int_\Omega\left(-|\boldsymbol{v}_1|^2\mathrm{div}\boldsymbol{v}+\left(\int_\xi^1\mathrm{div}\boldsymbol{v}\ \mathrm{d}\xi'\right)\frac{\partial|\boldsymbol{v}_1|^2}{\partial\xi}\right)\\
=&-\frac{1}{2}\int_\Omega\left(|\boldsymbol{v}_1|^2\left(\mathrm{div}\boldsymbol{v}+\frac{\partial W\,(\boldsymbol{v})}{\partial\xi}\right)\right)\\
&+\int_\Omega|\boldsymbol{v}_1|^2\left(\int_\xi^1\mathrm{div}\boldsymbol{v}\ \mathrm{d}\xi'\right)|_{\xi=0,1}\\
=&0.
\end{aligned}$$

Similarly, we prove $b_2(\boldsymbol{v},T_1,T_1)=b_3(\boldsymbol{v},q_1,q_1)=0$.

$$\begin{aligned}
b_4(\boldsymbol{U},\boldsymbol{U},\boldsymbol{U})=&\int_\Omega\left\{\left[\int_\xi^1\frac{bP}{p}\mathrm{grad}(Tcq)\mathrm{d}\xi'\right]\cdot\boldsymbol{v}-\frac{bP}{p}cqW(\boldsymbol{v})T\right\}\\
=&\int_\Omega\left\{\left[-\int_\xi^1\frac{bP}{p}(Tcq)\mathrm{d}\xi'\right]\cdot\mathrm{div}\boldsymbol{v}-\frac{bP}{p}cqW(\boldsymbol{v})T\right\}\\
=&\int_\Omega\left\{\left[-\int_\xi^1\frac{bP}{p}(Tcq)\mathrm{d}\xi'\right]\cdot\frac{\partial(\int_\xi^1\mathrm{div}\boldsymbol{v}\ \mathrm{d}\xi')}{\partial\xi}-\frac{bP}{p}cqW(\boldsymbol{v})T\right\}\\
=&\int_\Omega\left(\frac{bP}{p}TcqW(\boldsymbol{v})-\frac{bP}{p}cqW(\boldsymbol{v})T\right)=0.
\end{aligned}$$

(ii) For any $\boldsymbol{U} = (\boldsymbol{v}, T, q)$, $\boldsymbol{U}_1 = (\boldsymbol{v}_1, T_1, q_1) \in D(A)$, $\boldsymbol{U}_2 = (\boldsymbol{v}_2, T_2, q_2) \in D(A) \cap V^{(3)}$,

$$\left| \int_\Omega \left(\left(\int_\xi^1 \operatorname{div}\boldsymbol{v} \, \mathrm{d}\xi' \right) \frac{\partial \boldsymbol{v}_1}{\partial \xi} \right) \cdot \boldsymbol{v}_2 \right| \le \int_\Omega \left| \boldsymbol{v}_1 \cdot \left(\frac{\partial \boldsymbol{v}_2}{\partial \xi} \int_\xi^1 \operatorname{div}\boldsymbol{v} \, \mathrm{d}\xi' + \boldsymbol{v}_2 \operatorname{div}\boldsymbol{v} \right) \right|$$

$$\le |\boldsymbol{v}\|_{V_1} |\boldsymbol{v}_1|_2 \left| \frac{\partial \boldsymbol{v}_2}{\partial \xi} \right|_{L^\infty} \le \|\boldsymbol{U}\| |\boldsymbol{U}_1|_2 \left| \frac{\partial \boldsymbol{U}_2}{\partial \xi} \right|_{L^\infty} \le C \|\boldsymbol{U}\| |\boldsymbol{U}_1|_2 \|\boldsymbol{U}_2\|_{(3)},$$

where C is a positive constant. In the same way, we obtain $|b(\boldsymbol{U}, \boldsymbol{U}, \boldsymbol{U}_2)| \le c_7 \|\boldsymbol{U}\| |\boldsymbol{U}|_2 \|\boldsymbol{U}_2\|_{(3)}$.

For the linear terms in equations (3.1.17)-(3.1.19), we define a bilinear functional $e : V \times V \to \mathbb{R}$ and its corresponding operator $\widetilde{E} : H \to H$ by

$$\begin{aligned} e(\boldsymbol{U}, \boldsymbol{U}_1) &= (\widetilde{E}\boldsymbol{U}, \boldsymbol{U}_1)_H \\ &= \int_\Omega \left[\frac{f}{R_0} (\boldsymbol{k} \times \boldsymbol{v}) \cdot \boldsymbol{v}_1 + \left(\int_\xi^1 \frac{bP}{p} \operatorname{grad} T \mathrm{d}\xi' \right) \cdot \boldsymbol{v}_1 - \frac{bP}{p} W(\boldsymbol{v}) T_1 \right]. \end{aligned}$$

Lemma 3.1.4.

(i) For any $\boldsymbol{U}$, $\boldsymbol{U}_1 \in V$,

$$|e(\boldsymbol{U}, \boldsymbol{U}_1)| \le C \|\boldsymbol{U}\| \|\boldsymbol{U}_1\|, \tag{3.1.27}$$

where C is a positive constant.

(ii) For any $\boldsymbol{U}$, $\boldsymbol{U} \in V$, we have $e(\boldsymbol{U}, \boldsymbol{U}) = 0$.

Proof. The first part of the lemma is obvious, so we omit the detail of the proof. The proof of the second part is similar to the proof of $b_4(\boldsymbol{U}, \boldsymbol{U}, \boldsymbol{U}) = 0$.

3.1.2.3 *Weak Formulation of Problem IBVP*

In this part, we will introduce a weak formulation of problem IBVP by eliminating the geopotential Φ_s in equation (3.1.17). This method is similar to eliminating the pressure term in obtaining the global existence of weak solutions of Navier-Stokes system.

Firstly, we make the boundary condition (3.1.21) of T, q homogeneous. By

$$\frac{\partial T'}{\partial \xi} = \alpha_s (\overline{T_s} - T'), \quad \frac{\partial q'}{\partial \xi} = \beta_s (\overline{q_s} - q'),$$

we get

$$T' = \overline{T_s}(1 - \exp(-\alpha_s \xi)), \quad q' = \overline{q_s}(1 - \exp(-\beta_s \xi)).$$

Let

$$T'_\varepsilon = T' \psi_\varepsilon(\xi), \; q'_\varepsilon = q' \psi_\varepsilon(\xi),$$

where $0 < \varepsilon < \frac{1}{2}$, and

$$\psi_\varepsilon(\xi) := \begin{cases} 1, & 1-\varepsilon \le \xi \le 1, \\ \text{increasing}, & 1-2\varepsilon \le \xi \le 1-\varepsilon, \\ 0, & 0 \le \xi \le 1-2\varepsilon. \end{cases}$$

Then, by letting $\widetilde{\boldsymbol{U}} = (\boldsymbol{v}, \widetilde{T}, \widetilde{q}) = \boldsymbol{U} - \boldsymbol{U}'_\varepsilon = (\boldsymbol{v}, T, q) - (0, T'_\varepsilon, q'_\varepsilon)$, we know that the problem IBVP can be rewritten as the following equations

$$\begin{aligned} &\frac{\partial \boldsymbol{v}}{\partial t} + \boldsymbol{\nabla}_{\boldsymbol{v}} \boldsymbol{v} + W(\boldsymbol{v}) \frac{\partial \boldsymbol{v}}{\partial \xi} + \frac{f}{R_0} \boldsymbol{k} \times \boldsymbol{v} + \operatorname{grad} \Phi_s + \int_\xi^1 \frac{bP}{p} \operatorname{grad}(1 + c\widetilde{q}) \widetilde{T} \, \mathrm{d}\xi' \\ &+ \int_\xi^1 \frac{bP}{p} \operatorname{grad}(cq'_\varepsilon \widetilde{T} + cT'_\varepsilon \widetilde{q}) \, \mathrm{d}\xi' - \frac{1}{Re_1} \Delta \boldsymbol{v} - \frac{1}{Re_2} \frac{\partial}{\partial \xi} \left[\left(\frac{p\overline{T_0}}{P\overline{T}} \right)^2 \frac{\partial \boldsymbol{v}}{\partial \xi} \right] \\ = \; &\widetilde{f_1} = f_1 - \int_\xi^1 \frac{bP}{p} \operatorname{grad}(1 + cq'_\varepsilon) T'_\varepsilon \, \mathrm{d}\xi', \end{aligned} \tag{3.1.28}$$

$$\begin{aligned} &a_1 \left(\frac{\partial \widetilde{T}}{\partial t} + \boldsymbol{\nabla}_{\boldsymbol{v}} \widetilde{T} + W(\boldsymbol{v}) \frac{\partial \widetilde{T}}{\partial \xi} \right) + a_1 \left(\boldsymbol{\nabla}_{\boldsymbol{v}} T'_\varepsilon + W(\boldsymbol{v}) \frac{\partial T'_\varepsilon}{\partial \xi} \right) - \frac{bP}{p} (1 + c\widetilde{q}) W(\boldsymbol{v}) \\ &- \frac{bP}{p} (cq'_\varepsilon) W(\boldsymbol{v}) - \frac{1}{Rt_1} \Delta \widetilde{T} - \frac{1}{Rt_2} \frac{\partial}{\partial \xi} \left[\left(\frac{p\overline{T_0}}{P\overline{T}} \right)^2 \frac{\partial \widetilde{T}}{\partial \xi} \right] \\ = \; &\widetilde{f_2} = f_2 + \frac{1}{Rt_1} \Delta T'_\varepsilon + \frac{1}{Rt_2} \frac{\partial}{\partial \xi} \left[\left(\frac{p\overline{T_0}}{P\overline{T}} \right)^2 \frac{\partial T'_\varepsilon}{\partial \xi} \right], \end{aligned} \tag{3.1.29}$$

$$\begin{aligned} &\frac{\partial \widetilde{q}}{\partial t} + \boldsymbol{\nabla}_{\boldsymbol{v}} \widetilde{q} + W(\boldsymbol{v}) \frac{\partial \widetilde{q}}{\partial \xi} + \boldsymbol{\nabla}_{\boldsymbol{v}} q'_\varepsilon + W(\boldsymbol{v}) \frac{\partial q'_\varepsilon}{\partial \xi} - \frac{1}{Rt_3} \Delta \widetilde{q} - \frac{1}{Rt_4} \frac{\partial}{\partial \xi} \left[(\frac{p\overline{T_0}}{P\overline{T}})^2 \frac{\partial \widetilde{q}}{\partial \xi} \right] \\ = \; &\widetilde{f_3} = f_3 + \frac{1}{Rt_3} \Delta q'_\varepsilon + \frac{1}{Rt_4} \frac{\partial}{\partial \xi} \left[\left(\frac{p\overline{T_0}}{P\overline{T}} \right)^2 \frac{\partial q'_\varepsilon}{\partial \xi} \right], \end{aligned} \tag{3.1.30}$$

$$\int_0^1 \operatorname{div} \boldsymbol{v} \mathrm{d}\xi = 0, \tag{3.1.31}$$

with boundary conditions and initial conditions

$$\xi = 1 (p = P): \; \boldsymbol{v} = 0, \; \frac{\partial \widetilde{T}}{\partial \xi} + \alpha_s \widetilde{T} = 0, \; \frac{\partial \widetilde{q}}{\partial \xi} + \beta_s \widetilde{q} = 0, \tag{3.1.32}$$

$$\xi = 0 (p = p_0): \; \boldsymbol{v} = 0, \; \frac{\partial \widetilde{T}}{\partial \xi} = 0, \; \frac{\partial \widetilde{q}}{\partial \xi} = 0, \tag{3.1.33}$$

$$\widetilde{U}_0 = (\boldsymbol{v}_0, \widetilde{T_0}, \widetilde{q_0}). \tag{3.1.34}$$

Now let's introduce the weak formulation of problem IBVP.

Problem 3.1.5. For $\widetilde{f} = (\widetilde{f}_1, \widetilde{f}_2, \widetilde{f}_3)$, $\widetilde{\boldsymbol{U}}_0 = (\boldsymbol{v}_0, \widetilde{T_0}, \widetilde{q_0}) \in H$ given, find $\widetilde{\boldsymbol{U}} = (\boldsymbol{v}, \widetilde{T}, \widetilde{q})$ such that

$$\widetilde{\boldsymbol{U}} \in L^2(0, \mathcal{T}; V) \cap L^\infty(0, \mathcal{T}; H), \ \forall \mathcal{T} > 0, \tag{3.1.35}$$

$$\begin{aligned} &\frac{\mathrm{d}}{\mathrm{d}t}(\widetilde{\boldsymbol{U}}, \boldsymbol{U}_1)_H + (A\widetilde{\boldsymbol{U}}, \boldsymbol{U}_1)_H + ((B\widetilde{\boldsymbol{U}}, \widetilde{\boldsymbol{U}}), \boldsymbol{U}_1)_H + ((B\widetilde{\boldsymbol{U}}, \boldsymbol{U}'_\varepsilon), \boldsymbol{U}_1)_H \\ &+ ((F\widetilde{\boldsymbol{U}}, \boldsymbol{U}'_\varepsilon), \boldsymbol{U}_1)_H + (\widetilde{E}\widetilde{\boldsymbol{U}}, \boldsymbol{U}_1)_H = (\widetilde{f}, \boldsymbol{U}_1)_H \ \text{ in } (C_0^\infty(0, \mathcal{T}))', \ \forall \boldsymbol{U}_1 \in D(A) \end{aligned} \tag{3.1.36}$$

$$\widetilde{\boldsymbol{U}}|_{t=0} = \widetilde{\boldsymbol{U}}_0 \text{ in } V^{(-3)}, \tag{3.1.37}$$

where

$$((F\widetilde{\boldsymbol{U}}, \boldsymbol{U}'_\varepsilon), \boldsymbol{U}_1)_H = \int_\Omega \left[\left(\int_\xi^1 \frac{bP}{p} \operatorname{grad}(cq'_\varepsilon \widetilde{T} + cT'_\varepsilon \widetilde{q}) \, \mathrm{d}\xi'\right) \cdot \boldsymbol{v}_1 - \frac{bP}{p}(cq'_\varepsilon) W(\boldsymbol{v}) T_1\right].$$

3.1.2.4 *The Proof of the Global Existence of Weak Solutions*

In this section, we prove the global existence of weak solutions for problem IBVP by proving the global existence of weak solutions to the problem 3.1.5. In order to do this, we need some lemmas.

Lemma 3.1.6. (cf. [140, Lemma 2.1])

(i) If $\boldsymbol{v} \in H_1$, $\int_\Omega \boldsymbol{v}\boldsymbol{v}_1 = 0$, $\forall \boldsymbol{v}_1 \in C_0^\infty(T\Omega|TS^2)$, then

$$\boldsymbol{v} = \operatorname{grad} \Phi_s, \ \ \Phi_s \in (C_0^\infty(S^2))'.$$

(ii) Let $H_1^\perp$ be the orthogonal complement of H_1 in $L^2(T\Omega|TS^2)$. Then

$$H_1^\perp = \left\{\boldsymbol{v}; \ \boldsymbol{v} \in L^2(T\Omega|TS^2), \ \boldsymbol{v} = \operatorname{grad} l, \ l \in H^1(S^2)\right\}, \tag{3.1.38}$$

$$H_1 = \left\{\boldsymbol{v}; \ \boldsymbol{v} \in L^2(T\Omega|TS^2), \ \int_0^1 \operatorname{div} \boldsymbol{v} \, \mathrm{d}\xi = 0\right\}, \tag{3.1.39}$$

$$V_1 = \left\{\boldsymbol{v}; \ \boldsymbol{v} \in H_0^1(T\Omega|TS^2), \ \int_0^1 \operatorname{div} \boldsymbol{v} \, \mathrm{d}\xi = 0\right\}. \tag{3.1.40}$$

According to the above lemma, we know that if $\widetilde{\boldsymbol{U}} = (\boldsymbol{v}, \widetilde{T}, \widetilde{q})$ is a solution of problem 3.1.5, then there exists a unique (up to a constant) $\Phi_s \in (C_0^\infty(S^2))'$ such that $(\boldsymbol{v}, T, q, \Phi_s)$ is a weak solution of problem IBVP. On the other hand, if $(\boldsymbol{v}, T, q, \Phi_s)$ is a solution of problem IBVP ($(\boldsymbol{v}, T, q, \Phi_s)$ is sufficiently smooth), then $\widetilde{\boldsymbol{U}} = (\boldsymbol{v}, \widetilde{T}, \widetilde{q})$ is a solution of problem 3.1.5.

In order to solve problem 3.1.5 by Faedo-Galerkin method, we need the following three lemmas.

Lemma 3.1.7. The eigenvalue problem

$$A\boldsymbol{U} = \mu \boldsymbol{U}, \quad \boldsymbol{U} \in V,$$

exists a series of eigenvalues $0 < \mu_1 < \mu_2 \le \mu_3 \le \cdots \le \mu_n \le \cdots$, which are finite multiplicity, and $\mu_n \to +\infty$. The first eigenvalue is simple, and the corresponding eigenfunction is positive. Moreover, the eigenfunction sequence $\{\phi_n\}$ is an orthogonal basis of space V.

Proof. In fact, Lemma 3.1.7 is a corollary of Lemma 3.1.2.

Lemma 3.1.8. For any $\delta > 0$, there is $0 < \varepsilon < 1/2$ such that

$$|((F\widetilde{\boldsymbol{U}}, \boldsymbol{U}'_\varepsilon), \widetilde{\boldsymbol{U}})_H| \le \delta \|\widetilde{\boldsymbol{U}}\|^2, \tag{3.1.41}$$

$$|((B\widetilde{\boldsymbol{U}}, \boldsymbol{U}'_\varepsilon), \widetilde{\boldsymbol{U}})_H| \le \delta \|\widetilde{\boldsymbol{U}}\|^2, \quad \forall \widetilde{\boldsymbol{U}} \in V. \tag{3.1.42}$$

Proof. According to the definition of F, for any $U_1 \in V$,

$$\begin{aligned}
&\left|\left(\left(F\widetilde{\boldsymbol{U}}, \boldsymbol{U}'_\varepsilon\right), \boldsymbol{U}_1\right)_H\right| \\
=&\left|\int_\Omega \left[\left(\int_\xi^1 \frac{bP}{p} \mathrm{grad}\left(cq'_\varepsilon \widetilde{T} + cT'_\varepsilon \widetilde{q}\right) \mathrm{d}\xi'\right) \cdot \boldsymbol{v}_1 - \frac{bP}{p}\left(cq'_\varepsilon\right) W(\boldsymbol{v})\, T_1\right]\right| \\
\le&\left|\int_\Omega \left(\int_\xi^1 \frac{bP}{p} cq'_\varepsilon \mathrm{grad}\widetilde{T}\, \mathrm{d}\xi'\right) \cdot \boldsymbol{v}_1\right| + \left|\int_\Omega \left(\int_\xi^1 \frac{bP}{p} c\widetilde{T} \mathrm{grad} q'_\varepsilon\, \mathrm{d}\xi'\right) \cdot \boldsymbol{v}_1\right| \\
&+\left|\int_\Omega \left(\int_\xi^1 \frac{bP}{p} cT'_\varepsilon \mathrm{grad}\widetilde{q}\, \mathrm{d}\xi'\right) \cdot \boldsymbol{v}_1\right| + \left|\int_\Omega \left(\int_\xi^1 \frac{bP}{p} c\widetilde{q} \mathrm{grad} T'_\varepsilon\, \mathrm{d}\xi'\right) \cdot \boldsymbol{v}_1\right| \\
&+\left|\int_\Omega \frac{bP}{p} cq'_\varepsilon \left(\int_\xi^1 \mathrm{div}\boldsymbol{v}\, \mathrm{d}\xi'\right) T_1\right| \\
\le& I_1 + I_2 + I_3 + I_4 + I_5.
\end{aligned}$$

By the definition of q'_ε,

$$\begin{aligned}
I_1 =& \left|\int_\Omega \left(\int_\xi^1 \frac{bP}{p} cq'_\varepsilon \mathrm{grad}\widetilde{T}\, \mathrm{d}\xi'\right) \cdot \boldsymbol{v}_1\right| \\
\le& \left|\int_{S^2\times[1-2\varepsilon]} \left(\int_{1-2\varepsilon}^1 \frac{bP}{p} cq'_\varepsilon \mathrm{grad}\widetilde{T} \mathrm{d}\xi'\right) \cdot \boldsymbol{v}_1\right| \\
\le& c_8 |q'_\varepsilon|_{L^\infty(S^2)} \left|\int_{S^2\times[1-2\varepsilon]} \left(\int_{1-2\varepsilon}^1 \mathrm{grad}\widetilde{T} \mathrm{d}\xi'\right) \cdot \boldsymbol{v}_1\right| \\
\le& 4c_8\varepsilon^2 |q'_\varepsilon|_{L^\infty(S^2)} \left|\int_{S^2} \left(\int_{1-2\varepsilon}^1 |\mathrm{grad}\widetilde{T}|^2\, \mathrm{d}\xi'\right)^{\frac{1}{2}} \left(\int_{1-2\varepsilon}^1 |\boldsymbol{v}_1|^2\, \mathrm{d}\xi'\right)^{\frac{1}{2}} \mathrm{d}S^2\right| \\
\le& 4c_8\varepsilon^2 |q'_\varepsilon|_{L^\infty(S^2)} \|T\|_{V_2} \|\boldsymbol{v}_1\|_{V_1} \le c_9\varepsilon \|\boldsymbol{U}\| \|\boldsymbol{U}_1\|.
\end{aligned}$$

Similarly, we prove that

$$I_i \le c_{8+i}\varepsilon \|\boldsymbol{U}\| \|\boldsymbol{U}_1\|, \; i = 2, 3, 4, 5.$$

Therefore,

$$|((F\widetilde{U}, U'_\varepsilon), \widetilde{U})_H| \le \delta \|\widetilde{U}\|^2.$$

Since the proof of (3.1.42) is similar to the proof of $I_1 \le c_9\varepsilon \|U\| \|U_1\|$, we omit the details of the proof.

Lemma 3.1.9. (Lions-Magenes, cf. [138]) Suppose that $g \in L^\infty(0, \mathcal{T}; E)$, and $g(t)$ is weakly continuous in E_0, $g \in C_\omega(0, \mathcal{T}; E_0)$, that is, for any function $\phi \in (E_0)'$, function $\langle g(t), \phi\rangle$ belongs to $C[0, \mathcal{T}]$, where E, $E_0(E \subset E_0)$ are Banach spaces. Then, when $0 \le t \le \mathcal{T}$, $g(t) \in E$ and $g(t)$ is weakly continuous in E.

Now we can state the main result of this section.

Theorem 3.1.10. On the assumptions of problem 3.1.5, for any $\mathcal{T} > 0$, there is at least one solution $\widetilde{U} = (v, \widetilde{T}, \widetilde{q})$ on the time interval $[0, \mathcal{T}]$ to problem 3.1.5, and $\widetilde{U} = (v, \widetilde{T}, \widetilde{q})$ satisfies

$$\widetilde{U}_t \in L^2(0, \mathcal{T}; V^{(-3)}), \tag{3.1.43}$$

$$\widetilde{U} \in C_\omega(0, \mathcal{T}; H). \tag{3.1.44}$$

Moreover, $\widetilde{U}$ satisfies the energy inequality

$$\begin{aligned} &- \int_0^\infty |\widetilde{U}(t)|_2^2 \psi'(s) \mathrm{d}s + \frac{1}{R_{\max}} \int_0^\infty \|\widetilde{U}(s)\|^2 \psi(s) \mathrm{d}s \\ \le& 2\int_0^\infty (\widetilde{f}, \widetilde{U}(s))_H \psi(s) \mathrm{d}s, \end{aligned} \tag{3.1.45}$$

where $\psi(s) \in C_0^\infty(]0, \mathcal{T}[)$, $\psi(s) \ge 0$.

Proof. We shall prove Theorem 3.1.10 with Faedo-Galerkin method. Since the proof is similar to the proof of the global existence of weak solutions to Navier-Stokes system in reference [137, Theorem 6.1], we only give the outline of the proof.

Firstly, we look for the approximate solutions $\widetilde{U}_m(x,t)$ for problem 3.1.5, $\widetilde{U}_m(x,t) = \sum_{i=1}^m \alpha_{i,m}(t)\phi_i(x)$, and $\phi_i(x)$ is defined in Lemma 3.1.7. The function $\widetilde{U}_m$ satisfies

$$\begin{aligned} &\frac{\mathrm{d}}{\mathrm{d}t}(\widetilde{U}_m, \phi_i(x))_H + (A\widetilde{U}_m, \phi_i(x))_H + ((B\widetilde{U}_m, \widetilde{U}_m), \phi_i(x))_H \\ &+((B\widetilde{U}_m, U'_\varepsilon), \phi_i(x))_H + ((F\widetilde{U}_m, U'_\varepsilon), \phi_i(x))_H + (\widetilde{E}\widetilde{U}_m, \phi_i(x))_H \\ =& (\widetilde{f}, \phi_i(x))_H, \quad i = 1, 2, \cdots, m, \end{aligned} \tag{3.1.46}$$

$$\widetilde{U}_m|_{t=0} = \widetilde{U}_{0,m}, \tag{3.1.47}$$

where $\widetilde{\boldsymbol{U}}_{0,m} \to \widetilde{\boldsymbol{U}}_0$ in H. As is in [137], according to (3.1.46) and Lemma 3.1.8, we have

$$\frac{1}{2}\frac{\mathrm{d}}{\mathrm{d}t}|\widetilde{\boldsymbol{U}}_m(s)|_2^2 + \frac{1}{R_{\max}}\|\widetilde{\boldsymbol{U}}_m(s)\|^2 \leq \frac{1}{2R_{\max}}\|\widetilde{\boldsymbol{U}}_m(s)\|^2 + (\widetilde{f}, \widetilde{\boldsymbol{U}}_m(s))_H. \tag{3.1.48}$$

Integrating (3.1.48) from 0 to t, $t \in]0, \mathcal{T}]$, we obtain

$$|\widetilde{\boldsymbol{U}}_m(t)|_2^2 + \frac{1}{R_{\max}}\int_0^t \|\widetilde{\boldsymbol{U}}_m(s)\|^2 \mathrm{d}s \leq |\widetilde{\boldsymbol{U}}_{0,m}|_2^2 + 2\int_0^t (\widetilde{f}, \widetilde{\boldsymbol{U}}_m(s))_H \mathrm{d}s, \quad t \in]0, \mathcal{T}]. \tag{3.1.49}$$

By Gronwall Inequality and (3.1.48), we have

$$|\widetilde{\boldsymbol{U}}_m(t)|_2^2 + c_{14}\int_0^t \|\widetilde{\boldsymbol{U}}_m(t)\|^2 \mathrm{d}s \leq |\widetilde{\boldsymbol{U}}_{0,m}|_2^2 + c_{15}\int_0^t \|\widetilde{f}(s)\|_{V'} \mathrm{d}s, \quad t \in]0, \mathcal{T}]. \tag{3.1.50}$$

Therefore $\{\widetilde{\boldsymbol{U}}_m\}$ is bounded in space $L^2(0, \mathcal{T}; V) \cap L^\infty(0, \mathcal{T}; H)$. Going if necessary to a subsequence denoted again by $\{\widetilde{\boldsymbol{U}}_m\}$, $\widetilde{\boldsymbol{U}}_m \to \widetilde{\boldsymbol{U}}(m \to \infty)$ weakly in $L^2(0, \mathcal{T}; V)$ and *weakly in $L^\infty(0, \mathcal{T}; H)$. Due to a compactness theorem (see [137]), we can extract a subsequence denoted again by $\{\widetilde{\boldsymbol{U}}_m\}$ such that $\widetilde{\boldsymbol{U}}_m \to \widetilde{\boldsymbol{U}}$ $(m \to \infty)$ strongly in $L^2(0, \mathcal{T}; H)$. By passing to the limit in (3.1.46), we conclude that $\widetilde{\boldsymbol{U}}$ is a weak solution to problem 3.1.5. According to Lemmas 3.1.2, 3.1.3 and 3.2.4, we know that $\{(\widetilde{\boldsymbol{U}}_m)_t\}$ is a bounded set in $L^2(0, \mathcal{T}; V^{(-3)})$, so $\widetilde{\boldsymbol{U}}_t \in L^2(0, \mathcal{T}; V^{(-3)})$. At last, according to Lemma 3.1.9 and $\widetilde{\boldsymbol{U}} \in C(0, \mathcal{T}; V^{(-3)})$, we have $\widetilde{\boldsymbol{U}} \in C_\omega(0, \mathcal{T}; H)$.

Next we prove the energy inequality. From $\widetilde{\boldsymbol{U}}_m \to \widetilde{\boldsymbol{U}}(m \to \infty)$ strongly in $L^2(0, \mathcal{T}; H)$, it follows that $|\widetilde{\boldsymbol{U}}_m|_2 \to |\widetilde{\boldsymbol{U}}|_2(m \to \infty)$ strongly in $L^2(0, \mathcal{T})$. Going if necessary to a subsequence, $|\widetilde{\boldsymbol{U}}_m|_2 \to |\widetilde{\boldsymbol{U}}|_2(m \to \infty)$ a.e. in $[0, \mathcal{T}]$. Let $\psi(s) \in C_0^\infty(]0, \mathcal{T}[)$, $\psi(s) \geq 0$. By (3.1.50) and Lebesgue Dominated Theorem, we have

$$\int_0^\infty |\widetilde{\boldsymbol{U}}_m(t)|_2^2 \psi'(s)\mathrm{d}s \to \int_0^\infty |\widetilde{\boldsymbol{U}}(t)|_2^2 \psi'(s)\mathrm{d}s, \quad m \to \infty. \tag{3.1.51}$$

$\widetilde{\boldsymbol{U}}_m \to \widetilde{\boldsymbol{U}}(m \to \infty)$ weakly in $L^2(0, \mathcal{T}; V)$ implies that $\widetilde{\boldsymbol{U}}_m \psi^{\frac{1}{2}}(s) \to \widetilde{\boldsymbol{U}} \psi^{\frac{1}{2}}(s)(m \to \infty)$ weakly in $L^2(0, \mathcal{T}; V)$. By the lower weak semicontinuity of the norm, we get

$$\int_0^\infty \|\widetilde{\boldsymbol{U}}(s)\|^2 \psi(s)\mathrm{d}s \leq \liminf_{m \to \infty} \int_0^\infty \|\widetilde{\boldsymbol{U}}_m(s)\|^2 \psi(s)\mathrm{d}s. \tag{3.1.52}$$

Using (3.1.46) and (3.1.47), we have

$$-\int_0^\infty |\widetilde{\boldsymbol{U}}_m(t)|_2^2 \psi'(s)\mathrm{d}s + \frac{1}{R_{\max}}\int_0^\infty \|\widetilde{\boldsymbol{U}}_m(s)\|^2 \psi(s)\mathrm{d}s$$

$$\leq 2\int_0^{\infty}(\widetilde{f},\widetilde{\boldsymbol{U}}_m(s))_H\psi(s)\mathrm{d}s. \tag{3.1.53}$$

Combining (3.1.51) with (3.1.52), we pass to the limit in (3.1.53) and obtain the energy inequality. The proof of Theorem 3.1.10 is complete.

3.1.3 *Trajectory and Global Attractors for the Moist Atmospheric Equations*

This subsection is divided into two parts. In the first part, we give some preliminaries about the trajectory and global attractors. In the second part, we prove the existence of the trajectory and global attractors of system (3.1.18)-(3.1.33).

3.1.3.1 *Some Preliminaries about the Trajectory Attractors*

Let's recall the trajectory attractor theory of Vishik and Chepyzhov, see [37, 206].

Let $E,\ E_0$ be two Banach spaces such that $E \subseteq E_0$. An autonomous evolution equation is

$$\frac{\partial u}{\partial t} = G(u), \tag{3.1.54}$$

where G is a differential operator. Denote a special family of solutions of equation (3.1.54) as $\mathcal{K}^+, \mathcal{K}^+ \subset C(\mathbb{R}_+;\ E_0) \cap L^\infty(\mathbb{R}_+;\ E)$. $\mathcal{K}^+$ is called the **trajectory space of equation** (3.1.54), and its elements are called the **trajectories of equation** (3.1.54). $\mathcal{K}^+$ is **translation invariant** in the following sense: if $\forall u(s) \in \mathcal{K}^+,\ h \in \mathbb{R}_+$, then $u(s+h) \in \mathcal{K}^+$.

The action of the translation operator $T(t)(t \geq 0)$ on the space $C(\mathbb{R}_+;\ E_0) \cap L^\infty(\mathbb{R}_+;\ E)$ is defined by

$$T(t)u(s) = u(t+s),\ \forall t \geq 0,\ u \in C(\mathbb{R}_+;\ E_0) \cap L^\infty(\mathbb{R}_+;\ E).$$

According to this definition, we have $T(t_1+t_2) = T(t_1)T(t_2)$, for $t_1, t_2 \geq 0$, and $T(0)$ is the identity operator on the space $C(\mathbb{R}_+;\ E_0) \cap L^\infty(\mathbb{R}_+;\ E)$. The semigroup $\{T(t)\} = \{T(t); t \geq 0\}$ is called the **time translation group** on space $C(\mathbb{R}_+;\ E_0) \cap L^\infty(\mathbb{R}_+;\ E)$.

We introduce a topology in the trajectory space $\mathcal{K}^+$. The sequence $\{f_n(s)\}$ ($\{f_n(s)\} \subset C(\mathbb{R}_+;\ E_0)$) converges to $f(s)(f(s) \in C(\mathbb{R}_+;\ E_0))$ in the topology space $C_{\mathrm{loc}}(\mathbb{R}_+;\ E_0)$ if

$$\max_{s\in[0,\mathcal{T}]} \|f_n(s) - f(s)\|_{E_0} \to 0, \quad \text{as} \quad n \to \infty,\ \forall \mathcal{T} > 0. \tag{3.1.55}$$

The topology of $\mathcal{K}^+$ is induced by the topological space $C_{\text{loc}}(\mathbb{R}_+;\ E_0)$. Obviously, the translation semigroup $\{T(t)\}$ is continuous in space $C_{\text{loc}}(\mathbb{R}_+;\ E_0)$. Especially, the translation semigroup $\{T(t)\}$ is continuous in space $\mathcal{K}^+$. A set Ξ is said to be bounded in space $\mathcal{K}^+$, if

$$\|u\|_{L^\infty(\mathbb{R}_+;\ E)} = \operatorname*{ess\,sup}_{s\geq 0} \|u(s)\|_E \leq c_{16}, \quad \forall u \in \Xi. \tag{3.1.56}$$

Definition 3.1.11. A set $\Lambda \subset C(\mathbb{R}_+;\ E_0) \cap L^\infty(\mathbb{R}_+;\ E)$ is said to be attracting for the trajectory space $\mathcal{K}^+$ in the topology of $C_{\text{loc}}(\mathbb{R}_+;\ E_0)$, if for any bounded set $\Xi \subseteq \mathcal{K}^+$ and $\mathcal{T} \geq 0$, the following relationship holds:

$$\text{dist}_{C(0,\mathcal{T};E_0)}(\Pi_{\mathcal{T}} T(t)\Xi, \Pi_{\mathcal{T}}\Lambda) \to 0, \text{ when } t \to \infty, \tag{3.1.57}$$

where $\Pi_{\mathcal{T}}$ is the restriction operator to the interval $[0, \mathcal{T}]$ defined by the following if $u \in C(\mathbb{R}_+; E_0) \cap L^\infty(\mathbb{R}_+; E)$, then

$$\Pi_{\mathcal{T}} u \in C(0, \mathcal{T}; E_0) \cap L^\infty(0, \mathcal{T}; E), \Pi_{\mathcal{T}} u(s) = u(s) \text{ for } s \in [0, \mathcal{T}],$$

and

$$\text{dist}_{C(0,\mathcal{T};E_0)}(\Pi_{\mathcal{T}} T(t)\Xi, \Pi_{\mathcal{T}}\Lambda) = \sup_{a\in\Xi} \inf_{b\in\Lambda} \max_{s\in[0,\mathcal{T}]} \|a(s+t) - b(s)\|_{E_0}.$$

Definition 3.1.12. A set $\mathcal{H} \subseteq \mathcal{K}^+$ is called the **trajectory attractor** in the trajectory space $\mathcal{K}^+$ with respect to the topology of $C_{\text{loc}}(\mathbb{R}_+;\ E_0)$ if we have the following properties:

(i) $\mathcal{H}$ is a compact set in $C_{\text{loc}}(\mathbb{R}_+;\ E_0)$, and bounded in $L^\infty(\mathbb{R}_+;\ E)$;
(ii) $\mathcal{H}$ is strictly invariant with respect to $\{T(t)\}$, that is,

$$T(t)\mathcal{H} = \mathcal{H}, \ \forall t \geq 0;$$

(iii) $\mathcal{H}$ is an attracting set in $\mathcal{K}^+$ in the topology of $C_{\text{loc}}(\mathbb{R}_+;\ E_0)$.

Theorem 3.1.13. (cf. [206]) Suppose that the trajectory space $\mathcal{K}^+$ is translation invariant. Suppose also that for $\mathcal{K}^+$ there exists an attracting set Λ such that $\Lambda \subseteq \mathcal{K}^+$, Λ is a compact subset in $C_{\text{loc}}(\mathbb{R}_+;\ E_0)$ and bounded in $L^\infty(\mathbb{R}_+;\ E)$. Then in $\mathcal{K}^+$ there exists a trajectory attractor $\mathcal{H} \subseteq \Lambda$, and $\mathcal{H}$ is unique in $\mathcal{K}^+$. Moreover

$$\mathcal{H} = \cap_{\mathcal{T}\geq 0}\overline{\cup_{t\geq\mathcal{T}} T(t)\Lambda}, \tag{3.1.58}$$

where $\overline{\cup_{t\geq\mathcal{T}} T(t)\Lambda}$ is the closure of the set $\cup_{t\geq\mathcal{T}} T(t)\Lambda$ in space $C_{\text{loc}}(\mathbb{R}_+;\ E_0)$.

Now let's recall the definition of the global attractors based on the trajectory attractors. Firstly, let's introduce some notations. For a set $\Xi(t) \subseteq E$, $t \geq 0$, we define $\Xi(t) \subseteq E$ as follows

$$\Xi(t) = \{u(t);\ u \in \Xi\} \subseteq E.$$

In a similar way, for the trajetory attractor $\mathcal{H}$, we define a set

$$\mathcal{H}(t) = \{u(t);\ u \in \mathcal{H}\} \subseteq E, \text{ for any } t \geq 0.$$

We note that $\mathcal{H}(t)$ is independent of time t.

Definition 3.1.14. A set $\mathcal{A} \subseteq E$ is called as the **global attractor** in E_0 of equation (3.1.54) if the following properties hold:

(i) $\mathcal{A}$ is compact in E_0 and bounded in E;

(ii) For any bounded trajectory set $\Xi \subset \mathcal{K}^+$ in $L^\infty(\mathbb{R}_+;\ E)$, the following relationship holds:

$$\text{dist}_{E_0}(\Xi(t), \mathcal{A}) \to 0, \quad \text{as} \quad t \to \infty; \tag{3.1.59}$$

(iii) $\mathcal{A}$ is the minimal set satisfying condition (i) and (ii), that is, $\mathcal{A}$ belongs to any attracting set which is compact in E_0 and bounded in E.

Theorem 3.1.15. (cf. [206]) If the assumptions of Theorem 3.1.13 are satisfied, then there exists a global attractor $\mathcal{A}$ in E_0, and $\mathcal{A} = \mathcal{H}(0)$.

Remark 3.1.16. Definition 3.1.14 generalizes (E, E_0)—global attractors which are usually based on the semigroup corresponding to the Cauchy problem of equation (3.1.54) under the assumption that the solution of the Cauchy problem of equation (3.1.54) is unique (see [8, 97, 201]). Suppose that for any u_0, there exists a unique trajectory $u \in \mathcal{K}^+$ such that $u(0) = u_0$. Then the global attractor is identical to the usual (E, E_0)—global attractors. We shall explain this point in Theorem 3.1.18.

Under the assumption that the Cauchy problem

$$\frac{\partial u}{\partial t} = G(u), \quad u(0) = u_0, \tag{3.1.60}$$

has a unique solution, following the standard approach, we introduce an operator semigroup $\{S(t); t \geq 0\}$ in the space E corresponding to the problem (3.1.60)

$$S(t)u_0 = u(t), \ t \geq 0. \tag{3.1.61}$$

Definition 3.1.17. A set $\mathcal{A}_1 \subseteq E$ is called a (E, E_0)—global attractor of semigroup $\{S(t)\}$ acting on E if the following properties hold:

(i) $\mathcal{A}_1$ is compact in E_0 and bounded in E;

(ii) $S(t)\mathcal{A}_1 = \mathcal{A}_1, \ \forall t \geq 0$;

(iii) The bounded (in E) sets $\Xi_0 \subset E$ satisfy:

$$\text{dist}_{E_0}(S(t)\Xi_0, \mathcal{A}_1) \to 0, \text{ as } t \to \infty. \tag{3.1.62}$$

Theorem 3.1.18. (cf. [206]). Suppose that the assumptions of Theorem 3.1.15 are satisfied, and the semigroup $\{S(t)\}$ is bounded (for any bounded (in E) set $\Xi_0 \subset E$, the set $\overline{\cup_{t\geq 0} S(t)\Xi_0}$ is bounded in E). Then the (E_0, E)—global attractor $\mathcal{A}_1$ of equation (3.1.54) exists, moreover $\mathcal{A}_1 = \mathcal{A} = \mathcal{H}(0)$, where $\mathcal{A}$ is the E_0—global attractor $\mathcal{A}_1$ of equation (3.1.54).

3.1.3.2 *Existence of Trajectory and Global Attractors*

Firstly, we construct the trajectory space $\mathcal{K}^+$ of system (3.1.28)-(3.1.33). According to Theorem 3.1.10, if $\widetilde{\boldsymbol{U}}$ is a weak solution to problem 3.1.5, then $\widetilde{\boldsymbol{U}} \in L^2(0,\mathcal{T};V) \cap L^\infty(0,\mathcal{T};H)$, and $\widetilde{\boldsymbol{U}}$ satisfies the energy inequality

$$\begin{aligned}&-\int_0^\infty |\widetilde{\boldsymbol{U}}(t)|_2^2 \psi'(s)\mathrm{d}s + \frac{1}{R_{\max}}\int_0^\infty \|\widetilde{\boldsymbol{U}}(s)\|^2 \psi(s)\mathrm{d}s \\ \leq &2\int_0^\infty (\widetilde{f},\widetilde{\boldsymbol{U}}(s))_H \psi(s)\mathrm{d}s, \end{aligned} \tag{3.1.63}$$

where $\psi(s) \in C_0^\infty(]0,\mathcal{T}[)$, $\psi(s) \geq 0$. The above inequality can be interpreted as follows

$$\frac{1}{2}\frac{\mathrm{d}}{\mathrm{d}t}|\widetilde{\boldsymbol{U}}(s)|_2^2 + \frac{1}{R_{\max}}\|\widetilde{\boldsymbol{U}}(s)\|^2 \leq \frac{1}{2R_{\max}}\|\widetilde{\boldsymbol{U}}(s)\|^2 + (\widetilde{f},\widetilde{\boldsymbol{U}}(s))_H, \ s \in]0,\mathcal{T}]. \tag{3.1.64}$$

Definition 3.1.19. The trajectory space $\mathcal{K}^+$ of system (3.1.28)-(3.1.33) is consists of the following functions

$$\widetilde{\boldsymbol{U}} \in L^2_{\mathrm{loc}}(\mathbb{R}_+;V) \cap L^\infty(\mathbb{R}_+;H),$$

where $\widetilde{\boldsymbol{U}}$ satisfies the following conditions: for any $\mathcal{T} > 0$, the function $\Pi_{\mathcal{T}}\widetilde{\boldsymbol{U}}$ is a weak solution of problem 3.1.5 on the time interval $[0,\mathcal{T}]$, and also satisfies energy inequality (3.1.63).

Lemma 3.1.20. (cf. [207]) Suppose that Y is a Banach space, and $E \subset\subset E_0 \subset Y$, where $\subset\subset$ denotes a compact embedding. Then we have the following embedding

$$W^{\infty,p}(0,\mathcal{T};E,Y) \subset\subset C(0,\mathcal{T};E_0),$$

where $W^{\infty,p}(0,\mathcal{T};E,Y) = \{u(s);\ s \in (0,\mathcal{T}), u \in L^\infty(0,\mathcal{T};E),\ u_t \in L^p(0,\mathcal{T};Y)\}$, $p > 1$, with the norm

$$\|u\|_{W^{\infty,p}} = \mathrm{ess}\sup_{\mathcal{T}\geq s\geq 0}\|u\|_E + \left(\int_0^{\mathcal{T}} \|u_t(s)\|_Y^p \mathrm{d}s\right)^{\frac{1}{p}}.$$

Proposition 3.1.21. If $\mathcal{K}^+$ is a trajectory space in definition 3.1.19, then

(i) For any $\widetilde{\boldsymbol{U}}_0 \in H$, there exists a trajectory $\widetilde{\boldsymbol{U}}(t) \in \mathcal{K}^+$ such that $\widetilde{\boldsymbol{U}}(0) = \widetilde{\boldsymbol{U}}_0$;

(ii) $\mathcal{K}^+ \subset C(\mathbb{R}_+;V^{(-3)}) \cap L^\infty(\mathbb{R}_+;H)$;

(iii) $\mathcal{K}^+$ is trajectory invariant, that is, $T(t)\mathcal{K}^+ \subseteq \mathcal{K}^+$.

Proof. According to definition 3.1.19, we know that (i),(ii) are corollaries of Theorem 3.1.10 and Lemma 3.1.20. Since equations (3.1.28)-(3.1.30) are autonomous, $\mathcal{K}^+$ is trajectory invariant for the translation semigroup $\{T(t)\}$.

Proposition 3.1.22. For any $\widetilde{\boldsymbol{U}} \in \mathcal{K}^+$, the following inequality holds:

$$\begin{aligned}
&\|T(t)\widetilde{\boldsymbol{U}}\|_{L^\infty(\mathbb{R}_+;H)} + \|T(t)\widetilde{\boldsymbol{U}}\|_{L^2(0,1;V)} + \|T(t)\widetilde{\boldsymbol{U}}_t\|_{L^2(0,1;V^{(-3)})}\\
=&\operatorname{ess\,sup}_{s\geq t} |\widetilde{\boldsymbol{U}}(s)| + \left(\int_t^{t+1} \|\widetilde{\boldsymbol{U}}(s)\|^2 \mathrm{d}s\right)^{\frac{1}{2}} + \left(\int_t^{t+1} \|\widetilde{\boldsymbol{U}}_t(s)\|^2_{V^{(-3)}} \mathrm{d}s\right)^{\frac{1}{2}}\\
\leq& c_{22}\|\widetilde{\boldsymbol{U}}\|^2_{L^\infty(0,1;H)} \exp(-c_{23}t) + c_{24}\|\widetilde{\boldsymbol{U}}\|_{L^\infty(0,1;H)} \exp(-c_{23}t) + c_{25}, \quad \forall t \geq 0.
\end{aligned} \tag{3.1.65}$$

In order to prove Proposition 3.1.22, we need the following general Gronwall Lemma.

Lemma 3.1.23. (cf. [37]) Let $y(s),\ \varphi(s) \in L^1_{\mathrm{loc}}(0,\infty)$, and

$$-\int_0^\infty y(s)\psi'(s)\mathrm{d}s + \alpha\int_0^\infty y(s)\psi(s)\mathrm{d}s \leq \int_0^\infty \varphi(s)\psi(s)\mathrm{d}s, \tag{3.1.66}$$

for any $\psi(s) \in C_0^\infty(\mathbb{R}_+),\ \psi(s) \geq 0$, where $\alpha \in \mathbb{R}$. Then

$$y(t)\mathrm{e}^{\alpha t} - y(\tau)\mathrm{e}^{\alpha \mathcal{T}} \leq \int_{\mathcal{T}}^t \varphi(s)\mathrm{e}^{\alpha s}\mathrm{d}s,$$

for any $t,\ \mathcal{T} \in \mathbb{R}_+ \setminus \widetilde{Q},\ t \geq \mathcal{T}$, where $\mu(\widetilde{Q}) = 0$, that is, the Lebesgue measure of $\widetilde{Q}$ is zero.

The proof of Proposition 3.1.22.

(1) According to the definitions of spaces $V,\ H,\ \mathcal{K}^+$, we have

$$c_{17}|\widetilde{\boldsymbol{U}}(s)|_2^2 \leq \|\widetilde{\boldsymbol{U}}(s)\|^2,\ \forall \widetilde{\boldsymbol{U}} \in \mathcal{K}^+,\ s \geq 0. \tag{3.1.67}$$

By (3.1.63), we obtain

$$\begin{aligned}
&-\int_0^\infty |\widetilde{\boldsymbol{U}}(s)|_2^2\psi'(s)\mathrm{d}s + \frac{c_{17}}{2R_{\max}}\int_0^\infty |\widetilde{\boldsymbol{U}}(s)|_2^2\psi(s)\mathrm{d}s\\
\leq&\int_0^\infty (2R_{\max}\|\widetilde{f}\|^2_{V'} - \frac{1}{2R_{\max}}(\|\widetilde{\boldsymbol{U}}(s)\|^2 - c_{17}|\widetilde{\boldsymbol{U}}(s)|_2^2))\psi(s)\mathrm{d}s.
\end{aligned} \tag{3.1.68}$$

Applying Lemma 3.1.23, we have the following inequality

$$\begin{aligned}
&|\widetilde{\boldsymbol{U}}(t)|_2^2\mathrm{e}^{c_{18}t} - |\widetilde{\boldsymbol{U}}(\mathcal{T})|_2^2\mathrm{e}^{c_{18}\mathcal{T}}\\
\leq&\int_{\mathcal{T}}^t (2R_{\max}\|\widetilde{f}\|^2_{V'} - \frac{1}{2R_{\max}}(\|\widetilde{\boldsymbol{U}}(s)\|^2 - c_{17}|\widetilde{\boldsymbol{U}}(s)|_2^2))\mathrm{e}^{c_{18}s}\mathrm{d}s,
\end{aligned} \tag{3.1.69}$$

for any $t,\ \mathcal{T} \in \mathbb{R}_+ \setminus \widetilde{Q},\ t \geq \mathcal{T},\ c_{18} = \dfrac{c_{17}}{2R_{\max}}$, where $\mu(\widetilde{Q}) = 0$. Combining (3.1.67) with (3.1.69), we obtain

$$|\widetilde{\boldsymbol{U}}(t)|_2^2\mathrm{e}^{c_{18}t} - |\widetilde{\boldsymbol{U}}(\mathcal{T})|_2^2\mathrm{e}^{c_{18}\mathcal{T}} \leq \int_{\mathcal{T}}^t 2R_{\max}\|\widetilde{f}\|_{V'}\mathrm{e}^{c_{18}s}\mathrm{d}s. \tag{3.1.70}$$

By (3.1.70), we have

$$\|T(t)\widetilde{\boldsymbol{U}}\|_{L^\infty(\mathbb{R}_+;H)} \le \|\widetilde{\boldsymbol{U}}\|_{L^\infty(0,1;H)} \exp(-c_{18}t) + c_{19}, \quad \forall t \ge 0. \qquad (3.1.71)$$

(2) Combining (3.1.69) with (3.1.71), we get

$$\begin{aligned}
&\frac{1}{2R_{\max}} \int_t^{t+1} (\|\widetilde{\boldsymbol{U}}(s)\|^2 - c_{17}|\widetilde{\boldsymbol{U}}(s)|_2^2) \mathrm{e}^{c_{18}s} \mathrm{d}s \\
\le & c_{19}(\mathrm{e}^{c_{18}(t+1)} - \mathrm{e}^{c_{18}t}) + |\widetilde{\boldsymbol{U}}(t)|_2^2 \mathrm{e}^{c_{18}t} \\
\le & c_{19}(\mathrm{e}^{c_{18}(t+1)} - \mathrm{e}^{c_{18}t}) + \|\widetilde{\boldsymbol{U}}\|^2_{L^\infty(0,1;H)} + c_{19}\mathrm{e}^{c_{18}t} \\
\le & \|\widetilde{\boldsymbol{U}}\|^2_{L^\infty(0,1;H)} + c_{19}\mathrm{e}^{c_{18}(t+1)},
\end{aligned}$$

that is,

$$\begin{aligned}
&\frac{1}{2R_{\max}} \int_t^{t+1} \|\widetilde{\boldsymbol{U}}(s)\|^2 \mathrm{e}^{c_{18}s} \mathrm{d}s \\
\le & c_{18} \int_t^{t+1} |\widetilde{\boldsymbol{U}}(s)|_2^2 \mathrm{e}^{c_{18}s} \mathrm{d}s + \|\widetilde{\boldsymbol{U}}\|^2_{L^\infty(0,1;H)} + c_{19}\mathrm{e}^{c_{18}(t+1)}. \qquad (3.1.72)
\end{aligned}$$

By (3.1.71), we get

$$c_{18} \int_t^{t+1} |\widetilde{\boldsymbol{U}}(s)|_2^2 \mathrm{e}^{c_{18}s} \mathrm{d}s \le c_{18}\|\widetilde{\boldsymbol{U}}\|^2_{L^\infty(0,1;H)} + c_{19}(\mathrm{e}^{c_{18}(t+1)} - \mathrm{e}^{c_{18}t}). \qquad (3.1.73)$$

Combining (3.1.72) with (3.1.73), we have

$$\frac{1}{2R_{\max}} \int_t^{t+1} \|\widetilde{\boldsymbol{U}}(s)\|^2 \mathrm{e}^{c_{18}s} \mathrm{d}s \le (c_{18}+1)\|\widetilde{\boldsymbol{U}}\|^2_{L^\infty(0,1;H)} + c_{19}(2\mathrm{e}^{c_{18}(t+1)} - \mathrm{e}^{c_{18}t}).$$

Thus

$$\frac{1}{2R_{\max}} \int_t^{t+1} \|\widetilde{\boldsymbol{U}}(s)\|^2 \mathrm{d}s \le (c_{18}+1)\|\widetilde{\boldsymbol{U}}\|^2_{L^\infty(0,1;H)} \mathrm{e}^{-c_{18}t} + c_{19}(2\mathrm{e}^{c_{18}} - 1). \qquad (3.1.74)$$

(3) By Lemmas 3.1.3, 3.1.4, 3.1.8, assumptions of problem 3.1.5 and (3.1.36), we have

$$\|\widetilde{\boldsymbol{U}}_t(s)\|_{V^{(-3)}} \le \|\widetilde{\boldsymbol{U}}(s)\| |\widetilde{\boldsymbol{U}}(s)|_2 + c_{20}\|\widetilde{\boldsymbol{U}}(s)\| + \|\widetilde{f}\|_{V'}.$$

So,

$$\begin{aligned}
&\left(\int_t^{t+1} \|\widetilde{\boldsymbol{U}}_t(s)\|^2_{V^{(-3)}} \mathrm{d}s\right)^{\frac{1}{2}} \\
\le & \left(\int_t^{t+1} \|\widetilde{\boldsymbol{U}}(s)\|^2 |\widetilde{\boldsymbol{U}}(s)|_2^2 \mathrm{d}s\right)^{\frac{1}{2}} + c_{21} + c_{21}\left(\int_t^{t+1} \|\widetilde{\boldsymbol{U}}(s)\|^2 \mathrm{d}s\right)^{\frac{1}{2}}
\end{aligned}$$

$$\leq c_{21}\|\widetilde{U}\|_{L^\infty(t,t+1;H)}\left(\int_t^{t+1}\|\widetilde{U}(s)\|^2\mathrm{d}s\right)^{\frac{1}{2}}$$
$$+c_{21}+c_{21}\left(\int_t^{t+1}\|\widetilde{U}(s)\|^2\mathrm{d}s\right)^{\frac{1}{2}},\quad \forall t\geq 0. \tag{3.1.75}$$

Therefore, by (3.1.71), (3.1.74), (3.1.75), we know that there exist two positive constants c_{20} and c_{21} such that

$$\|T(t)\widetilde{U}\|_{L^\infty(\mathbb{R}_+;H)}+\|T(t)\widetilde{U}\|_{L^2(0,1;V)}+\|T(t)\widetilde{U}_t\|_{L^2(0,1;V^{(-3)})}$$
$$\leq c_{22}\|\widetilde{U}\|^2_{L^\infty(0,1;H)}\exp(-c_{23}t)+c_{24}\|\widetilde{U}\|_{L^\infty(0,1;H)}\exp(-c_{23}t)+c_{25},\quad \forall t\geq 0.$$

The proof is complete.

Proposition 3.1.24. If $\{\widetilde{U}_n\}\subset\mathcal{K}^+$ is a bounded sequence in $L^\infty(\mathbb{R}_+;H)$, and for some $\widetilde{U}\in C(\mathbb{R}_+;V^{(-3)})$, the following relationship holds,

$$\widetilde{U}_n\to\widetilde{U}\quad\text{in } C_{\text{loc}}(\mathbb{R}_+;V^{(-3)}),\qquad \text{as } n\to\infty,$$

Then $\widetilde{U}\in\mathcal{K}^+$.

Proof. We only give the outline of the proof. Since $\{\widetilde{U}_n\}\subset\mathcal{K}^+$ is bounded in $L^\infty(\mathbb{R}_+;H)$, by Proposition 3.1.22, we know that, if necessary going to a subsequence $\{\widetilde{U}_n\}$,

$$(\widetilde{U}_n)_t+A\widetilde{U}_n+(B\widetilde{U}_n,\widetilde{U}_n)+(B\widetilde{U}_n,U')+(F\widetilde{U}_n,U')+\widetilde{E}\widetilde{U}_n-\widetilde{f}$$
$$\rightharpoonup(\widetilde{U})_t+A\widetilde{U}+(B\widetilde{U},\widetilde{U})+(B\widetilde{U},U')+(F\widetilde{U},U')+\widetilde{E}\widetilde{U}-\widetilde{f}$$

weakly in $L^2(0,\mathcal{T};V^{(-3)})$, for any $\mathcal{T}>0$. Thus $\widetilde{U}$ is a weak solution of problem 3.1.5.

In order to prove $\widetilde{U}\in\mathcal{K}^+$, we shall check that $\widetilde{U}$ satisfies the energy inequality (3.1.63). According to $\{\widetilde{U}_n\}\subset\mathcal{K}^+$, for any n, $\widetilde{U}_n$ satisfies

$$-\int_0^\infty|\widetilde{U}_n(t)|_2^2\psi'(s)\mathrm{d}s+\frac{1}{R_{\max}}\int_0^\infty\|\widetilde{U}_n(s)\|^2\psi(s)\mathrm{d}s$$
$$\leq 2\int_0^\infty(\widetilde{f},\widetilde{U}_n(s))_H\psi(s)\mathrm{d}s,$$

where $\psi(s)\in C_0^\infty(]0,\mathcal{T}[),\psi(s)\geq 0$. Since $\{\widetilde{U}_n\}\subset\mathcal{K}^+$ is bounded in $L^\infty(\mathbb{R}_+;H)$, by Proposition 3.1.22, if necessary going to a subsequence, we obtain that $\widetilde{U}_m\to\widetilde{U}(m\to\infty)$ weakly in $L^2(0,\mathcal{T};V)$, $*$-weakly in $L^\infty(0,\mathcal{T};H)$. Similarly to the proof of (3.1.53), we can take limit in the above inequality, thus get the energy inequality

$$-\int_0^\infty|\widetilde{U}(t)|_2^2\psi'(s)\mathrm{d}s+\frac{1}{R_{\max}}\int_0^\infty\|\widetilde{U}(s)\|^2\psi(s)\mathrm{d}s\leq 2\int_0^\infty(\widetilde{f},\widetilde{U}(s))_H\psi(s)\mathrm{d}s.$$

The proof of Proposition 3.1.24 is complete.

Now we can give the other main result of this section.

Theorem 3.1.25. If the assumptions of problem 3.1.5 are satisfied, then for the system (3.1.28)-(3.1.33) there exists a trajectory attractor $\mathcal{H} \subseteq \mathcal{K}^+$, and $\mathcal{H}$ is unique in $\mathcal{K}^+$. Moreover $\mathcal{A} = \mathcal{H}(0)$ is the global attractor of system (3.1.28)-(3.1.33) in $V^{(-3)}$.

Proof. In order to prove Theorem 3.1.25, by using Theorem 3.1.13 and Theorem 3.1.15, we construct an attracting set Λ of the time translation semigroup $\{T(t)\}$, where $\Lambda \subseteq \mathcal{K}^+$, and Λ is compact in $C_{\text{loc}}(\mathbb{R}_+;\ V^{(-3)})$, and bounded in $L^\infty(\mathbb{R}_+;\ H)$. Let

$$\Lambda = \{\widetilde{\boldsymbol{U}} \in \mathcal{K}^+;\ \operatorname*{ess\,sup}_{t\geq 0}\{\|\widetilde{\boldsymbol{U}}\|_{L^\infty(t,t+1;H)} + \|\widetilde{\boldsymbol{U}}_t\|_{L^2(t,t+1;V^{(-3)})}\} \leq 3c_{25}\}. \tag{3.1.76}$$

We claim that Λ is an attracting set satisfying the conditions in Theorem 3.1.13. Indeed, suppose that $\Xi \subseteq \mathcal{K}^+$, and Ξ is bounded in $L^\infty(\mathbb{R}_+; H)$, that is

$$\|\widetilde{\boldsymbol{U}}\|_{L^\infty(\mathbb{R}_+;\ H)} = \operatorname*{ess\,sup}_{s\geq 0}\|\widetilde{\boldsymbol{U}}\|_E \leq c_{26}, \quad \forall u \in \Xi.$$

By Proposition 3.1.22, we know that there exists $t_1 > 0$ such that $T(t)\Xi \subseteq \Lambda$ for any $t \geq t_1$. By the definition of Λ, we obtain that Λ is bounded in $L^\infty(\mathbb{R}_+;\ H)$. According to (3.1.76), for any $\mathcal{T} > 0$, $\Pi_{\mathcal{T}}\Lambda$ is bounded in $W^{\infty,2}(0,\mathcal{T}; H, V^{(-3)})$. Applying Lemma 3.1.20, we know that $\Pi_{\mathcal{T}}\Lambda$ is compact in $C(0,\mathcal{T};\ V^{(-3)})$, that is, the closure of Λ is compact in $C_{\text{loc}}(\mathbb{R}_+; V^{(-3)})$. Next we prove that Λ is closed in $C_{\text{loc}}(\mathbb{R}_+; V^{(-3)})$. Suppose that there exists a sequence $\{\widetilde{\boldsymbol{U}}_n\} \subset \Lambda$, that satisfies

$$\widetilde{\boldsymbol{U}}_n \to \widetilde{\boldsymbol{U}} \ \text{ in } C_{\text{loc}}(\mathbb{R}_+; V^{(-3)}), \text{ as } \quad n \to \infty.$$

On the other hand, by (3.1.76), $\{\widetilde{\boldsymbol{U}}_n\}$ is bounded in $L^\infty(\mathbb{R}_+; H)$. Applying Proposition 3.1.24, we obtain $\widetilde{\boldsymbol{U}} \in \mathcal{K}^+$. By the lower weak semicontinuity of the norm and

$$\|\widetilde{\boldsymbol{U}}_n\|_{L^\infty(t,t+1;H)} + \|(\widetilde{\boldsymbol{U}}_n)_t\|_{L^2(t,t+1;V^{(-3)})} \leq 3c_{25},$$

we have

$$\|\widetilde{\boldsymbol{U}}\|_{L^\infty(t,t+1;H)} + \|\widetilde{\boldsymbol{U}}_t\|_{L^2(t,t+1;V^{(-3)})} \leq 3c_{25}.$$

Thus $\widetilde{\boldsymbol{U}} \in \Lambda$, that is, Λ is compact in $C_{\text{loc}}(\mathbb{R}_+;\ V^{(-3)})$. Applying Theorem 3.1.13 and Theorem 3.1.15, we get Theorem 3.1.25 and the trajectory attractor

$$\mathcal{H} = \cap_{\mathcal{T}\geq 0}\overline{\cup_{t\geq \mathcal{T}}T(t)\Lambda},$$

where $\overline{\cup_{t\geq T}T(t)\Lambda}$ is the closure of set $\cup_{t\geq T}T(t)\Lambda$ in space $C_{\text{loc}}(\mathbb{R}_+;V^{(-3)})$. The proof is complete.

Remark 3.1.26. In Lemma 3.1.20, we can let $E_0 = V^{(-\delta)}, 0 < \delta \leq 3$, $E = H, Y = V^{(-3)}$, where E_0 is in Theorem 3.1.18, $V^{(-\delta)} = (V^{(\delta)})'$, and the definition of $V^{(\delta)}$ is similar to the definition of $V^{(3)}$. Then, we can replace $V^{(-3)}$ in Propositions 3.1.21 and 3.1.24 by $V^{(-\delta)}$. Thus we get the trajectory attractor $\mathcal{H}$ of system (3.1.28)-(3.1.33) (here $\mathcal{H} \subseteq \mathcal{K}^+ \subset C(\mathbb{R}_+;V^{(-\delta)}) \cap L^\infty(\mathbb{R}_+;H))$ and the global attractor in the space $V^{(-\delta)}$.

3.2 Long-Time Behavior of the Strong Solutions of the Primitive Equations of the Large-Scale Moist Atmosphere

In this section, we mainly concern with the global existence, uniqueness, stability and long-time behavior of the strong solutions to the initial boundary value problem of the primitive equations of the large-scale atmosphere. The initial boundary value problem of the primitive equations is denoted as IBVP, which will be given in section 3.2.1. Our main results are Theorem 3.2.2, Theorem 3.2.3, Proposition 3.2.4 and Theorem 3.2.5. Firstly, we get the global well-posedness of problem IBVP. Secondly, through studying the long-time behavior of strong solutions, we prove that the H^1-norm of strong solutions is uniformly bounded in time t, and to that the semigroup $\{S(t)\}_{t\geq 0}$ corresponding to the boundary value problem of the primitive equations possesses a bounded absorbing set B_ρ in V (the definition of s-pace V will be given in section 3.2.2). With the bounded absorbing set B_ρ, we construct a weakly compact global attractor $\mathcal{A}$. Because the global well-posedness of the three-dimensional incompressible Navier-Stokes system is still open, with Theorem 3.2.2 and Theorem 3.2.3, we strictly proved mathematically that the large-scale moist atmospheric primitive equations are more simpler than the three-dimensional incompressible Navier-Stokes system. This is consistent with the physical point of view (the large-scale moist atmospheric primitive equations are derived from the atmospheric equations under the hydrostatic approximation).

For studying the long-time behavior of the strong solutions of problem IBVP, we make three key estimates. First, before studying the long-time behavior of the strong solutions, we make estimates of L^3-norm of the baroclinic flow $\tilde{v}$ and temperature T. Without these estimates, we only obtain the global well-posedness of problem IBVP. Second, we make estimates about L^4-norm of the baroclinic flow, temperature and the mixing ratio of the vapor in the air. If we make estimates about L^6-norm of $\tilde{v}$, T, q only like in [26], we can't study the long-time behavior of the strong solutions of problem IBVP. At last, we have to make estimates about the vertical derivations $\partial_\xi T$, $\partial_\xi q$ before we prove that H_1-norm of v, T, q is uniformly bounded in time.

The arrangement of this section is as follows. In subsection 3.2.1, we give the primitive equations of the large-scale moist atmosphere. Our main results are formulated in subsection 3.2.2. In subsection 3.2.3, the *a priori* estimates of local strong solutions are given. We shall prove the main results of this section in the last two parts.

3.2.1 *The Primitive Equations of the Large-Scale Moist Atmosphere*

The model which we consider in this section is the dimensionless three-dimensional viscous primitive equations of the large-scale moist atmosphere in the pressure coordinate frame

$$\frac{\partial \boldsymbol{v}}{\partial t} + \boldsymbol{\nabla}_{\boldsymbol{v}} \boldsymbol{v} + \omega \frac{\partial \boldsymbol{v}}{\partial \xi} + \frac{f}{R_0} \boldsymbol{k} \times \boldsymbol{v} + \operatorname{grad} \Phi - \frac{1}{Re_1} \Delta \boldsymbol{v} - \frac{1}{Re_2} \frac{\partial^2 \boldsymbol{v}}{\partial \xi^2} = 0, \tag{3.2.1}$$

$$\operatorname{div} \boldsymbol{v} + \frac{\partial \omega}{\partial \xi} = 0, \tag{3.2.2}$$

$$\frac{\partial \Phi}{\partial \xi} + \frac{bP}{p}(1 + aq)T = 0, \tag{3.2.3}$$

$$\frac{\partial T}{\partial t} + \boldsymbol{\nabla}_{\boldsymbol{v}} T + \omega \frac{\partial T}{\partial \xi} - \frac{bP}{p}(1 + aq)\omega - \frac{1}{Rt_1} \Delta T - \frac{1}{Rt_2} \frac{\partial^2 T}{\partial \xi^2} = Q_1, \tag{3.2.4}$$

$$\frac{\partial q}{\partial t} + \boldsymbol{\nabla}_{\boldsymbol{v}} q + \omega \frac{\partial q}{\partial \xi} - \frac{1}{Rt_3} \Delta q - \frac{1}{Rt_4} \frac{\partial^2 q}{\partial \xi^2} = Q_2. \tag{3.2.5}$$

The space domain of (3.2.1)-(3.2.5) is

$$\Omega = S^2 \times (0, 1),$$

where S^2 is two-dimensional unit sphere. The boundary conditions of (3.2.1)-(3.2.5) are given by

$$\xi = 1 (p = P): \ \frac{\partial \boldsymbol{v}}{\partial \xi} = 0, \ \omega = 0, \ \frac{\partial T}{\partial \xi} = \alpha_s (\overline{T}_s - T), \ \frac{\partial q}{\partial \xi} = \beta_s (\overline{q}_s - q), \tag{3.2.6}$$

$$\xi = 0 (p = p_0): \ \frac{\partial \boldsymbol{v}}{\partial \xi} = 0, \ \omega = 0, \ \frac{\partial T}{\partial \xi} = 0, \ \frac{\partial q}{\partial \xi} = 0, \tag{3.2.7}$$

where α_s, β_s are positive constants, $\overline{T}_s$ the temperature of the environmental air, and $\overline{q}_s$ the given mixing ratio of the water vapor on the earth surface. For simplicity and without loss of generality, we assume that $\overline{T}_s = 0$ and $\overline{q}_s = 0$. If $\overline{T}_s \neq 0$ or $\overline{q}_s \neq 0$, we can make the boundary conditions of T, q homogeneous (see [88]).

Remark 3.2.1. The difference between the boundary value problem (3.2.1)-(3.2.7) of the three-dimensional viscous primitive equations of the large-scale moist atmosphere and (3.1.8)-(3.1.14) is the boundary condition of $\boldsymbol{v}$. As for the viscous term, for convenience, we take the above form.

Integrating (3.2.2) with respect to ξ, and using boundary conditions (3.2.6) and (3.2.7), we have

$$\omega(t; \theta, \varphi, \xi) = W(\boldsymbol{v})(t; \theta, \varphi, \xi) = \int_\xi^1 \operatorname{div} \boldsymbol{v}(t; \theta, \varphi, \xi') \, \mathrm{d}\xi', \tag{3.2.8}$$

$$\int_0^1 \operatorname{div} \boldsymbol{v} \, \mathrm{d}\xi = 0. \tag{3.2.9}$$

Suppose that Φ_s is a certain unknown function at the isobaric surface $\xi = 1$. Integrating (3.2.3) with respect to ξ, we get

$$\Phi(t, \theta, \varphi, \xi) = \Phi_s(t, \theta, \varphi) + \int_\xi^1 \frac{bP}{p}(1 + aq) T \, \mathrm{d}\xi'. \tag{3.2.10}$$

In this section, for convenience, **we suppose that constants Re_1, Re_2, Rt_1, Rt_2, Rt_3, Rt_4 are all equal to 1**, which do not change our results. Then equations (3.2.1)-(3.2.5) can be rewritten as

$$\begin{aligned} &\frac{\partial \boldsymbol{v}}{\partial t} + \nabla_{\boldsymbol{v}} \boldsymbol{v} + W(\boldsymbol{v}) \frac{\partial \boldsymbol{v}}{\partial \xi} + \frac{f}{R_0} \boldsymbol{k} \times \boldsymbol{v} + \operatorname{grad} \Phi_s + \int_\xi^1 \frac{bP}{p} \operatorname{grad}[(1 + aq) T] \, \mathrm{d}\xi' \\ &- \Delta \boldsymbol{v} - \frac{\partial^2 \boldsymbol{v}}{\partial \xi^2} = 0, \end{aligned} \tag{3.2.11}$$

$$\frac{\partial T}{\partial t} + \nabla_{\boldsymbol{v}} T + W(\boldsymbol{v}) \frac{\partial T}{\partial \xi} - \frac{bP}{p}(1 + aq) W(\boldsymbol{v}) - \Delta T - \frac{\partial^2 T}{\partial \xi^2} = Q_1, \tag{3.2.12}$$

$$\frac{\partial q}{\partial t} + \nabla_{\boldsymbol{v}} q + W(\boldsymbol{v}) \frac{\partial q}{\partial \xi} - \Delta q - \frac{\partial^2 q}{\partial \xi^2} = Q_2, \tag{3.2.13}$$

$$\int_0^1 \operatorname{div} \boldsymbol{v} \, \mathrm{d}\xi = 0, \tag{3.2.14}$$

where the definitions of $\text{grad}[(1+aq)T]$, and $\text{grad}\,\Phi_s$ are as in the above section. The boundary conditions of equations (3.2.11)-(3.2.14) are given by

$$\xi = 1: \ \frac{\partial \boldsymbol{v}}{\partial \xi} = 0, \ \frac{\partial T}{\partial \xi} = -\alpha_s T, \ \frac{\partial q}{\partial \xi} = -\beta_s q, \tag{3.2.15}$$

$$\xi = 0: \ \frac{\partial \boldsymbol{v}}{\partial \xi} = 0, \ \frac{\partial T}{\partial \xi} = 0, \ \frac{\partial q}{\partial \xi} = 0; \tag{3.2.16}$$

and the initial condition is given by

$$U|_{t=0} = (\boldsymbol{v}|_{t=0}, T|_{t=0}, q|_{t=0}) = U_0 = (\boldsymbol{v}_0, T_0, q_0). \tag{3.2.17}$$

We call (3.2.11)-(3.2.17) as the initial boundary value problem of the new formulation of the three-dimensional viscous primitive equations of the large-scale moist atmosphere, which is denoted as IBVP.

Next, we define the baroclinic flow $\tilde{\boldsymbol{v}}$ and the barotrophic flow $\bar{\boldsymbol{v}}$, and find the equations satisfied $\tilde{\boldsymbol{v}}$ and $\bar{\boldsymbol{v}}$. Integrating (3.2.11) from 0 to 1 with respect to ξ, with the boundary conditions (3.2.15)-(3.2.16), we have

$$\begin{aligned}&\frac{\partial \bar{\boldsymbol{v}}}{\partial t} + \int_0^1 \left(\nabla_{\boldsymbol{v}} \boldsymbol{v} + W(\boldsymbol{v}) \frac{\partial \boldsymbol{v}}{\partial \xi} \right) \mathrm{d}\xi + \frac{f}{R_0} \boldsymbol{k} \times \bar{\boldsymbol{v}} + \text{grad}\,\Phi_s \\ &+ \int_0^1 \int_\xi^1 \frac{bP}{p} \text{grad}[(1+aq)T] \mathrm{d}\xi' \mathrm{d}\xi - \Delta \bar{\boldsymbol{v}} = 0 \ \text{ in } S^2,\end{aligned} \tag{3.2.18}$$

where $\bar{\boldsymbol{v}} = \displaystyle\int_0^1 \boldsymbol{v} \mathrm{d}\xi$.

Denote the baroclinic flow as

$$\tilde{\boldsymbol{v}} = \boldsymbol{v} - \bar{\boldsymbol{v}}.$$

We notice that

$$\bar{\tilde{\boldsymbol{v}}} = \int_0^1 \tilde{\boldsymbol{v}} \mathrm{d}\xi = 0, \quad \nabla \cdot \bar{\boldsymbol{v}} = 0.$$

By integration by parts and with the above equalities, we get

$$\int_0^1 W(\boldsymbol{v}) \frac{\partial \boldsymbol{v}}{\partial \xi} \mathrm{d}\xi = \int_0^1 \boldsymbol{v} \,\text{div} \boldsymbol{v} \mathrm{d}\xi = \int_0^1 \tilde{\boldsymbol{v}} \,\text{div} \tilde{\boldsymbol{v}} \mathrm{d}\xi, \tag{3.2.19}$$

$$\int_0^1 \nabla_{\boldsymbol{v}} \boldsymbol{v} \mathrm{d}\xi = \int_0^1 \nabla_{\tilde{\boldsymbol{v}}} \tilde{\boldsymbol{v}} \mathrm{d}\xi + \nabla_{\bar{\boldsymbol{v}}} \bar{\boldsymbol{v}}. \tag{3.2.20}$$

According to (3.2.18)-(3.2.20), we obtain

$$\frac{\partial \bar{\boldsymbol{v}}}{\partial t} + \nabla_{\bar{\boldsymbol{v}}} \bar{\boldsymbol{v}} + \overline{\tilde{\boldsymbol{v}} \,\text{div} \tilde{\boldsymbol{v}} + \nabla_{\tilde{\boldsymbol{v}}} \tilde{\boldsymbol{v}}} + \frac{f}{R_0} \boldsymbol{k} \times \bar{\boldsymbol{v}} + \text{grad}\,\Phi_s$$

$$+\int_0^1\int_\xi^1 \frac{bP}{p}\mathrm{grad}[(1+aq)T]\mathrm{d}\xi'\mathrm{d}\xi - \Delta\bar{\boldsymbol{v}} = 0 \ \text{ in } S^2. \tag{3.2.21}$$

Subtracting (3.2.21) from (3.2.11), we know that the baroclinic flow $\tilde{\boldsymbol{v}}$ satisfies the following equation and boundary conditions

$$\frac{\partial\tilde{\boldsymbol{v}}}{\partial t} - \Delta\tilde{\boldsymbol{v}} - \frac{\partial^2\tilde{\boldsymbol{v}}}{\partial\xi^2} + \boldsymbol{\nabla}_{\tilde{\boldsymbol{v}}}\tilde{\boldsymbol{v}} + \left(\int_\xi^1 \mathrm{div}\tilde{\boldsymbol{v}}\mathrm{d}\xi'\right)\frac{\partial\tilde{\boldsymbol{v}}}{\partial\xi} + \boldsymbol{\nabla}_{\tilde{\boldsymbol{v}}}\bar{\boldsymbol{v}} + \boldsymbol{\nabla}_{\bar{\boldsymbol{v}}}\tilde{\boldsymbol{v}} - \overline{(\tilde{\boldsymbol{v}}\mathrm{div}\tilde{\boldsymbol{v}} + \boldsymbol{\nabla}_{\tilde{\boldsymbol{v}}}\tilde{\boldsymbol{v}})}$$
$$+\frac{f}{R_0}\boldsymbol{k}\times\tilde{\boldsymbol{v}} + \int_\xi^1 \frac{bP}{p}\mathrm{grad}[(1+aq)T]\mathrm{d}\xi' - \int_0^1\int_\xi^1 \frac{bP}{p}\mathrm{grad}[(1+aq)T]\mathrm{d}\xi'\mathrm{d}\xi$$
$$= 0 \ \text{ in } \Omega, \tag{3.2.22}$$

$$\xi = 1: \frac{\partial\tilde{\boldsymbol{v}}}{\partial\xi} = 0, \quad \xi = 0: \frac{\partial\tilde{\boldsymbol{v}}}{\partial\xi} = 0. \tag{3.2.23}$$

3.2.2 *Main Results*

Before stating the main results of this section, we define some function spaces.

The definitions of spaces $L^p(\Omega)$, $L^2(T\Omega|TS^2)$, $C^\infty(S^2)$, $C^\infty(T\Omega|TS^2)$ and $H^m(\Omega)$ are as in the above section. The definitions of the horizontal divergence div, horizontal gradient $\boldsymbol{\nabla} = \mathrm{grad}$, horizontal covariant derivative $\boldsymbol{\nabla}_{\boldsymbol{v}}$ and the horizontal Laplace-Beltrami operator Δ of the scalar and vector functions are also as in the above section.

Next we define the working spaces of problem IBVP. Let

$$\widetilde{\mathcal{V}_1} := \left\{\boldsymbol{v};\ \boldsymbol{v}\in C^\infty(T\Omega|TS^2),\ \frac{\partial\boldsymbol{v}}{\partial\xi}\bigg|_{\xi=0} = 0,\ \frac{\partial\boldsymbol{v}}{\partial\xi}\bigg|_{\xi=1} = 0,\ \int_0^1 \mathrm{div}\boldsymbol{v}\ \mathrm{d}\xi = 0\right\},$$
$$\widetilde{\mathcal{V}_2} := \left\{T;\ T\in C^\infty(\Omega),\ \frac{\partial T}{\partial\xi}\bigg|_{\xi=0} = 0,\ \frac{\partial T}{\partial\xi}\bigg|_{\xi=1} = -\alpha_s T\right\};$$
$$\widetilde{\mathcal{V}_3} := \left\{q;\ q\in C^\infty(\Omega),\ \frac{\partial q}{\partial\xi}\bigg|_{\xi=0} = 0,\ \frac{\partial q}{\partial\xi}\bigg|_{\xi=1} = -\beta_s q\right\},$$

V_1 = the closure of $\widetilde{\mathcal{V}_1}$ with respect to norm $\|\cdot\|_1$, V_2 and V_3 respectively closures of $\widetilde{\mathcal{V}_2}$ and $\widetilde{\mathcal{V}_3}$ with respect to norm $\|\cdot\|_1$,
H_1 = the closure of $\widetilde{\mathcal{V}_1}$ with respect to norm $|\cdot|_2$, $H_2 = L^2(\Omega)$,
$V = V_1\times V_2\times V_3$, $H = H_1\times H_2\times H_2$.
The inner product and norm of V_1, V_2, V_3, H, V are given by

$$(\boldsymbol{v},\boldsymbol{v}_1)_{V_1} = \int_\Omega\left(\boldsymbol{\nabla}_{e_\theta}\boldsymbol{v}\cdot\boldsymbol{\nabla}_{e_\theta}\boldsymbol{v}_1 + \boldsymbol{\nabla}_{e_\varphi}\boldsymbol{v}\cdot\boldsymbol{\nabla}_{e_\varphi}\boldsymbol{v}_1 + \frac{\partial\boldsymbol{v}}{\partial\xi}\frac{\partial\boldsymbol{v}_1}{\partial\xi} + \boldsymbol{v}\cdot\boldsymbol{v}_1\right),$$
$$\|\boldsymbol{v}\| = (\boldsymbol{v},\boldsymbol{v})_{V_1}^{\frac{1}{2}}, \quad \forall\boldsymbol{v},\ \boldsymbol{v}_1\in V_1,$$

$$(T,T_1)_{V_2} = \int_\Omega \left(\text{grad}T \cdot \text{grad}T_1 + \frac{\partial T}{\partial \xi}\frac{\partial T_1}{\partial \xi} + TT_1\right),$$
$$\|T\| = (T,T)_{V_2}^{\frac{1}{2}}, \quad \forall T,\ T_1 \in V_2,$$
$$(q,q_1)_{V_3} = \int_\Omega \left(\text{grad}q \cdot \text{grad}q_1 + \frac{\partial q}{\partial \xi}\frac{\partial q_1}{\partial \xi} + qq_1\right),$$
$$\|q\| = (q,q)_{V_3}^{\frac{1}{2}}, \quad \forall q,\ q_1 \in V_2,$$
$$(\boldsymbol{U},\boldsymbol{U}_1)_H = (\boldsymbol{v},\boldsymbol{v}_1) + (T,T_1) + (q,q_1),$$
$$(\boldsymbol{U},\boldsymbol{U}_1)_V = (\boldsymbol{v},\boldsymbol{v}_1)_{V_1} + (T,T_1)_{V_2} + (q,q_1)_{V_3},$$
$$\|\boldsymbol{U}\| = (\boldsymbol{U},\boldsymbol{U})_V^{\frac{1}{2}},\ |\boldsymbol{U}|_2 = (\boldsymbol{U},\boldsymbol{U})_H^{\frac{1}{2}},\ \forall \boldsymbol{U} = (\boldsymbol{v},T,q),\ \boldsymbol{U}_1 = (\boldsymbol{v}_1,T_1,q_1) \in V,$$
where $(\cdot,\cdot)$ is L^2 inner product of $L^2(\Omega)$.

Next we give the main results of this section.

Theorem 3.2.2. Let $Q_1,\ Q_2 \in H^1(\Omega)$, $\boldsymbol{U}_0 = (\boldsymbol{v}_0,T_0,q_0) \in V$. Then, for any $\mathcal{T} > 0$ given, there exists a strong solution $\boldsymbol{U}$ of the system (3.2.11)-(3.2.17) on the interval $[0,\mathcal{T}]$, where the definition of the strong solutions of system (3.2.11)-(3.2.17) will be given in section 3.2.3.1.

Theorem 3.2.3. If $Q_1,\ Q_2 \in H^1(\Omega)$, $\boldsymbol{U}_0 = (\boldsymbol{v}_0,T_0,q_0) \in V$, then, for any $\mathcal{T} > 0$ given, the strong solution $\boldsymbol{U}$ of system (3.2.11)-(3.2.17) on the interval $[0,\mathcal{T}]$ is unique. Moreover the strong solution $\boldsymbol{U}$ is dependent continuously on the initial data.

Proposition 3.2.4. If $Q_1,\ Q_2 \in H^1(\Omega)$, $\boldsymbol{U}_0 = (\boldsymbol{v}_0,T_0,q_0) \in V$, then the strong solution of system (3.2.11)-(3.2.17) satisfies $\boldsymbol{U} \in L^\infty(0,\infty;V)$. Moreover, the corresponding semigroup $\{S(t)\}_{t\geq 0}$ of system (3.2.11)-(3.2.16) possesses a bounded absorbing set B_δ in V, that is, for every bounded $B \subset V$, there exists $t_0(B) > 0$ large enough such that for any $t \geq t_0$, $S(t)B \subset B_\delta$, where $B_\delta = \{\boldsymbol{U}; \|\boldsymbol{U}\| \leq \delta\}$, and δ is a positive constant dependent on $\|Q_1\|_1$, $\|Q_2\|_1$.

Theorem 3.2.5. (3.2.11)-(3.2.16) possess a global attractor $\mathcal{A} = \cap_{s\geq 0}\overline{\cup_{t\geq s}T(t)B_\rho}$ that captures all the trajectories, where the closure is taken with respect to the weak topology of V. The global attractor $\mathcal{A}$ has the following properties:

(i) (Weak compactness) $\mathcal{A}$ is bounded in V, and weakly closed;

(ii) (Invariant) for any $t \geq 0$, $S(t)\mathcal{A} = \mathcal{A}$;

(iii) (Attracting) for any bounded set in V, as $t \to +\infty$, the set family $S(t)B$ converges to $\mathcal{A}$ with respect to the weak topology of V, that is, $\lim\limits_{t\to+\infty} d_V^w(S(t)B,\ \mathcal{A}) = 0$, where the distance d_V^w is induced by the weak topology of V.

Remark 3.2.6. The global attractor $\mathcal{A}$ has the following additional properties:

(i) By Rellich-Kondrachov Compact Embedding Theorem (see [3]), we know that for any $1 \leq p < 6$, the sets $S(t)B$ converge to $\mathcal{A}$ with respect to $L^p(\Omega) \times L^p(\Omega) \times L^p(\Omega) \times L^p(\Omega)$-norm.

(ii) The global attractor $\mathcal{A}$ is unique, and is connected with respect to the weak topology of V.

Remark 3.2.7. In comparison with the three-dimensional incompressible Navier-Stokes equation, the three-dimensional viscous primitive equations of the large-scale moist atmosphere have not the time derivative of the vertical velocity ω. Thus we can't prove the bounded absorbing set B_ρ in V is bounded in $H^2(\Omega) \times H^2(\Omega) \times H^2(\Omega) \times H^2(\Omega)$ as three-dimensional incompressible Navier-Stokes equation. For the three-dimensional incompressible Navier-Stokes equation, if there exists a bounded absorbing set in $H_0^1(\Omega) \times H_0^1(\Omega) \times H_0^1(\Omega)$, then one can prove that this bounded absorbing set is bounded in $H^2(\Omega) \times H^2(\Omega) \times H^2(\Omega)$.

3.2.3 *Uniform Estimates of Local Strong Solutions in Time*

First, we give some lemmas to be used later.

Lemma 3.2.8. Let $\boldsymbol{v} = (v_\theta, v_\varphi)$, $\boldsymbol{v}_1 = ((v_1)_\theta, (v_1)_\varphi) \in C^\infty(T\Omega|TS^2)$, $p \in C^\infty(S^2)$. Then we have

(i)

$$\int_{S^2} p\mathrm{div}\boldsymbol{v} = -\int_{S^2} \boldsymbol{\nabla} p \cdot \boldsymbol{v},$$

in particular, $\displaystyle\int_{S^2} \mathrm{div}\boldsymbol{v} = 0$;

(ii)

$$\int_\Omega (-\Delta \boldsymbol{v}) \cdot \boldsymbol{v}_1 = \int_\Omega (\boldsymbol{\nabla}_{e_\theta} \boldsymbol{v} \cdot \boldsymbol{\nabla}_{e_\theta} \boldsymbol{v}_1 + \boldsymbol{\nabla}_{e_\varphi} \boldsymbol{v} \cdot \boldsymbol{\nabla}_{e_\varphi} \boldsymbol{v}_1 + \boldsymbol{v} \cdot \boldsymbol{v}_1).$$

Proof. we can use Stokes Theorem and the definitions of the horizontal divergence div and gradient operators and (see [201]) to prove (i). By direct calculation, we obtain the second part.

Lemma 3.2.9 (Interpolation Inequalities). Let Ω_1 be the bounded region in $\mathbb{R}^n$, whose boundary $\partial\Omega_1$ satisfies: $\partial\Omega_1 \in C^m$. Then, for any $\boldsymbol{u} \in W^{m,r}(\Omega_1) \cap L^q(\Omega_1)$, $0 \leq l \leq m$,

(i) When $m - l - \dfrac{n}{r}$ is not a nonnegative integer,

$\|D^l \boldsymbol{u}\|_{L^p(\Omega_1)} \leq c\|\boldsymbol{u}\|^\alpha_{W^{m,r}(\Omega_1)} \|\boldsymbol{u}\|^{1-\alpha}_{L^q(\Omega_1)}$, where l, p, α, m, r, q satisfy

$$\frac{l}{m} \le \alpha \le 1, 1 \le r, q \le \infty, \frac{1}{p} - \frac{l}{n} = \alpha\left(\frac{1}{r} - \frac{m}{n}\right) + (1-\alpha)\frac{1}{q};$$

(ii) When $m - l - \frac{n}{r}$ is a nonnegative integer,

$\|D^l \boldsymbol{u}\|_{L^p(\Omega_1)} \le c\|\boldsymbol{u}\|^{\alpha}_{W^{m,r}(\Omega_1)}\|\boldsymbol{u}\|^{1-\alpha}_{L^q(\Omega_1)}$, where l, p, α, m, r, q satisfy

$$\frac{l}{m} \le \alpha < 1, 1 < r < \infty, 1 < q < \infty.$$

In particular, the following interpolation inequalities hold

(iii) For each $\boldsymbol{u} \in H^1(S^2)$ (see [138] for definitions of $H^1(S^2)$, $L^p(S^2)$),

$$\|\boldsymbol{u}\|_{L^4(S^2)} \le c\|\boldsymbol{u}\|^{\frac{1}{2}}_{L^2(S^2)}\|\boldsymbol{u}\|^{\frac{1}{2}}_{H^1(S^2)}, \tag{3.2.24}$$

$$\|\boldsymbol{u}\|_{L^6(S^2)} \le c\|\boldsymbol{u}\|^{\frac{2}{3}}_{L^4(S^2)}\|\boldsymbol{u}\|^{\frac{1}{3}}_{H^1(S^2)}, \tag{3.2.25}$$

$$\|\boldsymbol{u}\|_{L^8(S^2)} \le c\|\boldsymbol{u}\|^{\frac{1}{2}}_{L^4(S^2)}\|\boldsymbol{u}\|^{\frac{1}{2}}_{H^1(S^2)}. \tag{3.2.26}$$

(iv) For any $\boldsymbol{u} \in H^1(\Omega)$,

$$\|\boldsymbol{u}\|_{L^4(\Omega)} \le c\|\boldsymbol{u}\|^{\frac{1}{4}}_{L^2(\Omega)}\|\boldsymbol{u}\|^{\frac{3}{4}}_{H^1(\Omega)}. \tag{3.2.27}$$

Proof. The proof of (i), (ii) is similar to that of $\Omega_1 = \mathbb{R}^n$. For the proof of the details, the reader can refer to [3, 76]. For the proof of (3.2.24)-(3.2.26), see [138, Chapter 1].

Lemma 3.2.10. For any $h \in C^\infty(S^2)$, $\boldsymbol{v} \in C^\infty(T\Omega|TS^2)$, we have

$$\int_{S^2} \nabla_{\boldsymbol{v}} h + \int_{S^2} h \text{div}\boldsymbol{v} = \int_{S^2} \text{div}(h\boldsymbol{v}) = 0.$$

Proof. By direct calculation, we can prove Lemma 3.2.10.

Lemma 3.2.11. Suppose $\boldsymbol{v}$, $\boldsymbol{v}_1 \in V_1$, $T \in V_2$, and $q \in V_3$. Then we have

(i) $\int_\Omega \left(\nabla_{\boldsymbol{v}} \boldsymbol{v}_1 + \left(\int_\xi^1 \text{div}\boldsymbol{v} d\xi'\right)\frac{\partial \boldsymbol{v}_1}{\partial \xi}\right) \cdot \boldsymbol{v}_1 = 0,$

(ii) $\int_\Omega \left(\nabla_{\boldsymbol{v}} T + \left(\int_\xi^1 \text{div}\boldsymbol{v} d\xi'\right)\frac{\partial T}{\partial \xi}\right) T = 0,$

(iii) $\int_\Omega \left(\nabla_{\boldsymbol{v}} q + \left(\int_\xi^1 \text{div}\boldsymbol{v} d\xi'\right)\frac{\partial q}{\partial \xi}\right) q = 0,$

(iv) $\int_\Omega \left(\int_\xi^1 \frac{bP}{p}\text{grad}[(1+aq)T]d\xi' \cdot \boldsymbol{v} - \frac{bP}{p}(1+aq)W(\boldsymbol{v})T\right) = 0.$

For the detail of the proof of Lemma 3.2.11, see [88, Lemma 3.2].

Lemma 3.2.12 (Minkowski Inequality). Let (X, μ), (Y, ν) be two measure spaces, and $f(x, y)$ be a measurable function about $\mu \times \nu$ on

$X \times Y$. If for a.e. $y \in Y$, $f(\cdot, y) \in L^p(X, \mu)$, $1 \le p \le \infty$, and $\int_Y \|f(\cdot, y)\|_{L^p(X,\mu)} \mathrm{d}\nu(y) < \infty$, then

$$\Big\| \int_Y f(\cdot, y) \mathrm{d}\nu(y) \Big\|_{L^p(X,\mu)} \le \int_Y \|f(\cdot, y)\|_{L^p(X,\mu)} \mathrm{d}\nu(y).$$

Lemma 3.2.13 (The Uniform Gronwall Lemma). Let ϕ, ψ, φ be three positive local integrable functions defined on $[t_0, +\infty)$, such that φ' is a positive local integrable function defined on $[t_0, +\infty)$, and satisfy:

$$\frac{\mathrm{d}\varphi}{\mathrm{d}t} \le \phi\varphi + \psi, \text{ for any } t \ge t_0,$$

$$\int_t^{t+r} \phi(s)\mathrm{d}s \le a_1, \ \int_t^{t+r} \psi(s)\mathrm{d}s \le a_2, \ \int_t^{t+r} \varphi(s)\mathrm{d}s \le a_3, \text{ for any } t \ge t_0,$$

where r, a_1, a_2, a_3 are positive constants. Then

$$\varphi(t + r) \le (\frac{a_3}{r} + a_2) \exp(a_1), \ \forall t \ge t_0.$$

For the detail of the proof of Lemma 3.2.13, see [202, p.91].

3.2.3.1 *Local Existence of the Strong Solutions*

In this part, we recall the local existence of strong solutions of the three-dimensional viscous primitive equations of the large-scale moist atmosphere.

Definition 3.2.14. Let $\boldsymbol{U}_0 = (\boldsymbol{v}_0, T_0, q_0) \in V$, and $\mathcal{T}$ be a given positive time. $\boldsymbol{U} = (\boldsymbol{v}, T, q)$ is called a strong solution of system (3.2.11)-(3.2.17) in $[0, \mathcal{T}]$, if $\boldsymbol{U}$ satisfies (3.2.11)-(3.2.13) in weak sense such that

$$\begin{aligned}
&\boldsymbol{v} \in C([0, \mathcal{T}]; V_1) \cap L^2(0, \mathcal{T}; (H^2(\Omega))^2), \\
&T \in C([0, \mathcal{T}]; V_2) \cap L^2(0, \mathcal{T}; H^2(\Omega)), \\
&q \in C([0, \mathcal{T}]; V_3) \cap L^2(0, \mathcal{T}; H^2(\Omega)), \\
&\frac{\partial \boldsymbol{v}}{\partial t} \in L^2(0, \mathcal{T}; (L^2(TR|TS^2))), \\
&\frac{\partial T}{\partial t}, \ \frac{\partial q}{\partial t} \in L^2(0, \mathcal{T}; L^2(\Omega)).
\end{aligned}$$

Proposition 3.2.15. Let Q_1, $Q_2 \in H^1(\Omega)$, $\boldsymbol{U}_0 = (\boldsymbol{v}_0, T_0, q_0) \in V$. Then, there exists $\mathcal{T}^* > 0$, $\mathcal{T}^* = \mathcal{T}^*(\|\boldsymbol{U}_0\|)$, and there exists a strong solution $\boldsymbol{U}$ of the system (3.2.11)-(3.2.17) on $[0, \mathcal{T}^*]$.

Proof. Since the proof of Proposition 3.2.15 is similar to the reference of [87, 203] that for the primitive equations of the large-scale oceans, here we omit its detail.

To prove the global existence of the strong solutions of system (3.2.11)-(3.2.17), we should make *a priori* estimates of the H^1-norm of the local strong solutions obtained in Proposition 3.2.15. We shall prove that if $\mathcal{T}^* < \infty$, then the H^1-norm of strong solution $\boldsymbol{U}(t)$ is uniformly bounded in $[0, \mathcal{T}^*]$.

3.2.3.2 *A Prior Estimates of Local Strong Solutions*

L^2 estimates of $\boldsymbol{v}$, T, q. Taking L^2 inner product of equation (3.2.11) with $\boldsymbol{v}$, we get

$$\begin{aligned}&\frac{1}{2}\frac{\mathrm{d}|\boldsymbol{v}|_2^2}{\mathrm{d}t} + \int_\Omega \left(|\nabla_{e_\theta}\boldsymbol{v}|^2 + |\nabla_{e_\varphi}\boldsymbol{v}|^2 + |\boldsymbol{v}|^2\right) + \int_\Omega \left|\frac{\partial \boldsymbol{v}}{\partial \xi}\right|^2 \\ &= -\int_\Omega \left(\nabla_{\boldsymbol{v}}\boldsymbol{v} + W(\boldsymbol{v})\frac{\partial \boldsymbol{v}}{\partial \xi} + \frac{f}{R_0}\boldsymbol{k}\times\boldsymbol{v} + \operatorname{grad}\Phi_s\right)\cdot\boldsymbol{v} \\ &\quad - \int_\Omega \left[\int_\xi^1 \frac{bP}{p}\operatorname{grad}\left((1+aq)T\right)\right]\cdot\boldsymbol{v}.\end{aligned}$$

By integration by parts and (3.2.14), according to Lemma 3.2.11 and $\left(\frac{f}{R_0}\boldsymbol{k}\times\boldsymbol{v}\right)\cdot\boldsymbol{v} = 0$, we obtain

$$\begin{aligned}&\frac{1}{2}\frac{\mathrm{d}|\boldsymbol{v}|_2^2}{\mathrm{d}t} + \int_\Omega \left(|\nabla_{e_\theta}\boldsymbol{v}|^2 + |\nabla_{e_\varphi}\boldsymbol{v}|^2 + |\boldsymbol{v}|^2 + \left|\frac{\partial \boldsymbol{v}}{\partial \xi}\right|^2\right) \\ &= -\int_\Omega \left[\int_\xi^1 \frac{bP}{p}\operatorname{grad}\left((1+aq)T\right)\right]\cdot\boldsymbol{v}.\end{aligned} \tag{3.2.28}$$

Taking $L^2(\Omega)$ inner product of equation (3.2.12) with T, according to Lemma 3.2.11, we get

$$\frac{1}{2}\frac{\mathrm{d}|T|_2^2}{\mathrm{d}t} + \int_\Omega |\nabla T|^2 + \int_\Omega \left|\frac{\partial T}{\partial \xi}\right|^2 + \alpha_s |T|_{\xi=1}|_2^2 = \int_\Omega \frac{bP}{p}(1+aq)TW(\boldsymbol{v}) + \int_\Omega Q_1 T. \tag{3.2.29}$$

In a similar way, we have

$$\frac{1}{2}\frac{\mathrm{d}|q|_2^2}{\mathrm{d}t} + \int_\Omega |\nabla q|^2 + \int_\Omega \left|\frac{\partial q}{\partial \xi}\right|^2 + \beta_s |q|_{\xi=1}|_2^2 = \int_\Omega qQ_2. \tag{3.2.30}$$

Applying Lemma 3.2.11, we get

$$-\int_\Omega \left[\int_\xi^1 \frac{bP}{p}\operatorname{grad}((1+aq)T)\mathrm{d}\xi'\right]\cdot\boldsymbol{v} + \int_\Omega \frac{bP}{p}(1+cq)TW(\boldsymbol{v}) = 0.$$

Thus, according to (3.2.28)-(3.2.30), we obtain

$$\frac{1}{2}\frac{\mathrm{d}(|\boldsymbol{v}|_2^2+|T|_2^2+|q|_2^2)}{\mathrm{d}t}+\int_\Omega(|\boldsymbol{\nabla}_{e_\theta}\boldsymbol{v}|^2+|\boldsymbol{\nabla}_{e_\varphi}\boldsymbol{v}|^2+|\boldsymbol{v}|^2)+\int_\Omega\left|\frac{\partial\boldsymbol{v}}{\partial\xi}\right|^2$$
$$+\int_\Omega|\boldsymbol{\nabla}T|^2+\int_\Omega\left|\frac{\partial T}{\partial\xi}\right|^2+\alpha_s|T|_{\xi=1}|_2^2+\int_\Omega|\boldsymbol{\nabla}q|^2+\int_\Omega\left|\frac{\partial q}{\partial\xi}\right|^2+\beta_s|q|_{\xi=1}|_2^2$$
$$=\int_\Omega Q_1T+\int_\Omega qQ_2. \tag{3.2.31}$$

By using $T(\theta,\varphi,\xi)=-\int_\xi^1\frac{\partial T}{\partial\xi'}\mathrm{d}\xi'+T|_{\xi=1}$, Hölder Inequality and Cauchy-Schwarz Inequality, we have

$$|T|_2^2\le 2\left|\frac{\partial T}{\partial\xi}\right|_2^2+2|T|_{\xi=1}|_2^2.$$

Similarly,

$$|q|_2^2\le 2\left|\frac{\partial q}{\partial\xi}\right|_2^2+2|q|_{\xi=1}|_2^2.$$

By Young Inequality, we have

$$\left|\int_\Omega Q_1T\right|\le\varepsilon|T|_2^2+c|Q_1|_2^2,\quad \left|\int_\Omega qQ_2\right|\le\varepsilon|q|_2^2+c|Q_2|_2^2,$$

where ε is a positive constant small enough. Thus, we derive from (3.2.31)

$$\frac{\mathrm{d}(|\boldsymbol{v}|_2^2+|T|_2^2+|q|_2^2)}{\mathrm{d}t}+\int_\Omega(|\boldsymbol{\nabla}_{e_\theta}\boldsymbol{v}|^2+|\boldsymbol{\nabla}_{e_\varphi}\boldsymbol{v}|^2+|\boldsymbol{v}|^2)+\int_\Omega\left|\frac{\partial\boldsymbol{v}}{\partial\xi}\right|^2$$
$$+\int_\Omega|\boldsymbol{\nabla}T|^2+\int_\Omega\left|\frac{\partial T}{\partial\xi}\right|^2+\alpha_s|T|_{\xi=1}|_2^2+\int_\Omega|\boldsymbol{\nabla}q|^2+\int_\Omega\left|\frac{\partial q}{\partial\xi}\right|^2+\beta_s|q|_{\xi=1}|_2^2$$
$$\le c\int_\Omega(Q_1^2+Q_2^2). \tag{3.2.32}$$

By Gronwall Inequality, we have

$$|\boldsymbol{v}|_2^2+|T|_2^2+|q|_2^2\le \mathrm{e}^{-c_0t}(|\boldsymbol{v}_0|_2^2+|T_0|_2^2+|q_0|_2^2)+c(|Q_1|_2^2+|Q_2|_2^2)\le E_0, \tag{3.2.33}$$

where $c_0=\min\{\frac{1}{2},\frac{\alpha_s}{2},\frac{\beta_s}{2}\}>0,\ t\ge 0$, and E_0 is a positive constant. By Minkowski Inequality and Hölder Inequality, for any $t\ge 0$, we have

$$\|\bar{\boldsymbol{v}}(t)\|_{L^2(S^2)}^2\le|\boldsymbol{v}(t)|_2^2\le \mathrm{e}^{-c_0t}(|\boldsymbol{v}_0|_2^2+|T_0|_2^2+|q_0|_2^2)+c(|Q_1|_2^2+|Q_2|_2^2)\le E_0. \tag{3.2.34}$$

Combining (3.2.32) with (3.2.33), we have

$$c_1\int_t^{t+r}\left[\int_\Omega\left(|\boldsymbol{\nabla}_{e_\theta}\boldsymbol{v}|^2+|\boldsymbol{\nabla}_{e_\varphi}\boldsymbol{v}|^2+\left|\frac{\partial\boldsymbol{v}}{\partial\xi}\right|^2\right)+\int_\Omega\left(|\boldsymbol{\nabla}T|^2+\left|\frac{\partial T}{\partial\xi}\right|^2+|T|^2\right)\right.$$

$$+\int_\Omega\left(|\nabla q|^2+\left|\frac{\partial q}{\partial\xi}\right|^2+|q|^2\right)+|T|_{\xi=1}|_2^2+|q|_{\xi=1}|_2^2\Bigg]+|U(t)|_2^2$$
$$\leq 2e^{-c_0t}\left(|\boldsymbol{v}_0|_2^2+|T_0|_2^2+|q_0|_2^2\right)+c(|Q_1|_2^2+|Q_2|_2^2)(2+r)\leq E_1, \qquad (3.2.35)$$

where $c_1=\min\{\frac{1}{3},\frac{\alpha_s}{2},\frac{\beta_s}{2}\}$, $t\geq 0$, $1\geq r>0$ is given, E_1 is a positive constant, and $\int_t^{t+r}\cdot ds$ is remarked as $\int_t^{t+r}\cdot$. Since

$$\int_{S^2}(|\nabla_{e_\theta}\bar{\boldsymbol{v}}|^2+|\nabla_{e_\varphi}\bar{\boldsymbol{v}}|^2)\leq\int_\Omega(|\nabla_{e_\theta}\boldsymbol{v}|^2+|\nabla_{e_\varphi}\boldsymbol{v}|^2),$$

we derive from (3.2.35)

$$c_1\int_t^{t+r}\int_{S^2}\left(|\nabla_{e_\theta}\bar{\boldsymbol{v}}|^2+|\nabla_{e_\varphi}\bar{\boldsymbol{v}}|^2\right)+\|\bar{\boldsymbol{v}}\|_{L^2}^2\leq E_1,\forall t\geq 0. \qquad (3.2.36)$$

L^4 estimates of q. Taking $L^2(\Omega)$ inner product of equation (3.2.13) with $|q|^2q$, we have

$$\frac{1}{4}\frac{d|q|_4^4}{dt}+3\int_\Omega|\nabla q|^2q^2+3\int_\Omega\left|\frac{\partial q}{\partial\xi}\right|^2q^2+\beta_s\int_{S^2}|q|_{\xi=1}|^4$$
$$=\int_\Omega Q_2|q|^2q-\int_\Omega\left[\nabla_{\boldsymbol{v}}q+\left(\int_\xi^1\mathrm{div}\boldsymbol{v}d\xi'\right)\frac{\partial q}{\partial\xi}\right]|q|^2q.$$

According to Lemma 3.2.10, we get

$$\int_\Omega\left(\nabla_{\boldsymbol{v}}q+\left(\int_\xi^1\mathrm{div}\boldsymbol{v}d\xi'\right)\frac{\partial q}{\partial\xi}\right)|q|^2q$$
$$=\frac{1}{4}\int_\Omega\nabla_{\boldsymbol{v}}q^4+\int_{S^2}\left[\int_0^1(\int_\xi^1\mathrm{div}\boldsymbol{v}d\xi')d\left(\frac{1}{4}q^4\right)\right]$$
$$=\frac{1}{4}\int_\Omega(\nabla_{\boldsymbol{v}}q^4+q^4\mathrm{div}\boldsymbol{v})=0.$$

By the above two equalities, we obtain

$$\frac{1}{4}\frac{d|q|_4^4}{dt}+3\int_\Omega|\nabla q|^2q^2+3\int_\Omega\left|\frac{\partial q}{\partial\xi}\right|^2q^2+\beta_s\int_{S^2}|q|_{\xi=1}|^4=\int_\Omega Q_2|q|^2q.$$

Since $q^4(\theta,\varphi,\xi)=-\int_\xi^1\frac{\partial q^4}{\partial\xi'}d\xi'+q^4|_{\xi=1}$, using Hölder Inequality and Cauchy-Schwarz Inequality, we get

$$|q|_4^4\leq 4\int_{S^2}\left(\int_0^1\left(\int_\xi^1|q|^3\left|\frac{\partial q}{\partial\xi'}\right|\right)\right)+|q|_{\xi=1}|_4^4$$

$$\leq c\left(\int_\Omega |q|^2\left|\frac{\partial q}{\partial \xi}\right|^2\right)+\frac{1}{2}\int_\Omega q^4+|q|_{\xi=1}|_4^4.$$

Since $\left|\int_\Omega Q_2|q|^2 q\right| \leq c|Q_2|_4^4+\varepsilon|q|_4^4$, by choosing ε small enough, we have

$$\frac{\mathrm{d}|q|_4^4}{\mathrm{d}t}+c_2|q|_4^4 \leq c|Q_2|_4^4,$$

where c_2 is a positive constant. According to Gronwall Inequality, we have

$$|q(t)|_4^4 \leq \mathrm{e}^{-c_2 t}|q_0|_4^4+c|Q_2|_4^4 \leq E_2, \tag{3.2.37}$$

where $t \geq 0$, and E_2 is a positive constant. Moreover,

$$c_1\int_t^{t+r}|q|_{\xi=1}|_4^4 \leq 2E_2, \text{ to any } t \geq 0. \tag{3.2.38}$$

In making L^3 and L^4 estimates of T by anisotropic estimates, we need the following lemma.

Lemma 3.2.16. If $\boldsymbol{v} \in V_1$, and $T \in V_2$, then

(i) $\left\|\frac{bP}{p}\int_\xi^1 \operatorname{div}\boldsymbol{v}\mathrm{d}\xi'\right\|_{L^2_{(\theta,\varphi)}L^\infty_\xi} \leq |\operatorname{div}\boldsymbol{v}|_2,$

(ii) $\|T^n\|_{L^2_{(\theta,\varphi)}L^1_\xi}=\left\|\left(\int_{S^2}|T^n|^2\right)^{\frac{1}{2}}\right\|_{L^1_\xi} \leq \begin{cases} c|T|_2\|T\|, & \text{when } n=2, \\ c|T|_4^2\|T\|, & \text{when } n=3, \end{cases}$

where $\left\|\frac{bP}{p}\int_\xi^1 \operatorname{div}\boldsymbol{v}\mathrm{d}\xi'\right\|_{L^2_{(\theta,\varphi)}L^\infty_\xi}=\left\|\left[\int_{S^2}(\frac{bP}{p}\int_\xi^1 \operatorname{div}\boldsymbol{v}\mathrm{d}\xi')^2\right]^{\frac{1}{2}}\right\|_{L^\infty_\xi}.$

Proof. With Hölder Inequality, we can prove (i). According to (3.2.24), (3.2.25) and the Hölder Inequality, we can obtain (ii).

L^3 estimates of T. Taking the $L^2(\Omega)$ inner product of equation (3.2.12) with $|T|T$, we get

$$\begin{aligned}
&\frac{1}{3}\frac{\mathrm{d}|T|_3^3}{\mathrm{d}t}+2\int_\Omega|\boldsymbol{\nabla}T|^2|T|+2\int_\Omega\left|\frac{\partial T}{\partial\xi}\right|^2|T|+\alpha_s\int_{S^2}|T|_{\xi=1}|^3\\
=&\int_\Omega Q_1|T|T-\int_\Omega\left[\boldsymbol{\nabla}_{\boldsymbol{v}}T+\left(\int_\xi^1\operatorname{div}\boldsymbol{v}\right)\frac{\partial T}{\partial\xi}\right]|T|T\\
&+\int_\Omega\frac{bP}{p}(1+aq)|T|T\left(\int_\xi^1\operatorname{div}\boldsymbol{v}\right).
\end{aligned}$$

Applying Hölder Inequality and Young Inequality, we have

$$\left|\int_\Omega Q_1|T|T\right| \le c|Q_1|_3^3 + \varepsilon|T|_3^3.$$

By Hölder Inequality and Lemma 3.2.16, we obtain

$$\begin{aligned}\left|\int_\Omega \frac{bP}{p}\left(\int_\xi^1 \mathrm{div}\boldsymbol{v}\mathrm{d}\xi'\right)|T|T\right| \le& \left\|\frac{bP}{p}\int_\xi^1 \mathrm{div}\boldsymbol{v}\mathrm{d}\xi'\right\|_{L^2_{(\theta,\varphi)}L^\infty_\xi}\|T^2\|_{L^2_{(\theta,\varphi)}L^1_\xi}\\ \le& c\int_\Omega(|\nabla_{\boldsymbol{e}_\varphi}\boldsymbol{v}|^2 + |\nabla_{\boldsymbol{e}_\varphi}\boldsymbol{v}|^2) + c|T|_2^2\|T\|^2.\end{aligned}$$

By Hölder Inequality, Young Inequality and

$$\|u\|_{L^{\frac{16}{3}}(S^2)} \le c\|u\|_{L^2(S^2)}^{\frac{3}{8}}\|u\|_{H^1(S^2)}^{\frac{5}{8}}, \quad \text{for any} \quad u \in H^1(S^2),$$

we have

$$\begin{aligned}&\left|\int_\Omega \frac{abP}{p}q\left(\int_\xi^1 \mathrm{div}\boldsymbol{v}\right)|T|T\right|\\ \le& c\int_0^1\left\{\left(\int_{S^2}q^4\right)^{\frac{1}{4}}\left[\int_{S^2}\left(|T|^{\frac{3}{2}}\right)^{\frac{16}{3}}\right]^{\frac{1}{4}}\right\}\left\|\frac{bP}{p}\int_\xi^1 \mathrm{div}\boldsymbol{v}\right\|_{L^2_{(\theta,\varphi)}L^\infty_\xi}\\ \le& c|q|_4|T|_3^{\frac{3}{4}}\left(\int_\Omega |T||\nabla T|^2 + |T|_3^3\right)^{\frac{5}{12}}|\mathrm{div}\boldsymbol{v}|_2\\ \le& c|q|_4^{\frac{12}{7}}\left(1+|T|_3^3\right)\left[1+\int_\Omega(|\nabla_{\boldsymbol{e}_\theta}\boldsymbol{v}|^2 + |\nabla_{\boldsymbol{e}_\varphi}\boldsymbol{v}|^2)\right] + \varepsilon\left(\int_\Omega |T||\nabla T|^2 + |T|_3^3\right).\end{aligned}$$

Thus, we get

$$\begin{aligned}&\frac{\mathrm{d}|T|_3^3}{\mathrm{d}t} + 2\int_\Omega |\nabla T|^2|T| + 2\int_\Omega \left|\frac{\partial T}{\partial\xi}\right|^2|T| + \alpha_s\int_{S^2}|T|_{\xi=1}|^3\\ \le& c|q|_4^{\frac{12}{7}}[1+\int_\Omega(|\nabla_{\boldsymbol{e}_\theta}\boldsymbol{v}|^2 + |\nabla_{\boldsymbol{e}_\varphi}\boldsymbol{v}|^2)]|T|_3^3 + c|Q_1|_3^3\\ &+ c(1+|q|_4^{\frac{12}{7}})\left[1+\int_\Omega(|\nabla_{\boldsymbol{e}_\theta}\boldsymbol{v}|^2 + |\nabla_{\boldsymbol{e}_\varphi}\boldsymbol{v}|^2)\right] + c|T|_2^2\|T\|^2. \qquad (3.2.39)\end{aligned}$$

By using Lemma 3.2.13, (3.2.35), (3.2.37) and $|T|_3^3 \le c|T|_2^{\frac{3}{2}}\|T\|^{\frac{3}{2}}$, we obtain

$$\begin{aligned}|T(t+r)|_3^3 \le& c\Big(|Q_1|_3^3 + \left(1+E_2^{\frac{3}{7}}\right)(1+E_1) + E_1^2\\ &+ \frac{(E_0E_1)^{\frac{3}{4}}}{r}\Big)\exp\left(cE_2^{\frac{3}{7}}(1+E_1)\right) = E_3, \qquad (3.2.40)\end{aligned}$$

where E_3 is a positive constant, and $t \ge 0$. By Gronwall Inequality, we have

$$|T(t)|_3^3 \le c[|Q_1|_3^3 + (1+E_2^{\frac{3}{7}})(1+E_1) + |T_0|_3^3]\exp(cE_2^{\frac{3}{7}}(1+E_1)),$$

$$\text{for any } 0 \le t < r. \tag{3.2.41}$$

L^4 estimates of T. Taking the $L^2(\Omega)$ inner product of equation (3.2.12) with $|T|^2T$, we have

$$\begin{aligned}
&\frac{1}{4}\frac{\mathrm{d}|T|_4^4}{\mathrm{d}t} + 3\int_\Omega |\nabla T|^2 T^2 + 3\int_\Omega \left|\frac{\partial T}{\partial \xi}\right|^2 T^2 + \alpha_s \int_{S^2} |T|_{\xi=1}|^4 \\
=& \int_\Omega Q_1 |T|^2 T - \int_\Omega \left[\nabla_{\boldsymbol{v}} T + \left(\int_\xi^1 \mathrm{div}\boldsymbol{v}\right)\frac{\partial T}{\partial \xi}\right]|T|^2 T \\
&+ \int_\Omega \frac{bP}{p}(1+aq)|T|^2 T\left(\int_\xi^1 \mathrm{div}\boldsymbol{v}\right).
\end{aligned}$$

By Hölder Inequality, Minkowski Inequality and Lemma 3.2.16, we get

$$\left|\int_\Omega \frac{bP}{p}\left(\int_\xi^1 \mathrm{div}\boldsymbol{v}\mathrm{d}\xi'\right)|T|^2T\right| \le c\int_\Omega (|\nabla_{e_\varphi}\boldsymbol{v}|^2 + |\nabla_{e_\varphi}\boldsymbol{v}|^2) + c|T|_4^4\|T\|^2.$$

Applying Hölder Inequality, Young Inequality and $\|u\|_{L^6(S^2)} \le c\|u\|_{L^2(S^2)}^{\frac{1}{3}}\|u\|_{H^1(S^2)}^{\frac{2}{3}}$ for any $u \in H^1(S^2)$, we obtain

$$\begin{aligned}
&\left|\int_\Omega \frac{abP}{p} q\left(\int_\xi^1 \mathrm{div}\boldsymbol{v}\mathrm{d}\xi'\right)|T|^2T\right| \\
\le& c(|q|_4^4 + |T|_4^4)\int_\Omega (|\nabla_{e_\theta}\boldsymbol{v}|^2 + |\nabla_{e_\varphi}\boldsymbol{v}|^2) + \varepsilon\left(\int_\Omega |T|^2|\nabla T|^2 + |T|_4^4\right),
\end{aligned}$$

where ε is small enough. Thus, we have

$$\begin{aligned}
&\frac{\mathrm{d}|T|_4^4}{\mathrm{d}t} + 3\int_\Omega |\nabla T|^2T^2 + 3\int_\Omega |\frac{\partial T}{\partial \xi}|^2 T^2 + \alpha_s\int_{S^2} |T|_{\xi=1}|^4 \\
\le\ & c[\|T\|^2 + \int_\Omega (|\nabla_{e_\theta}\boldsymbol{v}|^2 + |\nabla_{e_\varphi}\boldsymbol{v}|^2)]|T|_4^4 + c|Q_1|_4^4 \\
&+ c(1+|q|_4^4)\int_\Omega (|\nabla_{e_\theta}\boldsymbol{v}|^2 + |\nabla_{e_\varphi}\boldsymbol{v}|^2).
\end{aligned} \tag{3.2.42}$$

By using Lemma 3.2.13, (3.2.25), (3.2.37), (3.2.40) and $|T|_4^4 \le c|T|_3^2\|T\|^2$, we obtain

$$|T(t+2r)|_4^4 \le c\left(|Q_1|_4^4 + E_1E_2 + E_1 + \frac{E_1E_3^{\frac{2}{3}}}{r}\right)\exp(cE_1) = E_4, \tag{3.2.43}$$

where E_4 is a positive constant, and $t \ge 0$. By Gronwall Inequality, we derive from (3.2.42)

$$|T(t)|_4^4 \le c(|Q_1|_4^4 + E_1E_2 + E_1 + |T_0|_4^4)\exp(cE_1) = C_1, \tag{3.2.44}$$

where $C_1 = C_1(\|U_0\|, \|Q_1\|_1, \|Q_2\|_1) > 0$, and $0 \le t < 2r$. Combining (3.2.42) with (3.2.43), we obtain

$$c_1 \int_{t+2r}^{t+3r} |T|_{\xi=1}|_4^4 \le E_4^2 + E_4, \tag{3.2.45}$$

for any $t \ge 0$.

L^3 estimates of $\tilde{v}$. Taking L^2 inner product of equation (3.2.22) with $|\tilde{v}|\tilde{v}$, and obtain

$$\frac{1}{3}\frac{\mathrm{d}|\tilde{v}|_3^3}{\mathrm{d}t} + \int_\Omega \left[(|\nabla_{e_\theta}\tilde{v}|^2 + |\nabla_{e_\varphi}\tilde{v}|^2)|\tilde{v}| + \frac{4}{9}|\nabla_{e_\theta}|\tilde{v}|^{\frac{3}{2}}|^2 + \frac{4}{9}|\nabla_{e_\varphi}|\tilde{v}|^{\frac{3}{2}}|^2 + |\tilde{v}|^3\right]$$
$$+ \int_\Omega \left(|\tilde{v}_\xi|^2|\tilde{v}| + \frac{4}{9}|\partial_\xi|\tilde{v}|^{\frac{3}{2}}|^2\right)$$
$$= \int_\Omega \overline{(\tilde{v}\mathrm{div}\tilde{v} + \nabla_{\tilde{v}}\tilde{v})} \cdot |\tilde{v}|\tilde{v} - \int_\Omega (\frac{f}{R_0}k \times \tilde{v}) \cdot |\tilde{v}|\tilde{v}$$
$$- \int_\Omega \left(\nabla_{\tilde{v}}\tilde{v} + \left(\int_\xi^1 \mathrm{div}\tilde{v}\mathrm{d}\xi'\right)\frac{\partial\tilde{v}}{\partial\xi}\right) \cdot |\tilde{v}|\tilde{v} - \int_\Omega (\nabla_{\bar{v}}\tilde{v}) \cdot |\tilde{v}|\tilde{v} - \int_\Omega |\tilde{v}|\tilde{v} \cdot \nabla_{\tilde{v}}\bar{v}$$
$$- \int_\Omega \left[\int_\xi^1 \frac{bP}{p}\mathrm{grad}((1+aq)T) - \int_0^1\int_\xi^1 \frac{bP}{p}\mathrm{grad}((1+aq)T)\right] \cdot |\tilde{v}|\tilde{v},$$

where $\tilde{v}_\xi = \partial_\xi\tilde{v}$. Applying Lemma 3.2.10 and integration by parts, we get

$$\int_\Omega \left(\nabla_{\tilde{v}}\tilde{v} + \left(\int_\xi^1 \mathrm{div}\tilde{v}\mathrm{d}\xi'\right)\frac{\partial\tilde{v}}{\partial\xi}\right) \cdot |\tilde{v}|\tilde{v}$$
$$= \frac{1}{3}\left[\int_\Omega \nabla_{\tilde{v}}|\tilde{v}|^3 + \int_{S^2}\left(\int_0^1 \left(\int_\xi^1 \mathrm{div}\tilde{v}\mathrm{d}\xi'\right)\mathrm{d}|\tilde{v}|^3\right)\right]$$
$$= \frac{1}{3}\int_\Omega (\nabla_{\tilde{v}}|\tilde{v}|^3 + |\tilde{v}|^3\mathrm{div}\tilde{v})$$
$$= \frac{1}{3}\int_\Omega \mathrm{div}\left(|\tilde{v}|^3\tilde{v}\right) = 0.$$

According to Lemma 3.2.10, we have

$$\int_\Omega \mathrm{div}\left(|\tilde{v}|^3\bar{v}\right) = \int_\Omega \nabla_{\bar{v}}|\tilde{v}|^3 + |\tilde{v}|^3\mathrm{div}\bar{v} = 0.$$

Since $\mathrm{div}\bar{v} = 0$,

$$\int_\Omega (\nabla_{\bar{v}}\tilde{v}) \cdot |\tilde{v}|\tilde{v} = \frac{1}{3}\int_\Omega \nabla_{\bar{v}}|\tilde{v}|^3 = 0.$$

By Lemma 3.2.10, we have

$$\int_\Omega \mathrm{div}\left(\left(|\tilde{v}|\tilde{v}\cdot\bar{v}\right)\tilde{v}\right) = \int_\Omega \nabla_{\tilde{v}}\left(|\tilde{v}|\tilde{v}\cdot\bar{v}\right) + \int_\Omega |\tilde{v}|\tilde{v}\cdot\bar{v}\mathrm{div}\tilde{v}$$

$$= \int_\Omega \left(|\tilde{\boldsymbol{v}}|\tilde{\boldsymbol{v}} \cdot \boldsymbol{\nabla}_{\tilde{\boldsymbol{v}}}\bar{\boldsymbol{v}} + \bar{\boldsymbol{v}} \cdot \boldsymbol{\nabla}_{\tilde{\boldsymbol{v}}}\left(|\tilde{\boldsymbol{v}}|\tilde{\boldsymbol{v}}\right)\right) + \int_\Omega |\tilde{\boldsymbol{v}}|\tilde{\boldsymbol{v}} \cdot \bar{\boldsymbol{v}}\mathrm{div}\tilde{\boldsymbol{v}} = 0,$$

and

$$-\int_\Omega |\tilde{\boldsymbol{v}}|\tilde{\boldsymbol{v}} \cdot \boldsymbol{\nabla}_{\tilde{\boldsymbol{v}}}\bar{\boldsymbol{v}} = \int_\Omega \bar{\boldsymbol{v}} \cdot \boldsymbol{\nabla}_{\tilde{\boldsymbol{v}}}\left(|\tilde{\boldsymbol{v}}|\tilde{\boldsymbol{v}}\right) + \int_\Omega |\tilde{\boldsymbol{v}}|\tilde{\boldsymbol{v}} \cdot \bar{\boldsymbol{v}}\mathrm{div}\tilde{\boldsymbol{v}}.$$

By integration by parts, we obtain

$$\begin{aligned}
&\int_\Omega \left(\int_0^1 \left(\tilde{\boldsymbol{v}}\mathrm{div}\tilde{\boldsymbol{v}} + \boldsymbol{\nabla}_{\tilde{\boldsymbol{v}}}\tilde{\boldsymbol{v}}\right)\mathrm{d}\xi\right) \cdot |\tilde{\boldsymbol{v}}|\tilde{\boldsymbol{v}} \\
&= \int_\Omega \left(\int_0^1 \tilde{\boldsymbol{v}}_{e_\theta}\tilde{\boldsymbol{v}}\mathrm{d}\xi\right) \cdot \boldsymbol{\nabla}_{\boldsymbol{e}_\theta}\left(|\tilde{\boldsymbol{v}}|\tilde{\boldsymbol{v}}\right) + \int_\Omega \left(\int_0^1 \tilde{\boldsymbol{v}}_{e_\varphi}\tilde{\boldsymbol{v}}\mathrm{d}\xi\right) \cdot \boldsymbol{\nabla}_{\boldsymbol{e}_\varphi}\left(|\tilde{\boldsymbol{v}}|\tilde{\boldsymbol{v}}\right).
\end{aligned}$$

Moreover,

$$\int_\Omega \left(\frac{f}{R_0}\boldsymbol{k} \times \tilde{\boldsymbol{v}}\right) \cdot |\tilde{\boldsymbol{v}}|\tilde{\boldsymbol{v}} = 0.$$

According to Lemma 3.2.8, we get

$$\begin{aligned}
&-\int_\Omega \left[\int_\xi^1 \frac{bP}{p}\mathrm{grad}\left((1+aq)T\right)\mathrm{d}\xi' - \int_0^1\int_\xi^1 \frac{bP}{p}\mathrm{grad}\left((1+aq)T\right)\mathrm{d}\xi'\mathrm{d}\xi\right] \cdot |\tilde{\boldsymbol{v}}|\tilde{\boldsymbol{v}} \\
&= \int_\Omega \left[\int_\xi^1 \frac{bP}{p}(1+aq)T\mathrm{d}\xi' - \int_0^1\int_\xi^1 \frac{bP}{p}(1+aq)T\mathrm{d}\xi'\mathrm{d}\xi\right] \cdot \mathrm{div}\left(|\tilde{\boldsymbol{v}}|\tilde{\boldsymbol{v}}\right).
\end{aligned}$$

From the above relationships, we have

$$\begin{aligned}
&\frac{1}{3}\frac{\mathrm{d}|\tilde{\boldsymbol{v}}|_3^3}{\mathrm{d}t} + \int_\Omega \left[\left(|\boldsymbol{\nabla}_{\boldsymbol{e}_\theta}\tilde{\boldsymbol{v}}|^2 + |\boldsymbol{\nabla}_{\boldsymbol{e}_\varphi}\tilde{\boldsymbol{v}}|^2\right)|\tilde{\boldsymbol{v}}| + \frac{4}{9}|\boldsymbol{\nabla}_{\boldsymbol{e}_\theta}|\tilde{\boldsymbol{v}}|^{\frac{3}{2}}|^2 + \frac{4}{9}|\boldsymbol{\nabla}_{\boldsymbol{e}_\varphi}|\tilde{\boldsymbol{v}}|^{\frac{3}{2}}|^2 + |\tilde{\boldsymbol{v}}|^3\right] \\
&+ \int_\Omega \left(|\tilde{\boldsymbol{v}}_\xi|^2|\tilde{\boldsymbol{v}}| + \frac{4}{9}|\partial_\xi|\tilde{\boldsymbol{v}}|^{\frac{3}{2}}|^2\right) \\
&= \int_\Omega \left(\bar{\boldsymbol{v}} \cdot \boldsymbol{\nabla}_{\tilde{\boldsymbol{v}}}(|\tilde{\boldsymbol{v}}|\tilde{\boldsymbol{v}}) + |\tilde{\boldsymbol{v}}|\tilde{\boldsymbol{v}} \cdot \bar{\boldsymbol{v}}\mathrm{div}\tilde{\boldsymbol{v}}\right) \\
&+ \int_\Omega \left[\left(\int_0^1 \tilde{\boldsymbol{v}}_\theta\tilde{\boldsymbol{v}}\mathrm{d}\xi\right) \cdot \boldsymbol{\nabla}_{\boldsymbol{e}_\theta}\left(|\tilde{\boldsymbol{v}}|\tilde{\boldsymbol{v}}\right) + \left(\int_0^1 \tilde{\boldsymbol{v}}_\varphi\tilde{\boldsymbol{v}}\mathrm{d}\xi\right) \cdot \boldsymbol{\nabla}_{\boldsymbol{e}_\varphi}(|\tilde{\boldsymbol{v}}|\tilde{\boldsymbol{v}})\right] \\
&+ \int_\Omega \left[\int_\xi^1 \frac{bP}{p}(1+aq)T - \int_0^1\int_\xi^1 \frac{bP}{p}(1+aq)T\right]\mathrm{div}\left(|\tilde{\boldsymbol{v}}|\tilde{\boldsymbol{v}}\right). \qquad (3.2.46)
\end{aligned}$$

By Hölder Inequality and (3.2.46), we get

$$\begin{aligned}
&\frac{1}{3}\frac{\mathrm{d}|\tilde{\boldsymbol{v}}|_3^3}{\mathrm{d}t} + \int_\Omega \left[\left(|\boldsymbol{\nabla}_{\boldsymbol{e}_\theta}\tilde{\boldsymbol{v}}|^2 + |\boldsymbol{\nabla}_{\boldsymbol{e}_\varphi}\tilde{\boldsymbol{v}}|^2\right)|\tilde{\boldsymbol{v}}| + \frac{4}{9}\left(|\boldsymbol{\nabla}_{\boldsymbol{e}_\theta}|\tilde{\boldsymbol{v}}|^{\frac{3}{2}}|^2 + |\boldsymbol{\nabla}_{\boldsymbol{e}_\varphi}|\tilde{\boldsymbol{v}}|^{\frac{3}{2}}|^2\right)\right. \\
&\left.+|\tilde{\boldsymbol{v}}|^3\right] + \int_\Omega \left(|\tilde{\boldsymbol{v}}_\xi|^2|\tilde{\boldsymbol{v}}| + \frac{4}{9}|\partial_\xi|\tilde{\boldsymbol{v}}|^{\frac{3}{2}}|^2\right)
\end{aligned}$$

$$
\begin{aligned}
&\leq c\int_{S^2}(\overline{|T|}+\overline{|qT|})\left[\int_0^1|\tilde{\boldsymbol{v}}|(|\boldsymbol{\nabla}_{\boldsymbol{e}_\theta}\tilde{\boldsymbol{v}}|^2+|\boldsymbol{\nabla}_{\boldsymbol{e}_\varphi}\tilde{\boldsymbol{v}}|^2)^{\frac{1}{2}}\right]\\
&\quad+c\int_{S^2}|\bar{\boldsymbol{v}}|\int_0^1|\tilde{\boldsymbol{v}}|^2\left(|\boldsymbol{\nabla}_{\boldsymbol{e}_\theta}\tilde{\boldsymbol{v}}|^2+|\boldsymbol{\nabla}_{\boldsymbol{e}_\varphi}\tilde{\boldsymbol{v}}|^2\right)^{\frac{1}{2}}\\
&\quad+c\int_{S^2}\left(\int_0^1|\tilde{\boldsymbol{v}}|^2\right)\left[\int_0^1|\tilde{\boldsymbol{v}}|\left(|\boldsymbol{\nabla}_{\boldsymbol{e}_\theta}\tilde{\boldsymbol{v}}|^2+|\boldsymbol{\nabla}_{\boldsymbol{e}_\varphi}\tilde{\boldsymbol{v}}|^2\right)^{\frac{1}{2}}\right]\\
&\leq c\|\bar{\boldsymbol{v}}\|_{L^4(S^2)}\left(\int_\Omega|\tilde{\boldsymbol{v}}|(|\boldsymbol{\nabla}_{\boldsymbol{e}_\theta}\tilde{\boldsymbol{v}}|^2+|\boldsymbol{\nabla}_{\boldsymbol{e}_\varphi}\tilde{\boldsymbol{v}}|^2)\right)^{\frac{1}{2}}\left(\int_{S^2}\left(\int_0^1|\tilde{\boldsymbol{v}}|^3\mathrm{d}\xi\right)^2\right)^{\frac{1}{4}}\\
&\quad+c\left(\int_\Omega|\tilde{\boldsymbol{v}}|(|\boldsymbol{\nabla}_{\boldsymbol{e}_\theta}\tilde{\boldsymbol{v}}|^2+|\boldsymbol{\nabla}_{\boldsymbol{e}_\varphi}\tilde{\boldsymbol{v}}|^2)\right)^{\frac{1}{2}}\cdot\left(\int_{S^2}\left(\int_0^1|\tilde{\boldsymbol{v}}|^2\mathrm{d}\xi\right)^{\frac{5}{2}}\right)^{\frac{1}{2}}\\
&\quad+c(\|\overline{|T|}\|_{L^4(S^2)}+\|\overline{|qT|}\|_{L^4(S^2)})|\tilde{\boldsymbol{v}}|_2^{\frac{1}{2}}\left[\int_\Omega|\tilde{\boldsymbol{v}}|(|\boldsymbol{\nabla}_{\boldsymbol{e}_\theta}\tilde{\boldsymbol{v}}|^2+|\boldsymbol{\nabla}_{\boldsymbol{e}_\varphi}\tilde{\boldsymbol{v}}|^2)\right]^{\frac{1}{2}}.
\end{aligned}
\tag{3.2.47}
$$

By using Minkowski Inequality, Hölder Inequality and (3.2.24), we have

$$
\begin{aligned}
\left[\int_{S^2}\left(\int_0^1|\tilde{\boldsymbol{v}}|^3\mathrm{d}\xi\right)\right]^{\frac{1}{2}}&\leq\int_0^1\left[\int_{S^2}\left(|\tilde{\boldsymbol{v}}|^{\frac{3}{2}}\right)^4\right]^{\frac{1}{2}}\mathrm{d}\xi\\
&\leq c|\tilde{\boldsymbol{v}}|_3^{\frac{3}{2}}\left(\int_0^1\left(\|\boldsymbol{\nabla}|\tilde{\boldsymbol{v}}|^{\frac{3}{2}}\|_{L^2}^2+\||\tilde{\boldsymbol{v}}|^{\frac{3}{2}}\|_{L^2}^2\right)\mathrm{d}\xi\right)^{\frac{1}{2}}.
\end{aligned}
$$

By using Minkowski Inequality, Hölder Inequality and the inequality

$$
\|u\|_{L^5(S^2)}\leq c\|u\|_{L^3(S^2)}^{\frac{3}{5}}\|u\|_{H^1(S^2)}^{\frac{2}{5}},\quad \text{for any } u\in H^1(S^2),
$$

we have

$$
\begin{aligned}
\int_{S^2}\left(\int_0^1|\tilde{\boldsymbol{v}}|^2\mathrm{d}\xi\right)^{\frac{5}{2}}&\leq\left(\int_0^1\left(\int_{S^2}|\tilde{\boldsymbol{v}}|^5\right)^{\frac{2}{5}}\mathrm{d}\xi\right)^{\frac{5}{2}}\\
&\leq\left(\int_0^1\|\tilde{\boldsymbol{v}}\|_{L^3}^{\frac{6}{5}}\|\tilde{\boldsymbol{v}}\|_{H^1}^{\frac{4}{5}}\mathrm{d}\xi\right)^{\frac{5}{2}}\leq c\|\tilde{\boldsymbol{v}}\|^2|\tilde{\boldsymbol{v}}|_3^3.
\end{aligned}
$$

According to (3.2.24),

$$
\|\bar{\boldsymbol{v}}\|_{L^4(S^2)}\leq\|\bar{\boldsymbol{v}}\|_{L^2(S^2)}^{\frac{1}{2}}\|\bar{\boldsymbol{v}}\|_{H^1(S^2)}^{\frac{1}{2}}.
$$

By Minkowski Inequality, (3.2.24), (3.2.26) and Hölder Inequality, we get

$$
\|\overline{|T|}\|_{L^4(S^2)}=\left(\int_{S^2}\left(\int_0^1|T|\mathrm{d}\xi\right)^4\right)^{\frac{1}{4}}\leq|T|_4,
$$

and

$$\|\overline{|qT|}\|_{L^4(S^2)} \leq \left(\int_{S^2}\left(\int_0^1 |q|^2 \mathrm{d}\xi\right)^2 \left(\int_0^1 |T|^2 \mathrm{d}\xi\right)^2\right)^{\frac{1}{4}} \leq c|q|_4^{\frac{1}{2}}\|q\|^{\frac{1}{2}}|T|_4^{\frac{1}{2}}\|T\|^{\frac{1}{2}}.$$

Thus, we obtain

$$\frac{\mathrm{d}|\tilde{\boldsymbol{v}}|_3^3}{\mathrm{d}t} + \int_\Omega \left[(|\boldsymbol{\nabla}_{\boldsymbol{e}_\theta}\tilde{\boldsymbol{v}}|^2 + |\boldsymbol{\nabla}_{\boldsymbol{e}_\varphi}\tilde{\boldsymbol{v}}|^2)|\tilde{\boldsymbol{v}}| + \frac{4}{9}|\boldsymbol{\nabla}_{\boldsymbol{e}_\theta}|\tilde{\boldsymbol{v}}|^{\frac{3}{2}}|^2 + \frac{4}{9}|\boldsymbol{\nabla}_{\boldsymbol{e}_\varphi}|\tilde{\boldsymbol{v}}|^{\frac{3}{2}}|^2 + |\tilde{\boldsymbol{v}}|^3\right]$$

$$+ \int_\Omega \left(|\tilde{\boldsymbol{v}}_\xi|^2|\tilde{\boldsymbol{v}}| + \frac{4}{9}|\partial_\xi|\tilde{\boldsymbol{v}}|^{\frac{3}{2}}|^2\right)$$

$$\leq c(\|\bar{\boldsymbol{v}}\|_{L^2}^2\|\bar{\boldsymbol{v}}\|_{H^1}^2 + \|\tilde{\boldsymbol{v}}\|^2)|\tilde{\boldsymbol{v}}|_3^3 + c|T|_4^4 + c|T|_4^2\|T\|^2 + c(1 + |q|_4^2\|q\|^2)|\tilde{\boldsymbol{v}}|_2^2.$$

By using Lemma 3.2.13, (3.2.34)-(3.2.37), (3.2.43) and $|\tilde{\boldsymbol{v}}|_3^3 \leq |\tilde{\boldsymbol{v}}|_2^{\frac{3}{2}}\|\tilde{\boldsymbol{v}}\|_2^{\frac{3}{2}}$, we get

$$\begin{aligned} &|\tilde{\boldsymbol{v}}(t+3r)|_3^3 \\ \leq& c(E_4 + E_4^{\frac{1}{2}}E_1 + 2E_0(1 + E_2^{\frac{1}{2}}E_1) + \frac{(4E_0E_1)^{\frac{3}{4}}}{r}) \exp c(E_0E_1 + 2E_1) \leq E_5, \end{aligned} \tag{3.2.48}$$

where E_5 is a positive constant, and $t \geq 0$.

L^4 estimates of $\tilde{\boldsymbol{v}}$. Taking L^2 inner product of equation (3.2.22) with $|\tilde{\boldsymbol{v}}|^2\tilde{\boldsymbol{v}}$, like (3.2.46), we have

$$\begin{aligned} &\frac{1}{4}\frac{\mathrm{d}|\tilde{\boldsymbol{v}}|_4^4}{\mathrm{d}t} + \int_\Omega \left[\left(|\boldsymbol{\nabla}_{\boldsymbol{e}_\theta}\tilde{\boldsymbol{v}}|^2 + |\boldsymbol{\nabla}_{\boldsymbol{e}_\varphi}\tilde{\boldsymbol{v}}|^2\right)|\tilde{\boldsymbol{v}}|^2 + \frac{1}{2}|\boldsymbol{\nabla}_{\boldsymbol{e}_\theta}|\tilde{\boldsymbol{v}}|^2|^2 + \frac{1}{2}|\boldsymbol{\nabla}_{\boldsymbol{e}_\varphi}|\tilde{\boldsymbol{v}}|^2|^2 + |\tilde{\boldsymbol{v}}|^4\right] \\ &+ \int_\Omega \left(|\tilde{\boldsymbol{v}}_\xi|^2|\tilde{\boldsymbol{v}}|^2 + \frac{1}{2}|\partial_\xi|\tilde{\boldsymbol{v}}|^2|^2\right) \\ =& \int_\Omega \left[\bar{\boldsymbol{v}}\cdot\boldsymbol{\nabla}_{\tilde{\boldsymbol{v}}}\left(|\tilde{\boldsymbol{v}}|^2\tilde{\boldsymbol{v}}\right) + |\tilde{\boldsymbol{v}}|^2\tilde{\boldsymbol{v}}\cdot\bar{\boldsymbol{v}}\mathrm{div}\tilde{\boldsymbol{v}}\right] \\ &+ \int_\Omega \left[\left(\int_0^1 \tilde{\boldsymbol{v}}_\theta\tilde{\boldsymbol{v}}\mathrm{d}\xi\right)\cdot\boldsymbol{\nabla}_{\boldsymbol{e}_\theta}\left(|\tilde{\boldsymbol{v}}|^2\tilde{\boldsymbol{v}}\right) + \left(\int_0^1 \tilde{\boldsymbol{v}}_\varphi\tilde{\boldsymbol{v}}\mathrm{d}\xi\right)\cdot\boldsymbol{\nabla}_{\boldsymbol{e}_\varphi}\left(|\tilde{\boldsymbol{v}}|^2\tilde{\boldsymbol{v}}\right)\right] \\ &+ \int_\Omega \left[\int_\xi^1 \frac{bP}{p}(1+aq)T\mathrm{d}\xi' - \int_0^1\int_\xi^1 \frac{bP}{p}(1+aq)T\mathrm{d}\xi'\mathrm{d}\xi\right]\mathrm{div}\left(|\tilde{\boldsymbol{v}}|^2\tilde{\boldsymbol{v}}\right). \end{aligned}$$

By Hölder Inequality, we get

$$\frac{1}{4}\frac{\mathrm{d}|\tilde{\boldsymbol{v}}|_4^4}{\mathrm{d}t} + \int_\Omega \left[\left(|\boldsymbol{\nabla}_{\boldsymbol{e}_\theta}\tilde{\boldsymbol{v}}|^2 + |\boldsymbol{\nabla}_{\boldsymbol{e}_\varphi}\tilde{\boldsymbol{v}}|^2\right)|\tilde{\boldsymbol{v}}|^2 + \frac{1}{2}|\boldsymbol{\nabla}_{\boldsymbol{e}_\theta}|\tilde{\boldsymbol{v}}|^2|^2 + \frac{1}{2}|\boldsymbol{\nabla}_{\boldsymbol{e}_\varphi}|\tilde{\boldsymbol{v}}|^2|^2 + |\tilde{\boldsymbol{v}}|^4\right]$$

$$+\int_{\Omega}\left(|\tilde{\boldsymbol{v}}_{\xi}|^2|\tilde{\boldsymbol{v}}|^2+\frac{1}{2}|\partial_{\xi}|\tilde{\boldsymbol{v}}|^2|^2\right)$$

$$\leq c\|\bar{\boldsymbol{v}}\|_{L^4}\left[\int_{\Omega}|\tilde{\boldsymbol{v}}|^2\left(|\nabla_{e_\theta}\tilde{\boldsymbol{v}}|^2+|\nabla_{e_\varphi}\tilde{\boldsymbol{v}}|^2\right)\right]^{\frac{1}{2}}\left[\int_{S^2}\left(\int_0^1|\tilde{\boldsymbol{v}}|^4\mathrm{d}\xi\right)^2\right]^{\frac{1}{4}}$$

$$+c\left[\int_{\Omega}|\tilde{\boldsymbol{v}}|^2\left(|\nabla_{e_\theta}\tilde{\boldsymbol{v}}|^2+|\nabla_{e_\varphi}\tilde{\boldsymbol{v}}|^2\right)\right]^{\frac{1}{2}}\left[\int_{S^2}\left(\int_0^1|\tilde{\boldsymbol{v}}|^2\mathrm{d}\xi\right)^3\right]^{\frac{1}{2}}$$

$$+c(\||\overline{T}|\|_{L^4(S^2)}+\||\overline{qT}|\|_{L^4(S^2)})\left[\int_{\Omega}|\tilde{\boldsymbol{v}}|^2(|\nabla_{e_\theta}\tilde{\boldsymbol{v}}|^2+|\nabla_{e_\varphi}\tilde{\boldsymbol{v}}|^2)\right]^{\frac{1}{2}}$$

$$\cdot\left[\int_{S^2}\left(\int_0^1|\tilde{\boldsymbol{v}}|^2\mathrm{d}\xi\right)^2\right]^{\frac{1}{4}}.$$

Applying Minkowski Inequality, Hölder Inequality and (3.2.24), we obtain

$$\left(\int_{S^2}\left(\int_0^1|\tilde{\boldsymbol{v}}|^4\right)^2\right)^{\frac{1}{2}}\leq\int_0^1\left(\int_{S^2}(|\tilde{\boldsymbol{v}}|^2)^4\right)^{\frac{1}{2}}$$

$$\leq c|\tilde{\boldsymbol{v}}|_4^2\left(\int_0^1(\|\nabla|\tilde{\boldsymbol{v}}|^2\|_{L^2}^2+\||\tilde{\boldsymbol{v}}|^2\|_{L^2}^2)\right)^{\frac{1}{2}}.$$

Similar to the above inequality,

$$\int_{S^2}\left(\int_0^1|\tilde{\boldsymbol{v}}|^2\mathrm{d}\xi\right)^3\leq\left(\int_0^1\left(\int_{S^2}|\tilde{\boldsymbol{v}}|^6\right)^{\frac{1}{3}}\mathrm{d}\xi\right)^3\leq c\|\tilde{\boldsymbol{v}}\|^2|\tilde{\boldsymbol{v}}|_4^4.$$

By Young Inequality, we obtain

$$\frac{\mathrm{d}|\tilde{\boldsymbol{v}}|_4^4}{\mathrm{d}t}+\int_{\Omega}\left[\left(|\nabla_{e_\theta}\tilde{\boldsymbol{v}}|^2+|\nabla_{e_\varphi}\tilde{\boldsymbol{v}}|^2\right)|\tilde{\boldsymbol{v}}|^2+\frac{1}{2}|\nabla_{e_\theta}|\tilde{\boldsymbol{v}}|^2|^2+\frac{1}{2}|\nabla_{e_\varphi}|\tilde{\boldsymbol{v}}|^2|^2+|\tilde{\boldsymbol{v}}|^4\right]$$

$$+\int_{\Omega}\left(|\tilde{\boldsymbol{v}}_{\xi}|^2|\tilde{\boldsymbol{v}}|^2+\frac{1}{2}|\partial_{\xi}|\tilde{\boldsymbol{v}}|^2|^2\right)$$

$$\leq c\left(\|\bar{\boldsymbol{v}}\|_{L^2}^2\|\bar{\boldsymbol{v}}\|_{H^1}^2+|T|_4^2+\|\tilde{\boldsymbol{v}}\|^2+|q|_4^2\|q\|^2\right)|\tilde{\boldsymbol{v}}|_4^4+c|T|_4^2+c|T|_4^2\|T\|^2. \tag{3.2.49}$$

Applying Lemma 3.2.13, (3.2.34)-(3.2.36), (3.2.37), (3.2.43), (3.2.48) and $|\tilde{\boldsymbol{v}}|_4^4\leq|\tilde{\boldsymbol{v}}|_3^2\|\tilde{\boldsymbol{v}}\|^2$, we derive from (3.2.49)

$$|\tilde{\boldsymbol{v}}(t+4r)|_4^4$$

$$\leq c\left(E_4^{\frac{1}{2}}+E_1E_4^{\frac{1}{2}}+\frac{E_5^{\frac{2}{3}}E_1}{r}\right)\exp c(E_0E_1+E_1+E_4^{\frac{1}{2}}+E_2^{\frac{1}{2}}E_1)\leq E_6, \tag{3.2.50}$$

where E_6 is a positive constant, and $t \geq 0$. By using (3.2.49) and (3.2.50), we have

$$\int_{t+4r}^{t+5r} \left[\int_\Omega \left(\left(|\nabla_{e_\theta} \tilde{v}|^2 + |\nabla_{e_\varphi} \tilde{v}|^2 \right) |\tilde{v}|^2 + \frac{1}{2} |\nabla_{e_\theta} |\tilde{v}|^2|^2 + \frac{1}{2} |\nabla_{e_\varphi} |\tilde{v}|^2|^2 + |\tilde{v}|^4 \right) \right.$$
$$\left. + \int_\Omega \left(|\tilde{v}_\xi|^2 |\tilde{v}|^2 + \frac{1}{2} |\partial_\xi |\tilde{v}|^2|^2 \right) \right] \leq E_6^2 + E_6 = E_7. \qquad (3.2.51)$$

By Gronwall Inequality and (3.2.49), we obtain

$$|\tilde{v}(t)|_4^4 \leq C_2, \qquad (3.2.52)$$

where $C_2 = C_2(\|U_0\|, \|Q_1\|_1, \|Q_2\|_1) > 0$, and $0 \leq t < 4r$.

H^1 **estimates of** $\bar{v}$**.** Taking L^2 inner product of equation (3.2.21) with $-\Delta \bar{v}$, we get

$$\frac{1}{2} \frac{\mathrm{d}\|\bar{v}\|_{H^1}^2}{\mathrm{d}t} + \|\Delta \bar{v}\|_{L^2}^2 = \int_{S^2} \left[\nabla_{\bar{v}} \bar{v} + \int_0^1 (\tilde{v} \operatorname{div} \tilde{v} + \nabla_{\tilde{v}} \tilde{v}) \mathrm{d}\xi \right] \cdot \Delta \bar{v}$$
$$+ \int_{S^2} (\operatorname{grad} \Phi_s + \frac{f}{R_0} k \times \bar{v}) \cdot \Delta \bar{v} + \int_{S^2} \left[\int_0^1 \int_\xi^1 \frac{bP}{p} \operatorname{grad}((1 + aq)T) \right] \cdot \Delta \bar{v}.$$

Applying Hölder Inequality, (3.2.24) and Young Inequality, we have

$$\left| \int_{S^2} (\nabla_{\bar{v}} \bar{v} \cdot \Delta \bar{v}) \right| \leq c \|\bar{v}\|_{L^4} \left[\int_{S^2} (|\nabla_{e_\theta} \bar{v}|^2 + |\nabla_{e_\varphi} \bar{v}|^2)^2 \right]^{\frac{1}{4}} \|\Delta \bar{v}\|_{L^2}$$
$$\leq c(\|\bar{v}\|_{L^2}^2 + \|\bar{v}\|_{H^1}^2 + \|\bar{v}\|_{L^2}^2 \|\bar{v}\|_{H^1}^2) \|\bar{v}\|_{H^1}^2 + \varepsilon \|\Delta \bar{v}\|_{L^2}^2.$$

By Hölder Inequality and Minkowski Inequality, we obtain

$$\left| \int_{S^2} \left(\int_0^1 (\tilde{v} \operatorname{div} \tilde{v} + \nabla_{\tilde{v}} \tilde{v}) \mathrm{d}\xi \cdot \Delta \bar{v} \right) \right| \leq c \int_\Omega |\tilde{v}|^2 (|\nabla_{e_\theta} \tilde{v}|^2 + |\nabla_{e_\varphi} \tilde{v}|^2) + \varepsilon \|\Delta \bar{v}\|_{L^2}^2.$$

Using Lemma 3.2.8 and $\operatorname{div} \bar{v} = 0$, we get

$$\int_{S^2} \operatorname{grad} \Phi_s \cdot \Delta \bar{v} = 0, \qquad \int_{S^2} \left[\int_0^1 \int_\xi^1 \frac{bP}{p} \operatorname{grad}((1 + aq)T) \right] \cdot \Delta \bar{v} = 0.$$

Choosing ε small enough, we derive from the above equalities

$$\frac{\mathrm{d}\|\bar{v}\|_{H^1}^2}{\mathrm{d}t} + \|\Delta \bar{v}\|_{L^2}^2$$
$$\leq c(\|\bar{v}\|_{L^2}^2 + \|\bar{v}\|_{H^1}^2 + \|\bar{v}\|_{L^2}^2 \|\bar{v}\|_{H^1}^2) \|\bar{v}\|_{H^1}^2 + c \int_\Omega |\tilde{v}|^2 \left(|\nabla_{e_\theta} \tilde{v}|^2 + |\nabla_{e_\varphi} \tilde{v}|^2 \right). \qquad (3.2.53)$$

Applying Lemma 3.2.13, (3.2.34)-(3.2.36) and (3.2.51), we get

$$\|\bar{v}(t + 5r)\|_{H^1}^2 \leq c \left(\frac{E_1}{r} + E_7 \right) \exp c(E_0 E_1 + E_1) \leq E_8, \qquad (3.2.54)$$

where E_8 is a positive constant. By Gronwall Inequality, we derive from (3.2.53)

$$\|\bar{\boldsymbol{v}}(t)\|_{H^1}^2 \le C_3, \tag{3.2.55}$$

where $C_3 = C_3(\|U_0\|,\ \|Q_1\|_1,\ \|Q_2\|_1) > 0$, and $0 \le t < 5r$.

L^2 estimates of $\boldsymbol{v}_\xi$. By taking the derivative of equation (3.2.11) with respect to ξ, we get

$$\frac{\partial \boldsymbol{v}_\xi}{\partial t} - \Delta \boldsymbol{v}_\xi - \frac{\partial^2 \boldsymbol{v}_\xi}{\partial \xi^2} + \boldsymbol{\nabla}_{\boldsymbol{v}} \boldsymbol{v}_\xi + \left(\int_\xi^1 \mathrm{div}\boldsymbol{v}\mathrm{d}\xi'\right) \frac{\partial \boldsymbol{v}_\xi}{\partial \xi} + \boldsymbol{\nabla}_{\boldsymbol{v}_\xi} \boldsymbol{v}$$

$$-(\mathrm{div}\boldsymbol{v})\frac{\partial \boldsymbol{v}}{\partial \xi} + \frac{f}{R_0}\boldsymbol{k} \times \boldsymbol{v}_\xi - \frac{bP}{p}\mathrm{grad}[(1+aq)T] = 0.$$

Taking L^2 inner product of the above equation with $\boldsymbol{v}_\xi$, we have

$$\frac{1}{2}\frac{\mathrm{d}|\boldsymbol{v}_\xi|_2^2}{\mathrm{d}t} + \int_\Omega \left(|\boldsymbol{\nabla}_{e_\theta} \boldsymbol{v}_\xi|^2 + |\boldsymbol{\nabla}_{e_\varphi} \boldsymbol{v}_\xi|^2 + |\boldsymbol{v}_\xi|^2 + |\frac{\partial \boldsymbol{v}_\xi}{\partial \xi}|^2\right)$$

$$= -\int_\Omega \left[\boldsymbol{\nabla}_{\boldsymbol{v}} \boldsymbol{v}_\xi + \left(\int_\xi^1 \mathrm{div}\boldsymbol{v}\right)\frac{\partial \boldsymbol{v}_\xi}{\partial \xi}\right] \cdot \boldsymbol{v}_\xi - \int_\Omega \left[\boldsymbol{\nabla}_{\boldsymbol{v}_\xi}\boldsymbol{v} - (\mathrm{div}\boldsymbol{v})\frac{\partial \boldsymbol{v}}{\partial \xi}\right] \cdot \boldsymbol{v}_\xi$$

$$-\int_\Omega \left(\frac{f}{R_0}\boldsymbol{k} \times \boldsymbol{v}_\xi\right) \cdot \boldsymbol{v}_\xi + \int_\Omega \frac{bP}{p}\mathrm{grad}\left[(1+aq)\,T\right] \cdot \boldsymbol{v}_\xi.$$

By Lemma 3.2.10 and integration by parts, we obtain

$$\int_\Omega \left[\boldsymbol{\nabla}_{\boldsymbol{v}} \boldsymbol{v}_\xi + \left(\int_\xi^1 \mathrm{div}\boldsymbol{v}\mathrm{d}\xi'\right)\frac{\partial \boldsymbol{v}_\xi}{\partial \xi}\right] \cdot \boldsymbol{v}_\xi = 0.$$

By integration by parts, Hölder Inequality, (3.2.27) and Young Inequality, we have

$$-\int_\Omega \left(\boldsymbol{\nabla}_{\boldsymbol{v}_\xi}\boldsymbol{v} - (\mathrm{div}\boldsymbol{v})\frac{\partial \boldsymbol{v}}{\partial \xi}\right) \cdot \boldsymbol{v}_\xi \le c\int_\Omega |\boldsymbol{v}||\boldsymbol{v}_\xi|(|\boldsymbol{\nabla}_{e_\theta} \boldsymbol{v}_\xi|^2 + |\boldsymbol{\nabla}_{e_\varphi} \boldsymbol{v}_\xi|^2)^{\frac{1}{2}}$$

$$\le c|\boldsymbol{v}|_4 |\boldsymbol{v}_\xi|_2^{\frac{1}{4}} \|\boldsymbol{v}_\xi\|^{\frac{3}{4}} \left(\int_\Omega (|\boldsymbol{\nabla}_{e_\theta} \boldsymbol{v}_\xi|^2 + |\boldsymbol{\nabla}_{e_\varphi} \boldsymbol{v}_\xi|^2)\right)^{\frac{1}{2}} \le \varepsilon\|\boldsymbol{v}_\xi\|^2 + c|\boldsymbol{v}|_4^8 |\boldsymbol{v}_\xi|_2^2.$$

By Lemma 3.2.8, Hölder Inequality and Young Inequality, we have

$$\int_\Omega \frac{bP}{p}\mathrm{grad}[(1+aq)T] \cdot \boldsymbol{v}_\xi$$

$$= -\int_\Omega \frac{bP}{p}(1+aq)T\mathrm{div}\boldsymbol{v}_\xi \le c|T|_2^2 + c|q|_4^2|T|_4^2 + \varepsilon\|\boldsymbol{v}_\xi\|^2.$$

Choosing ε small enough, we obtain

$$\frac{\mathrm{d}|\boldsymbol{v}_\xi|_2^2}{\mathrm{d}t} + \int_\Omega (|\boldsymbol{\nabla}_{e_\theta} \boldsymbol{v}_\xi|^2 + |\boldsymbol{\nabla}_{e_\varphi} \boldsymbol{v}_\xi|^2 + |\boldsymbol{v}_\xi|^2) + \int_\Omega |\frac{\partial \boldsymbol{v}_\xi}{\partial \xi}|^2$$

$$\leq c(|\bar{\boldsymbol{v}}|_{H^1}^8 + |\tilde{\boldsymbol{v}}|_4^8)|\boldsymbol{v}_\xi|_2^2 + c|T|_2^2 + c|q|_4^4 + c|T|_4^4. \tag{3.2.56}$$

Applying Lemma 3.2.13, (3.2.35), (3.2.37), (3.2.43), (3.2.50) and (3.2.54), we have

$$|\boldsymbol{v}_\xi(t+6r)|_2^2 \leq c\left(E_0 + \frac{E_1}{r} + E_2 + E_4\right)\exp c(E_6^2 + E_8^4) \leq E_9, \tag{3.2.57}$$

where E_9 is a positive constant, and $t \geq 0$. Combining (3.2.56) with (3.2.57), we have

$$c_1 \int_{t+6r}^{t+7r} \|\boldsymbol{v}_\xi\|^2 \leq E_9^2 + E_9 = E_{10}. \tag{3.2.58}$$

By Gronwall Inequality, we derive from (3.2.56)

$$|\boldsymbol{v}_\xi(t)|_2^2 \leq C_4, \tag{3.2.59}$$

where $C_4 = C_4(\|U_0\|, \|Q_1\|_1, \|Q_2\|_1) > 0$, and $0 \leq t < 6r$.

L^2 estimates of T_ξ and q_ξ. Taking the derivatives of equations (3.2.12) and (3.2.13) with respect to ξ, we get

$$\frac{\partial T_\xi}{\partial t} - \Delta T_\xi - \frac{\partial^2 T_\xi}{\partial \xi^2} + \nabla_{\boldsymbol{v}} T_\xi + W(\boldsymbol{v})\frac{\partial T_\xi}{\partial \xi} + \nabla_{\boldsymbol{v}_\xi} T - (\mathrm{div}\boldsymbol{v})\frac{\partial T}{\partial \xi}$$
$$+ \frac{bP}{p}(1+aq)\mathrm{div}\boldsymbol{v} - \frac{abP}{p} q_\xi W(\boldsymbol{v}) + \frac{bP(P-p_0)}{p^2}(1+aq)W(\boldsymbol{v}) = Q_{1\xi},$$
$$\frac{\partial q_\xi}{\partial t} - \Delta q_\xi - \frac{\partial^2 q_\xi}{\partial \xi^2} + \nabla_{\boldsymbol{v}} q_\xi + W(\boldsymbol{v})\frac{\partial q_\xi}{\partial \xi} + \nabla_{\boldsymbol{v}_\xi} q - (\mathrm{div}\boldsymbol{v})\frac{\partial q}{\partial \xi} = Q_{2\xi}.$$

Taking $L^2(\Omega)$ inner product of the first equation above with T_ξ, we have

$$\frac{1}{2}\frac{\mathrm{d}|T_\xi|_2^2}{\mathrm{d}t} + \int_\Omega |\nabla T_\xi|^2 + \int_\Omega |T_{\xi\xi}|^2 - \int_{S^2}(T_\xi|_{\xi=1} \cdot T_{\xi\xi}|_{\xi=1})$$
$$= -\int_\Omega \left(\nabla_{\boldsymbol{v}} T_\xi + W(\boldsymbol{v})\frac{\partial T_\xi}{\partial \xi}\right) T_\xi - \int_\Omega \left(\nabla_{\boldsymbol{v}_\xi} T - (\mathrm{div}\boldsymbol{v})\frac{\partial T}{\partial \xi}\right) T_\xi + \int_\Omega Q_{1\xi} T_\xi$$
$$+ \int_\Omega \left[-\frac{bP}{p}(1+aq)(\mathrm{div}\boldsymbol{v}) - \frac{bP(P-p_0)}{p^2}(1+aq)W(\boldsymbol{v}) + \frac{abP}{p} q_\xi W(\boldsymbol{v})\right] T_\xi.$$

By integration by parts, Hölder Inequality, (3.2.27), Poincaré Inequality and Young Inequality, we get

$$\left|\int_\Omega \left[\nabla_{\boldsymbol{v}_\xi} T - \mathrm{div}(\boldsymbol{v})\frac{\partial T}{\partial \xi}\right] T_\xi\right|$$
$$\leq c\int_\Omega \left[(|\nabla_{e_\theta} \boldsymbol{v}_\xi|^2 + |\nabla_{e_\varphi} \boldsymbol{v}_\xi|^2)^{\frac{1}{2}} |T||T_\xi| + |\boldsymbol{v}_\xi||T||\nabla T_\xi| + |\boldsymbol{v}||\nabla T_\xi||T_\xi|\right]$$
$$\leq \varepsilon(|T_{\xi\xi}|_2^2 + |\nabla T_\xi|_2^2) + c\left[|\boldsymbol{v}_{\xi\xi}|_2^2 + \int_\Omega (|\nabla_{e_\theta} \boldsymbol{v}_\xi|^2 + |\nabla_{e_\varphi} \boldsymbol{v}_\xi|^2)\right] + c|T|_4^8 |\boldsymbol{v}_\xi|_2^2$$

$$+ c(|T|_4^8 + |\boldsymbol{v}|_4^8)|T_\xi|_2^2.$$

In a similar way, there is

$$\left|\int_\Omega \left[-\frac{bP}{p}(1+aq)(\mathrm{div}\boldsymbol{v}) - \frac{bP(P-p_0)}{p^2}(1+aq)W(\boldsymbol{v}) + \frac{abP}{p}q_\xi W(\boldsymbol{v})\right] T_\xi\right|$$
$$\leq \varepsilon(|\nabla q_\xi|_2^2 + |\nabla T_\xi|_2^2) + c|\nabla q|_2^2 + c\|\boldsymbol{v}\|^2 + c(|\boldsymbol{v}|_4^2 + |q|_4^2)|T_\xi|_4^2$$
$$+ c|\boldsymbol{v}|_4^2(|q_\xi|_4^2 + |q|_4^2) + c|T_\xi|_2^2$$
$$\leq \varepsilon(|\nabla T_\xi|_2^2 + |T_{\xi\xi}|_2^2) + \varepsilon(|\nabla q_\xi|_2^2 + |q_{\xi\xi}|_2^2) + c(|\boldsymbol{v}|_4^8 + |q|_4^8 + 1)|T_\xi|_2^2$$
$$+ c|\boldsymbol{v}|_4^8|q_\xi|_2^2 + c(|\nabla q|_2^2 + \|\boldsymbol{v}\|^2) + c|\boldsymbol{v}|_4^2|q|_4^2.$$

Taking the trace on $\xi = 1$ of equation (3.2.12), we get

$$T_{\xi\xi}|_{\xi=1} = \frac{\partial T|_{\xi=1}}{\partial t} + (\nabla_{\boldsymbol{v}} T)|_{\xi=1} - \Delta T|_{\xi=1} - Q_1|_{\xi=1}.$$

By (3.2.15), we have

$$- \int_{S^2} (T_\xi|_{\xi=1} T_{\xi\xi}|_{\xi=1})$$
$$= \alpha_s \int_{S^2} T|_{\xi=1} \left[\frac{\partial T|_{\xi=1}}{\partial t} + (\nabla_{\boldsymbol{v}} T)|_{\xi=1} - \Delta T|_{\xi=1} - Q_1|_{\xi=1}\right]$$
$$= \alpha_s \left(\frac{1}{2}\frac{\mathrm{d}|T|_{\xi=1}|_2^2}{\mathrm{d}t} + |\nabla T|_{\xi=1}|_2^2\right) + \alpha_s \int_{S^2} T|_{\xi=1}\left[(\nabla_{\boldsymbol{v}} T)|_{\xi=1} - Q_1|_{\xi=1}\right].$$

Applying Lemma 3.2.10, we obtain

$$- \alpha_s \int_{S^2} T|_{\xi=1}\left((\nabla_{\boldsymbol{v}} T)|_{\xi=1} - Q_1|_{\xi=1}\right)$$
$$= \frac{\alpha_s}{2}\int_{S^2} T^2|_{\xi=1}\mathrm{div}\boldsymbol{v}|_{\xi=1} + \alpha_s \int_{S^2} (TQ_1)|_{\xi=1}$$
$$= \frac{\alpha_s}{2}\int_{S^2} T^2|_{\xi=1}\left(\int_\xi^1 \mathrm{div}\boldsymbol{v}_\xi \mathrm{d}\xi' + \mathrm{div}\boldsymbol{v}\right) + \alpha_s \int_{S^2} T|_{\xi=1}Q_1|_{\xi=1}$$
$$\leq c|T|_{\xi=1}|_4^4 + c\|\boldsymbol{v}_\xi\|^2 + c\|\boldsymbol{v}\|^2 + c|T|_{\xi=1}|_2^2 + c|Q_1|_{\xi=1}|_2^2.$$

We derive from the above equalities

$$\frac{1}{2}\frac{\mathrm{d}\left(|T_\xi|_2^2 + \alpha_s|T|_{\xi=1}|_2^2\right)}{\mathrm{d}t} + \int_\Omega |\nabla T_\xi|^2 + \int_\Omega |T_{\xi\xi}|^2 + \alpha_s|\nabla T|_{\xi=1}|_2^2$$
$$\leq 2\varepsilon\left(|T_{\xi\xi}|_2^2 + |\nabla T_\xi|_2^2\right) + \varepsilon\left(|\nabla q_\xi|_2^2 + |q_{\xi\xi}|_2^2\right) + c\left(1 + |T|_4^8 + |\boldsymbol{v}|_4^8 + |q|_4^8\right)|T_\xi|_2^2$$
$$+ c|\boldsymbol{v}|_4^8|q_\xi|_2^2 + c\|\boldsymbol{v}_\xi\|^2 + c\|\boldsymbol{v}\|^2 + c|\nabla q|_2^2 + c|T|_4^8|\boldsymbol{v}_\xi|_2^2 + c|q|_4^2|\boldsymbol{v}|_4^2$$
$$+ c|T|_{\xi=1}|_4^4 + c|T|_{\xi=1}|_2^2 + c|Q_1|_{\xi=1}|_2^2 + c|Q_1|_2^2 + c|Q_{1\xi}|_2^2.$$

Similar to the above equality,

$$\frac{1}{2}\frac{\mathrm{d}\left(|q_\xi|_2^2+\beta_s|q|_{\xi=1}|_2^2\right)}{\mathrm{d}t}+\int_\Omega|\boldsymbol{\nabla} q_\xi|^2+\int_\Omega|q_{\xi\xi}|^2+\beta_s|\boldsymbol{\nabla} q|_{\xi=1}|_2^2$$
$$\le\varepsilon\left(|\boldsymbol{\nabla} q_\xi|_2^2+|q_{\xi\xi}|_2^2\right)+c\left(|\boldsymbol{v}|_4^8+|T|_4^8\right)|q_\xi|_2^2+c\|\boldsymbol{v}_\xi\|^2+c\|\boldsymbol{v}\|^2+c|q|_4^8|\boldsymbol{v}_\xi|_2^2$$
$$+c|q|_{\xi=1}|_4^4+c|q|_{\xi=1}|_2^2+c|Q_2|_{\xi=1}|_2^2+c|Q_2|_2^2+c|Q_{2\xi}|_2^2.$$

According to the above two inequalities, choosing ε small enough, we obtain

$$\frac{\mathrm{d}\left(|T_\xi|_2^2+|q_\xi|_2^2+\beta_s|q|_{\xi=1}|_2^2+\alpha_s|T|_{\xi=1}|_2^2\right)}{\mathrm{d}t}+\int_\Omega|\boldsymbol{\nabla} T_\xi|^2+\int_\Omega|T_{\xi\xi}|^2T|_{\xi=1}|_2^2$$
$$+\alpha_s|\boldsymbol{\nabla}+\int_\Omega|\boldsymbol{\nabla} q_\xi|^2+\int_\Omega|q_{\xi\xi}|^2+\beta_s|\boldsymbol{\nabla} q|_{\xi=1}|_2^2$$
$$\le c\left(1+|T|_4^8+|\boldsymbol{v}|_4^8+|q|_4^8\right)\left(|T_\xi|_2^2+|q_\xi|_2^2\right)+c\|\boldsymbol{v}_\xi\|^2+c\|\boldsymbol{v}\|^2+c\|q\|^2$$
$$+c\left(|T|_4^8+|q|_4^8\right)|\boldsymbol{v}_\xi|_2^2+c|q|_4^2|\boldsymbol{v}|_4^2+c|T|_{\xi=1}|_4^4+c|T|_{\xi=1}|_2^2+c|q|_{\xi=1}|_4^4$$
$$+c|q|_{\xi=1}|_2^2+c(|Q_1|_{\xi=1}|_2^2+|Q_2|_{\xi=1}|_2^2)$$
$$+c\left(|Q_1|_2^2+|Q_{1\xi}|_2^2+|Q_2|_2^2+|Q_{2\xi}|_2^2\right). \tag{3.2.60}$$

By using Lemma 3.2.13, (3.2.35), (3.2.37), (3.2.38), (3.2.43), (3.2.45), (3.2.50), (3.2.54), (3.2.57) and (3.2.58), we get

$$|T_\xi(t+7r)|_2^2+|q_\xi(t+7r)|_2^2\le E_{11}, \tag{3.2.61}$$

where

$$E_{11}=c[\frac{E_1}{r}+E_1+E_2+E_4+E_4^2+E_2^{\frac{1}{2}}(E_6^{\frac{1}{2}}+E_8)+(E_4^2+E_2^2)E_9+E_{10}$$
$$+c\|Q_1\|_1^2+c\|Q_2\|_1^2]\cdot\exp c(1+E_2^2+E_4^2+E_6^2+E_8^4).$$

Combining (3.2.60) with (3.2.61), we have

$$c_1\int_{t+7r}^{t+8r}(\|T_\xi\|^2+\|q_\xi\|^2)\le E_{11}^2+2E_{11}+E_1=E_{12}. \tag{3.2.62}$$

According to Gronwall Inequality, we derive from (3.2.60)

$$|T_\xi(t)|_2^2+|q_\xi(t)|_2^2\le C_5, \tag{3.2.63}$$

where $C_5=C_5(\|U_0\|,\ \|Q_1\|_1,\ \|Q_2\|_1)>0$, and $0\le t<7r$.

H^1 **estimate of** $\boldsymbol{v}$**,** T**,** q**.** Taking L^2 inner product of equation (3.2.11) with $-\Delta\boldsymbol{v}$, we get

$$\frac{1}{2}\frac{\mathrm{d}\int_\Omega(|\boldsymbol{\nabla}_{e_\theta}\boldsymbol{v}|^2+|\boldsymbol{\nabla}_{e_\varphi}\boldsymbol{v}|^2+|\boldsymbol{v}|^2)}{\mathrm{d}t}+\int_\Omega|\Delta\boldsymbol{v}|_2^2$$

$$+\int_\Omega (|\nabla_{e_\theta} v_\xi|^2 + |\nabla_{e_\varphi} v_\xi|^2 + |v_\xi|^2)$$
$$= \int_\Omega (\nabla_v v + W(v) v_\xi) \cdot \Delta v + \int_\Omega \left[\int_\xi^1 \frac{bP}{p} \text{grad}((1+aq)T)\right] \cdot \Delta v$$
$$+ \int_\Omega \left(\frac{f}{R_0} k \times v + \text{grad}\, \Phi_s\right) \cdot \Delta v.$$

By Hölder Inequality, (3.2.27) and Young Inequality, we have

$$\left|\int_\Omega \nabla_v v \cdot \Delta v\right| \le \int_\Omega |v| \left(|\nabla_{e_\theta} v|^2 + |\nabla_{e_\varphi} v|^2\right)^{\frac{1}{2}} |\Delta v|$$
$$\le c(|v|_4^8 + |v|_4^2) \int_\Omega (|\nabla_{e_\theta} v|^2 + |\nabla_{e_\varphi} v|^2)$$
$$+ 2\varepsilon(|\Delta v|_2^2 + \int_\Omega (|\nabla_{e_\theta} v_\xi|^2 + |\nabla_{e_\varphi} v_\xi|^2)).$$

By Hölder Inequality, Minkowski Inequality, Young Inequality and (3.2.24), we obtain

$$\left|\int_\Omega W(v) v_\xi \cdot \Delta v\right| \le \int_{S^2} \left[\int_0^1 (|\nabla_{e_\theta} v|^2 + |\nabla_{e_\varphi} v|^2)^{\frac{1}{2}} \mathrm{d}\xi \int_0^1 |v_\xi||\Delta v| \mathrm{d}\xi\right]$$
$$\le c \left\{\int_0^1 \left[\int_{S^2} (|\nabla_{e_\theta} v|^2 + |\nabla_{e_\varphi} v|^2)\right]^{\frac{1}{2}}\right.$$
$$\left. \cdot \left[\int_{S^2} (|\nabla_{e_\theta} v|^2 + |\nabla_{e_\varphi} v|^2 + |\Delta v|^2)\right]^{\frac{1}{2}} \mathrm{d}\xi\right\}$$
$$\cdot c \left\{\int_0^1 \left(\int_{S^2} |v_\xi|^2\right)^{\frac{1}{2}} \cdot \left[\int_{S^2} (|\nabla_{e_\theta} v_\xi|^2 + |\nabla_{e_\varphi} v_\xi|^2 + |v_\xi|^2)\right]^{\frac{1}{2}} \mathrm{d}\xi\right\} + \varepsilon|\Delta v|_2^2$$
$$\le 2\varepsilon|\Delta v|_2^2 + c[2|v_\xi|_2^2 + |v_\xi|_2^4 + (|v_\xi|_2^2 + 1) \int_\Omega (|\nabla_{e_\theta} v_\xi|^2 + |\nabla_{e_\varphi} v_\xi|^2)]$$
$$\cdot \int_\Omega (|\nabla_{e_\theta} v|^2 + |\nabla_{e_\varphi} v|^2),$$

$$\left|\int_\Omega \int_\xi^1 \frac{bP}{p} \text{grad}[(1+aq)T] \mathrm{d}\xi' \cdot \Delta v\right|$$
$$\le c \left[\int_\Omega (\int_0^1 |q|^2 \mathrm{d}\xi)^2\right]^{\frac{1}{2}} \left[\int_\Omega (\int_0^1 |\nabla T|^2 \mathrm{d}\xi)^2\right]^{\frac{1}{2}}$$
$$+ c \left[\int_\Omega (\int_0^1 |T|^2 \mathrm{d}\xi)^2\right]^{\frac{1}{2}} \left[\int_\Omega (\int_0^1 |\nabla q|^2 \mathrm{d}\xi)^2\right]^{\frac{1}{2}} + c|\nabla T|_2^2 + \varepsilon|\Delta v|_2^2$$
$$\le c|T|_4^2 \int_0^1 \left[\|\nabla q\|_{L^2} (\|\nabla q\|_{L^2}^2 + \|\Delta q\|_{L^2}^2)^{\frac{1}{2}}\right] \mathrm{d}\xi + c|\nabla T|_2^2 + \varepsilon|\Delta v|_2^2$$

$$
\begin{aligned}
&+ c|q|_4^2 \int_0^1 \left[\|\nabla T\|_{L^2} (\|\nabla T\|_{L^2}^2 + \|\Delta T\|_{L^2}^2)^{\frac{1}{2}} \right] \mathrm{d}\xi \\
&\leq \varepsilon(|\Delta \boldsymbol{v}|_2^2 + |\Delta T|_2^2 + |\Delta q|_2^2) + c|q|_4^2 |\nabla T|_2^2 + c|T|_4^2 |\nabla q|_2^2 + c|q|_4^4 |\nabla T|_2^2 \\
&+ c|T|_4^4 |\nabla q|_2^2 + c|\nabla T|_2^2.
\end{aligned}
$$

Thus, we have

$$
\begin{aligned}
&\frac{1}{2} \frac{\mathrm{d} \int_\Omega (|\nabla_{e_\theta} \boldsymbol{v}|^2 + |\nabla_{e_\varphi} \boldsymbol{v}|^2 + |\boldsymbol{v}|^2)}{\mathrm{d}t} + |\Delta \boldsymbol{v}|_2^2 \\
&+ \int_\Omega (|\nabla_{e_\theta} \boldsymbol{v}_\xi|^2 + |\nabla_{e_\varphi} \boldsymbol{v}_\xi|^2 + |\boldsymbol{v}_\xi|^2) \\
&\leq c[|\boldsymbol{v}|_4^8 + |\boldsymbol{v}|_4^2 + 2|\boldsymbol{v}_\xi|_2^2 + |\boldsymbol{v}_\xi|_2^4 + (|\boldsymbol{v}_\xi|_2^2 + 1)\|\boldsymbol{v}_\xi\|^2] \int_\Omega (|\nabla_{e_\theta} \boldsymbol{v}|^2 + |\nabla_{e_\varphi} \boldsymbol{v}|^2) \\
&\quad + c(1 + |q|_4^2 + |q|_4^4)|\nabla T|_2^2 + c(|T|_4^2 + |T|_4^4)|\nabla q|_2^2 + 2\varepsilon \int_\Omega (|\nabla_{e_\theta} \boldsymbol{v}_\xi|^2 \\
&\quad + |\nabla_{e_\varphi} \boldsymbol{v}_\xi|^2) + 5\varepsilon |\Delta \boldsymbol{v}|_2^2 + \varepsilon |\Delta T|_2^2 + \varepsilon |\Delta q|_2^2 + c|\nabla T|_2^2. \qquad (3.2.64)
\end{aligned}
$$

Taking $L^2(\Omega)$ inner product of equation (3.2.12) with $-\Delta T$, we get

$$
\begin{aligned}
&\frac{1}{2} \frac{\mathrm{d}|\nabla T|_2^2}{\mathrm{d}t} + |\Delta T|_2^2 + (|\nabla T_\xi|_2^2 + \alpha_s |\nabla T|_{\xi=1}|_2^2) \\
&= \int_\Omega \left(\nabla_{\boldsymbol{v}} T + W(\boldsymbol{v}) \frac{\partial T}{\partial \xi} \right) \Delta T - \int_\Omega \frac{bP}{p}(1 + aq) W(\boldsymbol{v}) \Delta T - \int_\Omega Q_1 \Delta T.
\end{aligned}
$$

In a similar way, we obtain

$$
\begin{aligned}
&\left| \int_\Omega \Delta T \nabla_{\boldsymbol{v}} T \right| \leq c(|\boldsymbol{v}|_4^8 + |\boldsymbol{v}|_4^2)|\nabla T|_2^2 + 2\varepsilon(|\Delta T|_2^2 + |\nabla T_\xi|_2^2), \\
&\left| \int_\Omega W(\boldsymbol{v}) T_\xi \Delta T \right| \leq \varepsilon |\Delta T|_2^2 + \varepsilon |\Delta \boldsymbol{v}|_2^2 + c[2|T_\xi|_2^2 + |T_\xi|_2^4 \\
&\qquad + (|T_\xi|_2^2 + 1) \int_\Omega |\nabla T_\xi|^2] \int_\Omega (|\nabla_{e_\theta} \boldsymbol{v}|^2 + |\nabla_{e_\varphi} \boldsymbol{v}|^2).
\end{aligned}
$$

By Hölder Inequality, Minkowski Inequality, Young Inequality and (3.2.24), we have

$$
\begin{aligned}
&\left| \int_\Omega \frac{bP}{p}(1 + aq) W(\boldsymbol{v}) \Delta T \right| \\
&\leq c \left(\int_\Omega |\mathrm{div} \boldsymbol{v}|^2 \right)^{\frac{1}{2}} |\Delta T|_2 + c|q|_4 \left(\int_\Omega \left(\int_0^1 |\mathrm{div} \boldsymbol{v}|^2 \mathrm{d}\xi \right)^2 \right)^{\frac{1}{4}} |\Delta T|_2 \\
&\leq \varepsilon |\Delta T|_2^2 + \varepsilon |\Delta \boldsymbol{v}|_2^2 + c(1 + |q|_4^2 + |q|_4^4) \int_\Omega (|\nabla_{e_\theta} \boldsymbol{v}|^2 + |\nabla_{e_\varphi} \boldsymbol{v}|^2).
\end{aligned}
$$

So we obtain

$$\frac{1}{2}\frac{\mathrm{d}|\nabla T|_2^2}{\mathrm{d}t} + |\Delta T|_2^2 + (|\nabla T_\xi|_2^2 + \alpha_s |\nabla T|_{\xi=1}|_2^2)$$
$$\leq 5\varepsilon|\Delta T|_2^2 + 2\varepsilon|\nabla T_\xi|_2^2 + 2\varepsilon|\Delta \boldsymbol{v}|_2^2 + c[1 + |q|_4^2 + |q|_4^4 + 2|T_\xi|_2^2 + |T_\xi|_2^4$$
$$+ (|T_\xi|_2^2 + 1)\int_\Omega |\nabla T_\xi|^2] \cdot \int_\Omega (|\nabla_{e_\theta}\boldsymbol{v}|^2 + |\nabla_{e_\varphi}\boldsymbol{v}|^2) + c(|\boldsymbol{v}|_4^2 + |\boldsymbol{v}|_4^8)|\nabla T|_2^2$$
$$+ c|Q_1|_2^2. \tag{3.2.65}$$

Similar to (3.2.65),

$$\frac{1}{2}\frac{\mathrm{d}|\nabla q|_2^2}{\mathrm{d}t} + |\Delta q|_2^2 + (|\nabla q_\xi|_2^2 + \beta_s |\nabla q|_{\xi=1}|_2^2)$$
$$\leq 4\varepsilon|\Delta q|_2^2 + 2\varepsilon|\nabla q_\xi|_2^2 + \varepsilon|\Delta \boldsymbol{v}|_2^2 + c(|\boldsymbol{v}|_4^2 + |\boldsymbol{v}|_4^8)|\nabla q|_2^2 + c[|q_\xi|_2^2 + |q_\xi|_2^4$$
$$+ (|q_\xi|_2^2 + 1)|\nabla q_\xi|_2^2](|\nabla_{e_\theta}\boldsymbol{v}|_2^2 + |\nabla_{e_\varphi}\boldsymbol{v}|_2^2) + c|Q_2|_2^2. \tag{3.2.66}$$

According to (3.2.64)-(3.2.66), choosing ε small enough, we get

$$\frac{\mathrm{d}\left[\int_\Omega (|\nabla_{e_\theta}\boldsymbol{v}|^2 + |\nabla_{e_\varphi}\boldsymbol{v}|^2 + |\boldsymbol{v}|^2) + |\nabla T|_2^2 + |\nabla q|_2^2\right]}{\mathrm{d}t} + |\Delta \boldsymbol{v}|_2^2 + |\Delta T|_2^2$$
$$+ |\Delta q|_2^2 + \int_\Omega (|\nabla_{e_\theta}\boldsymbol{v}_\xi|^2 + |\nabla_{e_\varphi}\boldsymbol{v}_\xi|^2 + |\boldsymbol{v}_\xi|^2) + (|\nabla T_\xi|_2^2 + \alpha_s|\nabla T|_{\xi=1}|_2^2)$$
$$+ (|\nabla q_\xi|_2^2 + \beta_s|\nabla q|_{\xi=1}|_2^2)$$
$$\leq c\Big[1 + |q|_4^2 + |q|_4^4 + |T|_4^2 + |T|_4^4 + |\boldsymbol{v}|_4^2 + |\boldsymbol{v}|_4^8 + 2|\boldsymbol{v}_\xi|_2^2 + |\boldsymbol{v}_\xi|_2^4 + (|\boldsymbol{v}_\xi|_2^2$$
$$+ 1)\|\boldsymbol{v}_\xi\|^2 + 2|T_\xi|_2^2 + |T_\xi|_2^4 + (|T_\xi|_2^2 + 1)\int_\Omega |\nabla T_\xi|^2 + 2|q_\xi|_2^2 + |q_\xi|_2^4$$
$$+(|q_\xi|_2^2 + 1)\int_\Omega |\nabla q_\xi|^2\Big]\left[\int_\Omega (|\nabla_{e_\theta}\boldsymbol{v}|^2 + |\nabla_{e_\varphi}\boldsymbol{v}|^2) + |\nabla T|_2^2 + |\nabla q|_2^2\right]$$
$$+ c|Q_1|_2^2 + c|Q_2|_2^2. \tag{3.2.67}$$

Applying Lemma 3.2.13, (3.2.35), (3.2.37), (3.2.43), (3.2.50), (3.2.54), (3.2.57), (3.2.58), (3.2.61) and (3.2.62), we obtain

$$|\nabla_{e_\theta}\boldsymbol{v}(t+8r)|_2^2 + |\nabla_{e_\varphi}\boldsymbol{v}(t+8r)|_2^2 + |\nabla T(t+8r)|_2^2 + |\nabla q(t+8r)|_2^2 \leq E_{13}, \tag{3.2.68}$$

here

$$E_{13} = c(\frac{E_1}{r} + |Q_1|_2^2 + |Q_2|_2^2)\cdot \exp c[1 + E_2 + E_4 + E_6^2 + E_8^4 + E_9^2$$
$$+(E_9 + 1)E_{10} + E_{11}^2 + (E_{11} + 1)E_{12}].$$

By Gronwall Inequality, we derive from (3.2.67)

$$|\nabla_{e_\theta}\boldsymbol{v}(t)|_2^2 + |\nabla_{e_\varphi}\boldsymbol{v}(t)|_2^2 + |\nabla T(t)|_2^2 + |\nabla q(t)|_2^2 \leq C_6, \tag{3.2.69}$$

where $C_6 = C_6(\|U_0\|,\ \|Q_1\|_1,\ \|Q_2\|_1) > 0$, and $0 \leq t < 8r$.

3.2.4 *Global Existence and Uniqueness of Strong Solutions*

Proof of Theorem 3.2.2. Applying Proposition 3.2.15, we prove Theorem 3.2.2 by contradiction. In fact, let U be a strong solution of system (3.2.11)-(3.2.17) on the maximal interval $[0,\mathcal{T}^*]$. If $\mathcal{T}^* < +\infty$, then $\limsup_{t\to\mathcal{T}^{*-}} \|U\| = +\infty$. By (3.2.35), (3.2.57), (3.2.59), (3.2.61), (3.2.63), (3.2.68) and (3.2.69), we know $\limsup_{t\to\mathcal{T}^{*-}} \|U\| = +\infty$, which is impossible. Thus, Theorem 3.2.2 is proved.

Proof of Theorem 3.2.3. Suppose that $(\boldsymbol{v}_1, T_1, q_1)$ and $(\boldsymbol{v}_2, T_2, q_2)$ are two solutions of system (3.2.11)-(3.2.17) in interval $[0,\mathcal{T}]$, which are respectively corresponding to Φ_{s_1}, Φ_{s_2} and initial data $((\boldsymbol{v}_0)_1, (T_0)_1, (q_0)_1)$, $((\boldsymbol{v}_0)_2, (T_0)_2, (q_0)_2)$. Define $\boldsymbol{v} = \boldsymbol{v}_1 - \boldsymbol{v}_2$, $T = T_1 - T_2$, $q = q_1 - q_2$, $\Phi_s = \Phi_{s_1} - \Phi_{s_2}$. Then $\boldsymbol{v}$, T, q, Φ_s satisfy the following system

$$\frac{\partial \boldsymbol{v}}{\partial t} - \Delta \boldsymbol{v} - \frac{\partial^2 \boldsymbol{v}}{\partial \xi^2} + \nabla_{\boldsymbol{v}_1}\boldsymbol{v} + \nabla_{\boldsymbol{v}}\boldsymbol{v}_2 + W(\boldsymbol{v}_1)\frac{\partial \boldsymbol{v}}{\partial \xi} + W(\boldsymbol{v})\frac{\partial \boldsymbol{v}_2}{\partial \xi} + \frac{f}{R_0}\boldsymbol{k}\times\boldsymbol{v}$$
$$+ \operatorname{grad}\Phi_s + \int_\xi^1 \frac{bP}{p}\operatorname{grad}T\,\mathrm{d}\xi' + \int_\xi^1 \frac{abP}{p}\operatorname{grad}(q_1 T)\,\mathrm{d}\xi'$$
$$+ \int_\xi^1 \frac{abP}{p}\operatorname{grad}(qT_2)\,\mathrm{d}\xi' = 0, \tag{3.2.70}$$

$$\frac{\partial T}{\partial t} - \Delta T - \frac{\partial^2 T}{\partial \xi^2} + \nabla_{\boldsymbol{v}_1}T + \nabla_{\boldsymbol{v}}T_2 + W(\boldsymbol{v}_1)\frac{\partial T}{\partial \xi} + W(\boldsymbol{v})\frac{\partial T_2}{\partial \xi} - \frac{bP}{p}W(\boldsymbol{v})$$
$$- \frac{abP}{p}q_1 W(\boldsymbol{v}) - \frac{abP}{p}qW(\boldsymbol{v}_2) = 0, \tag{3.2.71}$$

$$\frac{\partial q}{\partial t} - \Delta q - \frac{\partial^2 q}{\partial \xi^2} + \nabla_{\boldsymbol{v}_1}q + \nabla_{\boldsymbol{v}}q_2 + W(\boldsymbol{v}_1)\frac{\partial q}{\partial \xi} + W(\boldsymbol{v})\frac{\partial q_2}{\partial \xi} = 0, \tag{3.2.72}$$

$$(\boldsymbol{v}|_{t=0}, T|_{t=0}, q|_{t=0}) = ((\boldsymbol{v}_0)_1 - (\boldsymbol{v}_0)_2, (T_0)_1 - (T_0)_2, (q_0)_1 - (q_0)_2),$$
$$\xi = 1: \frac{\partial \boldsymbol{v}}{\partial \xi} = 0, \frac{\partial T}{\partial \xi} = -\alpha_s T, \frac{\partial q}{\partial \xi} = -\beta_s q, \ \xi = 0: \frac{\partial \boldsymbol{v}}{\partial \xi} = 0, \frac{\partial T}{\partial \xi} = 0, \frac{\partial q}{\partial \xi} = 0.$$

Taking $L^2(\Omega)\times L^2(\Omega)$ inner product of equation (3.2.70) with $\boldsymbol{v}$, we get

$$\frac{1}{2}\frac{\mathrm{d}|\boldsymbol{v}|_2^2}{\mathrm{d}t} + \int_\Omega (|\nabla_{e_\theta}\boldsymbol{v}|^2 + |\nabla_{e_\varphi}\boldsymbol{v}|^2 + |\boldsymbol{v}|^2) + \int_\Omega |\boldsymbol{v}_\xi|^2$$
$$= -\int_\Omega \left[\nabla_{\boldsymbol{v}_1}\boldsymbol{v} + W(\boldsymbol{v}_1)\frac{\partial \boldsymbol{v}}{\partial \xi}\right]\cdot\boldsymbol{v} - \int_\Omega \boldsymbol{v}\cdot\nabla_{\boldsymbol{v}}\boldsymbol{v}_2 - \int_\Omega W(\boldsymbol{v})\frac{\partial \boldsymbol{v}_2}{\partial \xi}\cdot\boldsymbol{v}$$
$$- \int_\Omega \left(\frac{f}{R_0}\boldsymbol{k}\times\boldsymbol{v} + \operatorname{grad}\Phi_s\right)\cdot\boldsymbol{v} - \int_\Omega \left(\int_\xi^1 \frac{bP}{p}\operatorname{grad}T\,\mathrm{d}\xi'\right)\cdot\boldsymbol{v}$$

$$-\int_{\Omega}\left[\int_{\xi}^{1}\frac{abP}{p}\text{grad}(q_1T)\text{d}\xi^{'}\right]\cdot\boldsymbol{v}-\int_{\Omega}\left[\int_{\xi}^{1}\frac{abP}{p}\text{grad}(qT_2)\text{d}\xi^{'}\right]\cdot\boldsymbol{v}.$$

By using Lemma 3.2.11, we have

$$\int_{\Omega}\left(\nabla_{\boldsymbol{v}_1}\boldsymbol{v}+W(\boldsymbol{v}_1)\frac{\partial\boldsymbol{v}}{\partial\xi}\right)\cdot\boldsymbol{v}=0.$$

Applying Lemma 3.2.10, Hölder Inequality, Young Inequality and (3.2.27), we obtain

$$\left|\int_{\Omega}\boldsymbol{v}\cdot\nabla_{\boldsymbol{v}}\boldsymbol{v}_2\right|=\left|\int_{\Omega}(\boldsymbol{v}_2\cdot\nabla_{\boldsymbol{v}}\boldsymbol{v}+\boldsymbol{v}_2\cdot\boldsymbol{v}\text{div}\boldsymbol{v})\right|$$

$$\leq c\int_{\Omega}|\boldsymbol{v}||\boldsymbol{v}_2|(|\nabla_{\boldsymbol{e}_\theta}\boldsymbol{v}|^2+|\nabla_{\boldsymbol{e}_\varphi}\boldsymbol{v}|^2)^{\frac{1}{2}}$$

$$\leq\varepsilon\int_{\Omega}(|\nabla_{\boldsymbol{e}_\theta}\boldsymbol{v}|^2+|\nabla_{\boldsymbol{e}_\varphi}\boldsymbol{v}|^2)+c|\boldsymbol{v}|_2^{\frac{1}{2}}|\boldsymbol{v}_2|_4^2\|\boldsymbol{v}\|^{\frac{3}{2}}\leq 2\varepsilon\|\boldsymbol{v}\|^2+c|\boldsymbol{v}_2|_4^8|\boldsymbol{v}|_2^2.$$

By Hölder Inequality, Young Inequality, Minkowski Inequality and (3.2.24), we get

$$\left|\int_{\Omega}W(\boldsymbol{v})\frac{\partial\boldsymbol{v}_2}{\partial\xi}\cdot\boldsymbol{v}\right|\leq\int_{S^2}\left[\int_0^1(|\nabla_{\boldsymbol{e}_\theta}\boldsymbol{v}|^2+|\nabla_{\boldsymbol{e}_\varphi}\boldsymbol{v}|^2)^{\frac{1}{2}}\,\text{d}\xi\int_0^1|\boldsymbol{v}_{2\xi}||\boldsymbol{v}|\text{d}\xi\right]$$

$$\leq\varepsilon\int_{\Omega}(|\nabla_{\boldsymbol{e}_\theta}\boldsymbol{v}|^2+|\nabla_{\boldsymbol{e}_\varphi}\boldsymbol{v}|^2)+c\int_0^1\left(\int_{S^2}|\boldsymbol{v}_{2\xi}|^4\right)^{\frac{1}{2}}\text{d}\xi\int_0^1\left(\int_{S^2}|\boldsymbol{v}|^4\right)^{\frac{1}{2}}\text{d}\xi$$

$$\leq\varepsilon\|\boldsymbol{v}\|^2+c\int_0^1\left[\|\boldsymbol{v}_{2\xi}\|_{L^2}\left(\int_{S^2}(|\nabla_{\boldsymbol{e}_\theta}\boldsymbol{v}_{2\xi}|^2+|\nabla_{\boldsymbol{e}_\varphi}\boldsymbol{v}_{2\xi}|^2)\right)^{\frac{1}{2}}\right]$$

$$\cdot\int_0^1\left[\|\boldsymbol{v}\|_{L^2}\left(\int_{S^2}(|\nabla_{\boldsymbol{e}_\theta}\boldsymbol{v}|^2+|\nabla_{\boldsymbol{e}_\varphi}\boldsymbol{v}|^2+|\boldsymbol{v}|^2)\right)^{\frac{1}{2}}\right]$$

$$\leq 2\varepsilon\|\boldsymbol{v}\|^2+c\left[(|\boldsymbol{v}_{2\xi}|_2^2+1)\int_{\Omega}(|\nabla_{\boldsymbol{e}_\theta}\boldsymbol{v}_{2\xi}|^2+|\nabla_{\boldsymbol{e}_\varphi}\boldsymbol{v}_{2\xi}|^2)+|\boldsymbol{v}_{2\xi}|_2^2\right]|\boldsymbol{v}|_2^2.$$

By Lemma 3.2.8, Hölder Inequality, Young Inequality, Minkowski Inequality and (3.2.24), we have

$$\left|\int_{\Omega}\left[\int_{\xi}^{1}\frac{abP}{p}\text{grad}\,(qT_2)\right]\cdot\boldsymbol{v}\right|$$

$$\leq c\int_0^1\left(\int_{S^2}|q|^4\right)^{\frac{1}{2}}\int_0^1\left(\int_{S^2}|T_2|^4\right)^{\frac{1}{2}}+\varepsilon\int_{\Omega}(|\nabla_{\boldsymbol{e}_\theta}\boldsymbol{v}|^2+|\nabla_{\boldsymbol{e}_\varphi}\boldsymbol{v}|^2)$$

$$\leq c\left(|T_2|_4^2+|T_2|_4^4\right)|q|_2^2+\varepsilon|\nabla q|_2^2+\varepsilon\int_{\Omega}\left(|\nabla_{\boldsymbol{e}_\theta}\boldsymbol{v}|^2+|\nabla_{\boldsymbol{e}_\varphi}\boldsymbol{v}|^2\right).$$

From the above relationships, we obtain

$$\frac{1}{2}\frac{\mathrm{d}|\boldsymbol{v}|_2^2}{\mathrm{d}t}+\int_\Omega\left(|\nabla_{e_\theta}\boldsymbol{v}|^2+|\nabla_{e_\varphi}\boldsymbol{v}|^2+|\boldsymbol{v}|^2\right)+\int_\Omega|\boldsymbol{v}_\xi|^2$$
$$\leq 5\varepsilon\|\boldsymbol{v}\|^2+\varepsilon|\nabla q|_2^2+c\left(|T_2|_4^2+|T_2|_4^4\right)|q|_2^2-\int_\Omega\left(\int_\xi^1\frac{bP}{p}\mathrm{grad}T\right)\cdot\boldsymbol{v}$$
$$-\int_\Omega\left(\int_\xi^1\frac{abP}{p}\mathrm{grad}\left(q_1T\right)\right)\cdot\boldsymbol{v}$$
$$+c\left[|\boldsymbol{v}_2|_4^8+|\boldsymbol{v}_{2\xi}|_2^2+\left(|\boldsymbol{v}_{2\xi}|_2^2+1\right)\int_\Omega\left(|\nabla_{e_\theta}\boldsymbol{v}_{2\xi}|^2+|\nabla_{e_\varphi}\boldsymbol{v}_{2\xi}|^2\right)\right]|\boldsymbol{v}|_2^2.$$

Similar to the above inequality,

$$\frac{1}{2}\frac{\mathrm{d}|T|_2^2}{\mathrm{d}t}+\int_\Omega|\nabla T|^2+\int_\Omega|T_\xi|^2+\alpha_s|T|_{\xi=1}|_2^2$$
$$\leq\int_\Omega\frac{bP}{p}T\left[W(\boldsymbol{v})\left(1+aq_1\right)+aqW(\boldsymbol{v}_2)\right]+3\varepsilon\|\boldsymbol{v}\|^2+3\varepsilon\|T\|^2$$
$$+c|T_2|_4^8\left(|T|_2^2+|\boldsymbol{v}|_2^2\right)+c\left[\left(|T_{2\xi}|_2^2+1\right)|\nabla T_{2\xi}|_2^2+|T_{2\xi}|_2^2\right]|T|_2^2,$$
$$\frac{1}{2}\frac{\mathrm{d}|q|_2^2}{\mathrm{d}t}+\int_\Omega|\nabla q|^2+\int_\Omega|q_\xi|^2+\beta_s|q|_{\xi=1}|_2^2$$
$$\leq 3\varepsilon\|\boldsymbol{v}\|^2+3\varepsilon\|q\|^2+c\left[\left(|q_{2\xi}|_2^2+1\right)|\nabla q_{2\xi}|_2^2+|q_{2\xi}|_2^2\right]|q|_2^2+c|q_2|_4^8\left(|q|_2^2+|\boldsymbol{v}|_2^2\right).$$

By integration by parts, we have

$$-\int_\Omega\left(\int_\xi^1\frac{bP}{p}\mathrm{grad}T\mathrm{d}\xi'\right)\cdot\boldsymbol{v}+\int_\Omega\frac{bP}{p}W(\boldsymbol{v})T=0,$$
$$-\int_\Omega\left(\int_\xi^1\frac{abP}{p}\mathrm{grad}\left(q_1T\right)\mathrm{d}\xi'\right)\cdot\boldsymbol{v}+\int_\Omega\frac{abP}{p}q_1W(\boldsymbol{v})T=0.$$

Thus, we obtain

$$\left|\int_\Omega\frac{abP}{p}qW(\boldsymbol{v}_2)T\right|=\left|\int_\Omega\int_\xi^1\frac{abP}{p}\mathrm{grad}\left(qT\right)\mathrm{d}\xi'\cdot\boldsymbol{v}_2\right|$$
$$\leq c|\boldsymbol{v}_2|_4|q|_4|\nabla T|_2+c|\boldsymbol{v}_2|_4|T|_4|\nabla q|_2\leq c|\boldsymbol{v}_2|_4^8\left(|q|_2^2+|T|_2^2\right)+2\varepsilon\|q\|^2+2\varepsilon\|T\|^2.$$

Choosing ε small enough, we have

$$\frac{\mathrm{d}\left(|\boldsymbol{v}|_2^2+|T|_2^2+|q|_2^2\right)}{\mathrm{d}t}+\int_\Omega\left(|\nabla_{e_\theta}\boldsymbol{v}|^2+|\nabla_{e_\varphi}\boldsymbol{v}|^2+|\boldsymbol{v}|^2\right)+\int_\Omega|\boldsymbol{v}_\xi|^2+\int_\Omega|\nabla T|^2$$
$$+\int_\Omega|T_\xi|^2+\alpha_s|T|_{\xi=1}|_2^2+\int_\Omega|\nabla q|^2+\int_\Omega|q_\xi|^2+\beta_s|q|_{\xi=1}|_2^2$$
$$\leq c\left[|\boldsymbol{v}_2|_4^8+|T_2|_4^8+|q_2|_4^8+|\boldsymbol{v}_{2\xi}|_2^2+\left(|\boldsymbol{v}_{2\xi}|_2^2+1\right)\int_\Omega\left(|\nabla_{e_\theta}\boldsymbol{v}_{2\xi}|^2+|\nabla_{e_\varphi}\boldsymbol{v}_{2\xi}|^2\right)\right]$$

$$\cdot |\boldsymbol{v}|_2^2 + c\left(|\boldsymbol{v}_2|_4^8 + |T_2|_4^8 + |T_{2\xi}|_2^2 + \left(|T_{2\xi}|_2^2 + 1\right)|\nabla T_{2\xi}|_2^2\right)|T|_2^2$$
$$+ c\left(|\boldsymbol{v}_2|_4^8 + |T_2|_4^2 + |T_2|_4^4 + |q_2|_4^8 + |q_{2\xi}|_2^2 + \left(|q_{2\xi}|_2^2 + 1\right)|\nabla q_{2\xi}|_2^2\right)|q|_2^2. \tag{3.2.73}$$

By Gronwall Inequality, Lemma 3.2.2 and (3.2.73), we prove Theorem 3.2.3.

3.2.5 *Some Preliminaries about the Infinite-Dimensional Dynamical System*

In this subsection, we introduce a very important concept in the infinite-dimensional dynamical system, that is, the global attractor based on semi-group, and recall two theorems about the existence of the global attractors.

Consider solutions of a differential equation

$$\frac{\mathrm{d}u(t)}{\mathrm{d}t} = F(u(t)) \tag{3.2.74}$$

with initial condition

$$u(0) = u_0 \tag{3.2.75}$$

and the asymptotic behavior of $u(t)$ as $t \to \infty$, where the unknown function $u = u(t)$ belongs to a linear space of a phase space H, and F is a map on H. There are two cases to consider:

(1) When $u = u(t) \in H = R^N$, (3.2.74) and (3.2.75) are called a finite-dimensional system.

(2) When $u = u(t) \in H = R^N$, where H is a infinite-dimensional Banach space, (3.2.74) and (3.2.75) are called an **infinite-dimensional dynamical system**.

Definition 3.2.17. Let E be a Banach space, and $S(t)$ be a semigroup operator, that is, $S(t) : E \to E, S(t+\tau) = S(t) \cdot S(\tau), \forall t, \tau \geq 0, S(0) = I$, which is the identity operator. If there exists a compact set $\mathcal{A} \subset E$, which satisfies the following properties:

(i) Invariant: that is,

$$S(t)\mathcal{A} = \mathcal{A}, \forall t \geq 0.$$

(ii) Attracting: $\mathcal{A}$ absorbs all bounded sets in E, that is, for any bounded set $B \subset E$,

$$\operatorname{dist}(S(t)B, \mathcal{A}) = \sup_{x \in B} \inf_{y \in \mathcal{A}} \|S(t)x - y\|_E \to 0, t \to \infty,$$

especially, as $t \to \infty$, all the trajectories $S(t)u_0$ started from u_0 converge to $\mathcal{A}$, i.e.,

$$\operatorname{dist}(S(t)u_0, \mathcal{A}) \to 0, t \to \infty,$$

then, the compact set $\mathcal{A}$ is called the **global attractor** of the semigroup $S(t)$.

In general, the structure of the global attractor is very complex, which includes not only the trivial u_0, $F(u_0) = 0$, which might be multiple solutions of the initial value problem of the nonlinear evolution equation

$$\frac{\mathrm{d}u(t)}{\mathrm{d}t} = F(u(t)), \quad u(0) = u_0,$$

but also periodic orbits, quasi-periodic orbits, fractal and stranger attractors, it might not be smooth, but has non-integer dimensions.

Before recalling the existence theorems of the global attractors, we introduce the concept of absorbing sets.

Theorem 3.2.18. For a bounded set $B_0 \subset E$, if there exists $t_0(B_0) > 0$ such that for any bounded set $B \subset E$,

$$S(t)B \subset B_0, \ \forall t \geq t_0,$$

then, B_0 is denoted as a **bounded absorbing set** in E.

Theorem 3.2.19 (cf.[8] or [201, Theorem I.1.1]). Let E be a Banach space, and $\{S(t), t \geq 0\}$ be a semigroup operator, $S(t) : E \to E$, $S(t+\tau) = S(t) \cdot S(\tau), t, \tau \geq 0, S(0) = I$, where I is the identity operator. Suppose that the semigroup operator $S(t)$ satisfies the following conditions.

(i) The semigroup operator $S(t)$ is uniformly bounded in E, that is, for any $R > 0$, there exists a constant $C(R)$, such that for $\|u\|_E \leq R$,

$$\|S(t)u\|_E \leq C(R), \ \forall t \in [0, \infty).$$

(ii) There exists a bounded absorbing set B_0 in E.

(iii) For any $t > 0$, $S(t)$ is a completely continuous operator,then the semigroup $S(t)$ has a completely global attractor $\mathcal{A}$.

Remark 3.2.20. If the bounded absorbing set B_0 in condition (ii) is replaced by a compact absorbing set B_0, then the complete continuity of $S(t)$ in condition (iii) is replaced by a continuous operator, and Theorem 3.2.19 is still valid.

Remark 3.2.21. We can prove that the above global attractor $\mathcal{A}$ is the ω limit set of the absorbing set B_0, that is,

$$\mathcal{A} = \omega(B_0) = \bigcup_{s \geq 0} \overline{\bigcup_{t \geq s} S(t)B_0},$$

where the closures are taken on E.

Another useful theorem for proving the existence of global attractors is as follows.

Theorem 3.2.22. Let E be a Banach space. If the semigroup operator $S(t)$ is continuous, there exists an open set $\mathcal{U} \subset E$ and a bounded set B in $\mathcal{U}$, such that B is absorbing in $\mathcal{U}$, and $S(t)$ satisfies one of the following two conditions:

(i) The operator $S(t)$ is uniformly compact for t large enough, that is, for any bounded set B, there exists $t = t_0(B)$, such that

$$\bigcup_{t \geq t_0} S(t)B \tag{3.2.76}$$

is relatively compact in E.

(ii) $S(t) = S_1(t) + S_2(t)$, where $S_1(\cdot)$ is uniformly compact for t large enough, which means $S(t)$ satisfies the condition (3.2.76), and operator $S_2(t)$ is a continuous mapping, $S_2(t) : E \to E$, and for every bounded set $B \subset E$, there exists

$$r_B(t) = \sup_{\varphi \in \bar{B}} \|S_2(t)\varphi\|_E \to 0, \tag{3.2.77}$$

then, the $\mathcal{A} = \omega(B)$, ω limit of $\mathcal{B}$, is a compact attractor, which also absorbs bounded sets of $\mathcal{U}$. It's also the maximal bounded attractor in $\mathcal{U}$. Moreover, if $\mathcal{U}$ is convex and connected, then $\mathcal{A}$ is connected.

In order to prove the existence of global attractors, one may verify the assumptions in Theorem 3.2.19 or Theorem 3.2.22. In applications, we mainly prove the following three conditions:

(1) the existence of the semigroup operator $S(t)$,

(2) the existence of a bounded or compact absorbing set,

(3)$S(t)(t > 0)$ is a completely continuous operator or satisfies conditions (3.2.76) or (3.2.77).

3.2.6 *The Existence of Global Attractors*

Proof of Proposition 3.2.4. Applying (3.2.35), (3.2.57), (3.2.59), (3.2.61), (3.2.63), (3.2.68) and (3.2.69), we know $\boldsymbol{U} \in L^\infty(0, \infty; V)$. By Theorem 3.2.2 and Theorem 3.2.3, we define a semigroup $\{S(t)\}_{t \geq 0}$ corresponding to the system (3.2.11)-(3.2.16), $S(t) : V \to V, S(t)\boldsymbol{U}_0 = \boldsymbol{U}(t)$. Applying (3.2.33), (3.2.35), (3.2.37), (3.2.40), (3.2.43), (3.2.48), (3.2.50), (3.2.51), (3.2.54), (3.2.57), (3.2.58), (3.2.61), (3.2.62) and (3.2.68), we prove that $\{S(t)\}_{t \geq 0}$ exists a bounded absorbing set B_δ in V, that is, for any $\boldsymbol{U}_0 \in V$, there exists t_0 big enough, such that $S(t)\boldsymbol{U}_0 \in B_\delta, \quad \forall\, t \geq t_0$, where $B_\delta = \{\boldsymbol{U};\ \boldsymbol{U} \in V,\ \|\boldsymbol{U}\| \leq \delta\}$, and δ is a positive constant depending on $\|Q_1\|_1$ and $\|Q_2\|_1$.

In order to prove Theorem 3.2.5, we need to use the following property of the semigroup $\{S(t)\}_{t\geq 0}$.

Proposition 3.2.23. For any $t \geq 0$, the mapping $S(t)$ is weakly continuous on V.

Proof of Proposition 3.2.23. Let $\{\boldsymbol{U}_n\}$ be a sequence in V, where $\boldsymbol{U}_n$ weakly converges to $\boldsymbol{U}$ in V. Then $\{\boldsymbol{U}_n\}$ is bounded in V. According to the *a priori* estimates in subsection 3.2.4, $\{S(t)\boldsymbol{U}_n\}$ is bounded in V, for any $t \geq 0$ with which we subtract a subsequence $S(t)\boldsymbol{U}_{n_k}$ weakly converging to u in V. Since the imbedding $V \hookrightarrow L^2(T\Omega|TS^2) \times L^2(\Omega) \times L^2(\Omega)$ is compact, $\boldsymbol{U}_{n_k}$ strongly converges to $\boldsymbol{U}$ in $L^2(T\Omega|TS^2) \times L^2(\Omega) \times L^2(\Omega)$. According to (3.2.73), we know that $S(t)\boldsymbol{U}_{n_k}$ strongly converges to $S(t)U$ in $L^2(T\Omega|TS^2) \times L^2(\Omega) \times L^2(\Omega)$. Thus, $u = S(t)\boldsymbol{U}$. Thus, the sequence $\{S(t)\boldsymbol{U}_n\}$ satisfies that $S(t)\boldsymbol{U}_{n_k}$ weakly converges to $S(t)\boldsymbol{U}$ in V.

Proof of Theorem 3.2.5. With Proposition 3.2.4 and Proposition 3.2.23, we note that the proof of Theorem 3.2.5 is similar to that of Theorem I.1.1 in [201]. We only replace "strongly converge in H" in Theorem I.1.1 into "weakly converge in V". Here the details of the proof of Theorem 3.2.5 is omitted.

3.3 The Global Well-Posedness of the Primitive Equations

In this section, we consider the well-posedness of the primitive equations of the large-scale dry atmosphere based on the results of the above two sections. The main results of this section are Theorem 3.3.2 and Theorem 3.3.3. Firstly, let's prove the global well-posedness of the primitive equations in weaker initial conditions compared to the above section. Next, we obtain the global existence of the smooth solutions of the primitive equations. Moreover, we obtain a compact global attractor of the dynamic system corresponding to the boundary value problem of the primitive equations. Here we omit the water vapor equation, but the results obtained here are still valid to the moist atmospheric primitive equations.

The arrangement of this section is as follows. In subsection 3.3.1, we give the main results. We shall prove the global well-posedness of the primitive equations in subsection 3.3.2. At last, we prove the global existence of the smooth solutions to the primitive equations.

3.3.1 Main Results

First, we present the initial boundary value problem IBVP of the primitive equations of the large-scale dry atmosphere:

$$\frac{\partial \boldsymbol{v}}{\partial t}+\boldsymbol{\nabla}_{\boldsymbol{v}}\boldsymbol{v}+W(\boldsymbol{v})\frac{\partial \boldsymbol{v}}{\partial \xi}+f\boldsymbol{k}\times\boldsymbol{v}+\mathrm{grad}\,\Phi_s+\int_\xi^1\frac{bP}{p}\mathrm{grad}T\,\mathrm{d}\xi'-\Delta\boldsymbol{v}-\frac{\partial^2\boldsymbol{v}}{\partial\xi^2}$$
$$=0, \tag{3.3.1}$$

$$\frac{\partial T}{\partial t}+\boldsymbol{\nabla}_{\boldsymbol{v}}T+W(\boldsymbol{v})\frac{\partial T}{\partial \xi}-\frac{bP}{p}W(\boldsymbol{v})-\Delta T-\frac{\partial^2 T}{\partial\xi^2}=Q, \tag{3.3.2}$$

$$\int_0^1\mathrm{div}\boldsymbol{v}\,\mathrm{d}\xi=0, \tag{3.3.3}$$

$$\xi=1:\ \frac{\partial \boldsymbol{v}}{\partial\xi}=0,\ \frac{\partial T}{\partial\xi}=-\alpha_s T, \tag{3.3.4}$$

$$\xi=0:\ \frac{\partial \boldsymbol{v}}{\partial\xi}=0,\ \frac{\partial T}{\partial\xi}=0, \tag{3.3.5}$$

$$U|_{t=0}=(\boldsymbol{v}|_{t=0},T|_{t=0})=U_0=(\boldsymbol{v}_0,T_0). \tag{3.3.6}$$

Let the viscosity coefficient be 1 for simplicity.

We firstly introduce a definition of weakly strong solutions of IBVP. Let $\bar{\boldsymbol{v}}=\int_0^1\boldsymbol{v}\mathrm{d}\xi$, and a baroclinic flow $\tilde{\boldsymbol{v}}$ be given by

$$\tilde{\boldsymbol{v}}=\boldsymbol{v}-\bar{\boldsymbol{v}}.$$

Then $\tilde{\boldsymbol{v}}$ satisfies the following boundary value problem:

$$\frac{\partial \tilde{\boldsymbol{v}}}{\partial t}+\boldsymbol{\nabla}_{\tilde{\boldsymbol{v}}}\tilde{\boldsymbol{v}}+\left(\int_\xi^1\mathrm{div}\tilde{\boldsymbol{v}}\mathrm{d}\xi'\right)\frac{\partial\tilde{\boldsymbol{v}}}{\partial\xi}+\boldsymbol{\nabla}_{\tilde{\boldsymbol{v}}}\bar{\boldsymbol{v}}+\boldsymbol{\nabla}_{\bar{\boldsymbol{v}}}\tilde{\boldsymbol{v}}-\overline{(\tilde{\boldsymbol{v}}\mathrm{div}\tilde{\boldsymbol{v}}+\boldsymbol{\nabla}_{\tilde{\boldsymbol{v}}}\tilde{\boldsymbol{v}})}+\frac{f}{R_0}\boldsymbol{k}\times\tilde{\boldsymbol{v}}$$
$$+\int_\xi^1\frac{bP}{p}\mathrm{grad}T\mathrm{d}\xi'-\int_0^1\int_\xi^1\frac{bP}{p}\mathrm{grad}T\mathrm{d}\xi'\mathrm{d}\xi-\Delta\tilde{\boldsymbol{v}}-\frac{\partial^2\tilde{\boldsymbol{v}}}{\partial\xi^2}=0\ \text{ in } \Omega, \tag{3.3.7}$$

$$\xi=1:\ \frac{\partial\tilde{\boldsymbol{v}}}{\partial\xi}=0,\ \xi=0:\ \frac{\partial\tilde{\boldsymbol{v}}}{\partial\xi}=0. \tag{3.3.8}$$

Let

$$V=V_1\times V_2,\quad H=H_1\times H_2.$$

The definitions of V_1, V_2, H_1, and H_2 are the same as those given in the last section. Then, we give the definition of the weakly strong solution of IBVP.

Definition 3.3.1. Suppose that $Q\in H^1(\Omega)$, $\boldsymbol{U}_0=(\boldsymbol{v}_0,T_0)$ satisfies conditions: $\boldsymbol{U}_0\in H$, $\tilde{\boldsymbol{v}}_0\in L^4(T\Omega|TS^2)$, $T_0\in L^4(\Omega)$, $\partial_\xi\boldsymbol{v}_0\in$

$L^2(T\Omega|TS^2)$, $\partial_\xi T_0 \in L^2(\Omega)$, and $\mathcal{T}$ is a given positive time. $\boldsymbol{U} = (\boldsymbol{v}, T)$ is called **the weakly strong solution** of the initial boundary value problem (3.3.1)-(3.3.6) in $[0, \mathcal{T}]$, if (v, T) satisfies (3.3.1)-(3.3.3) in weak sense, and

$$\boldsymbol{v} \in L^2(0,\mathcal{T};V_1) \cap L^\infty(0,\mathcal{T};H_1),$$
$$\tilde{\boldsymbol{v}} \in L^\infty(0,\mathcal{T};L^4(T\Omega|TS^2)),$$
$$\partial_\xi \boldsymbol{v} \in L^\infty(0,\mathcal{T};L^2(T\Omega|TS^2)) \cap L^2(0,\mathcal{T};H^1(T\Omega|TS^2)),$$
$$T \in L^\infty(0,\mathcal{T};L^4(\Omega)) \cap L^2(0,\mathcal{T};V_2),$$
$$\partial_\xi T \in L^\infty(0,\mathcal{T};L^2(\Omega)) \cap L^2(0,\mathcal{T};H^1(\Omega)).$$
$$\frac{\partial \boldsymbol{v}}{\partial t} \in L^2(0,V_1'),\ \frac{\partial T}{\partial t} \in L^2(0,\mathcal{T};V_2'),$$

where V_i' is the dual space of $V_i (i = 1, 2)$.

Now we give the main results of this section.

Theorem 3.3.2 (The global existence and uniqueness of weakly strong solution of IBVP). If $Q \in H^1(\Omega)$, $\boldsymbol{U}_0 = (\boldsymbol{v}_0, T_0)$ satisfies conditions: $\boldsymbol{U}_0 \in H$, $\tilde{\boldsymbol{v}}_0 \in L^4(T\Omega|TS^2)$, $T_0 \in L^4(\Omega)$, $\partial_\xi \boldsymbol{v}_0 \in L^2(T\Omega|TS^2)$, $\partial_\xi T_0 \in L^2(\Omega)$, then for any $\mathcal{T} > 0$ given, the initial boundary value problem (3.3.1)-(3.3.6) has a unique weakly strong solution $\boldsymbol{U}$ on $[0, \mathcal{T}]$.

Theorem 3.3.3 (The global existence of the smooth solutions of IBVP). Let $Q \in H^1(\Omega)$, and $\boldsymbol{U}_0 \in V \cap (H^2(T\Omega|TS^2) \times H^2(\Omega))$. Then, for any $\mathcal{T} > 0$ given, IBVP has a strong solution $\boldsymbol{U}$ on $[0, \mathcal{T}]$, such that $\boldsymbol{U} \in L^\infty(0, \mathcal{T}; V \cap (H^2(T\Omega|TS^2) \times H^2(\Omega)))$. Moreover, if $Q \in C^\infty(\Omega)$, $\boldsymbol{U}_0 = (\boldsymbol{v}_0, T_0) \in V \cap (C^\infty(T\Omega|TS) \times C^\infty(\Omega))$, then for any $\mathcal{T} > 0$ given, IBVP has a smooth solution $\boldsymbol{U}$ on $[0, \mathcal{T}]$.

Remark 3.3.4. Theorem 3.3.3 only gives the global existence of smooth solutions of the atmospheric primitive equations under the hydrostatic approximation and with the ideal boundary conditions. But, the upper and lower boundary conditions are very complex in applications. People still can't prove the global well-posedness problem of the atmospheric primitive equations under general conditions (such as containing vacuum, that is the density of the upper boundary is zero).

3.3.2 *The Global Well-Posedness of IBVP*

3.3.2.1 *The Global Existence of the Weakly Strong Solutions*

We use Faedo-Galerkin method to prove the global existence of the weak strong solutions of IBVP. Since the proof is similar to the proof of the global existence of the weak solutions of the three-dimensional incompressible Navier-Stokes equation, we only give some *a priori* estimate of approximate solutions.

As is in subsection 3.2.2, we get the following estimates:

$$c_1 \int_t^{t+r} [\|T\|^2 + |T|_{\xi=1}|_2^2] + |T(t)|_2^2 \leq 2\mathrm{e}^{-c_0 t}|T_0|_2^2 + 3c|Q|_2^2 \leq E_1, \quad (3.3.9)$$

$$\int_t^{t+r} (\|\boldsymbol{v}\|^2) + |\boldsymbol{v}(t)|_2^2 \leq 2\mathrm{e}^{-ct}|\boldsymbol{v}_0|_2^2 + cE_1 \leq E_2, \quad (3.3.10)$$

$$c_1 \int_t^{t+r} (\|\bar{\boldsymbol{v}}\|_{H^1}^2) + \|\bar{\boldsymbol{v}}(t)\|_{L^2}^2 \leq E_2, \quad (3.3.11)$$

which means

$$\int_t^{t+r} |\bar{\boldsymbol{v}}|_4^4 = \int_t^{t+r} \|\bar{\boldsymbol{v}}\|_{L^4}^4 \leq \int_t^{t+r} \|\bar{\boldsymbol{v}}\|_{L^2}^2 \|\bar{\boldsymbol{v}}\|_{H^1}^2 \leq cE_2^2, \quad (3.3.12)$$

$$|T(t)|_4^4 + c_1 \int_t^{t+r} |T|_{\xi=1}|_4^4 \leq 3E_3, \quad (3.3.13)$$

$$|\tilde{\boldsymbol{v}}(t)|_4^4 \leq E_4, \quad (3.3.14)$$

where $c_0 = \min\{\frac{1}{2}, \frac{\alpha_s}{2}\}$, $c_1 = \min\{1, \frac{1}{3}, \frac{\alpha_s}{2}\}$, $t \geq 0$, $1 \geq r > 0$ is given, E_i is a positive constant, and $\int_t^{t+r} \cdot$ represents $\int_t^{t+r} \cdot \mathrm{d}s$.

L^2 estimates of $\boldsymbol{v}_\xi$. Taking the derivative of equation (3.3.1) with respect to ξ, we have

$$\frac{\partial \boldsymbol{v}_\xi}{\partial t} - \Delta \boldsymbol{v}_\xi - \frac{\partial^2 \boldsymbol{v}_\xi}{\partial \xi^2} + \boldsymbol{\nabla}_{\boldsymbol{v}} \boldsymbol{v}_\xi + \left(\int_\xi^1 \mathrm{div}\boldsymbol{v}\right) \frac{\partial \boldsymbol{v}_\xi}{\partial \xi} + \boldsymbol{\nabla}_{\boldsymbol{v}_\xi} \boldsymbol{v} - (\mathrm{div}\boldsymbol{v}) \frac{\partial \boldsymbol{v}}{\partial \xi}$$
$$+ f\boldsymbol{k} \times \boldsymbol{v}_\xi - \frac{bP}{p} \mathrm{grad}T = 0. \quad (3.3.15)$$

Taking $L^2(\Omega) \times L^2(\Omega)$ inner product of the above equation with $\boldsymbol{v}_\xi$, we obtain

$$\frac{1}{2}\frac{\mathrm{d}|\boldsymbol{v}_\xi|_2^2}{\mathrm{d}t} + \int_\Omega (|\boldsymbol{\nabla}_{e_\theta} \boldsymbol{v}_\xi|^2 + |\boldsymbol{\nabla}_{e_\varphi} \boldsymbol{v}_\xi|^2 + |\boldsymbol{v}_\xi|^2) + \int_\Omega \left|\frac{\partial \boldsymbol{v}_\xi}{\partial \xi}\right|^2$$
$$= - \int_\Omega \left[\boldsymbol{\nabla}_{\boldsymbol{v}} \boldsymbol{v}_\xi + \left(\int_\xi^1 \mathrm{div}\boldsymbol{v}\right) \frac{\partial \boldsymbol{v}_\xi}{\partial \xi}\right] \cdot \boldsymbol{v}_\xi - \int_\Omega \left[\boldsymbol{\nabla}_{\boldsymbol{v}_\xi} \boldsymbol{v} - (\mathrm{div}\boldsymbol{v}) \frac{\partial \boldsymbol{v}}{\partial \xi}\right] \cdot \boldsymbol{v}_\xi$$
$$- \int_\Omega (f\boldsymbol{k} \times \boldsymbol{v}_\xi) \cdot \boldsymbol{v}_\xi + \int_\Omega \frac{bP}{p} \mathrm{grad}T \cdot \boldsymbol{v}_\xi.$$

By integration by parts, Hölder Inequality and Young Inequality, we get

$$- \int_\Omega \left[\boldsymbol{\nabla}_{\boldsymbol{v}_\xi} \boldsymbol{v} - (\mathrm{div}\boldsymbol{v}) \frac{\partial \boldsymbol{v}}{\partial \xi}\right] \cdot \boldsymbol{v}_\xi$$

$$\leq c\int_\Omega (|\tilde{\boldsymbol{v}}| + |\bar{\boldsymbol{v}}|)\,|\boldsymbol{v}_\xi|\left(|\boldsymbol{\nabla}_{e_\theta}\boldsymbol{v}_\xi|^2 + |\boldsymbol{\nabla}_{e_\varphi}\boldsymbol{v}_\xi|^2\right)^{\frac{1}{2}}$$

$$\leq c|\tilde{\boldsymbol{v}}|_4|\boldsymbol{v}_\xi|_4\left[\int_\Omega \left(|\boldsymbol{\nabla}_{e_\theta}\boldsymbol{v}_\xi|^2 + |\boldsymbol{\nabla}_{e_\varphi}\boldsymbol{v}_\xi|^2\right)\right]^{\frac{1}{2}}$$

$$+ c\int_{S^2} |\bar{\boldsymbol{v}}|\left[\int_0^1 |\boldsymbol{v}_\xi|\left(|\boldsymbol{\nabla}_{e_\theta}\boldsymbol{v}_\xi|^2 + |\boldsymbol{\nabla}_{e_\varphi}\boldsymbol{v}_\xi|^2\right)^{\frac{1}{2}}\right]$$

$$\leq c|\tilde{\boldsymbol{v}}|_4|\boldsymbol{v}_\xi|_2^{\frac{1}{4}}\|\boldsymbol{v}_\xi\|^{\frac{7}{4}} + c|\bar{\boldsymbol{v}}|_4\left[\int_0^1\left(\int_{S^2}|\boldsymbol{v}_\xi|^4\right)^{\frac{1}{2}}\right]^{\frac{1}{2}}\|\boldsymbol{v}_\xi\|$$

$$\leq \varepsilon\|\boldsymbol{v}_\xi\|^2 + c\left(|\tilde{\boldsymbol{v}}|_4^8 + \|\bar{\boldsymbol{v}}\|_{L^4}^4\right)|\boldsymbol{v}_\xi|_2^2.$$

Choosing ε small enough, we derive from the above two relationships that

$$\frac{\mathrm{d}|\boldsymbol{v}_\xi|_2^2}{\mathrm{d}t} + \int_\Omega \left(|\boldsymbol{\nabla}_{e_\theta}\boldsymbol{v}_\xi|^2 + |\boldsymbol{\nabla}_{e_\varphi}\boldsymbol{v}_\xi|^2 + |\boldsymbol{v}_\xi|^2\right) + \int_\Omega \left|\frac{\partial \boldsymbol{v}_\xi}{\partial \xi}\right|^2$$
$$\leq c\left(|\tilde{\boldsymbol{v}}|_4^8 + \|\bar{\boldsymbol{v}}\|_{L^4}^4\right)|\boldsymbol{v}_\xi|_2^2 + c|T|_2^2. \tag{3.3.16}$$

Applying the uniform Gronwall Inequality, (3.3.9), (3.3.10), (3.3.12) and (3.3.14), we deduce from the above equality that

$$|\boldsymbol{v}_\xi(t+r)|_2^2 \leq E_5, \tag{3.3.17}$$

where E_5 is a positive constant, and $t \geq 0$. According to (3.3.16) and (3.3.17), we have

$$c_1\int_{t+r}^{t+2r} \|\boldsymbol{v}_\xi\|^2 \leq E_5^2 + E_5 = E_6. \tag{3.3.18}$$

By Gronwall Inequality, we derive from (3.3.16)

$$|\boldsymbol{v}_\xi(t)|_2^2 \leq c, \text{ for } 0 \leq t < r. \tag{3.3.19}$$

L^2 estimates of T_ξ. Taking the derivative of equation (3.3.2) with respect to ξ, we get

$$\frac{\partial T_\xi}{\partial t} - \Delta T_\xi - \frac{\partial^2 T_\xi}{\partial \xi^2} + \boldsymbol{\nabla}_{\boldsymbol{v}}T_\xi + W(\boldsymbol{v})\frac{\partial T_\xi}{\partial \xi} + \boldsymbol{\nabla}_{\boldsymbol{v}_\xi}T - (\mathrm{div}\boldsymbol{v})\frac{\partial T}{\partial \xi}$$
$$+ \frac{bP}{p}\mathrm{div}\boldsymbol{v} + \frac{bP(P-p_0)}{p^2}W(\boldsymbol{v}) = Q_\xi. \tag{3.3.20}$$

Taking $L^2(\Omega)$ inner product of the above equation with T_ξ, we have

$$\frac{1}{2}\frac{\mathrm{d}|T_\xi|_2^2}{\mathrm{d}t} + \int_\Omega |\boldsymbol{\nabla}T_\xi|^2 + \int_\Omega |T_{\xi\xi}|^2 - \int_{S^2}(T_\xi|_{\xi=1}\cdot T_{\xi\xi}|_{\xi=1})$$

$$
\begin{aligned}
&= -\int_\Omega \left[\nabla_{\boldsymbol{v}} T_\xi + W(\boldsymbol{v})\frac{\partial T_\xi}{\partial \xi}\right] T_\xi - \int_\Omega \left[\nabla_{\boldsymbol{v}_\xi} T - (\mathrm{div}\boldsymbol{v})\frac{\partial T}{\partial \xi}\right] T_\xi + \int_\Omega Q_\xi T_\xi \\
&\quad + \int_\Omega \left[-\frac{bP}{p}(\mathrm{div}\boldsymbol{v}) - \frac{bP(P-p_0)}{p} W(\boldsymbol{v})\right] T_\xi.
\end{aligned}
$$

By integration by parts, Hölder Inequality, Poincaré Inequality and Young Inequality, we obtain

$$
\begin{aligned}
&\left|\int_\Omega \left[\nabla_{\boldsymbol{v}_\xi} T - \mathrm{div}(\boldsymbol{v})\frac{\partial T}{\partial \xi}\right] T_\xi\right| \\
\le & c\int_\Omega \left[(|\nabla_{e_\theta}\boldsymbol{v}_\xi|^2 + |\nabla_{e_\varphi}\boldsymbol{v}_\xi|^2)^{\frac{1}{2}}|T||T_\xi| + |\boldsymbol{v}_\xi||T||\nabla T_\xi| + (|\tilde{\boldsymbol{v}}| + |\bar{\boldsymbol{v}}|)|\nabla T_\xi||T_\xi|\right] \\
\le & \varepsilon(|T_{\xi\xi}|_2^2 + |\nabla T_\xi|_2^2) + c\left[|\boldsymbol{v}_{\xi\xi}|_2^2 + \int_\Omega (|\nabla_{e_\theta}\boldsymbol{v}_\xi|^2 + |\nabla_{e_\varphi}\boldsymbol{v}_\xi|^2)\right] + c|T|_4^8|v_\xi|_2^2 \\
&+ c(|T|_4^8 + |\tilde{\boldsymbol{v}}|_4^8 + \|\bar{\boldsymbol{v}}\|_{L^4}^4)|T_\xi|_2^2.
\end{aligned}
$$

Similar to the above inequality,

$$
\left|\int_\Omega \left[-\frac{bP}{p}(\mathrm{div}\boldsymbol{v}) - \frac{bP(P-p_0)}{p}W(\boldsymbol{v})\right] T_\xi\right| \le \varepsilon\left|\nabla T_\xi\right|_2^2 + c\left|T_\xi\right|_2^2 + c\|\boldsymbol{v}\|^2.
$$

Since the trace of equation (3.3.2) on $\xi = 1$ is

$$
T_{\xi\xi}|_{\xi=1} = \frac{\partial T|_{\xi=1}}{\partial t} + (\nabla_{\boldsymbol{v}} T)|_{\xi=1} - \Delta T|_{\xi=1} - Q|_{\xi=1},
$$

by the boundary condition (3.3.4), we have

$$
\begin{aligned}
&-\int_{S^2} (T_\xi|_{\xi=1} T_{\xi\xi}|_{\xi=1}) \\
=&\alpha_s \int_{S^2} T|_{\xi=1}\left[\frac{\partial T|_{\xi=1}}{\partial t} + (\nabla_{\boldsymbol{v}} T)|_{\xi=1} - \Delta T|_{\xi=1} - Q|_{\xi=1}\right] \\
=&\alpha_s(\frac{1}{2}\frac{\mathrm{d}|T(\xi=1)|_2^2}{\mathrm{d}t} + |\nabla T(\xi=1)|_2^2) + \alpha_s\int_{S^2} T|_{\xi=1}\left[(\nabla_{\boldsymbol{v}} T)|_{\xi=1} - Q|_{\xi=1}\right].
\end{aligned}
$$

By integration by parts, we have

$$
\begin{aligned}
&-\alpha_s \int_{S^2} T|_{\xi=1}((\nabla_{\boldsymbol{v}} T)|_{\xi=1} - Q|_{\xi=1}) \\
=&\frac{\alpha_s}{2}\int_{S^2} T^2|_{\xi=1}\mathrm{div}\boldsymbol{v}|_{\xi=1} + \alpha_s\int_{S^2}(TQ)|_{\xi=1} \\
=&\frac{\alpha_s}{2}\int_{S^2} T^2|_{\xi=1}\left(\int_\xi^1 \mathrm{div}\boldsymbol{v}_\xi \mathrm{d}\xi' + \mathrm{div}\boldsymbol{v}\right) + \alpha_s\int_{S^2} T|_{\xi=1}Q|_{\xi=1} \\
\le& c|T(\xi=1)|_4^4 + c\|\boldsymbol{v}_\xi\|^2 + c\|\boldsymbol{v}\|^2 + c|T(\xi=1)|_2^2 + c|Q(\xi=1)|_2^2.
\end{aligned}
$$

Thus, choosing ε small enough, we obtain

$$\frac{\mathrm{d}(|T_\xi|_2^2+\alpha_s|T(\xi=1)|_2^2)}{\mathrm{d}t}+\int_\Omega|\nabla T_\xi|^2+\int_\Omega|T_{\xi\xi}|^2+\alpha_s|\nabla T(\xi=1)|_2^2$$
$$\leq c(1+|T|_4^8+|\tilde{\boldsymbol{v}}|_4^8+\|\bar{\boldsymbol{v}}\|_{L^4}^4)|T_\xi|_2^2+c\|\boldsymbol{v}_\xi\|^2+c\|\boldsymbol{v}\|^2+c|T|_4^8|\boldsymbol{v}_\xi|_2^2$$
$$+c|T(\xi=1)|_4^4+c|T(\xi=1)|_2^2+c|Q(\xi=1)|_2^2+c|Q_\xi|_2^2. \tag{3.3.21}$$

Applying the uniform Gronwall Inequality, (3.3.9), (3.3.10), (3.3.12)-(3.3.14) and (3.3.18), we derive from (3.3.21) that

$$|T_\xi(t+2r)|_2^2\leq E_7, \tag{3.3.22}$$

where E_7 is a positive constant. Combining (3.3.21) with (3.3.22), we have

$$c_1\int_{t+2r}^{t+3r}\|T_\xi\|^2\leq E_7^2+2E_7+E_1=E_8. \tag{3.3.23}$$

By Gronwall Inequality and (3.3.21), we get

$$|T_\xi(t)|_2^2\leq c, \tag{3.3.24}$$

where $0\leq t<2r$.

According to the above *a priori* estimates, we obtain the following results.

Proposition 3.3.5 (long-time behavior of weakly strong solutions). If $\boldsymbol{U}$ is a weakly strong solution of initial boundary value problem (3.3.1)-(3.3.6), then $\boldsymbol{U}$ satisfies: $\boldsymbol{U}\in L^\infty(0,\infty;H)$, $\partial_\xi\boldsymbol{v}\in L^\infty(0,\infty;L^2(T\Omega|TS^2))$, $\tilde{\boldsymbol{v}}\in L^\infty(0,\infty;L^4(T\Omega|TS^2))$, $T\in L^\infty(0,\infty;L^4(\Omega))$, $\partial_\xi T\in L^\infty(0,\infty;L^2(\Omega))$.

3.3.2.2 *The Uniqueness of Weakly Strong Solution*

Suppose that $(\boldsymbol{v}_1,T_1)$ and $(\boldsymbol{v}_2,T_2)$ are two weakly strong solutions of system (3.3.1)-(3.3.6) in interval $[0,\mathcal{T}]$ respectively corresponding to Φ_{s_1}, Φ_{s_2}, and initial data $((\boldsymbol{v}_0)_1,(T_0)_1)$, $((\boldsymbol{v}_0)_2,(T_0)_2)$. Let $\boldsymbol{v}=\boldsymbol{v}_1-\boldsymbol{v}_2$, $T=T_1-T_2$, $\Phi_s=\Phi_{s_1}-\Phi_{s_2}$. Then $\boldsymbol{v}$, T, Φ_s satisfy the following system

$$\frac{\partial\boldsymbol{v}}{\partial t}-\Delta\boldsymbol{v}-\frac{\partial^2\boldsymbol{v}}{\partial\xi^2}+\nabla_{\boldsymbol{v}_1}\boldsymbol{v}+\nabla_{\boldsymbol{v}}\boldsymbol{v}_2+W(\boldsymbol{v}_1)\frac{\partial\boldsymbol{v}}{\partial\xi}+W(\boldsymbol{v})\frac{\partial\boldsymbol{v}_2}{\partial\xi}+f\boldsymbol{k}\times\boldsymbol{v}$$
$$+\operatorname{grad}\Phi_s+\int_\xi^1\frac{bP}{p}\operatorname{grad}T\mathrm{d}\xi'=0, \tag{3.3.25}$$

$$\frac{\partial T}{\partial t}-\Delta T-\frac{\partial^2T}{\partial\xi^2}+\nabla_{\boldsymbol{v}_1}T+\nabla_{\boldsymbol{v}}T_2+W(\boldsymbol{v}_1)\frac{\partial T}{\partial\xi}+W(\boldsymbol{v})\frac{\partial T_2}{\partial\xi}-\frac{bP}{p}W(\boldsymbol{v})=0, \tag{3.3.26}$$

$$\boldsymbol{v}|_{t=0}=(\boldsymbol{v}_0)_1-(\boldsymbol{v}_0)_2, \tag{3.3.27}$$

$$T|_{t=0} = (T_0)_1 - (T_0)_2, \tag{3.3.28}$$

$$\xi = 1: \ \frac{\partial \boldsymbol{v}}{\partial \xi} = 0, \ \frac{\partial T}{\partial \xi} = -\alpha_s T; \quad \mathrm{d}\xi = 0: \ \frac{\partial \boldsymbol{v}}{\partial \xi} = 0, \ \frac{\partial T}{\partial \xi} = 0. \tag{3.3.29}$$

Taking $L^2(\Omega) \times L^2(\Omega)$ inner product of equation (3.3.25) with $\boldsymbol{v}$, we get

$$\begin{aligned}
&\frac{1}{2}\frac{\mathrm{d}|\boldsymbol{v}|_2^2}{\mathrm{d}t} + \int_\Omega (|\nabla_{e_\theta}\boldsymbol{v}|^2 + |\nabla_{e_\varphi}\boldsymbol{v}|^2 + |\boldsymbol{v}|^2) + \int_\Omega |\boldsymbol{v}_\xi|^2 \\
=& -\int_\Omega \left[\nabla_{\boldsymbol{v}_1}\boldsymbol{v} + W(\boldsymbol{v}_1)\frac{\partial \boldsymbol{v}}{\partial \xi}\right]\cdot \boldsymbol{v} - \int_\Omega \boldsymbol{v}\cdot\left(\nabla_{\boldsymbol{v}}\boldsymbol{v}_2 + W(\boldsymbol{v})\frac{\partial \boldsymbol{v}_2}{\partial \xi}\right) \\
&- \int_\Omega \left(\frac{f}{R_0}\boldsymbol{k}\times\boldsymbol{v} + \mathrm{grad}\,\Phi_s\right)\cdot\boldsymbol{v} - \int_\Omega \left(\int_\xi^1 \frac{bP}{p}\mathrm{grad}T\right)\cdot\boldsymbol{v}.
\end{aligned}$$

By integration by parts, we have

$$\begin{aligned}
&\left|\int_\Omega \boldsymbol{v}\cdot\nabla_{\boldsymbol{v}}\boldsymbol{v}_2\right| \le c\int_\Omega (|\tilde{\boldsymbol{v}}_2| + |\bar{\boldsymbol{v}}_2|)|\boldsymbol{v}|(|\nabla_{e_\theta}\boldsymbol{v}|^2 + |\nabla_{e_\varphi}\boldsymbol{v}|^2)^{\frac{1}{2}} \\
\le& \varepsilon\int_\Omega (|\nabla_{e_\theta}\boldsymbol{v}|^2 + |\nabla_{e_\varphi}\boldsymbol{v}|^2) + c|\tilde{\boldsymbol{v}}_2|_4^2|\boldsymbol{v}|_4^2 + c\|\bar{\boldsymbol{v}}_2\|_{L^4}^2|\boldsymbol{v}|_2\|\boldsymbol{v}\|_2 \\
\le& \varepsilon\|\boldsymbol{v}\|_2^2 + c(|\tilde{\boldsymbol{v}}_2|_4^8 + \|\bar{\boldsymbol{v}}_2\|_{L^4}^4)|\boldsymbol{v}|_2^2.
\end{aligned}$$

By Hölder Inequality, Young Inequality and Minkowski Inequality, we get

$$\begin{aligned}
&\left|\int_\Omega W(\boldsymbol{v})\frac{\partial \boldsymbol{v}_2}{\partial \xi}\cdot\boldsymbol{v}\right| \le \int_{S^2}\left[\int_0^1 (|\nabla_{e_\theta}\boldsymbol{v}|^2 + |\nabla_{e_\varphi}\boldsymbol{v}|^2)^{\frac{1}{2}}\mathrm{d}\xi\int_0^1 |\boldsymbol{v}_{2\xi}||\boldsymbol{v}|\mathrm{d}\xi\right] \\
\le& 2\varepsilon\|\boldsymbol{v}\|^2 + c\left[(|\boldsymbol{v}_{2\xi}|_2^2 + 1)\int_\Omega (|\nabla_{e_\theta}\boldsymbol{v}_{2\xi}|^2 + |\nabla_{e_\varphi}\boldsymbol{v}_{2\xi}|^2) + |\boldsymbol{v}_{2\xi}|_2^2\right]|\boldsymbol{v}|_2^2.
\end{aligned}$$

Thus, we obtain

$$\begin{aligned}
&\frac{1}{2}\frac{\mathrm{d}|\boldsymbol{v}|_2^2}{\mathrm{d}t} + \int_\Omega (|\nabla_{e_\theta}\boldsymbol{v}|^2 + |\nabla_{e_\varphi}\boldsymbol{v}|^2 + |\boldsymbol{v}|^2) + \int_\Omega |\boldsymbol{v}_\xi|^2 \\
\le& -\int_\Omega \left(\int_\xi^1 \frac{bP}{p}\mathrm{grad}T\mathrm{d}\xi'\right)\cdot\boldsymbol{v} + 4\varepsilon\|\boldsymbol{v}\|^2 \\
&+ c\left[|\tilde{\boldsymbol{v}}_2|_4^8 + \|\bar{\boldsymbol{v}}_2\|_{L^4}^4 + |\boldsymbol{v}_{2\xi}|_2^2 + (|\boldsymbol{v}_{2\xi}|_2^2 + 1)\|\boldsymbol{v}_{2\xi}\|^2\right]|\boldsymbol{v}|_2^2.
\end{aligned} \tag{3.3.30}$$

Taking $L^2(\Omega)$ inner product of equation (3.3.26) with T, we get

$$\begin{aligned}
&\frac{1}{2}\frac{\mathrm{d}|T|_2^2}{\mathrm{d}t} + \int_\Omega |\nabla T|^2 + \int_\Omega |T_\xi|^2 + \alpha_s|T(\xi=1)|_2^2 \\
=& -\int_\Omega [\nabla_{\boldsymbol{v}_1}T + W(\boldsymbol{v}_1)\frac{\partial T}{\partial \xi}]T - \int_\Omega T\nabla_{\boldsymbol{v}}T_2 \\
&- \int_\Omega W(\boldsymbol{v})\frac{\partial T_2}{\partial \xi}T + \int_\Omega \frac{bP}{p}W(\boldsymbol{v})T.
\end{aligned}$$

Similar to (3.3.30),

$$\frac{1}{2}\frac{\mathrm{d}|T|_2^2}{\mathrm{d}t}+\int_\Omega|\nabla T|^2+\int_\Omega|T_\xi|^2+\alpha_s|T(\xi=1)|_2^2\le 3\varepsilon\|\boldsymbol{v}\|^2+3\varepsilon\|T\|^2$$
$$+c|T_2|_4^8(|T|_2^2+|\boldsymbol{v}|_2^2)+c\left[(|T_{2\xi}|_2^2+1)|\nabla T_{2\xi}|_2^2+|T_{2\xi}|_2^2\right]|T|_2^2+\int_\Omega\frac{bP}{p}W(\boldsymbol{v})T.$$

By integration by parts, we have

$$-\int_\Omega\left(\int_\xi^1\frac{bP}{p}\mathrm{grad}T\mathrm{d}\xi'\right)\cdot\boldsymbol{v}+\int_\Omega\frac{bP}{p}W(\boldsymbol{v})T=0.$$

Therefore, choosing ε small enough, we obtain

$$\frac{\mathrm{d}(|\boldsymbol{v}|_2^2+|T|_2^2)}{\mathrm{d}t}+\|\boldsymbol{v}\|^2+\int_\Omega|\nabla T|^2+\int_\Omega|T_\xi|^2+\alpha_s|T(\xi=1)|_2^2$$
$$\le c\left[|\tilde{\boldsymbol{v}}_2|_4^8+\|\bar{\boldsymbol{v}}_2\|_{L^4}^4+|T_2|_4^8+|\boldsymbol{v}_{2\xi}|_2^2+(|\boldsymbol{v}_{2\xi}|_2^2+1)\|\boldsymbol{v}_{2\xi}\|^2\right]|\boldsymbol{v}|_2^2$$
$$+c(1+|T_2|_4^8+|T_{2\xi}|_2^2+(|T_{2\xi}|_2^2+1)|\nabla T_{2\xi}|_2^2)|T|_2^2. \qquad (3.3.31)$$

According to *a priori* estimates in part 3.3.2.1 and (3.3.31), applying Gronwall Inequality, we prove the uniqueness of the weakly strong solution to the initial boundary value problem (3.3.1)-(3.3.6).

3.3.3 *The Global Existence of Smooth Solutions of IBVP*

3.3.3.1 *A Priori Estimates*

Here, we give the proof of the global existence of strong solutions to the initial boundary value problem (3.3.1)-(3.3.6), which can be obtained with the method in subsection 3.3.2.

Proposition 3.3.6. If $Q\in H^1(\Omega)$, $U_0=(\boldsymbol{v}_0,T_0)\in V$, then IBVP has a unique global strong solution $\boldsymbol{U}$, $\boldsymbol{U}\in L^\infty(0,\infty;V)$. Moreover, the corresponding semigroup $\{S(t)\}_{t\ge0}$ of the boundary value problem (3.3.1)-(3.3.5) has a bounded absorbing set B_ρ in V, that is, for any bounded set $B\subset V$, there is $t_0(B)>0$ big enough such that

$$S(t)B\subset B_\rho,\quad\forall\, t\ge t_0,$$

where $B_\rho=\{\boldsymbol{U};\|\boldsymbol{U}\|\le\rho\}$, and ρ is a constant dependent on $\|Q\|_1$.

Now, let's make *a priori* estimates of strong solutions.

L^3 estimates of $\boldsymbol{v}_\xi$. Taking $L^2(\Omega)\times L^2(\Omega)$ inner product of equation (3.3.15) with $|\boldsymbol{v}_\xi|\boldsymbol{v}_\xi$, we have

$$\frac{1}{3}\frac{\mathrm{d}|\boldsymbol{v}_\xi|_3^3}{\mathrm{d}t}+\int_\Omega\left[\left(|\nabla_{e_\theta}\boldsymbol{v}_\xi|^2+|\nabla_{e_\varphi}\boldsymbol{v}_\xi|^2\right)|\boldsymbol{v}_\xi|+\frac{4}{9}|\nabla_{e_\theta}|\boldsymbol{v}_\xi|^{\frac{3}{2}}|^2\right.$$

$$+\frac{4}{9}|\nabla_{e_\varphi}|v_\xi|^{\frac{3}{2}}|^2+|v_\xi|^3\Big]+\int_\Omega\left(|v_{\xi\xi}|^2|v_\xi|+\frac{4}{9}|\partial_\xi|v_\xi|^{\frac{3}{2}}|^2\right)$$
$$=-\int_\Omega(f\boldsymbol{k}\times v_\xi)\cdot|v_\xi|v_\xi+\int_\Omega\frac{bP}{p}\mathrm{grad}T\cdot|v_\xi|v_\xi$$
$$-\int_\Omega\left[\nabla_v v_\xi+\left(\int_\xi^1\mathrm{div}v\mathrm{d}\xi'\right)\frac{\partial v_\xi}{\partial\xi}\right]\cdot|v_\xi|v_\xi$$
$$-\int_\Omega\left[\nabla_{v_\xi}v-(\mathrm{div}v)\frac{\partial v}{\partial\xi}\right]\cdot|v_\xi|v_\xi.$$

By integration by parts, Hölder Inequality and Young Inequality, we get

$$-\int_\Omega\left[\nabla_{v_\xi}v-(\mathrm{div}v)\frac{\partial v}{\partial\xi}\right]\cdot|v_\xi|v_\xi\le c\int_\Omega|v||v_\xi|^2(|\nabla_{e_\theta}v_\xi|^2+|\nabla_{e_\varphi}v_\xi|^2)^{\frac{1}{2}}$$
$$\le c|v|_4\Big(\int_\Omega|v_\xi|^{\frac{3}{2}\cdot 2}\Big)^{\frac{1}{8}}\||v_\xi|^{\frac{3}{2}}\|^{\frac{3}{4}}\left[\int_\Omega(|\nabla_{e_\theta}v_\xi|^2+|\nabla_{e_\varphi}v_\xi|^2)|v_\xi|\right]^{\frac{1}{2}}$$
$$\le\varepsilon\||v_\xi|^{\frac{3}{2}}\|^2+\varepsilon\int_\Omega(|\nabla_{e_\theta}v_\xi|^2+|\nabla_{e_\varphi}v_\xi|^2)|v_\xi|+c|v|_4^8|v_\xi|_3^3,$$
$$\int_\Omega\frac{bP}{p}\mathrm{grad}T\cdot|v_\xi|v_\xi\le c|T|_4^2|v_\xi|_2+\varepsilon\int_\Omega(|\nabla_{e_\theta}v_\xi|^2+|\nabla_{e_\varphi}v_\xi|^2)|v_\xi|.$$

Choosing ε small enough, we obtain

$$\frac{\mathrm{d}|v_\xi|_3^3}{\mathrm{d}t}+\int_\Omega\Big[(|\nabla_{e_\theta}v_\xi|^2+|\nabla_{e_\varphi}v_\xi|^2)|v_\xi|+\frac{4}{9}|\nabla_{e_\theta}|v_\xi|^{\frac{3}{2}}|^2+\frac{4}{9}|\nabla_{e_\varphi}|v_\xi|^{\frac{3}{2}}|^2$$
$$+|v_\xi|^3\Big]+\int_\Omega(|v_{\xi\xi}|^2|v_\xi|+\frac{4}{9}|\partial_\xi|v_\xi|^{\frac{3}{2}}|^2)\le c|v|_4^8|v_\xi|_3^3+c|T|_4^2|v_\xi|_2. \tag{3.3.32}$$

Applying the uniform Gronwall Inequality, $|v_\xi|_3^3\le c|v_\xi|_2^{\frac{3}{2}}\|v_\xi\|^{\frac{3}{2}}$ and Proposition 3.3.5, we derive from the above inequality

$$|v_\xi(t+r)|_3^3\le F_1, \tag{3.3.33}$$

where F_1 is a positive constant, $t\ge 0$, and $r>0$ is given. Applying Gronwall Inequality, we derive from (3.3.32) for any $0\le t<r$,

$$|v_\xi(t)|_3^3\le c. \tag{3.3.34}$$

L^4 estimates of v_ξ. Taking $L^2(\Omega)\times L^2(\Omega)$ inner product of equation (3.3.15) with $|v_\xi|^2v_\xi$, we obtain

$$\frac{1}{4}\frac{\mathrm{d}|v_\xi|_4^4}{\mathrm{d}t}+\int_\Omega\Big[(|\nabla_{e_\theta}v_\xi|^2+|\nabla_{e_\varphi}v_\xi|^2)|v_\xi|^2+\frac{1}{2}|\nabla_{e_\theta}|v_\xi|^2|^2$$

$$
\begin{aligned}
&+\frac{1}{2}|\nabla_{e_\varphi}|v_\xi|^2|^2+|v_\xi|^4\Big]+\int_\Omega\left(|v_{\xi\xi}|^2|v_\xi|^2+\frac{1}{2}|\partial_\xi|v_\xi|^2|^2\right)\\
=&-\int_\Omega\left(fk\times v_\xi-\frac{bP}{p}\text{grad}T\right)\cdot|v_\xi|^2v_\xi\\
&-\int_\Omega\left[\nabla_v v_\xi+\left(\int_\xi^1 \text{div}v\text{d}\xi'\right)\frac{\partial v_\xi}{\partial\xi}\right]\cdot|v_\xi|^2v_\xi\\
&-\int_\Omega[\nabla_{v_\xi}v-(\text{div}v)\frac{\partial v}{\partial\xi}]\cdot|v_\xi|^2v_\xi.
\end{aligned}
$$

By integration by parts, Hölder Inequality and Young Inequality, we have

$$
\begin{aligned}
&-\int_\Omega\left[\nabla_{v_\xi}v-(\text{div}v)\frac{\partial v}{\partial\xi}\right]\cdot|v_\xi|^2v_\xi\\
\le& c\int_\Omega|v||v_\xi|^3\left(|\nabla_{e_\theta}v_\xi|^2+|\nabla_{e_\varphi}v_\xi|^2\right)^{\frac{1}{2}}\\
\le& c|v|_4\left(\int_\Omega|v_\xi|^{2\cdot 2}\right)^{\frac{1}{8}}\||v_\xi|^2\|^{\frac{3}{4}}\left[\int_\Omega\left(|\nabla_{e_\theta}v_\xi|^2+|\nabla_{e_\varphi}v_\xi|^2\right)|v_\xi|^2\right]^{\frac{1}{2}}\\
\le&\varepsilon\||v_\xi|^2\|^2+\varepsilon\int_\Omega\left(|\nabla_{e_\theta}v_\xi|^2+|\nabla_{e_\varphi}v_\xi|^2\right)|v_\xi|^2+c|v|_4^8|v_\xi|_4^4,
\end{aligned}
$$

$$
\int_\Omega\frac{bP}{p}\text{grad}T\cdot|v_\xi|^2v_\xi\le c|T|_4^4+c|v_\xi|_4^4+\varepsilon\int_\Omega\left(|\nabla_{e_\theta}v_\xi|^2+|\nabla_{e_\varphi}v_\xi|^2\right)|v_\xi|^2.
$$

Thus, choosing ε small enough, we obtain

$$
\begin{aligned}
&\frac{\text{d}|v_\xi|_4^4}{\text{d}t}+\int_\Omega[(|\nabla_{e_\theta}v_\xi|^2+|\nabla_{e_\varphi}v_\xi|^2)|v_\xi|^2+\frac{1}{2}|\nabla_{e_\theta}|v_\xi|^2|^2+\frac{1}{2}|\nabla_{e_\varphi}|v_\xi|^2|^2\\
&+|v_\xi|^4]+\int_\Omega(|v_{\xi\xi}|^2|v_\xi|^2+\frac{1}{2}|\partial_\xi|v_\xi|^2|^2)\le c|T|_4^4+c(1+|v|_4^8)|v_\xi|_4^4.
\end{aligned}
\tag{3.3.35}
$$

By the uniform Gronwall Inequality, $|v_\xi|_4^4\le c|v_\xi|_3^2\|v_\xi\|^2$ and Proposition 3.3.5, we derive from the above inequality

$$
|v_\xi(t+2r)|_4^4+\int_{t+2r}^{t+3r}\||v_\xi|^2\|^2\le F_2, \tag{3.3.36}
$$

where F_2 is a positive constant, $t\ge 0$, and $r>0$ is given. According to (3.3.35), applying Gronwall Inequality, we have

$$
|v_\xi(t)|_4^4\le c, \tag{3.3.37}
$$

where $0\le t<2r$.

L^3 estimates of T_ξ. Taking $L^2(\Omega)$ inner product of equation (3.3.20) with $|T_\xi|T_\xi$, we get

$$\frac{1}{3}\frac{\mathrm{d}|T_\xi|_3^3}{\mathrm{d}t}+2\int_\Omega|\nabla T_\xi|^2|T_\xi|+2\int_\Omega|T_{\xi\xi}|^2|T_\xi|-\int_{S^2}[(|T_\xi|T_\xi)|_{\xi=1}\cdot T_{\xi\xi}|_{\xi=1}]$$
$$=-\int_\Omega\left[\nabla_{\boldsymbol{v}}T_\xi+W(\boldsymbol{v})\frac{\partial T_\xi}{\partial\xi}\right]|T_\xi|T_\xi-\int_\Omega\left[\nabla_{\boldsymbol{v}_\xi}T-(\mathrm{div}\boldsymbol{v})\frac{\partial T}{\partial\xi}\right]|T_\xi|T_\xi$$
$$+\int_\Omega Q_\xi|T_\xi|T_\xi+\int_\Omega\left[-\frac{bP}{p}(\mathrm{div}\boldsymbol{v})-\frac{bP(P-p_0)}{p}W(\boldsymbol{v})\right]|T_\xi|T_\xi.$$

By integration by parts, Hölder Inequality, Poincaré Inequality and Young Inequality, we obtain

$$\left|\int_\Omega\left[\nabla_{\boldsymbol{v}_\xi}T-(\mathrm{div}\boldsymbol{v})\frac{\partial T}{\partial\xi}\right]|T_\xi|T_\xi\right|$$
$$\leq c|\boldsymbol{v}_\xi|_4|\nabla T|_2\left(\int_\Omega|T_\xi|^{\frac{3}{2}\cdot\frac{16}{3}}\right)^{\frac{1}{4}}+c|\boldsymbol{v}|_4\left(\int_\Omega|T_\xi|^{\frac{3}{2}\cdot4}\right)^{\frac{1}{4}}\left(\int_\Omega|\nabla T_\xi|^2|T_\xi|\right)^{\frac{1}{2}}$$
$$\leq\varepsilon\left(\int_\Omega|\nabla T_\xi|^2|T_\xi|+\int_\Omega|T_{\xi\xi}|^2|T_\xi|\right)+c(1+|\boldsymbol{v}|_4^8)|T_\xi|_3^3+c|\boldsymbol{v}_\xi|_4^8\|T\|^8.$$

By Hölder Inequality and Young Inequality, we have

$$\left|\int_\Omega Q_\xi|T_\xi|T_\xi\right|\leq c|Q_\xi|_2\left(\int_\Omega|T_\xi|^{\frac{3}{2}\cdot\frac{8}{3}}\right)^{\frac{1}{2}}\leq c|Q_\xi|_2\left(\||T_\xi|^{\frac{3}{2}}|_2^{\frac{5}{3}}\||T_\xi|^{\frac{3}{2}}\|\right)^{\frac{1}{2}}$$
$$\leq\varepsilon\left(\int_\Omega|\nabla T_\xi|^2|T_\xi|+\int_\Omega|T_{\xi\xi}|^2|T_\xi|\right)+c|T_\xi|_3^3+c|Q_\xi|_2^3.$$

By integration by parts, Hölder Inequality, Minkowski Inequality, Poincaré Inequality and Young Inequality, we get

$$\left|\int_\Omega\left[-\frac{bP}{p}(\mathrm{div}\boldsymbol{v})-\frac{bP(P-p_0)}{p}W(\boldsymbol{v})\right]|T_\xi|T_\xi\right|\leq c|\boldsymbol{v}|_4^2|T_\xi|_2+\varepsilon\int_\Omega|\nabla T_\xi|^2|T_\xi|.$$

According to (3.3.4) and $T_{\xi\xi}|_{\xi=1}=\frac{\partial T|_{\xi=1}}{\partial t}+(\nabla_{\boldsymbol{v}}T)|_{\xi=1}-\Delta T|_{\xi=1}-Q|_{\xi=1}$, we have

$$-\int_{S^2}[(|T_\xi|T_\xi)|_{\xi=1}\cdot T_{\xi\xi}|_{\xi=1}]$$
$$=\alpha_s^2\left[\frac{1}{3}\frac{\mathrm{d}|T(\xi=1)|_3^3}{\mathrm{d}t}+2\int_{S^2}(|\nabla T|^2|T|)|_{\xi=1}\right]+\alpha_s^2\int_{S^2}[|T|T(\nabla_{\boldsymbol{v}}T-Q)]|_{\xi=1}.$$

By integration by parts, we obtain

$$-\alpha_s^2\int_{S^2}[|T|T(\nabla_{\boldsymbol{v}}T-Q)]|_{\xi=1}$$

$$
\begin{aligned}
=&\frac{\alpha_s^2}{3}\int_{S^2} T^3|_{\xi=1}\mathrm{div}\boldsymbol{v}|_{\xi=1}+\alpha_s^2\int_{S^2}(|T|TQ)|_{\xi=1}\\
=&\frac{\alpha_s^2}{3}\int_{S^2} T^3|_{\xi=1}\left(\int_\xi^1 \mathrm{div}\boldsymbol{v}_\xi \mathrm{d}\xi'+\mathrm{div}\boldsymbol{v}\right)+\alpha_s^2\int_{S^2}(|T|TQ)|_{\xi=1}\\
\le& c|T|_{\xi=1}|_6^6+c\|\boldsymbol{v}_\xi\|^2+c\|\boldsymbol{v}\|^2+c|T|_{\xi=1}|_4^4+c|Q|_{\xi=1}|_2^2.
\end{aligned}
$$

Thus, choosing ε small enough, we get

$$
\begin{aligned}
&\frac{\mathrm{d}(|T_\xi|_3^3+\alpha_s^2|T|_{\xi=1}|_3^3)}{\mathrm{d}t}+\int_\Omega |\boldsymbol{\nabla} T_\xi|^2|T_\xi|+\int_\Omega |T_{\xi\xi}|^2|T_\xi|\\
&+\alpha_s^2\int_{S^2}(|\boldsymbol{\nabla} T|^2|T|)|_{\xi=1}\\
\le& c(1+|\boldsymbol{v}|_4^8)|T_\xi|_3^3+c|\boldsymbol{v}_\xi|_4^8\|T\|^8+c|Q_\xi|_2^3+c|\boldsymbol{v}|_4^2|T_\xi|_2+c\|\boldsymbol{v}_\xi\|^2\\
&+c\|\boldsymbol{v}\|^2+c|T|_{\xi=1}|_6^6+c|T|_{\xi=1}|_4^4+c|Q|_{\xi=1}|_2^2.
\end{aligned}\tag{3.3.38}
$$

By the uniform Gronwall Inequality, $|T_\xi|_3^3 \le c|T_\xi|_2^{\frac{3}{2}}\|T_\xi\|^{\frac{3}{2}}$, (3.3.36) and Proposition 3.3.6, we derive from the above inequality

$$
|T_\xi(t+3r)|_3^3\le F_3, \tag{3.3.39}
$$

where F_3 is a positive constant. In the proof of (3.3.39), we apply $\int_{t+2r}^{t+3r}|T|_{\xi=1}|_6^6\le c$, which can be obtained by L^6 estimates of T. According to (3.3.38), applying Gronwall Inequality, we have

$$
|T_\xi(t)|_3^3\le c, \tag{3.3.40}
$$

where $0\le t<3r$.

L^4 estimates of T_ξ. Taking $L^2(\Omega)$ inner product of equation (3.3.20) with $|T_\xi|^2T_\xi$, we have

$$
\begin{aligned}
&\frac{1}{4}\frac{\mathrm{d}|T_\xi|_4^4}{\mathrm{d}t}+3\int_\Omega|\boldsymbol{\nabla} T_\xi|^2|T_\xi|^2+3\int_\Omega|T_{\xi\xi}|^2|T_\xi|^2-\int_{S^2}\left[(|T_\xi|^2T_\xi)|_{\xi=1}\cdot T_{\xi\xi}|_{\xi=1}\right]\\
=&-\int_\Omega\left[\boldsymbol{\nabla}_{\boldsymbol{v}}T_\xi+W(\boldsymbol{v})\frac{\partial T_\xi}{\partial\xi}\right]|T_\xi|^2T_\xi-\int_\Omega\left[\boldsymbol{\nabla}_{\boldsymbol{v}_\xi}T-\mathrm{div}\boldsymbol{v}\frac{\partial T}{\partial\xi}\right]|T_\xi|^2T_\xi\\
&+\int_\Omega Q_\xi|T_\xi|^2T_\xi+\int_\Omega\left[-\frac{bP}{p}\mathrm{div}\boldsymbol{v}-\frac{bP(P-p_0)}{p}W(\boldsymbol{v})\right]|T_\xi|^2T_\xi.
\end{aligned}
$$

By integration by parts, Hölder Inequality, Poincaré Inequality and Young Inequality, we obtain

$$
\begin{aligned}
&\left|\int_\Omega\left[\boldsymbol{\nabla}_{\boldsymbol{v}_\xi}T-\mathrm{div}\boldsymbol{v}\frac{\partial T}{\partial\xi}\right]|T_\xi|^2T_\xi\right|\le c\int_\Omega(|\boldsymbol{v}_\xi||\boldsymbol{\nabla} T||T_\xi|^3+|\boldsymbol{v}||\boldsymbol{\nabla} T_\xi||T_\xi|^3)\\
\le& c|\boldsymbol{\nabla} T|_2\left(\int_\Omega|\boldsymbol{v}_\xi|^{2\cdot4}\right)^{\frac{1}{8}}\left(\int_\Omega|T_\xi|^{2\cdot4}\right)^{\frac{3}{8}}
\end{aligned}
$$

$$
\begin{aligned}
&+ c|\boldsymbol{v}|_4 \left(\int_\Omega |T_\xi|^{2\cdot 4}\right)^{\frac{1}{4}} \left(\int_\Omega |\nabla T_\xi|^2 |T_\xi|^2\right)^{\frac{1}{2}} \\
\le& \varepsilon \left(\int_\Omega |\nabla T_\xi|^2|T_\xi|^2 + \int_\Omega |T_{\xi\xi}|^2|T_\xi|^2\right) + c\left(\|T\|^{\frac{16}{3}} + |\boldsymbol{v}|_4^8\right)|T_\xi|_4^4 + c|\boldsymbol{v}_\xi|_4^4 \\
&+ c\left\||\boldsymbol{v}_\xi|^2\right\|^2, \\
&\left|\int_\Omega Q_\xi |T_\xi|^2 T_\xi\right| \\
\le& c|Q_\xi|_2 \left(\int_\Omega |T_\xi|^{2\cdot 3}\right)^{\frac{1}{2}} \\
\le& \varepsilon \int_\Omega (|\nabla T_\xi|^2 + |T_{\xi\xi}|^2)|T_\xi|^2 + c|T_\xi|_4^4 + c|Q_\xi|_2^4, \\
&\left|\int_\Omega \left[-\frac{bP}{p}\mathrm{div}\boldsymbol{v} - \frac{bP(P-p_0)}{p}W(\boldsymbol{v})\right]|T_\xi|^2 T_\xi\right| \\
\le& c|\boldsymbol{v}|_4^4 + c|T_\xi|_4^4 + \varepsilon \int_\Omega |\nabla T_\xi|^2 |T_\xi|^2.
\end{aligned}
$$

Similar to (3.3.38),

$$
\begin{aligned}
&\frac{\mathrm{d}(|T_\xi|_4^4 + \alpha_s^3 |T|_{\xi=1}|_4^4)}{\mathrm{d}t} + \int_\Omega |\nabla T_\xi|^2 |T_\xi|^2 + \int_\Omega |T_{\xi\xi}|^2 |T_\xi|^2 \\
&+ \alpha_s^3 \int_{S^2} |\nabla T|_{\xi=1}|^2 |T|_{\xi=1}|^2 \\
\le& c(1 + \|T\|^{\frac{16}{3}} + |\boldsymbol{v}|_4^8)|T_\xi|_4^4 + c|\boldsymbol{v}_\xi|_4^4 + c\||\boldsymbol{v}_\xi|^2\|^2 \\
&+ c|Q_\xi|_2^4 + c|\boldsymbol{v}|_4^4 + c\|\boldsymbol{v}_\xi\|^2 + c\|\boldsymbol{v}\|^2 + c|T|_{\xi=1}|_8^8 + c|T|_{\xi=1}|_6^6 + c|Q|_{\xi=1}|_2^2.
\end{aligned}
\tag{3.3.41}
$$

With the uniform Gronwall Inequality, $|T_\xi|_4^4 \le c|T_\xi|_3^2 \|T_\xi\|^2$, (3.3.36), (3.3.39) and Proposition 3.3.6, we derive from (3.3.41)

$$
|T_\xi(t+4r)|_4^4 \le F_4, \tag{3.3.42}
$$

where F_4 is a positive constant. In the proof of (3.3.42), we apply $\int_{t+3r}^{t+4r} |T|_{\xi=1}|_8^8 \le c$, which is derived from L^8 estimates of T. By Gronwall Inequality, we derive from (3.3.41)

$$
|T_\xi(t)|_4^4 \le c, \tag{3.3.43}
$$

where $0 \le t < 4r$.

L^2 estimates of $\boldsymbol{v}_t$. Taking the derivative of (3.3.1) with respect to t, we get

$$
\frac{\partial \boldsymbol{v}_t}{\partial t} - \Delta \boldsymbol{v}_t - \frac{\partial^2 \boldsymbol{v}_t}{\partial \xi^2} + \nabla_{\boldsymbol{v}} \boldsymbol{v}_t + \left(\int_\xi^1 \mathrm{div}\boldsymbol{v}\right)\frac{\partial \boldsymbol{v}_t}{\partial \xi} + \nabla_{\boldsymbol{v}_t}\boldsymbol{v} + \left(\int_\xi^1 \mathrm{div}\boldsymbol{v}_t\right)\frac{\partial \boldsymbol{v}}{\partial \xi}
$$

$$+ f\boldsymbol{k} \times \boldsymbol{v}_t + \operatorname{grad} \Phi_{st} + \int_\xi^1 \frac{bP}{p} \operatorname{grad} T_t = 0.$$

Taking $L^2(\Omega) \times L^2(\Omega)$ inner product of the above equation with $\boldsymbol{v}_t$, by integration by parts, we have

$$\frac{1}{2}\frac{\mathrm{d}|\boldsymbol{v}_t|_2^2}{\mathrm{d}t} + \|\boldsymbol{v}_t\|^2 = -\int_\Omega \left[\boldsymbol{\nabla}_{\boldsymbol{v}_t}\boldsymbol{v} + \left(\int_\xi^1 \operatorname{div}\boldsymbol{v}_t\right)\frac{\partial \boldsymbol{v}}{\partial \xi} + \int_\xi^1 \frac{bP}{p}\operatorname{grad} T_t\right] \cdot \boldsymbol{v}_t.$$

By integration by parts, Hölder Inequality, Poincaré Inequality and Young Inequality, we obtain

$$\begin{aligned}
&-\int_\Omega \left[\boldsymbol{\nabla}_{\boldsymbol{v}_t}\boldsymbol{v} + \left(\int_\xi^1 \operatorname{div}\boldsymbol{v}_t\right)\frac{\partial \boldsymbol{v}}{\partial \xi}\right] \cdot \boldsymbol{v}_t \\
&\leq c\int_\Omega (|\boldsymbol{v}| + |\boldsymbol{v}_\xi|)|\boldsymbol{v}_t|(|\boldsymbol{\nabla}_{e_\theta}\boldsymbol{v}_t|^2 + |\boldsymbol{\nabla}_{e_\varphi}\boldsymbol{v}_t|^2)^{\frac{1}{2}} \\
&\leq c(|\boldsymbol{v}|_4 + |\boldsymbol{v}_\xi|_4)|\boldsymbol{v}_t|_2^{\frac{1}{4}}\|\boldsymbol{v}_t\|^{\frac{3}{4}}\left[\int_\Omega (|\boldsymbol{\nabla}_{e_\theta}\boldsymbol{v}_t|^2 + |\boldsymbol{\nabla}_{e_\varphi}\boldsymbol{v}_t|^2)\right]^{\frac{1}{2}} \\
&\leq \varepsilon\|\boldsymbol{v}_t\|^2 + c(|\boldsymbol{v}|_4^8 + |\boldsymbol{v}_\xi|_4^8)|\boldsymbol{v}_t|_2^2.
\end{aligned}$$

Thus, choosing ε small enough, we have

$$\frac{\mathrm{d}|\boldsymbol{v}_t|_2^2}{\mathrm{d}t} + \|\boldsymbol{v}_t\|^2 \leq c(|\boldsymbol{v}|_4^8 + |\boldsymbol{v}_\xi|_4^8)|\boldsymbol{v}_t|_2^2 + c|T_t|_2^2. \tag{3.3.44}$$

According to the above inequality, applying the uniform Gronwall Inequality, (3.3.36) and Proposition 3.3.6, we derived from (3.3.44)

$$|\boldsymbol{v}_t(t+5r)|_2^2 + \int_{t+5r}^{t+6r} \|\boldsymbol{v}_t\|^2 \leq F_5, \tag{3.3.45}$$

where F_5 is a positive constant, and $t \geq 0$. In the proof of (3.3.45), we apply $\int_{t+4r}^{t+5r} |v_t|_2^2 \leq c$, which is obtained by taking $L^2(\Omega) \times L^2(\Omega)$ inner product of (3.3.1) with $\boldsymbol{v}_t$. we derive from (3.3.44). With Gronwall Inequality, we have

$$|\boldsymbol{v}_t(t)|_2^2 \leq c, \tag{3.3.46}$$

where $0 \leq t < 5r$.

L^2 estimates of T_t. Taking the derivative of (3.3.2) with respect to t, we get

$$\frac{\partial T_t}{\partial t} - \Delta T_t - \frac{\partial^2 T_t}{\partial \xi^2} + \boldsymbol{\nabla}_{\boldsymbol{v}} T_t + \left(\int_\xi^1 \operatorname{div}\boldsymbol{v}\right)\frac{\partial T_t}{\partial \xi} + \boldsymbol{\nabla}_{\boldsymbol{v}_t} T + \left(\int_\xi^1 \operatorname{div}\boldsymbol{v}_t\right)\frac{\partial T}{\partial \xi}$$

$$-\frac{bP}{p}\left(\int_\xi^1 \mathrm{div}\boldsymbol{v}_t\right)=0.$$

Taking $L^2(\Omega)$ inner product of the above equation with T_t, then by intergration by parts, we have

$$\begin{aligned}&\frac{1}{2}\frac{\mathrm{d}|T_t|_2^2}{dt}+\int_\Omega|\boldsymbol{\nabla}T_t|^2+\int_\Omega|T_{t\xi}|^2+\alpha_s\int_{S^2}|T_t(\xi=1)|^2\\ =&-\int_\Omega\left[\boldsymbol{\nabla}_{\boldsymbol{v}_t}T+\left(\int_\xi^1\mathrm{div}\boldsymbol{v}_t\right)\frac{\partial T}{\partial\xi}\right]T_t+\int_\Omega\frac{bP}{p}\left(\int_\xi^1\mathrm{div}\boldsymbol{v}_t\right)T_t.\end{aligned}$$

By intergration by parts, Hölder Inequality, Poincaré Inequality and Young Inequality, we obtain

$$\begin{aligned}&\left|\int_\Omega\left[\boldsymbol{\nabla}_{\boldsymbol{v}_t}T+\left(\int_\xi^1\mathrm{div}\boldsymbol{v}_t\right)\frac{\partial T}{\partial\xi}\right]T_t\right|\\ \leq&c(|T|_4+|T_\xi|_4)|T_t|_4\|\boldsymbol{v}_t\|+c|T|_4|\boldsymbol{v}_t|_4|\boldsymbol{\nabla}T_t|_2\\ \leq&\varepsilon\|T_t\|^2+c(|T|_4^8+|T_\xi|_4^8)|T_t|_2^2+c\|\boldsymbol{v}_t\|^2+c|T|_4^8|\boldsymbol{v}_t|_2^2,\\ &\left|\int_\Omega\frac{bP}{p}\left(\int_\xi^1\mathrm{div}\boldsymbol{v}_t\right)T_t\right|\leq c|T_t|_2^2+c\|\boldsymbol{v}_t\|_2^2.\end{aligned}$$

Thus, choosing ε small enough, we have

$$\begin{aligned}&\frac{\mathrm{d}|T_t|_2^2}{dt}+\int_\Omega|\boldsymbol{\nabla}T_t|^2+\int_\Omega|T_{t\xi}|^2+\alpha_s\int_{S^2}|T_t(\xi=1)|^2\\ \leq&c(1+|T|_4^8+|T_\xi|_4^8)|T_t|_2^2+c\|\boldsymbol{v}_t\|^2+c|T|_4^8|\boldsymbol{v}_t|_2^2.\end{aligned}\tag{3.3.47}$$

Applying the uniform Gronwall Inequality, (3.3.42), (3.3.45) and Proposition 3.3.6, we derive from the above inequality

$$|T_t(t+6r)|_2^2\leq F_6,\tag{3.3.48}$$

where F_6 is a positive constant. In the proof of (3.3.48), we apply $\int_{t+5r}^{t+6r}|T_t|_2^2\leq c$, which is obtained by taking $L^2(\Omega)$ inner product of (3.3.2) with T_t. By Gronwall Inequality, we derive from (3.3.47)

$$|T_t(t)|_2^2\leq c,\tag{3.3.49}$$

where $0\leq t<6r$.

3.3.3.2 *The Global Existence of Smooth Solution*

Firstly, we give a proposition about the regularity of a linear elliptic problem, which will be used in the proof of the global existence of the smooth solutions of IBVP.

Proposition 3.3.7. Suppose that $\boldsymbol{v} \in V_1$, $\Phi_s \in L^2(S^2)$ are solutions of the following Stokes problem,

$$-\Delta \boldsymbol{v} - \frac{\partial^2 \boldsymbol{v}}{\partial \xi^2} + \operatorname{grad} \Phi_s = g, \tag{3.3.50}$$

$$\int_0^1 \operatorname{div} \boldsymbol{v} \, \mathrm{d}\xi = 0, \tag{3.3.51}$$

$$\xi = 1: \frac{\partial \boldsymbol{v}}{\partial \xi} = 0; \quad \xi = 0: \frac{\partial \boldsymbol{v}}{\partial \xi} = 0. \tag{3.3.52}$$

If $g \in W^{m,\alpha}(T\Omega|TS^2)$, then $\boldsymbol{v} \in V_1 \cap W^{2+m,\alpha}(T\Omega|TS^2)$, $\Phi_s \in W^{1+m,\alpha}(S^2)$, where $1 < \alpha < +\infty$, and $m \geq 0$.

Proof of Proposition 3.3.7. Integrating equation (3.3.50) from 0 to 1 with respect to ξ, and applying the boundary condition (3.3.52), we get

$$-\Delta \int_0^1 \boldsymbol{v} \, \mathrm{d}\xi + \operatorname{grad} \Phi_s = \int_0^1 f \, \mathrm{d}\xi.$$

Then, with the regularity results of Stokes problem and regularity of elliptic equation in S^2, we prove Proposition 3.3.7.

Next we give the proof of Theorem 3.3.3.

Proof of Theorem 3.3.3. According to the results of the *a priori* estimates in part 3.3.3.1, we know that the initial boundary value problem (3.3.1)-(3.3.6) has a unique strong solution $\boldsymbol{U} = (\boldsymbol{v}, T)$, and $\boldsymbol{v}_\xi \in L^\infty(0, \mathcal{T}; L^4(T\Omega|TS^2)), T_\xi \in L^\infty(0, \mathcal{T}; L^4(\Omega)), \boldsymbol{v}_t \in L^\infty(0, \mathcal{T}; L^2(T\Omega|TS^2)), T_t \in L^\infty(0, \mathcal{T}; L^2(\Omega))$. If

$$g = -\frac{\partial \boldsymbol{v}}{\partial t} - \nabla_{\boldsymbol{v}} \boldsymbol{v} - \left(\int_\xi^1 \operatorname{div} \boldsymbol{v} \, \mathrm{d}\xi' \right) \frac{\partial \boldsymbol{v}}{\partial \xi} - f \boldsymbol{k} \times \boldsymbol{v} - \int_\xi^1 \frac{bP}{p} \operatorname{grad} T \, \mathrm{d}\xi',$$

then

$$-\Delta \boldsymbol{v} - \frac{\partial^2 \boldsymbol{v}}{\partial \xi^2} + \operatorname{grad} \Phi_s = g. \tag{3.3.53}$$

Since $\boldsymbol{v} \in L^\infty(0, \mathcal{T}; V_1)$ and $\boldsymbol{v}_\xi \in L^\infty(0, \mathcal{T}; L^4(T\Omega|TS^2))$, for any $u \in L^4(T\Omega|TS^2)$, we have

$$\int_\Omega \left(\int_\xi^1 \operatorname{div} \boldsymbol{v} \, \mathrm{d}\xi' \right) \frac{\partial \boldsymbol{v}}{\partial \xi} \cdot u \leq c \|\boldsymbol{v}\| |\boldsymbol{v}_\xi|_4 |u|_4,$$

which implies

$$\left(\int_\xi^1 \text{div}\boldsymbol{v}\ d\xi'\right)\frac{\partial \boldsymbol{v}}{\partial \xi} \in L^\infty(0,\mathcal{T};L^{\frac{4}{3}}T\Omega|TS^2)).$$

Since $\boldsymbol{U} = (\boldsymbol{v},T)$ is a strong solution of (3.3.1)-(3.3.6), and $\boldsymbol{v}_t \in L^\infty(0,\mathcal{T};L^2)$,

$$\left(-\frac{\partial \boldsymbol{v}}{\partial t} - \boldsymbol{\nabla}_{\boldsymbol{v}}\boldsymbol{v} - f\boldsymbol{k}\times\boldsymbol{v} - \int_\xi^1 \frac{bP}{p}\text{grad}T\ d\xi'\right) \in L^\infty(0,\mathcal{T};L^{\frac{4}{3}}(T\Omega|TS^2)).$$

Thus,

$$g \in L^\infty(0,\mathcal{T};L^{\frac{4}{3}}(T\Omega|TS^2)). \tag{3.3.54}$$

According to Proposition 3.3.7, we know

$$\boldsymbol{v} \in L^\infty(0,\mathcal{T};V_1\cap W^{2,\frac{4}{3}}(T\Omega|TS^2)).$$

Because $W^{2,\frac{4}{3}}(T\Omega|TS^2) \subset W^{1,\frac{12}{5}}(T\Omega|TS^2)$, we have

$$\boldsymbol{v} \in L^\infty(0,\mathcal{T};V_1\cap W^{1,\frac{12}{5}}(T\Omega|TS^2)).$$

Next we shall use Proposition 3.3.7 repeatedly to improve (3.3.54). Since $\boldsymbol{v} \in L^\infty(0,\mathcal{T};V_1\cap W^{1,\frac{12}{5}}(T\Omega|TS^2)), \boldsymbol{v}_\xi \in L^\infty(0,\mathcal{T};L^4(T\Omega|\ TS^2))$, for any $u \in L^4(T\Omega|TS^2)$, we have

$$\int_\Omega\left(\int_\xi^1 \text{div}\boldsymbol{v}\ d\xi'\right)\frac{\partial \boldsymbol{v}}{\partial \xi}\cdot u \le c\|\boldsymbol{v}\|_{W^{1,\frac{12}{5}}}|\boldsymbol{v}_\xi|_4|u|_3,$$

which implies

$$\left(\int_\xi^1 \text{div}\boldsymbol{v}\ d\xi'\right)\frac{\partial \boldsymbol{v}}{\partial \xi} \in L^\infty(0,\mathcal{T};L^{\frac{3}{2}}(T\Omega|TS^2)),$$

Thus,

$$g \in L^\infty(0,\mathcal{T};L^{\frac{3}{2}}(T\Omega|TS^2)).$$

Applying Proposition 3.3.7 and Sobolev Imbedding Theorem, we obtain

$$\boldsymbol{v} \in L^\infty(0,\mathcal{T};V_1\cap W^{1,3}(T\Omega|TS^2)).$$

Since

$$\int_\Omega\left(\int_\xi^1 \text{div}\boldsymbol{v}\ d\xi'\right)\frac{\partial \boldsymbol{v}}{\partial \xi}\cdot u \le c\|\boldsymbol{v}\|_{W^{1,3}}|\boldsymbol{v}_\xi|_4|u|_{\frac{12}{5}},$$

and $W^{2,\frac{12}{7}}(T\Omega|TS^2) \subset W^{1,4}(T\Omega|TS^2)$, we have

$$\int_\Omega\left(\int_\xi^1 \text{div}\boldsymbol{v}\ d\xi'\right)\frac{\partial \boldsymbol{v}}{\partial \xi}\cdot u \le c\|\boldsymbol{v}\|_{W^{1,4}}|\boldsymbol{v}_\xi|_4|u|_2.$$

Thus,

$$g \in L^\infty(0, \mathcal{T}; L^2(T\Omega|TS^2)).$$

According to Proposition 3.3.7, we have

$$\boldsymbol{v} \in L^\infty(0, \mathcal{T}; V_1 \cap W^{2,2}(T\Omega|TS^2)).$$

Applying the elliptic regularity theory, we get

$$T \in L^\infty(0, \mathcal{T}; V_2 \cap W^{2,2}(\Omega)).$$

By Proposition 3.3.7, we prove the existence of smooth solutions of the initial boundary value problem (3.3.1)-(3.3.6). Here we omit the details.

According to the *a priori* estimates in part 3.3.3.1 and the proof of Theorem 3.3.3, we get the following proposition.

Proposition 3.3.8. Suppose that $Q \in H^1(\Omega)$, $\boldsymbol{U}_0 = (\boldsymbol{v}_0, T_0) \in V \cap (H^2(T\Omega|TS^2) \times H^2(\Omega))$. Then the system (3.3.1)-(3.3.6) has a unique strong solution $\boldsymbol{U}$, $\boldsymbol{U} \in L^\infty(0, \infty; V \cap (H^2(T\Omega|TS^2) \times H^2(\Omega)))$. Moreover, the corresponding solution semigroup $\{S(t)\}_{t\geq 0}$ of (3.3.1)-(3.3.5) has a bounded absorbing set B_ρ in $V \cap (H^2(T\Omega|TS^2) \times H^2(\Omega))$, that is, for any $B \subset V$ given, there exists a big enough $t_0(B) > 0$, such that for $t \geq t_0$, $S(t)B \subset B_\rho$, where $B_\delta = \{\boldsymbol{U}; \boldsymbol{U} \in V, \ \|\boldsymbol{U}\| \leq \delta\}$, and δ is a positive constant dependent on $\|Q\|_1$.

Applying Proposition 3.3.8, we can get the following conclusion.

Corollary 3.3.9. (3.3.1)-(3.3.5) have a global attractor $\mathcal{A} = \cap_{s\geq 0}\overline{\cup_{t\geq s} S(t)B_\rho}$, where the closures are taken with respect to the topology V. The global attractor $\mathcal{A}$ has the following properties:

(i) (compactness) $\mathcal{A}$ is compact in V;

(ii) (invariant) for any $t \geq 0$, $S(t)\mathcal{A} = \mathcal{A}$;

(iii) (absorbing) for any bounded set in V, when $t \to +\infty$, the sets $S(t)B$ converge to $\mathcal{A}$ with respect to the topology of V, that is,

$$\lim_{t\to+\infty} d_V(S(t)B, \ \mathcal{A}) = 0,$$

where the distance d_V is induced by the topology of V.

3.4 Global Well-Posedness of Primitive Equations of the Oceans

In this section, we shall prove the global well-posedness of the primitive equations of the large-scale oceans, in which the proof of local existence of the strong solutions is given by Guillén-González et al. in [87], and the proof of the global existence of the strong solutions is given by Cao and Titi in [26]. In order to prove the existence of the global attractor of the infinite dimensional dynamic system corresponding to the primitive equations, we use the methods in section 3.3. Here we do not give the details of the proof.

The arrangement of this section is as follows. In the first part, we give the viscous primitive equations of the large-scale oceans. The work spaces and the main results of this section are given in the second part. The last two parts are used to prove the global well-posedness of the primitive equations of the large-scale oceans.

3.4.1 *The Viscous Primitive Equations of the Large-Scale Oceans*

In this section, let's recall the model considered in [26]. In rectangular coordinates, the viscous three-dimensional primitive equations of the large-scale oceans are written as (about the details of deduction of this model, see [172, 188] and the references therein):

$$\frac{\partial \boldsymbol{v}}{\partial t} + (\boldsymbol{v} \cdot \boldsymbol{\nabla})\boldsymbol{v} + w\frac{\partial \boldsymbol{v}}{\partial z} + f\boldsymbol{k} \times \boldsymbol{v} + \boldsymbol{\nabla} p - \frac{1}{Re_1}\Delta \boldsymbol{v} - \frac{1}{Re_2}\frac{\partial^2 \boldsymbol{v}}{\partial z^2} = 0, \tag{3.4.1}$$

$$\frac{\partial p}{\partial z} + T = 0, \tag{3.4.2}$$

$$\text{div}\boldsymbol{v} + \frac{\partial w}{\partial z} = 0, \tag{3.4.3}$$

$$\frac{\partial T}{\partial t} + \boldsymbol{v} \cdot \boldsymbol{\nabla} T + w\frac{\partial T}{\partial z} - \frac{1}{Rt_1}\Delta T - \frac{1}{Rt_2}\frac{\partial^2 T}{\partial z^2} = Q, \tag{3.4.4}$$

where the unknown functions are $\boldsymbol{v}$, w, p, T, $\boldsymbol{v} = (v^{(1)}, v^{(2)})$ is the horizontal velocity, w is the vertical velocity, p is pressure, T is temperature, $f = f_0 + \beta y$ is the Coriolis parameter, $\boldsymbol{k}$ is the vertical unit vector, Re_1, Re_2 are Reynolds numbers, Rt_1, Rt_2 are horizontal and vertical heat diffusion coefficients, Q is a given function in Ω (the definition of Ω will be given later), and

$$\boldsymbol{\nabla} = (\partial_x,\ \partial_y),\ \Delta = \partial_x^2 + \partial_y^2,\ \text{div}\boldsymbol{v} = \partial_x v^{(1)} + \partial_y v^{(2)}.$$

Assume that the space domain of equation (3.4.1)-(3.4.4) is the following cylindrical domain

$$\Omega = \{(x, y, z):\ (x, y) \in M,\quad z \in (-h(x, y), 0)\},$$

where M is a smooth bounded domain in $\mathbb{R}^2$. For simplicity, suppose $h = 1$, that is, suppose $\Omega = M \times (-1, 0)$. For a general function $h(x, y)$, in order to obtain the results like those in this section, we need the regularity conditions of $h(x,y)$, such as $h(x,y) \in C^3(\overline{M})$. The boundary conditions of equation (3.4.1)-(3.4.4) are given by

$$\frac{\partial \boldsymbol{v}}{\partial z} = 0,\ w = 0,\ \frac{\partial T}{\partial z} = -\alpha_s T \qquad \text{on } M \times \{0\} = \Gamma_u, \tag{3.4.5}$$

$$\frac{\partial \boldsymbol{v}}{\partial z} = 0,\ w = 0,\ \frac{\partial T}{\partial z} = 0 \qquad \text{on } M \times \{-1\} = \Gamma_b, \tag{3.4.6}$$

$$\boldsymbol{v} \cdot \boldsymbol{n} = 0,\ \frac{\partial \boldsymbol{v}}{\partial \boldsymbol{n}} \times \boldsymbol{n} = 0,\ \frac{\partial T}{\partial \boldsymbol{n}} = 0 \qquad \text{on } \partial M \times [-1,0] = \Gamma_l, \tag{3.4.7}$$

where α_s is a positive constant, and $\boldsymbol{n}$ is the external normal vector of the side boundary Γ_l.

Remark 3.4.1. Here the salinity equation is omitted. If the salinity equation are taken into account, and the boundary conditions $\frac{\partial \boldsymbol{v}}{\partial z}|_{z=0} = -\alpha_s T$, $\frac{\partial T}{\partial z}|_{z=0} = 0$ are replaced by $\frac{\partial \boldsymbol{v}}{\partial z}|_{z=0} = \tau$, $\frac{\partial T}{\partial z}|_{z=0} = -\alpha_s(T - T^*)$, where τ, T^* are smooth enough, then the results in this section can be established.

Now, we use the boundary conditions (3.4.5)-(3.4.7) to simplify equations (3.4.1)-(3.4.4). Integrating (3.4.3) from -1 to z with respect to z, with boundary conditions (3.4.5), (3.4.6), we get

$$w(t;x,y,z) = W(\boldsymbol{v})(t;x,y,z) = -\int_{-1}^{z} \operatorname{div}\boldsymbol{v}(t;x,y,z')\,\mathrm{d}z', \tag{3.4.8}$$

$$\int_{-1}^{0} \operatorname{div}\boldsymbol{v}\,\mathrm{d}z = 0. \tag{3.4.9}$$

Suppose that p_b is an unknown function on $M \times \{-1\}$. Integrating (3.4.2) from -1 to z with respect to z, we have

$$p(t;x,y,z) = p_b(t;x,y) - \int_{-1}^{z} T\,\mathrm{d}z'. \tag{3.4.10}$$

In this section, for simplicity, suppose that Re_1, Re_2, Rt_1, Rt_2, are all 1. According to (3.4.8)-(3.4.10), equations (3.4.1)-(3.4.4) are rewritten as

$$\frac{\partial \boldsymbol{v}}{\partial t} + (\boldsymbol{v}\cdot\boldsymbol{\nabla})\boldsymbol{v} + W(\boldsymbol{v})\frac{\partial \boldsymbol{v}}{\partial z} + f\boldsymbol{k}\times\boldsymbol{v} + \boldsymbol{\nabla} p_b - \int_{-1}^{z} \boldsymbol{\nabla} T\mathrm{d}z' - \Delta\boldsymbol{v} - \frac{\partial^2 \boldsymbol{v}}{\partial z^2} = 0, \tag{3.4.11}$$

$$\frac{\partial T}{\partial t} + (\boldsymbol{v}\cdot\boldsymbol{\nabla})T + W(\boldsymbol{v})\frac{\partial T}{\partial z} - \Delta T - \frac{\partial^2 T}{\partial z^2} = Q, \tag{3.4.12}$$

$$\int_{-1}^{0} \mathrm{div}\boldsymbol{v}\ \mathrm{d}z = 0. \tag{3.4.13}$$

The boundary conditions of equation (3.4.11)-(3.4.13) are

$$\frac{\partial \boldsymbol{v}}{\partial z} = 0, \quad \frac{\partial T}{\partial z} = -\alpha_s T \qquad \text{on } \Gamma_u, \tag{3.4.14}$$

$$\frac{\partial \boldsymbol{v}}{\partial z} = 0, \quad \frac{\partial T}{\partial z} = 0 \qquad \text{on } \Gamma_b, \tag{3.4.15}$$

$$\boldsymbol{v}\cdot\boldsymbol{n} = 0, \quad \frac{\partial \boldsymbol{v}}{\partial \boldsymbol{n}} \times \boldsymbol{n} = 0, \quad \frac{\partial T}{\partial \boldsymbol{n}} = 0 \qquad \text{on } \Gamma_l. \tag{3.4.16}$$

The initial condition is

$$\boldsymbol{U}|_{t=0} = (\boldsymbol{v}|_{t=0}, T|_{t=0}) = \boldsymbol{U}_0 = (\boldsymbol{v}_0, T_0). \tag{3.4.17}$$

We call (3.4.11)-(3.4.17) the initial boundary value problem of the new formulation of the viscous three-dimensional primitive equations of the large-scale oceans, which is denoted by IBVP.

3.4.2 *Main Results*

Firstly, we give some definitions of some function spaces. Define Lebesgue space $L^p(\Omega) := \{u; u : \Omega \to \mathbb{R}, \int_\Omega |u|^p < +\infty\}$, with norm $|u|_p = (\int_\Omega |u|^p)^{\frac{1}{p}}$, $1 \le p < \infty$. In this section we denote $\int_\Omega \cdot \mathrm{d}\Omega$ and $\int_M \cdot \mathrm{d}M$ respectively as $\int_\Omega \cdot$, $\int_M \cdot$. $H^m(\Omega)$ is a general Sobolev space with norm

$$\|u\|_m = \left(\int_\Omega \left(\sum_{1\le k\le m} \sum_{i_j=1,2,3;j=1,\cdots,k} |\boldsymbol{\nabla}_{i_1}\cdots\boldsymbol{\nabla}_{i_k} u|^2 + |u|^2 \right)\right)^{\frac{1}{2}},$$

where $\boldsymbol{\nabla}_1 = \frac{\partial}{\partial x}$, $\boldsymbol{\nabla}_2 = \frac{\partial}{\partial y}$, $\boldsymbol{\nabla}_3 = \frac{\partial}{\partial z}$, and m is a positive integer. Now, we define the work spaces of problem IBVP. Let

$$\widetilde{\mathcal{V}_1} := \left\{ \boldsymbol{v} \in (C^\infty(\Omega))^2;\ \frac{\partial \boldsymbol{v}}{\partial z}\Big|_{z=0} = 0,\ \frac{\partial \boldsymbol{v}}{\partial z}\Big|_{z=-1} = 0,\ \boldsymbol{v}\cdot\boldsymbol{n}\Big|_{\Gamma_s} = 0,\right.$$
$$\left. \frac{\partial \boldsymbol{v}}{\partial \boldsymbol{n}} \times \boldsymbol{n}\,|_{\Gamma_s} = 0,\ \boldsymbol{\nabla}\cdot\boldsymbol{v} = 0 \right\},$$

$$\widetilde{\mathcal{V}_2} := \left\{ T \in C^\infty(\Omega),\ \frac{\partial T}{\partial z}|_{z=0} = -\alpha_s T,\ \frac{\partial T}{\partial z}|_{z=-1} = 0,\ \frac{\partial T}{\partial \boldsymbol{n}}|_{\Gamma_s} = 0 \right\},$$

V_1 = closure of $\widetilde{\mathcal{V}_1}$ with respect to norm $\|\cdot\|_1$ ($\|\boldsymbol{v}\|_m^2 = \|\boldsymbol{v}^{(1)}\|_m^2 + \|\boldsymbol{v}^{(2)}\|_m^2$),

V_2 = closure of $\widetilde{\mathcal{V}_2}$ with respect to norm $\|\cdot\|_1$,

H_1 = closure of $\widetilde{\mathcal{V}_1}$ with respect to norm $|\cdot|_2$,

$H_2 =$ closure of $\widetilde{\mathcal{V}_2}$ with respect to norm $|\cdot|_2$,

$V = V_1 \times V_2,\ H = H_1 \times H_2.$

The inner products and norms of $V_1,\ V_2,\ V$ are given by

$$(\boldsymbol{v}, \boldsymbol{v}_1)_{V_1} = \int_\Omega (\partial_x \boldsymbol{v} \cdot \partial_x \boldsymbol{v}_1 + \partial_y \boldsymbol{v} \cdot \partial_y \boldsymbol{v}_1 + \partial_z \boldsymbol{v} \cdot \partial_z \boldsymbol{v}_1 + \boldsymbol{v} \cdot \boldsymbol{v}_1),$$

$$\|\boldsymbol{v}\| = (\boldsymbol{v}, \boldsymbol{v})_{V_1}^{\frac{1}{2}},\quad \forall \boldsymbol{v},\ \boldsymbol{v}_1 \in V_1,$$

$$(T, T_1)_{V_2} = \int_\Omega \left(\nabla T \cdot \nabla T_1 + \frac{\partial T}{\partial z}\frac{\partial T_1}{\partial z} + T T_1 \right),$$

$$\|T\| = (T, T)_{V_2}^{\frac{1}{2}},\quad \forall T,\ T_1 \in V_2,$$

$$(\boldsymbol{U}, \boldsymbol{U}_1) = \left(v^{(1)}, v_1^{(1)}\right) + \left(v^{(2)}, v_1^{(2)}\right) + (T, T_1),$$

$$(\boldsymbol{U}, \boldsymbol{U}_1)_V = (\boldsymbol{v}, \boldsymbol{v}_1)_{V_1} + (T, T_1)_{V_2},$$

$$\|\boldsymbol{U}\| = (\boldsymbol{U}, \boldsymbol{U})_V^{\frac{1}{2}},\ |\boldsymbol{U}|_2 = (\boldsymbol{U}, \boldsymbol{U})^{\frac{1}{2}},\ \forall \boldsymbol{U} = (\boldsymbol{v}, T),\ \boldsymbol{U}_1 = (\boldsymbol{v}_1, T_1) \in V.$$

Here $\bar{\boldsymbol{v}} = \int_{-1}^0 \boldsymbol{v} dz$, and $(\cdot,\cdot)$ denotes L^2 inner product in $L^2(\Omega)$.

Before introducing the main results, we give the definition of the strong solution of problem (3.4.11)-(3.4.17).

Definition 3.4.2. Suppose that $Q \in H^1(\Omega)$, $\boldsymbol{U}_0 = (\boldsymbol{v}_0, T_0) \in V$, and $\mathcal{T}$ is positive constant given. We call $\boldsymbol{U} = (\boldsymbol{v}, T)$ **a strong solution** of problem (3.4.11)-(3.4.12) in $[0, \mathcal{T}]$, if it satisfies (3.4.11)-(3.4.12) in weak sense, and

$$\boldsymbol{v} \in L^\infty(0, \mathcal{T}; V_1) \cap L^2(0, \mathcal{T}; H^2(\Omega) \times H^2(\Omega)),$$

$$T \in L^\infty(0, \mathcal{T}; V_2) \cap L^2(0, \mathcal{T}; H^2(\Omega)),$$

$$\frac{\partial \boldsymbol{v}}{\partial t} \in L^2(0, \mathcal{T}; L^2(\Omega) \times L^2(\Omega)),\ \frac{\partial T}{\partial t} \in L^2(0, \mathcal{T}; L^2(\Omega)).$$

Remark 3.4.3. In [26], the definition of strong solution only requires

$$\frac{\partial \boldsymbol{v}}{\partial t} \in L^1(0, \mathcal{T}; L^2(\Omega) \times L^2(\Omega)),\ \frac{\partial T}{\partial t} \in L^1(0, \mathcal{T}; L^2(\Omega)).$$

Actually, when $Q \in H^1(\Omega)$, $\boldsymbol{U}_0 = (\boldsymbol{v}_0, T_0) \in V$, the strong solution satisfies

$$\frac{\partial \boldsymbol{v}}{\partial t} \in L^2(0, \mathcal{T}; L^2(\Omega) \times L^2(\Omega)),\ \frac{\partial T}{\partial t} \in L^2(0, \mathcal{T}; L^2(\Omega)).$$

Proposition 3.4.4. If $Q \in H^1(\Omega)$, $\boldsymbol{U}_0 = (\boldsymbol{v}_0, T_0) \in V$, then there exists $\mathcal{T}^* > 0$, $\mathcal{T}^* = \mathcal{T}^*(\|\boldsymbol{U}_0\|)$, such that the system (3.4.11)-(3.4.17) has a strong solution $\boldsymbol{U}$ in $[0, \mathcal{T}^*]$.

Theorem 3.4.5. If $Q \in H^1(\Omega)$, $\boldsymbol{U}_0 = (\boldsymbol{v}_0, T_0) \in V$, then, for any given $\mathcal{T} > 0$, the system (3.4.11)-(3.4.17) has a unique strong solution $\boldsymbol{U}$ in $[0, \mathcal{T}]$, which is continuously dependent on initial data.

3.4.3 The Local Existence of Strong Solutions

In order to prove Proposition 3.4.4, we consider the system (3.4.11)-(3.4.17). Suppose that Y is a solution of the following initial boundary value problem,

$$\begin{cases} \dfrac{\partial \boldsymbol{Y}}{\partial t} + \boldsymbol{\nabla} p_{b_1} - \Delta \boldsymbol{Y} - \dfrac{\partial^2 \boldsymbol{Y}}{\partial z^2} = 0, \\ \displaystyle\int_{-1}^{0} \boldsymbol{\nabla} \cdot \boldsymbol{Y} \mathrm{d}z = 0, \\ \dfrac{\partial \boldsymbol{Y}}{\partial z}|_{\Gamma_u, \Gamma_b} = 0, \ \boldsymbol{Y} \cdot \boldsymbol{n}|_{\Gamma_l} = 0, \ \dfrac{\partial \boldsymbol{Y}}{\partial \boldsymbol{n}} \times \boldsymbol{n}|_{\Gamma_l} = 0, \\ \boldsymbol{Y}|_{t=0} = \boldsymbol{v}_0. \end{cases}$$

If $\boldsymbol{v}_0 \in V_1$, for any $\mathcal{T} > 0$, then the above initial boundary value problem has a unique strong solution $\boldsymbol{Y}$,

$$\boldsymbol{Y} \in L^\infty(0, \mathcal{T}; V_1) \cap L^2(0, \mathcal{T}; (H^2(\Omega))^2). \tag{3.4.18}$$

For the details of proof of this conclusion, see [87]. Let $\boldsymbol{u}(t) = \boldsymbol{v}(t) - \boldsymbol{Y}$. If $\boldsymbol{U}(t) = (\boldsymbol{v}, T)$ is a strong solution of (3.4.11)-(3.4.17) in $[0, \mathcal{T}]$, then $(\boldsymbol{u}, T)$ is a strong solution of the following problem in $[0, \mathcal{T}]$,

$$\frac{\partial \boldsymbol{u}}{\partial t} + [(\boldsymbol{u} + \boldsymbol{Y}) \cdot \boldsymbol{\nabla}] (\boldsymbol{u} + \boldsymbol{Y}) + W(\boldsymbol{u} + \boldsymbol{Y}) \frac{\partial (\boldsymbol{u} + \boldsymbol{Y})}{\partial z} + f \boldsymbol{k} \times (\boldsymbol{u} + \boldsymbol{Y})$$

$$+ \boldsymbol{\nabla} p_{b_2} - \int_{-1}^{z} \boldsymbol{\nabla} T \mathrm{d}z' - \Delta \boldsymbol{u} - \frac{\partial^2 \boldsymbol{u}}{\partial z^2} = 0, \tag{3.4.19}$$

$$\frac{\partial T}{\partial t} + [(\boldsymbol{u} + \boldsymbol{Y}) \cdot \boldsymbol{\nabla}] T + W(\boldsymbol{u} + \boldsymbol{Y}) \frac{\partial T}{\partial z} - \Delta T - \frac{\partial^2 T}{\partial z^2} = Q, \tag{3.4.20}$$

$$\int_{-1}^{0} \boldsymbol{\nabla} \cdot \boldsymbol{u} \, \mathrm{d}z = 0, \tag{3.4.21}$$

$$(\boldsymbol{u}, T) \text{ satisfies boundary conditions (3.4.14)-(3.4.16)}, \tag{3.4.22}$$

$$(\boldsymbol{u}|_{t=0}, T|_{t=0}) = (0, T_0). \tag{3.4.23}$$

Proof of Proposition 3.4.4. We prove Proposition 3.4.4 by using Faedo-Galerkin method. Since the proof is similar to the local existence of strong solutions of the three-dimensional incompressible Navier-Stokes equation, here we only give the *a priori* estimates of approximate solutions. Suppose that $(\boldsymbol{u}_m, T_m)$ is an approximate solution of (3.4.19)-(3.4.23), $(\boldsymbol{u}_m, T_m) = \sum_{i=1}^{m} \alpha_{i,m}(t) \phi_i(x)$, where $\{\phi_m\}$ is a complete orthogonal basis of space V. Then $(\boldsymbol{u}_m, T_m)$ satisfies

$$\int_{\Omega} h_m \cdot \frac{\partial \boldsymbol{u}_m}{\partial t}$$

$$+\int_\Omega h_m\cdot\left\{[(\boldsymbol{u}_m+\boldsymbol{Y})\cdot\boldsymbol{\nabla}](\boldsymbol{u}_m+\boldsymbol{Y})+W(\boldsymbol{u}_m+\boldsymbol{Y})\frac{\partial(\boldsymbol{u}_m+\boldsymbol{Y})}{\partial z}\right\}$$
$$+\int_\Omega h_m\cdot[f\boldsymbol{k}\times(\boldsymbol{u}_m+\boldsymbol{Y})]-\int_\Omega h_m\cdot\int_{-1}^z\boldsymbol{\nabla}T_m\mathrm{d}z'+\int_\Omega h_m\cdot A_1\boldsymbol{u}_m=0, \tag{3.4.24}$$

$$\int_\Omega q_m\frac{\partial T_m}{\partial t}+\int_\Omega q_m\left\{[(\boldsymbol{u}_m+\boldsymbol{Y})\cdot\boldsymbol{\nabla}]T_m+W(\boldsymbol{u}_m+\boldsymbol{Y})\frac{\partial T_m}{\partial z}\right\}$$
$$+\int_\Omega q_mA_2T_m=\int_\Omega q_mQ, \tag{3.4.25}$$
$$\boldsymbol{u}_m(0)=0,\ T_m(0)=T_{0m}\to T_0\quad\text{in } V_2,$$

where $h_m\in V_{1m}$, $q_m\in V_{2m}$, $V_{1m}\times V_{2m}=\mathrm{span}\{\phi_1,...,\phi_m\}$, and the definitions of operator A_1, A_2 are similar to those in Lemma 3.1.2. Now, we begin to make *a priori* estimates of approximate solutions.

L^2 estimates of T_m, $\boldsymbol{u}_m$. Letting $q_m=T_m$ in (3.4.25), by integration by parts, we get

$$\frac{\mathrm{d}|T_m|_2^2}{\mathrm{d}t}+c\|T_m\|^2\le c|Q|_2^2,$$

which implies that for any $\mathcal{T}>0$, T_m is uniformly bounded in $L^\infty(0,\mathcal{T};L^2(\Omega))\cap L^2(0,\mathcal{T};V_2)$ with respect to m. By Hölder Inequality, Sobolev Inequality and Young Inequality, we have

$$\begin{aligned}&-\int_\Omega\boldsymbol{u}_m\cdot[(\boldsymbol{Y}\cdot\boldsymbol{\nabla})\boldsymbol{u}_m+(\boldsymbol{u}_m\cdot\boldsymbol{\nabla})\boldsymbol{Y}+(\boldsymbol{Y}\cdot\boldsymbol{\nabla})\boldsymbol{Y}]\\ \le&\varepsilon\|\boldsymbol{u}_m\|^2+c(|\boldsymbol{Y}|_4^8+\|\boldsymbol{Y}\|^4)|\boldsymbol{u}_m|_2^2+c\|\boldsymbol{Y}\|^2,\\ &-\int_\Omega\boldsymbol{u}_m\cdot\left[W(\boldsymbol{u}_m)\frac{\partial\boldsymbol{Y}}{\partial z}+W(\boldsymbol{Y})\frac{\partial\boldsymbol{u}_m}{\partial z}+W(\boldsymbol{Y})\frac{\partial\boldsymbol{Y}}{\partial z}\right]\\ \le&\varepsilon\|\boldsymbol{u}_m\|^2+c\|\boldsymbol{Y}\|^2\|\boldsymbol{Y}\|_2^2|\boldsymbol{u}_m|_2^2+c\|\boldsymbol{Y}\|^2.\end{aligned}$$

Letting $h_m=\boldsymbol{u}_m$ in (3.4.24), we have

$$\int_\Omega\left[(\boldsymbol{u}_m\cdot\boldsymbol{\nabla})\boldsymbol{u}_m+W(\boldsymbol{u}_m)\frac{\partial\boldsymbol{u}_m}{\partial z}\right]\cdot\boldsymbol{u}_m=0.$$

By Hölder Inequality and Young Inequality, we get

$$\frac{\mathrm{d}|\boldsymbol{u}_m|_2^2}{\mathrm{d}t}+c\|\boldsymbol{u}_m\|^2\le c(|\boldsymbol{Y}|_4^8+\|\boldsymbol{Y}\|^4+\|\boldsymbol{Y}\|^2\|\boldsymbol{Y}\|_2^2)|\boldsymbol{u}_m|_2^2+c(\|\boldsymbol{Y}\|^2+|T_m|_2^2).$$

According to (3.4.18), we know that for any $\mathcal{T}>0$, $\boldsymbol{u}_m$ is uniformly bounded in $L^\infty(0,\mathcal{T};H_1)\cap L^2(0,\mathcal{T};V_1)$ with respect to m.

H^1 **estimates of $\boldsymbol{u}_m$, T_m.** Letting $h_m = A_1\boldsymbol{u}_m$ in (3.4.24), we have

$$\begin{aligned}
&\frac{1}{2}\frac{\mathrm{d}}{\mathrm{d}t}\|\boldsymbol{u}_m\|^2 + |A_1\boldsymbol{u}_m|_2^2 \\
&= -\int_\Omega ((\boldsymbol{u}_m+\boldsymbol{Y})\cdot\boldsymbol{\nabla}\boldsymbol{u}_m)\cdot A_1\boldsymbol{u}_m - \int_\Omega ((\boldsymbol{u}_m+\boldsymbol{Y})\cdot\boldsymbol{\nabla}\boldsymbol{Y})\cdot A_1\boldsymbol{u}_m \\
&\quad - \int_\Omega W(\boldsymbol{u}_m+\boldsymbol{Y})\partial_z\boldsymbol{u}_m\cdot A_1\boldsymbol{u}_m - \int_\Omega W(\boldsymbol{u}_m+\boldsymbol{Y})\partial_z\boldsymbol{Y}\cdot A_1\boldsymbol{u}_m \\
&\quad + \int_\Omega \{-[f\boldsymbol{k}\times(\boldsymbol{u}_m+\boldsymbol{Y})] + \int_{-1}^{z}\boldsymbol{\nabla}T_m\mathrm{d}z'\}\cdot A_1\boldsymbol{u}_m = \sum_{i=1}^{5} I_i.
\end{aligned}$$

By using $|\boldsymbol{u}_m|_4 \le C|\boldsymbol{u}_m|_2^{1/4}\|\boldsymbol{u}_m\|^{3/4}$ and (3.4.18), we get

$$\begin{aligned}
I_1 &\le |A_1\boldsymbol{u}_m|_2(|\boldsymbol{u}_m|_4 + |\boldsymbol{Y}|_4)|\boldsymbol{\nabla}\boldsymbol{u}_m|_4 \\
&\le C|A_1\boldsymbol{u}_m|_2(\|\boldsymbol{u}_m\|^{3/4}|\boldsymbol{u}_m|_2^{1/4} + \|\boldsymbol{Y}\|^{3/4}|\boldsymbol{Y}|_2^{1/4})\|\boldsymbol{\nabla}\boldsymbol{u}_m\|^{3/4}|\boldsymbol{\nabla}\boldsymbol{u}_m|_2^{1/4} \\
&\le C|A_1\boldsymbol{u}_m|_2^{7/4}(\|\boldsymbol{u}_m\| + \|\boldsymbol{Y}\|)\|\boldsymbol{u}_m\|^{1/4} \\
&\le \frac{1}{10}|A_1\boldsymbol{u}_m|_2^2 + C(\|\boldsymbol{u}_m\|^8 + \|\boldsymbol{Y}\|^8)\|\boldsymbol{u}_m\|^2.
\end{aligned}$$

In the same way,

$$\begin{aligned}
I_2 &\le |A_1\boldsymbol{u}_m|_2(|\boldsymbol{u}_m|_4 + |\boldsymbol{Y}|_4)|\boldsymbol{\nabla}\boldsymbol{Y}|_4 \\
&\le C|A_1\boldsymbol{u}_m|_2(\|\boldsymbol{u}_m\|^{3/4}|\boldsymbol{u}_m|_2^{1/4} + \|\boldsymbol{Y}\|^{3/4}|\boldsymbol{Y}|_2^{1/4})|\boldsymbol{\nabla}\boldsymbol{Y}|_2^{1/4}\|\boldsymbol{\nabla}\boldsymbol{Y}\|^{3/4} \\
&\le \frac{1}{10}|A_1\boldsymbol{u}_m|_2^2 + C\|\boldsymbol{Y}\|^{1/2}\|\boldsymbol{Y}\|_2^{2/3}\|\boldsymbol{u}_m\|^2 + C\|\boldsymbol{Y}\|^{5/2}\|\boldsymbol{Y}\|_2^{3/2}.
\end{aligned}$$

Let $I_3 = J_1 + J_2$, where $J_1 = -\int_\Omega (W(\boldsymbol{u}_m)\partial_z\boldsymbol{u}_m)\cdot A_1\boldsymbol{u}_m$, $J_2 = -\int_\Omega (W(\boldsymbol{Y})\partial_z\boldsymbol{u}_m)\cdot A_1\boldsymbol{u}_m$.
Applying

$$\begin{aligned}
&\|W(\boldsymbol{u}_m)\|_{L_z^\infty L_{x,y}^4} \le C(\Omega)|\boldsymbol{\nabla}\boldsymbol{u}_m|_2^{\frac{1}{2}}\|\boldsymbol{\nabla}\boldsymbol{u}_m\|^{\frac{1}{2}}, \\
&\|\boldsymbol{u}_m\|_{L_z^2L_{x,y}^4}^2 \le 4|\boldsymbol{u}_m|_2\|\boldsymbol{\nabla}\boldsymbol{u}_m\|,
\end{aligned}$$

we obtain

$$|J_1| \le |A_1\boldsymbol{u}_m|_2\|W(\boldsymbol{u}_m)\|_{L_z^\infty L_{x,y}^4}\|\partial_z\boldsymbol{u}_m\|_{L_z^2L_{x,y}^4} \le C|A_1\boldsymbol{u}_m|_2^2\|\boldsymbol{u}_m\|.$$

Similar to J_1,

$$\begin{aligned}
|J_2| &\le C|A_1\boldsymbol{u}_m|_2\|\boldsymbol{u}_m\|_2^{1/2}\|\boldsymbol{u}_m\|^{1/2}\|\boldsymbol{Y}\|_2^{1/2}\|\boldsymbol{Y}\|^{1/2} \\
&\le \frac{1}{10}|A_1\boldsymbol{u}_m|_2^2 + C\|\boldsymbol{u}_m\|^{1/2}\|\boldsymbol{Y}\|^{1/2}\|\boldsymbol{Y}\|_2^{1/2}.
\end{aligned}$$

Similar to I_3,

$$\begin{aligned}
I_4 &\le (\|W(\boldsymbol{u}_m)\|_{L_z^\infty L_{x,y}^4} + \|W(\boldsymbol{Y})\|_{L_z^\infty L_{x,y}^4})\|\partial_z \boldsymbol{Y}\|_{L_z^2 L_{x,y}^4}|A_1\boldsymbol{u}_m|_2 \\
&\le C(\|\boldsymbol{u}_m\|^{1/2}\|\boldsymbol{u}_m\|_2^{1/2} + \|\boldsymbol{Y}\|^{1/2}\|\boldsymbol{Y}\|_2^{1/2})\|\boldsymbol{Y}\|^{1/2}\|\boldsymbol{Y}\|_2^{1/2}|A_1\boldsymbol{u}_m|_2 \\
&\le \frac{1}{10}|A_1\boldsymbol{u}_m|_2^2 + C\|\boldsymbol{Y}\|^2\|\boldsymbol{Y}\|_2^2\|\boldsymbol{u}_m\|^2 + C\|\boldsymbol{Y}\|^2\|\boldsymbol{Y}\|_2^2.
\end{aligned}$$

At last, by Hölder Inequality and Young Inequality, we have

$$I_5 \le \frac{1}{10}|A_1\boldsymbol{u}_m|_2^2 + C(|\boldsymbol{u}_m|_2^2 + |\boldsymbol{Y}|_2^2 + \|T_m\|^2).$$

Thus, we get

$$\begin{aligned}
&\frac{\mathrm{d}}{\mathrm{d}t}\|\boldsymbol{u}_m\|^2 + |A_1\boldsymbol{u}_m|_2^2 \\
\le& C|A_1\boldsymbol{u}_m|_2^2\|\boldsymbol{u}_m\| + C\|\boldsymbol{u}_m\|^{10} \\
&+ C(\|\boldsymbol{Y}\|^8 + \|\boldsymbol{Y}\|^{1/2}\|\boldsymbol{Y}\|_2^{3/2} + \|\boldsymbol{Y}\|^2\|\boldsymbol{Y}\|_2^2)\|\boldsymbol{u}_m\|^2 \\
&+ C(\|\boldsymbol{Y}\|^{5/2}\|\boldsymbol{Y}\|_2^{3/2} + \|\boldsymbol{Y}\|^2\|\boldsymbol{Y}\|_2^2 + |\boldsymbol{u}_m|_2^2 + |\boldsymbol{Y}|_2^2 + \|T_m\|^2). \qquad (3.4.26)
\end{aligned}$$

Similar to the above inequality,

$$\begin{aligned}
\frac{\mathrm{d}\|T_m\|^2}{\mathrm{d}t} + \frac{1}{2}|A_2T_m|_2^2 \le& c(|\boldsymbol{u}_m|_2^2\|\boldsymbol{u}_m\|^6 + |\boldsymbol{Y}|_2^2\|\boldsymbol{Y}\|^6)\|T_m\|^2 \\
&+ c(\|\boldsymbol{Y}\|^2\|\boldsymbol{Y}\|_2^2 + \|\boldsymbol{u}_m\|^2\|\boldsymbol{u}_m\|_2^2)\|T_m\|^2 + |Q|_2^2. \qquad (3.4.27)
\end{aligned}$$

By using $\boldsymbol{u}_m(0) = 0$ and (3.4.26), we know that there exists $\mathcal{T}^* > 0$ independent with m, such that for any $t \in [0, \mathcal{T}^*]$, $\|\boldsymbol{u}_m(t)\|$ is small enough. According to (3.4.18), (3.4.26) and (3.4.27), applying the methods in proving the local existence of strong solutions to the three-dimensional incompressible Navier-Stokes equations, we prove Proposition 3.4.4.

3.4.4 *The Global Existence and Uniqueness of Strong Solutions*

***A Priori* estimates.** In proving the global existence of strong solutions of the primitive equations of the oceans, the key step is to make the L^p $(4 \le p \le 6)$ estimates of velocity. Thus, we need to analyze the velocity. Let $\boldsymbol{v} = \bar{\boldsymbol{v}} + \tilde{\boldsymbol{v}}$, where $\bar{\boldsymbol{v}}$ is barotrophic, and $\tilde{\boldsymbol{v}}$ is baroclinic,

$$\bar{\boldsymbol{v}} = \int_{-1}^{0} \boldsymbol{v}\,\mathrm{d}z, \qquad \tilde{\boldsymbol{v}} = \boldsymbol{v} - \bar{\boldsymbol{v}}.$$

According to (3.4.13), we have

$$\bar{\tilde{\boldsymbol{v}}} = \int_{-1}^{0} \tilde{\boldsymbol{v}} \mathrm{d}z = 0, \quad \boldsymbol{\nabla} \cdot \bar{\boldsymbol{v}} = 0. \tag{3.4.28}$$

Next, we give the equations of $\tilde{\boldsymbol{v}}$ and $\bar{\boldsymbol{v}}$. Integrating equation (3.4.11) from z to -1 with respect to z, and then using boundary conditions (3.4.14)-(3.4.16) and (3.4.28), we get

$$\begin{aligned} &\frac{\partial \bar{\boldsymbol{v}}}{\partial t} + (\bar{\boldsymbol{v}} \cdot \boldsymbol{\nabla})\bar{\boldsymbol{v}} + \overline{\tilde{\boldsymbol{v}}\mathrm{div}\tilde{\boldsymbol{v}} + (\tilde{\boldsymbol{v}} \cdot \boldsymbol{\nabla})\tilde{\boldsymbol{v}}} + f\boldsymbol{k} \times \bar{\boldsymbol{v}} + \boldsymbol{\nabla} p_s - \int_{-1}^{0}\int_{-1}^{z} \boldsymbol{\nabla} T \mathrm{d}z' \mathrm{d}z \\ &- \Delta \bar{\boldsymbol{v}} = 0 \text{ in } M, \end{aligned} \tag{3.4.29}$$

subtracting (3.4.29) from equation (3.4.11), we know that $\tilde{\boldsymbol{v}}$ satisfies the following equation,

$$\begin{aligned} &\frac{\partial \tilde{\boldsymbol{v}}}{\partial t} + (\tilde{\boldsymbol{v}} \cdot \boldsymbol{\nabla})\tilde{\boldsymbol{v}} + W(\tilde{\boldsymbol{v}})\frac{\partial \tilde{\boldsymbol{v}}}{\partial z} + (\tilde{\boldsymbol{v}} \cdot \boldsymbol{\nabla})\bar{\boldsymbol{v}} + (\bar{\boldsymbol{v}} \cdot \boldsymbol{\nabla})\tilde{\boldsymbol{v}} - \overline{(\tilde{\boldsymbol{v}}\mathrm{div}\tilde{\boldsymbol{v}} + (\tilde{\boldsymbol{v}} \cdot \boldsymbol{\nabla})\tilde{\boldsymbol{v}})} \\ &+ f\boldsymbol{k} \times \tilde{\boldsymbol{v}} - \int_{-1}^{z} \boldsymbol{\nabla} T \mathrm{d}z' + \int_{-1}^{0}\int_{-1}^{z} \boldsymbol{\nabla} T \mathrm{d}z' \mathrm{d}z - \Delta \tilde{\boldsymbol{v}} - \frac{\partial^2 \tilde{\boldsymbol{v}}}{\partial z^2} = 0 \ \text{ in } \Omega, \end{aligned} \tag{3.4.30}$$

with the boundary conditions,

$$\frac{\partial \tilde{\boldsymbol{v}}}{\partial z} = 0 \text{ on } \Gamma_u, \ \frac{\partial \tilde{\boldsymbol{v}}}{\partial z} = 0 \text{ on } \Gamma_b, \ \tilde{\boldsymbol{v}} \cdot \boldsymbol{n} = 0, \ \frac{\partial \tilde{\boldsymbol{v}}}{\partial \boldsymbol{n}} \times \boldsymbol{n} = 0 \text{ on } \Gamma_l. \tag{3.4.31}$$

Now we make *a priori* estimates of local solutions of system (3.4.11)-(3.4.17) obtained in Proposition 3.4.4, and prove the global existence of strong solutions.

L^2 estimates of $\boldsymbol{v}, T$. Taking $L^2(\Omega)$ inner product of equation (3.4.12) with T, we have

$$\begin{aligned} &\frac{1}{2}\frac{\mathrm{d}|T|_2^2}{\mathrm{d}t} + |\boldsymbol{\nabla} T|_2^2 + |T_z|_2^2 + \alpha_s|T(z=0)|_2^2 \\ &= \int_\Omega QT - \int_\Omega \left(\boldsymbol{v}\boldsymbol{\nabla} T - \left(\int_{-1}^{z} \boldsymbol{\nabla}\boldsymbol{v} \mathrm{d}z'\right)\frac{\partial T}{\partial z}\right) T. \end{aligned}$$

By integration by parts, we get

$$\frac{1}{2}\frac{\mathrm{d}|T|_2^2}{\mathrm{d}t} + |\boldsymbol{\nabla} T|_2^2 + |T_z|_2^2 + \alpha_s|T(z=0)|_2^2 = \int_\Omega QT \le |Q|_2|T|_2.$$

Using

$$|T|_2^2 \le 2|T_z|_2^2 + 2|T(z=0)|_2^2$$

and Cauchy-Schwarz Inequality, we have

$$\frac{\mathrm{d}|T|_2^2}{\mathrm{d}t} + 2|\nabla T|_2^2 + |T_z|_2^2 + \alpha_s |T(z=0)|_2^2 \leq 2(1+\frac{1}{\alpha_s})|Q|_2^2.$$

By Gronwall Inequality, we derive from the above inequalities

$$|T|_2^2 \leq \mathrm{e}^{-\frac{1}{2(1+1/\alpha_s)}} |T_0|_2^2 + (2+2/\alpha_s)^2 |Q|_2^2, \tag{3.4.32}$$

$$\int_0^t \left[|\nabla T(s)|_2^2 + |T_z(s)|_2^2 + \alpha_s |T(z=0)(s)|_2^2\right] \mathrm{d}s \leq 2(1+1/\alpha_s)|Q|_2^2 t + |T_0|_2^2. \tag{3.4.33}$$

Taking $L^2(\Omega) \times L^2(\Omega)$ inner product of equation (3.4.11) with $\boldsymbol{v}$, we get

$$\begin{aligned}\frac{1}{2}\frac{\mathrm{d}|\boldsymbol{v}|_2^2}{\mathrm{d}t} + |\nabla \boldsymbol{v}|_2^2 + |\boldsymbol{v}_z|_2^2 = &-\int_\Omega \left[(\boldsymbol{v}\cdot\nabla)\boldsymbol{v} - \left(\int_{-1}^z \nabla\cdot\boldsymbol{v}\mathrm{d}z'\right)\frac{\partial \boldsymbol{v}}{\partial z}\right]\cdot\boldsymbol{v} \\ &+ \int_\Omega \left(f\boldsymbol{k}\times\boldsymbol{v} + \nabla p_s - \nabla\left(\int_{-1}^z T\mathrm{d}z'\right)\right)\cdot\boldsymbol{v}.\end{aligned}$$

By integration by parts and div $\int_0^1 \boldsymbol{v}\mathrm{d}z = 0$, we have

$$\frac{1}{2}\frac{\mathrm{d}|\boldsymbol{v}|_2^2}{\mathrm{d}t} + |\nabla \boldsymbol{v}|_2^2 + |\boldsymbol{v}_z|_2^2 = -\int_\Omega \int_{-1}^z T\mathrm{d}z'(\nabla\cdot\boldsymbol{v}) \leq |T|_2 |\nabla \boldsymbol{v}|_2.$$

By Cauchy-Schwarz Inequality, (3.4.32) and the above inequality, we have

$$\frac{\mathrm{d}|\boldsymbol{v}|_2^2}{\mathrm{d}t} + |\nabla \boldsymbol{v}|_2^2 + |\boldsymbol{v}_z|_2^2 \leq |T|_2^2 \leq (|T_0|_2^2 + (2+2/\alpha_s)|Q|_2^2).$$

Using the inequality (see reference [76, Vol. I, p. 55])

$$|\boldsymbol{v}|_2^2 \leq C_M |\nabla \boldsymbol{v}|_2^2$$

and Gronwall Inequality, we get

$$|\boldsymbol{v}|_2^2 \leq \mathrm{e}^{-\frac{t}{C_M}}(|\bar{\boldsymbol{v}}_0|_2^2 + |\tilde{\boldsymbol{v}}_0|_2^2) + C_M\left[|T_0|_2^2 + (2+2/\alpha_s)^2|Q|_2^2\right], \tag{3.4.34}$$

$$\begin{aligned}&\int_0^t \left[|\nabla \boldsymbol{v}(s)|_2^2 + |\boldsymbol{v}_z(s)|_2^2\right] \mathrm{d}s \\ &\leq (|T_0|_2^2 + (2+2/\alpha_s)^2|Q|_2^2)t + 2(|\bar{\boldsymbol{v}}_0|_2^2 + |\tilde{\boldsymbol{v}}_0|_2^2).\end{aligned} \tag{3.4.35}$$

According to (3.4.32)-(3.4.35), we obtain

$$\begin{aligned}&|\boldsymbol{v}(t)|_2^2 + \int_0^t \left[|\nabla \boldsymbol{v}|_2^2 + |\boldsymbol{v}_z|_2^2\right]\mathrm{d}s + |T(t)|_2^2 \\ &+ \int_0^t \left[|\nabla T|_2^2 + |T_z|_2^2 + \alpha_s|T(z=0)|_2^2\right]\mathrm{d}s \leq K_1(t),\end{aligned} \tag{3.4.36}$$

where

$$\begin{aligned}K_1(t)=&2(1+1/\alpha_s)|Q|_2^2 t+3(|\bar{\boldsymbol{v}}_0|_2^2+|\tilde{\boldsymbol{v}}_0|_2^2)\\&+(1+C_M+t)\left[|T_0|_2^2+(2+2/\alpha_s)^2|Q|_2^2\right].\end{aligned}$$

L^6 estimates of $\tilde{\boldsymbol{v}}$, T. Taking $L^2(\Omega)\times L^2(\Omega)$ inner product of equation (3.4.30) with $|\tilde{\boldsymbol{v}}|^4\tilde{\boldsymbol{v}}$, we get

$$\begin{aligned}&\frac{1}{6}\frac{\mathrm{d}|\tilde{\boldsymbol{v}}|_6^6}{\mathrm{d}t}+\int_\Omega(|\nabla\tilde{\boldsymbol{v}}|^2|\tilde{\boldsymbol{v}}|^4+|\nabla|\tilde{\boldsymbol{v}}|^2|^2|\tilde{\boldsymbol{v}}|^2)+\int_\Omega(|\tilde{\boldsymbol{v}}_z|^2|\tilde{\boldsymbol{v}}|^4+|\partial_z|\tilde{\boldsymbol{v}}|^2|^2|\tilde{\boldsymbol{v}}|^2)\\
&=-\int_\Omega\left\{(\tilde{\boldsymbol{v}}\cdot\nabla)\tilde{\boldsymbol{v}}-\left(\int_{-1}^z\nabla\cdot\tilde{\boldsymbol{v}}\mathrm{d}z'\right)\frac{\partial\tilde{\boldsymbol{v}}}{\partial z}+(\tilde{\boldsymbol{v}}\cdot\nabla)\bar{\boldsymbol{v}}+(\bar{\boldsymbol{v}}\cdot\nabla)\tilde{\boldsymbol{v}}\right.\\
&\quad-\overline{[(\tilde{\boldsymbol{v}}\cdot\nabla)\tilde{\boldsymbol{v}}+(\nabla\cdot\tilde{\boldsymbol{v}})\tilde{\boldsymbol{v}}]}+f\boldsymbol{k}\times\tilde{\boldsymbol{v}}\\
&\quad\left.-\nabla\left(\int_{-1}^z T\mathrm{d}z'-\int_{-1}^0\int_{-1}^z T\mathrm{d}z'\mathrm{d}z\right)\right\}\cdot|\tilde{\boldsymbol{v}}|^4\tilde{\boldsymbol{v}}.\end{aligned}$$

As is similar to (3.2.46),

$$\begin{aligned}&\frac{1}{6}\frac{\mathrm{d}|\tilde{\boldsymbol{v}}|_6^6}{\mathrm{d}t}+\int_\Omega(|\nabla\tilde{\boldsymbol{v}}|^2|\tilde{\boldsymbol{v}}|^4+|\nabla|\tilde{\boldsymbol{v}}|^2|^2|\tilde{\boldsymbol{v}}|^2)+\int_\Omega(|\tilde{\boldsymbol{v}}_z|^2|\tilde{\boldsymbol{v}}|^4+|\partial_z|\tilde{\boldsymbol{v}}|^2|^2|\tilde{\boldsymbol{v}}|^2)\\
&\le C\int_M\left[|\bar{\boldsymbol{v}}|\int_{-1}^0|\nabla\tilde{\boldsymbol{v}}||\tilde{\boldsymbol{v}}|^5\mathrm{d}z\right]+C\int_M\left[\left(\int_{-1}^0|\tilde{\boldsymbol{v}}|^2\mathrm{d}z\right)\left(\int_{-1}^0|\nabla\tilde{\boldsymbol{v}}||\tilde{\boldsymbol{v}}|^4\mathrm{d}z\right)\right]\\
&\quad+C\int_M\left[|\bar{T}|\int_{-1}^0|\nabla\tilde{\boldsymbol{v}}||\tilde{\boldsymbol{v}}|^4\mathrm{d}z\right]\\
&\le C\int_M\left[|\bar{\boldsymbol{v}}|\left(\int_{-1}^0|\nabla\tilde{\boldsymbol{v}}|^2|\tilde{\boldsymbol{v}}|^4\mathrm{d}z\right)^{1/2}\left(\int_{-1}^0|\tilde{\boldsymbol{v}}|^6\mathrm{d}z\right)^{1/2}\right]\\
&\quad+C\int_M\left[\left(\int_{-1}^0|\tilde{\boldsymbol{v}}|^2\mathrm{d}z\right)\left(\int_{-1}^0|\nabla\tilde{\boldsymbol{v}}|^2|\tilde{\boldsymbol{v}}|^4\mathrm{d}z\right)^{1/2}\left(\int_{-1}^0|\tilde{\boldsymbol{v}}|^4\mathrm{d}z\right)^{1/2}\right]\\
&\quad+C\int_M\left[|\bar{T}|\left(\int_{-1}^0|\nabla\tilde{\boldsymbol{v}}|^2|\tilde{\boldsymbol{v}}|^4\mathrm{d}z\right)^{1/2}\left(\int_{-1}^0|\tilde{\boldsymbol{v}}|^4\mathrm{d}z\right)^{1/2}\right]\\
&\le C\|\bar{\boldsymbol{v}}\|_{L^4(M)}\left(\int_\Omega|\nabla\tilde{\boldsymbol{v}}|^2|\tilde{\boldsymbol{v}}|^4\right)^{1/2}\left(\int_M\left(\int_{-1}^0|\tilde{\boldsymbol{v}}|^6\mathrm{d}z\right)^2\right)^{1/4}\\
&\quad+C\left(\int_M\left(\int_{-1}^0|\tilde{\boldsymbol{v}}|^2\mathrm{d}z\right)^4\right)^{1/4}\left(\int_\Omega|\nabla\tilde{\boldsymbol{v}}|^2|\tilde{\boldsymbol{v}}|^4\right)^{1/2}\left(\int_M\left(\int_{-1}^0|\tilde{\boldsymbol{v}}|^4\mathrm{d}z\right)^2\right)^{1/4}\\
&\quad+C\|\bar{T}\|_{L^4(M)}\left(\int_\Omega|\nabla\tilde{\boldsymbol{v}}|^2|\tilde{\boldsymbol{v}}|^4\right)^{1/2}\left(\int_M\left(\int_{-1}^0|\tilde{\boldsymbol{v}}|^4\mathrm{d}z\right)^2\right)^{1/4}.\end{aligned}\tag{3.4.37}$$

By Minkowski Inequality, we have

$$\left(\int_M\left(\int_{-1}^0|\tilde{\boldsymbol{v}}|^6\mathrm{d}z\right)^2\right)^{1/2}\le C\int_{-1}^0\left(\int_M|\tilde{\boldsymbol{v}}|^{12}\mathrm{d}x\mathrm{d}y\right)^{1/2}\mathrm{d}z.$$

Using (3.2.24), where S^2 is replaced by M, we get

$$\int_M|\tilde{\boldsymbol{v}}|^{12}\le C\left(\int_M|\tilde{\boldsymbol{v}}|^6\right)\left(\int_M|\tilde{\boldsymbol{v}}|^4|\boldsymbol{\nabla}\tilde{\boldsymbol{v}}|^2\right)+\left(\int_M|\tilde{\boldsymbol{v}}|^6\right)^2.$$

According to the above two inequalities, and applying Cauchy-Schwarz Inequality, we obtain

$$\left(\int_M\left(\int_{-1}^0|\tilde{\boldsymbol{v}}|^6\mathrm{d}z\right)^2\right)^{1/2}\le C|\tilde{\boldsymbol{v}}|_6^3\left(\int_\Omega|\tilde{\boldsymbol{v}}|^4|\boldsymbol{\nabla}\tilde{\boldsymbol{v}}|^2\right)^{1/2}+|\tilde{\boldsymbol{v}}|_6^6. \tag{3.4.38}$$

Similarly, by Minkowski Inequality and $\|\boldsymbol{v}\|_{L^8(M)}\le C\|\boldsymbol{v}\|_{L^6(M)}^{\frac{3}{4}}\cdot\|\boldsymbol{v}\|_{H^1(M)}^{\frac{1}{4}}$, we also get

$$\left(\int_M\left(\int_{-1}^0|\tilde{\boldsymbol{v}}|^4\mathrm{d}z\right)^2\right)^{1/2}\le C\int_{-1}^0\left(\int_M|\tilde{\boldsymbol{v}}|^8\right)^{1/2}\mathrm{d}z$$

$$\le C\int_{-1}^0\|\tilde{\boldsymbol{v}}\|_{L^6(M)}^3(\|\boldsymbol{\nabla}\tilde{\boldsymbol{v}}\|_{L^2(M)}+\|\tilde{\boldsymbol{v}}\|_{L^2(M)})\mathrm{d}z\le C|\tilde{\boldsymbol{v}}|_6^3(|\boldsymbol{\nabla}\tilde{\boldsymbol{v}}|_2+|\tilde{\boldsymbol{v}}|_2), \tag{3.4.39}$$

$$\left(\int_M\left(\int_{-1}^0|\tilde{\boldsymbol{v}}|^2\mathrm{d}z\right)^4\right)^{1/4}\le C\int_{-1}^0\left(\int_M|\tilde{\boldsymbol{v}}|^8\right)^{1/4}\mathrm{d}z$$

$$\le C\int_{-1}^0\|\tilde{\boldsymbol{v}}\|_{L^6(M)}^{3/2}\left(\|\boldsymbol{\nabla}\tilde{\boldsymbol{v}}\|_{L^2(M)}^{1/2}+\|\tilde{\boldsymbol{v}}\|_{L^2(M)}^{1/2}\right)\mathrm{d}z$$

$$\le C|\tilde{\boldsymbol{v}}|_6^{3/2}\left(|\boldsymbol{\nabla}\tilde{\boldsymbol{v}}|_2^{1/2}+|\tilde{\boldsymbol{v}}|_2^{1/2}\right). \tag{3.4.40}$$

By (3.4.37)-(3.4.40), $\|\boldsymbol{v}\|_{L^4(M)}\le C\|\boldsymbol{v}\|_{L^2(M)}^{\frac{1}{2}}\|\boldsymbol{v}\|_{H^1(M)}^{\frac{1}{2}}$, Young Inequality and Cauchy-Schwarz Inequality, we obtain

$$\frac{\mathrm{d}|\tilde{\boldsymbol{v}}|_6^6}{\mathrm{d}t}+\int_\Omega(|\boldsymbol{\nabla}\tilde{\boldsymbol{v}}|^2|\tilde{\boldsymbol{v}}|^4+|\boldsymbol{\nabla}|\tilde{\boldsymbol{v}}|^2|^2|\tilde{\boldsymbol{v}}|^2)+\int_\Omega(|\tilde{\boldsymbol{v}}_z|^2|\tilde{\boldsymbol{v}}|^4+|\partial_z|\tilde{\boldsymbol{v}}|^2|^2|\tilde{\boldsymbol{v}}|^2)$$

$$\le c(\|\bar{\boldsymbol{v}}\|_{L^2(M)}^2\|\bar{\boldsymbol{v}}\|_{H^1(M)}^2+|\boldsymbol{\nabla}\tilde{\boldsymbol{v}}|_2^2+|\tilde{\boldsymbol{v}}|_2^2)|\tilde{\boldsymbol{v}}|_6^6+c\|\bar{T}\|_{L^2(M)}^2\|\bar{T}\|_{H^1(M)}^2.$$

Using Gronwall Inequality, we derive from (3.4.36) and the above inequality

$$|\tilde{\boldsymbol{v}}(t)|_6^6+\int_0^t\left(\int_\Omega|\boldsymbol{\nabla}\tilde{\boldsymbol{v}}|^2|\tilde{\boldsymbol{v}}|^4+\int_\Omega|\tilde{\boldsymbol{v}}_z|^2|\tilde{\boldsymbol{v}}|^4\right)\le K_2(t), \tag{3.4.41}$$

where

$$K_2(t) = \mathrm{e}^{K_1^2(t)}(\|\boldsymbol{v}_0\|^6 + K_1^2(t)).$$

Taking $L^2(\Omega)$ inner product of equation (3.4.12) with $|T|^4 T$, we get

$$\frac{1}{6}\frac{\mathrm{d}|T|_6^6}{\mathrm{d}t} + 5\int_\Omega |\nabla T|^2|T|^4 + 5\int_\Omega |T_z|^2|T|^4 + \alpha_s|T(z=0)|_6^6$$

$$= \int_\Omega Q|T|^4 T - \int_\Omega \left[(\boldsymbol{v}\cdot\nabla)T - \left(\int_{-1}^z \nabla\cdot\boldsymbol{v}\mathrm{d}z'\right)\frac{\partial T}{\partial z}\right]|T|^4 T.$$

By integration by parts, Hölder Inequality and Gronwall Inequality, we get

$$|T|_6^6 \le \|Q\|_1 t + \|T_0\|. \tag{3.4.42}$$

After getting the L^6 estimates of $\tilde{\boldsymbol{v}}$, we make the uniform estimates of H^1 norm of $\bar{\boldsymbol{v}}$ with respect of time. Because we have already obtained the uniform estimates of L^2 norm of $\bar{\boldsymbol{v}}$, we only need to make the uniform estimates of L^2 norm of $\nabla\bar{\boldsymbol{v}}$.

L^2 estimates of $\nabla\bar{\boldsymbol{v}}$. Taking $L^2(M)\times L^2(M)$ inner product of equation (3.4.29) with $-\Delta\bar{\boldsymbol{v}}$, we get

$$\frac{1}{2}\frac{\mathrm{d}\|\nabla\bar{\boldsymbol{v}}\|_{L^2(M)}^2}{\mathrm{d}t} + \|\Delta\bar{\boldsymbol{v}}\|_{L^2(M)}^2$$

$$= \int_M \left[(\bar{\boldsymbol{v}}\cdot\nabla)\bar{\boldsymbol{v}} + \int_{-1}^0 (\tilde{\boldsymbol{v}}\mathrm{div}\tilde{\boldsymbol{v}} + (\tilde{\boldsymbol{v}}\cdot\nabla)\tilde{\boldsymbol{v}})\mathrm{d}z\right]\cdot\Delta\bar{\boldsymbol{v}}$$

$$+ \int_M (\mathrm{grad}p_s + f\boldsymbol{k}\times\bar{\boldsymbol{v}})\cdot\Delta\bar{\boldsymbol{v}} + \int_M \left(-\int_{-1}^0\int_{-1}^z \nabla T\mathrm{d}z'\mathrm{d}z\right)\cdot\Delta\bar{\boldsymbol{v}}.$$

By Hölder Inequality, $\|\boldsymbol{v}\|_{L^4(M)} \le C\|\boldsymbol{v}\|_{L^2(M)}^{\frac{1}{2}}\|\boldsymbol{v}\|_{H^1(M)}^{\frac{1}{2}}$ and Young Inequality, we have

$$\left|\int_M ((\bar{\boldsymbol{v}}\cdot\nabla)\,\bar{\boldsymbol{v}}\cdot\Delta\bar{\boldsymbol{v}})\right| \le c\|\bar{\boldsymbol{v}}\|_{L^4(M)}\left(\int_M |\nabla\bar{\boldsymbol{v}}|^4\right)^{\frac{1}{4}}\|\Delta\bar{\boldsymbol{v}}\|_{L^2(M)}$$

$$\le c\|\bar{\boldsymbol{v}}\|_{L^2(M)}^{\frac{1}{2}}\|\bar{\boldsymbol{v}}\|_{H^1(M)}^{\frac{1}{2}}\left(\int_M |\nabla\bar{\boldsymbol{v}}|^2\right)^{\frac{1}{4}}\left(\int_M |\nabla\bar{\boldsymbol{v}}|^2 + \|\Delta\bar{\boldsymbol{v}}\|_{L^2(M)}^2\right)^{\frac{1}{4}}\|\Delta\bar{\boldsymbol{v}}\|_{L^2(M)}$$

$$\le c\left(\|\bar{\boldsymbol{v}}\|_{L^2(M)}^2 + \|\bar{\boldsymbol{v}}\|_{H^1(M)}^2 + \|\bar{\boldsymbol{v}}\|_{L^2(M)}^2\|\bar{\boldsymbol{v}}\|_{H^1(M)}^2\right)\|\nabla\bar{\boldsymbol{v}}\|_{L^2(M)}^2 + \varepsilon\|\Delta\bar{\boldsymbol{v}}\|_{L^2(M)}^2.$$

By Hölder Inequality and Young Inequality, we get

$$\left|\int_M\int_{-1}^0 (\tilde{\boldsymbol{v}}\mathrm{div}\tilde{\boldsymbol{v}} + (\tilde{\boldsymbol{v}}\cdot\nabla)\,\tilde{\boldsymbol{v}})\,\mathrm{d}z\cdot\Delta\bar{\boldsymbol{v}}\right| \le \int_M\int_{-1}^0 |\tilde{\boldsymbol{v}}||\nabla\tilde{\boldsymbol{v}}|\mathrm{d}z|\Delta\bar{\boldsymbol{v}}|$$

$$\le \left[\int_M\left(\int_{-1}^0 |\tilde{\boldsymbol{v}}||\nabla\tilde{\boldsymbol{v}}|\mathrm{d}z\right)^2\right]^{\frac{1}{4}}\left[\int_M\left(\int_{-1}^0 |\nabla\tilde{\boldsymbol{v}}|\mathrm{d}z\right)^2\right]^{\frac{1}{4}}\|\Delta\bar{\boldsymbol{v}}\|_{L^2(M)}$$

$$\le c\int_\Omega |\tilde{\boldsymbol{v}}|^4|\nabla\tilde{\boldsymbol{v}}|^2 + c|\nabla\tilde{\boldsymbol{v}}|_2^2 + \varepsilon\|\Delta\bar{\boldsymbol{v}}\|_{L^2(M)}^2.$$

By integration by parts, (3.4.13) and the boundary condition (3.4.16), we obtain

$$\int_M \mathrm{grad} p_s \cdot \Delta \bar{\boldsymbol{v}} = 0, \quad \int_M \left(-\int_{-1}^0 \int_{-1}^z \boldsymbol{\nabla} T \mathrm{d}z' \mathrm{d}z \right) \cdot \Delta \bar{\boldsymbol{v}} = 0.$$

$(f\boldsymbol{k} \times \bar{\boldsymbol{v}}) \cdot \Delta \bar{\boldsymbol{v}} = 0$ implies

$$\int_M (f\boldsymbol{k} \times \bar{\boldsymbol{v}}) \cdot \Delta \bar{\boldsymbol{v}} = 0.$$

Thus, choosing ε small enough, we get

$$\begin{aligned}
&\frac{\mathrm{d}\|\boldsymbol{\nabla} \bar{\boldsymbol{v}}\|^2_{L^2(M)}}{\mathrm{d}t} + \|\Delta \bar{\boldsymbol{v}}\|^2_{L^2(M)} \\
\leq & c(\|\bar{\boldsymbol{v}}\|^2_{L^2(M)} + \|\bar{\boldsymbol{v}}\|^2_{H^1(M)} + \|\bar{\boldsymbol{v}}\|^2_{L^2(M)} \|\bar{\boldsymbol{v}}\|^2_{H^1(M)}) \|\boldsymbol{\nabla} \bar{\boldsymbol{v}}\|^2_{L^2(M)} \\
& + c \int_\Omega |\tilde{\boldsymbol{v}}|^4 |\boldsymbol{\nabla} \tilde{\boldsymbol{v}}|^2 + c|\boldsymbol{\nabla} \tilde{\boldsymbol{v}}|^2_2.
\end{aligned}$$

According to (3.4.36) and (3.4.41), we derive from the above inequality

$$\|\boldsymbol{\nabla} \bar{\boldsymbol{v}}(t)\|^2_{L^2(M)} + \int_0^t \|\Delta \bar{\boldsymbol{v}}(s)\|^2_{L^2(M)} \mathrm{d}s \leq K_3(t), \tag{3.4.43}$$

where

$$K_3(t) = \mathrm{e}^{K_1^2(t)} (\|\boldsymbol{v}_0\|^2 + K_1(t) + K_2(t)).$$

In the following, we begin to estimate H^1 norm of $\boldsymbol{v}$, T. Firstly we make the uniform estimates of L^2 norm of $\boldsymbol{v}_z$ with respect to time. Then we make the uniform estimates of L^2 norm of $\boldsymbol{\nabla v}$. At last we give the uniform estimates of H^1 norm of T.

L^2 estimates of $\partial_z \boldsymbol{v}$. Derivating equation (3.4.11) with respect to z, we get the following equation

$$\begin{aligned}
&\frac{\partial \boldsymbol{v}_z}{\partial t} - \Delta \boldsymbol{v}_z - \frac{\partial^2 \boldsymbol{v}_z}{\partial z^2} + (\boldsymbol{v} \cdot \boldsymbol{\nabla}) \boldsymbol{v}_z + W(\boldsymbol{v}) \frac{\partial \boldsymbol{v}_z}{\partial z} + (\boldsymbol{v}_z \cdot \boldsymbol{\nabla}) \boldsymbol{v} \\
& - (\mathrm{div} \boldsymbol{v}) \boldsymbol{v}_z + f\boldsymbol{k} \times \boldsymbol{v}_z - \boldsymbol{\nabla} T = 0,
\end{aligned}$$

where $\boldsymbol{v}_z = \partial_z \boldsymbol{v}$. Taking $L^2(\Omega) \times L^2(\Omega)$ inner product of the above equation with $\boldsymbol{v}_z$, we have

$$\begin{aligned}
&\frac{1}{2} \frac{\mathrm{d}|\boldsymbol{v}_z|_2^2}{\mathrm{d}t} + \int_\Omega |\boldsymbol{\nabla} \boldsymbol{v}_z|^2 + \int_\Omega \left| \frac{\partial \boldsymbol{v}_z}{\partial z} \right|^2 \\
= & - \int_\Omega \left((\boldsymbol{v} \cdot \boldsymbol{\nabla}) \boldsymbol{v}_z + W(\boldsymbol{v}) \frac{\partial \boldsymbol{v}_z}{\partial z} \right) \cdot \boldsymbol{v}_z - \int_\Omega ((\boldsymbol{v}_z \cdot \boldsymbol{\nabla}) \boldsymbol{v} - (\mathrm{div} \boldsymbol{v}) \boldsymbol{v}_z) \cdot \boldsymbol{v}_z \\
& - \int_\Omega (f\boldsymbol{k} \times \boldsymbol{v}_z) \cdot \boldsymbol{v}_z - \int_\Omega \boldsymbol{\nabla} T \cdot \boldsymbol{v}_z.
\end{aligned}$$

By integration by parts, we get

$$\int_\Omega \left((\boldsymbol{v} \cdot \boldsymbol{\nabla})\, \boldsymbol{v}_z + W(\boldsymbol{v}) \frac{\partial \boldsymbol{v}_z}{\partial z} \right) \cdot \boldsymbol{v}_z = 0.$$

By integration by parts, Hölder Inequality, $|\boldsymbol{v}|_3 \le C|\boldsymbol{v}|_2^{\frac{1}{2}} \|\boldsymbol{v}\|_1^{\frac{1}{2}}$ and Young Inequality, we obtain

$$\begin{aligned}
&-\int_\Omega ((\boldsymbol{v}_z \cdot \boldsymbol{\nabla})\boldsymbol{v} - (\mathrm{div}\boldsymbol{v})\boldsymbol{v}_z) \cdot \boldsymbol{v}_z \le c \int_\Omega |\boldsymbol{v}||\boldsymbol{v}_z||\boldsymbol{\nabla} \boldsymbol{v}_z| \\
\le & c|\boldsymbol{v}|_6 |\boldsymbol{v}_z|_3 \left(\int_\Omega |\boldsymbol{\nabla} \boldsymbol{v}_z|^2 \right)^{\frac{1}{2}} \le c|\boldsymbol{v}|_6 |\boldsymbol{v}_z|_2^{\frac{1}{2}} \|\boldsymbol{v}_z\|^{\frac{1}{2}} \left(\int_\Omega |\boldsymbol{\nabla} \boldsymbol{v}_z|^2 \right)^{\frac{1}{2}} \\
\le & \varepsilon \left(|\boldsymbol{\nabla} \boldsymbol{v}_z|_2^2 + \left| \frac{\partial \boldsymbol{v}_z}{\partial z} \right|_2^2 \right) + c \left(|\boldsymbol{\nabla} \bar{\boldsymbol{v}}|_2^4 + |\tilde{\boldsymbol{v}}|_6^4 \right) |\boldsymbol{v}_z|_2^2.
\end{aligned}$$

Similarly, we get

$$-\int_\Omega \boldsymbol{\nabla} T \cdot \boldsymbol{v}_z = \int_\Omega T \mathrm{div} \boldsymbol{v}_z \le c|T|_2^2 + \varepsilon |\boldsymbol{\nabla} \boldsymbol{v}_z|_2^2.$$

Thus, choosing ε small enough, we have

$$\frac{\mathrm{d}|\boldsymbol{v}_z|_2^2}{\mathrm{d}t} + \int_\Omega |\boldsymbol{\nabla} \boldsymbol{v}_z|^2 + \int_\Omega \left| \frac{\partial \boldsymbol{v}_z}{\partial z} \right|^2 \le c(|\boldsymbol{\nabla} \bar{\boldsymbol{v}}|_2^4 + |\tilde{\boldsymbol{v}}|_6^4)|\boldsymbol{v}_z|_2^2 + c|T|_2^2.$$

According to (3.4.36), (3.4.41) and (3.4.43), we derive from the above inequality that

$$|\boldsymbol{v}_z(t)|_2^2 + \int_0^t \left(|\boldsymbol{\nabla} \boldsymbol{v}_z(s)|_2^2 + \left| \frac{\partial \boldsymbol{v}_z}{\partial z}(s) \right|_2^2 \right) \mathrm{d}s \le K_4(t), \tag{3.4.44}$$

where

$$K_4(t) = \mathrm{e}^{(K_3^2(t) + K_2^{\frac{2}{3}}(t))t} (\|\boldsymbol{v}_0\|^2 + K_1(t)).$$

L^2 estimates of $\boldsymbol{\nabla v}$. Taking $L^2(\Omega) \times L^2(\Omega)$ inner product of equation (3.4.11) with $-\Delta \boldsymbol{v}$, we get

$$\begin{aligned}
&\frac{1}{2} \frac{\mathrm{d}|\boldsymbol{\nabla} \boldsymbol{v}|_2^2}{\mathrm{d}t} + \int_\Omega |\Delta \boldsymbol{v}|^2 + \int_\Omega |\boldsymbol{\nabla} \boldsymbol{v}_z|^2 \\
= & \int_\Omega ((\boldsymbol{v} \cdot \boldsymbol{\nabla})\boldsymbol{v} + W(\boldsymbol{v})\boldsymbol{v}_z) \cdot \Delta \boldsymbol{v} - \int_\Omega \left(\int_{-1}^z \boldsymbol{\nabla} T \mathrm{d}z' \right) \cdot \Delta \boldsymbol{v} \\
& + \int_\Omega (f \boldsymbol{k} \times \boldsymbol{v} + \boldsymbol{\nabla} p_s) \cdot \Delta \boldsymbol{v}.
\end{aligned}$$

By Hölder Inequality, $|\boldsymbol{v}|_3 \le C|\boldsymbol{v}|_2^{\frac{1}{2}}\|\boldsymbol{v}\|_1^{\frac{1}{2}}$ and Young Inequality, we have

$$\begin{aligned}
&\left|\int_\Omega (\boldsymbol{v}\cdot\boldsymbol{\nabla})\boldsymbol{v}\cdot\Delta\boldsymbol{v}\right| \le \int_\Omega |\boldsymbol{v}||\boldsymbol{\nabla}\boldsymbol{v}||\Delta\boldsymbol{v}| \le c|\boldsymbol{v}|_6^2\left(\int_\Omega|\boldsymbol{\nabla}\boldsymbol{v}|^3\right)^{\frac{2}{3}} + \varepsilon|\Delta\boldsymbol{v}|_2^2\\
\le& c|\boldsymbol{v}|_6^2\left(\int_\Omega|\boldsymbol{\nabla}\boldsymbol{v}|^2\right)^{\frac{2}{4}}\left(\int_\Omega\left(|\boldsymbol{\nabla}\boldsymbol{v}|^2+|\boldsymbol{\nabla}\boldsymbol{v}_z|^2+|\Delta\boldsymbol{v}|^2\right)\right)^{\frac{2}{4}} + \varepsilon|\Delta\boldsymbol{v}|_2^2\\
\le& c\left(|\boldsymbol{v}|_6^2+|\boldsymbol{v}|_6^4\right)|\boldsymbol{\nabla}\boldsymbol{v}|_2^2 + 2\varepsilon\left(|\Delta\boldsymbol{v}|_2^2+|\boldsymbol{\nabla}\boldsymbol{v}_z|_2^2\right).
\end{aligned}\tag{3.4.45}$$

Similarly, we get

$$\begin{aligned}
&\left|\int_\Omega W(\boldsymbol{v})\,\boldsymbol{v}_z\cdot\Delta\boldsymbol{v}\right| \le \int_M\left(\int_{-1}^0|\boldsymbol{\nabla}\boldsymbol{v}|\mathrm{d}z\int_{-1}^0|\boldsymbol{v}_z||\Delta\boldsymbol{v}|\mathrm{d}z\right)\\
\le& \int_M\left[\left(\int_{-1}^0|\boldsymbol{\nabla}\boldsymbol{v}|^2\mathrm{d}z\right)^{\frac{1}{2}}\left(\int_{-1}^0|\boldsymbol{v}_z|^2\mathrm{d}z\right)^{\frac{1}{2}}\left(\int_{-1}^0|\Delta\boldsymbol{v}|^2\mathrm{d}z\right)^{\frac{1}{2}}\right]\\
\le& c\left[\int_{-1}^0\left(\int_M|\boldsymbol{\nabla}\boldsymbol{v}|^4\right)^{\frac{1}{2}}\mathrm{d}z\right]\left[\int_{-1}^0\left(\int_M|\boldsymbol{v}_z|^4\right)^{\frac{1}{2}}\mathrm{d}z\right] + \varepsilon|\Delta\boldsymbol{v}|_2^2\\
\le& c\left[\int_{-1}^0\left(\int_M|\boldsymbol{\nabla}\boldsymbol{v}|^2\right)^{\frac{1}{2}}\left(\int_M\left(|\boldsymbol{\nabla}\boldsymbol{v}|^2+|\Delta\boldsymbol{v}|^2\right)\right)^{\frac{1}{2}}\mathrm{d}z\right]\\
&\cdot c\left[\int_{-1}^0\left(\int_M|\boldsymbol{v}_z|^2\right)^{\frac{1}{2}}\left(\int_M\left(|\boldsymbol{\nabla}\boldsymbol{v}_z|^2+|\boldsymbol{v}_z|^2\right)\right)^{\frac{1}{2}}\mathrm{d}z\right] + \varepsilon|\Delta\boldsymbol{v}|_2^2\\
\le& c\left(|\boldsymbol{\nabla}\boldsymbol{v}|_2^2+|\boldsymbol{\nabla}\boldsymbol{v}|_2|\Delta\boldsymbol{v}|_2\right)|\boldsymbol{v}_z|_2\left(|\boldsymbol{v}_z|_2+|\boldsymbol{\nabla}\boldsymbol{v}_z|_2\right) + \varepsilon|\Delta\boldsymbol{v}|_2^2\\
\le& 2\varepsilon|\Delta\boldsymbol{v}|_2^2 + c\left[2|\boldsymbol{v}_z|_2^2+|\boldsymbol{v}_z|_2^4+\left(|\boldsymbol{v}_z|_2^2+1\right)|\boldsymbol{\nabla}\boldsymbol{v}_z|_2^2\right]|\boldsymbol{\nabla}\boldsymbol{v}|_2^2.
\end{aligned}\tag{3.4.46}$$

Thus, choosing ε small enough, we obtain

$$\begin{aligned}
&\frac{\mathrm{d}|\boldsymbol{\nabla}\boldsymbol{v}|_2^2}{\mathrm{d}t} + \int_\Omega|\Delta\boldsymbol{v}|^2 + \int_\Omega|\boldsymbol{\nabla}\boldsymbol{v}_z|^2\\
\le& c\left[|\boldsymbol{v}|_6^2+|\boldsymbol{v}|_6^4+2|\boldsymbol{v}_z|_2^2+|\boldsymbol{v}_z|_2^4+(|\boldsymbol{v}_z|_2^2+1)|\boldsymbol{\nabla}\boldsymbol{v}_z|_2^2)\right]|\boldsymbol{\nabla}\boldsymbol{v}|_2^2 + c|\boldsymbol{\nabla}T|_2^2.
\end{aligned}$$

Applying (3.4.36), (3.4.41), (3.4.43) and (3.4.44), we derive from the above inequality

$$|\boldsymbol{\nabla}\boldsymbol{v}(t)|_2^2 + \int_0^t(|\Delta\boldsymbol{v}(s)|_2^2+|\boldsymbol{\nabla}\boldsymbol{v}_z(s)|_2^2)\mathrm{d}s \le K_5(t), \tag{3.4.47}$$

where

$$K_5(t) = \mathrm{e}^{(1+K_2^{\frac{2}{3}}(t)+K_3^{\frac{2}{3}}(t)+K_4^2(t))t+(K_4(t)+1)K_4(t)}(\|\boldsymbol{v}_0\|^2+K_1(t)).$$

H^1 estimates of T. Taking $L^2(\Omega)$ inner product of equation (3.4.12) with $-\Delta T - T_{zz}$, we get

$$\begin{aligned}&\frac{1}{2}\frac{\mathrm{d}(|\nabla T|_2^2+|T_z|_2^2+\alpha_s|T|_{z=0}|_2^2)}{\mathrm{d}t}+|\Delta T|_2^2+(|\nabla T_z|_2^2+\alpha_s|\nabla T|_{z=0}|_2^2)\\&+|\nabla T_{zz}|_2^2\\&=\int_\Omega((\boldsymbol{v}\cdot\nabla)T+W(\boldsymbol{v})\frac{\partial T}{\partial z})(\Delta T+T_{zz})-\int_\Omega Q(\Delta T+T_{zz}).\end{aligned}$$

Like (3.4.45),

$$\left|\int_\Omega(\boldsymbol{v}\cdot\nabla)T(\Delta T+T_{zz})\right|\le c(|\boldsymbol{v}|_6^2+|\boldsymbol{v}|_6^4)|\nabla T|_2^2+2\varepsilon(|\Delta T|_2^2+|T_{zz}|_2^2+|\nabla T_z|_2^2).$$

Like (3.4.46),

$$\begin{aligned}&|\int_\Omega W(\boldsymbol{v})T_z(\Delta T+T_{zz})|\\&\le\int_M\left[\left(\int_{-1}^0|\nabla\boldsymbol{v}|^2\mathrm{d}z\right)^{\frac{1}{2}}\left(\int_{-1}^0|T_z|^2\mathrm{d}z\right)^{\frac{1}{2}}\left(\int_{-1}^0|\Delta T+T_{zz}|^2\mathrm{d}z\right)^{\frac{1}{2}}\right]\\&\le c\left[\int_{-1}^0\left(\int_M|\nabla\boldsymbol{v}|^4\right)^{\frac{1}{2}}\mathrm{d}z\right]\left[\int_{-1}^0\left(\int_M|T_z|^4\right)^{\frac{1}{2}}\mathrm{d}z\right]+\varepsilon\int_\Omega|\Delta T+T_{zz}|^2\\&\le c\left[\int_{-1}^0\left(\int_M|\nabla\boldsymbol{v}|^2\right)^{\frac{1}{2}}\left(\int_M(|\nabla\boldsymbol{v}|^2+|\Delta\boldsymbol{v}|^2)\right)^{\frac{1}{2}}\mathrm{d}z\right]\\&\quad\cdot c\left[\int_{-1}^0\left(\int_M|T_z|^2\right)^{\frac{1}{2}}\left(\int_M(|\nabla T_z|^2+|T_z|^2)\right)^{\frac{1}{2}}\mathrm{d}z\right]+\varepsilon\int_\Omega|\Delta T+T_{zz}|^2\\&\le\varepsilon\left(|\Delta T|_2^2+|\nabla T_z|_2^2+|T_{zz}|_2^2\right)+c(2|\nabla\boldsymbol{v}|_2^2+|\Delta\boldsymbol{v}|_2^2+|\nabla\boldsymbol{v}|_2^4\\&\quad+|\nabla\boldsymbol{v}|_2^2|\Delta\boldsymbol{v}|_2^2)|T_z|_2^2.\end{aligned}$$

Thus, choosing ε small enough, we obtain

$$\begin{aligned}&\frac{\mathrm{d}(|\nabla T|_2^2+|T_z|_2^2+\alpha_s|T|_{z=0}|_2^2)}{\mathrm{d}t}+|\Delta T|_2^2+(|\nabla T_z|_2^2+\alpha_s|\nabla T|_{z=0}|_2^2)\\&+|\nabla T_{zz}|_2^2\\&\le c(|\boldsymbol{v}|_6^2+|\boldsymbol{v}|_6^4)|\nabla T|_2^2+(2|\nabla\boldsymbol{v}|_2^2+|\Delta\boldsymbol{v}|_2^2+|\nabla\boldsymbol{v}|_2^4+|\nabla\boldsymbol{v}|_2^2|\Delta\boldsymbol{v}|_2^2)|T_z|_2^2\\&+c|Q|_2^2.\end{aligned}$$

By (3.4.36), (3.4.41), (3.4.43) and (3.4.44), we derive from the above in

$$|\nabla T(t)|_2^2+|T_z(t)|_2^2+\int_0^t(|\Delta T(s)|_2^2+|\nabla T_z(s)|_2^2+|\nabla T_{zz}|_2^2)\mathrm{d}s\le K_6(t), \tag{3.4.48}$$

where

$$K_6(t) = \mathrm{e}^{(1+K_2^{\frac{2}{3}}(t)+K_3^{\frac{2}{3}}(t)+K_5^2(t))t+(K_5(t)+1)K_5(t)}(\|T_0\|^2 + |Q|_2^2).$$

Proof of the global existence. According to Proposition 3.4.4, we use contradiction to prove the global existence of strong solutions. In fact, suppose that $\boldsymbol{U}$ is a strong solution of system (3.4.11)-(3.4.17) in the maximal interval $[0, \mathcal{T}^*]$. If $\mathcal{T}^* < +\infty$, then $\limsup\limits_{t \to \mathcal{T}^{*-}} \|\boldsymbol{U}\| = +\infty$. According to (3.4.36), (3.4.44), (3.4.47) and (3.4.48), we know that $\limsup\limits_{t \to \mathcal{T}^{*-}} \|\boldsymbol{U}\| = +\infty$, which is impossible. Thus, the global existence of strong solutions is proved.
Proof of the uniqueness. Suppose that $(\boldsymbol{v}_1, T_1)$ and $(\boldsymbol{v}_2, T_2)$ are two strong solutions of equations (3.4.11)-(3.4.17) in interval $[0, \mathcal{T}]$ corresponding to p_{s_1}, p_{s_2} and initial data $((\boldsymbol{v}_0)_1, (T_0)_1)$, $((\boldsymbol{v}_0)_2, (T_0)_2)$ respectively. Let $\boldsymbol{v} = \boldsymbol{v}_1 - \boldsymbol{v}_2$, $T = T_1 - T_2$, $p_s = p_{s_1} - p_{s_2}$. Then $\boldsymbol{v}$, T, p_s, satisfy the following system

$$\frac{\partial \boldsymbol{v}}{\partial t} - \Delta \boldsymbol{v} - \frac{\partial^2 \boldsymbol{v}}{\partial z^2} + (\boldsymbol{v}_1 \cdot \nabla)\boldsymbol{v} + (\boldsymbol{v} \cdot \nabla)\boldsymbol{v}_2 + W(\boldsymbol{v}_1)\frac{\partial \boldsymbol{v}}{\partial z} + W(\boldsymbol{v})\frac{\partial \boldsymbol{v}_2}{\partial z}$$

$$+ f\boldsymbol{k} \times \boldsymbol{v} + \nabla p_s - \int_{-1}^{z} \nabla T \mathrm{d}z' = 0, \tag{3.4.49}$$

$$\frac{\partial T}{\partial t} - \Delta T - \frac{\partial^2 T}{\partial z^2} + (\boldsymbol{v}_1 \cdot \nabla)T + (\boldsymbol{v} \cdot \nabla)T_2 + W(\boldsymbol{v}_1)\frac{\partial T}{\partial z} + W(\boldsymbol{v})\frac{\partial T_2}{\partial z} = 0, \tag{3.4.50}$$

$$\begin{aligned}
&\boldsymbol{v}|_{t=0} = (\boldsymbol{v}_0)_1 - (\boldsymbol{v}_0)_2,\\
&T|_{t=0} = (T_0)_1 - (T_0)_2,\\
&\frac{\partial \boldsymbol{v}}{\partial z} = 0, \quad \frac{\partial T}{\partial z} = -\alpha_s T \qquad \text{on } \Gamma_u,\\
&\frac{\partial \boldsymbol{v}}{\partial z} = 0, \quad \frac{\partial T}{\partial z} = 0 \qquad \text{on } \Gamma_b,\\
&\boldsymbol{v} \cdot \boldsymbol{n} = 0, \ \frac{\partial \boldsymbol{v}}{\partial \boldsymbol{n}} \times \boldsymbol{n} = 0, \ \frac{\partial T}{\partial \boldsymbol{n}} = 0 \qquad \text{on } \partial \Gamma_l.
\end{aligned}$$

Taking $L^2(\Omega) \times L^2(\Omega)$ inner product of equation (3.4.49) with $\boldsymbol{v}$, we get

$$\begin{aligned}
&\frac{1}{2}\frac{\mathrm{d}|\boldsymbol{v}|_2^2}{\mathrm{d}t} + \int_\Omega |\nabla \boldsymbol{v}|^2 + \int_\Omega |\boldsymbol{v}_z|^2\\
&= -\int_\Omega \left((\boldsymbol{v}_1 \cdot \nabla)\boldsymbol{v} + W(\boldsymbol{v}_1)\frac{\partial \boldsymbol{v}}{\partial z} \right) \cdot \boldsymbol{v} - \int_\Omega (\boldsymbol{v} \cdot \nabla)\boldsymbol{v}_2 \cdot \boldsymbol{v}\\
&\quad - \int_\Omega W(\boldsymbol{v})\frac{\partial \boldsymbol{v}_2}{\partial z} \cdot \boldsymbol{v} - \int_\Omega (f\boldsymbol{k} \times \boldsymbol{v} + \nabla p_s) \cdot \boldsymbol{v} + \int_\Omega \left(\int_{-1}^{z} \nabla T \mathrm{d}z' \right) \cdot \boldsymbol{v}.
\end{aligned}$$

By Hölder Inequality, Young Inequality and $|\boldsymbol{v}|_4 \le C|\boldsymbol{v}|_2^{\frac{1}{4}}\|\boldsymbol{v}\|_1^{\frac{3}{4}}$, we have

$$\begin{aligned}
&\left|\int_\Omega (\boldsymbol{v}\cdot\boldsymbol{\nabla})\,\boldsymbol{v}_2\cdot\boldsymbol{v}\right| = \left|\int_\Omega (\boldsymbol{v}_2\cdot(\boldsymbol{v}\cdot\boldsymbol{\nabla})\,\boldsymbol{v} + \boldsymbol{v}_2\cdot\boldsymbol{v}\mathrm{div}\boldsymbol{v})\right| \\
\le&\varepsilon\int_\Omega |\boldsymbol{\nabla}\boldsymbol{v}|^2 + c|\boldsymbol{v}|_4^2|\boldsymbol{v}_2|_4^2 \\
\le&\varepsilon\int_\Omega |\boldsymbol{\nabla}\boldsymbol{v}|^2 + c|\boldsymbol{v}_2|_4^2|\boldsymbol{v}|_2^{\frac{1}{2}}\|\boldsymbol{v}\|^{\frac{3}{2}} \\
\le&2\varepsilon\int_\Omega \left(|\boldsymbol{\nabla}\boldsymbol{v}|^2 + |\boldsymbol{v}_z|^2\right) + c\left(|\boldsymbol{v}_2|_4^2 + |\boldsymbol{v}_2|_4^8\right)|\boldsymbol{v}|_2^2.
\end{aligned}$$

Similar to (3.4.46),

$$\begin{aligned}
&\left|\int_\Omega W(\boldsymbol{v})\frac{\partial \boldsymbol{v}_2}{\partial z}\cdot\boldsymbol{v}\right| \\
\le&2\varepsilon\int_\Omega |\boldsymbol{\nabla}\boldsymbol{v}|^2 + c\left[(|\boldsymbol{v}_{2z}|_2^2+1)\int_\Omega |\boldsymbol{\nabla}\boldsymbol{v}_{2z}|^2 + |\boldsymbol{v}_{2z}|_2^4 + |\boldsymbol{v}_{2z}|_2^2\right]|\boldsymbol{v}|_2^2.
\end{aligned}$$

Thus, we obtain

$$\begin{aligned}
&\frac{1}{2}\frac{\mathrm{d}|\boldsymbol{v}|_2^2}{\mathrm{d}t} + \int_\Omega |\boldsymbol{\nabla}\boldsymbol{v}|^2 + \int_\Omega |\boldsymbol{v}_z|^2 \qquad (3.4.51)\\
\le&4\varepsilon\int_\Omega (|\boldsymbol{\nabla}\boldsymbol{v}|^2 + |\boldsymbol{v}_z|^2) + \varepsilon|\boldsymbol{\nabla}T|_2^2 \\
&+c\left[1 + |\boldsymbol{v}_2|_4^2 + |\boldsymbol{v}_2|_4^8 + (|\boldsymbol{v}_{2z}|_2^2+1)|\boldsymbol{\nabla}\boldsymbol{v}_{2z}|_2^2 + |\boldsymbol{v}_{2z}|_2^4 + |\boldsymbol{v}_{2z}|_2^2\right]|\boldsymbol{v}|_2^2.
\end{aligned}$$

Taking $L^2(\Omega)$ inner product of equation (3.4.50) with T, we get

$$\begin{aligned}
&\frac{1}{2}\frac{\mathrm{d}|T|_2^2}{\mathrm{d}t} + \int_\Omega |\boldsymbol{\nabla}T|^2 + \int_\Omega |T_z|^2 + \alpha_s|T|_{z=0}|_2^2 \\
=&-\int_\Omega \left((\boldsymbol{v}_1\cdot\boldsymbol{\nabla})T + W(\boldsymbol{v}_1)\frac{\partial T}{\partial z}\right)T - \int_\Omega T(\boldsymbol{v}\cdot\boldsymbol{\nabla})T_2 - \int_\Omega W(\boldsymbol{v})\frac{\partial T_2}{\partial z}T.
\end{aligned}$$

Like (3.4.51),

$$\begin{aligned}
&\frac{1}{2}\frac{\mathrm{d}|T|_2^2}{\mathrm{d}t} + \int_\Omega |\boldsymbol{\nabla}T|^2 + \int_\Omega |T_z|^2 + \alpha_s|T|_{z=0}|_2^2 \\
\le&3\varepsilon\int_\Omega (|\boldsymbol{\nabla}\boldsymbol{v}|^2 + |\boldsymbol{v}_z|^2) + 3\varepsilon\int_\Omega (|\boldsymbol{\nabla}T|^2 + |T_z|^2) + c(|T_2|_4^2 + |T_2|_4^8)|\boldsymbol{v}|_2^2 \\
&+c\left[|T_2|_4^2 + |T_2|_4^8 + (|T_{2z}|_2^2+1)|\boldsymbol{\nabla}T_{2z}|_2^2 + |T_{2z}|_2^4 + |T_{2z}|_2^2\right]|T|_2^2. \qquad (3.4.52)
\end{aligned}$$

Thus, choosing ε small enough, we derive from (3.4.51) and (3.4.52)

$$\frac{\mathrm{d}(|\boldsymbol{v}|_2^2 + |T|_2^2)}{\mathrm{d}t} + \int_\Omega |\boldsymbol{\nabla}\boldsymbol{v}|^2 + \int_\Omega |\boldsymbol{v}_z|^2 + \int_\Omega |\boldsymbol{\nabla}T|^2 + \int_\Omega |T_z|^2 + \alpha_s|T|_{z=0}|_2^2$$

$$\leq c[1 + |T_2|_4^2 + |T_2|_4^8 + |\boldsymbol{v}_2|_4^2 + |\boldsymbol{v}_2|_4^8 + (|\boldsymbol{v}_{2z}|_2^2 + 1)|\nabla \boldsymbol{v}_{2z}|_2^2 + |\boldsymbol{v}_{2z}|_2^4$$
$$+|\boldsymbol{v}_{2z}|_2^2]|\boldsymbol{v}|_2^2 + c\left[|T_2|_4^2 + |T_2|_4^8 + (|T_{2z}|_2^2 + 1)|\nabla T_{2z}|_2^2 + |T_{2z}|_2^4 + |T_{2z}|_2^2\right]|T|_2^2.$$

Applying Gronwall Inequality and the above result of the global existence of the strong solutions, we prove the uniqueness of strong solutions from the above inequality. Theorem 3.4.5 is proved.

Remark 3.4.6. Recently, there are some research works about the global well-posedness of the 3D primitive equations with Dirichlet boundary conditions, see [71] and [124]. In [27], Cao and Titi proved the global well-posedness of the 3D primitive equations with partial vertical turbulence mixing heat diffusion.

Chapter 4

Random Dynamical Systems of Atmosphere and Ocean

The effectiveness of the deterministic weather forecasting is bounded, which is mainly determined by the space scale and time scale of the forecast target. And it's impossible to accurately predict long-term weather and climate changes in a deterministic framework. To predict climate changes more objectively, after 1975, researchers proposed some stochastic climate models, set up Langevin equation describing weather changes and the corresponding Fokker-Planck equation. Efforts in this section can be seen in [74, 99, 129]. After 1980, researchers set up some simplified stochastic climate models, repealing the effect of the random force to the climate system changes, which can be seen in [86, 123, 134, 176, 187].

In the past decade, researchers began to pay attention to the study of stochastic climate models. Majda and his cooperators have done a lot of theoretical and numerical studies about the stochastic climate model mathematically, and obtained many important results, see [75, 155, 156, 157, 160]. Zhou pointed out in [222] that the fact that macroscopic micro-scale random forces stem from molecular thermal motion is an intrinsic property of the atmosphere itself; solar radiation as a main factor determining the atmospheric motions and changes, of which changes has randomness, is a random factor, which has an important effect to the climate changes; earth-atmosphere interaction is a time-varying nonlinear mutual feedback coupling process, which forms the complex random effect to the lower boundary of the atmosphere. So in some situations, researchers must think about the factors of random forces, solar radiation and the random force of the earth surface.

In this chapter, we mainly introduce the qualitative theory of the random dynamic system of the atmosphere and oceans. In section 4.1, we consider the two-dimensional quasi-geostrophic equation with random external force, friction and dissipation, where the random external force denotes the wind stress on the upper surface of the homogeneous fluid, see [109, 150]. In section 4.2, we study the global well-posedness of the ocean primitive equations of the large-scale oceans with random external forces and the existence of the global attractors of the corresponding infinite-dimensional random dynamic system for the first time, of which the conclusion is suitable for the primitive equations of the large-scale atmosphere with random external forces and solar radiation, see [89]. In section 4.3, we study the qualitative theory of the ocean primitive equations with random boundary conditions, where the random boundary conditions denote the random force effects of the atmosphere to oceans, which agrees with considering the atmospheric motion as a random process, see [92].

4.1 Random Attractors of Two-Dimensional Quasi-Geostrophic Dynamical System

In this section, we consider the two-dimensional quasi-geostrophic equation with random external force, friction term and dissipation (the explicit definition of this equation will be given in subsection 4.1.1):

$$\left(\frac{\partial}{\partial t}+\frac{\partial\psi}{\partial x}\frac{\partial}{\partial y}-\frac{\partial\psi}{\partial y}\frac{\partial}{\partial x}\right)(\Delta\psi-F\psi+\beta_0 y)=\frac{1}{Re}\Delta^2\psi-\frac{r}{2}\Delta\psi+f(x,y,t)\text{ in }D,\tag{4.1.1}$$

where D is an open domain smooth enough (for example C^2) in $\mathbb{R}^2$ except in section 4.1.5 where D is bounded, ψ is a stream function, $\frac{1}{Re}\Delta^2\psi$ is an viscous term, $-\frac{r}{2}\Delta\psi$ is a friction term, and $f(x,y,t)$ is an external force term, cf. [172, p. 234] or [173]. The actual fluid problems are mostly under the influence of different external forces and dissipation. Moreover, dissipation and external force are especially important in the long-time behavior. If we consider equation (4.1.1) as the simplified model describing large-scale oceanic motions, then since the main external force of oceanic motions is the atmospheric windstress, $f(x,y,t)$ is the curl of the atmospheric wind stress. Generally, the atmospheric external force field can be regarded as a random field. cf. e.g., [74, 162, 163, 186, 187]. Thus, we take $f(x,y,t)$ as a random force given later. The equation studied in [17, 20, 62] is

$$\left(\frac{\partial}{\partial t}+\frac{\partial\psi}{\partial x}\frac{\partial}{\partial y}-\frac{\partial\psi}{\partial y}\frac{\partial}{\partial x}\right)(\Delta\psi+\beta_0 y)=\frac{1}{Re}\Delta^2\psi-\frac{r}{2}\Delta\psi+f(x,y,t)\text{ in }D,$$

where D is a smooth enough domain in $\mathbb{R}^2$, and $f(x, y, t)$ is a random external force term.

Our interest is in the asymptotic behavior of the following dynamical system

$$\begin{cases} \left(\frac{\partial}{\partial t} + \frac{\partial \psi}{\partial x}\frac{\partial}{\partial y} - \frac{\partial \psi}{\partial y}\frac{\partial}{\partial x}\right)(\Delta\psi - F\psi + \beta_0 y) = \frac{1}{Re}\Delta^2\psi - \frac{r}{2}\Delta\psi + f(x, y, t), \\ \qquad\qquad\qquad\qquad\qquad\qquad\qquad\qquad\qquad\qquad\qquad\qquad \text{in } D, \\ \psi(x, y, t) = 0, \qquad\qquad\qquad\qquad\qquad\qquad\qquad\qquad\quad \text{on } \partial D, \\ \Delta\psi(x, y, t) = 0, \qquad\qquad\qquad\qquad\qquad\qquad\qquad\qquad \text{on } \partial D. \end{cases} \tag{4.1.2}$$

The understanding of the asymptotic behavior of the dynamical systems is one of the most important problems of modern mathematical physics. One way to attack the problem is to consider the global attractors of the dynamical systems (see, e.g., [201, 209, 210]). We aim to study the existence of the random attractors of system (4.1.2) under some appropriate assumptions of f. Our main results are Theorem 4.1.3, Theorem 4.1.12 and Theorem 4.1.20. First, we use the Banach fixed point method to prove the existence and uniqueness of the initial boundary value problem of the stochastic equation (4.1.1). Second, by studying the asymptotic behavior of the solutions, we obtain the existence of the random attractors of the stochastic quasi-geostrophic equation on bounded domains. Third, we also prove the existence of a random attractor for the stochastic 2D quasi-geostrophic on an unbounded domains. Our approach to proving the existence of the random attractors is inspired by [10, 51, 52, 72, 208].

The arrangement of this section is as follows. In section 4.1.1, we shall give the two-dimensional quasi-geostrophic equation with random external force term and dissipation. We shall, in section 4.1.2, prove the global well-posedness of the initial boundary value problem of equation (4.1.1). In section 4.1.3, we shall introduce the definitions of the random attractors and some preliminaries. In section 4.1.4, we shall prove the existence of the random attractors of the dissipative quasi-geostrophic dynamic system with the effect of random external force. At last, we shall show the existence of random attractors for the stochastic quasi-geostrophic dynamic system on unbounded domains.

4.1.1 *Model*

The stochastic two-dimensional quasi-geostrophic equation is

$$\left(\frac{\partial}{\partial t} + \frac{\partial \psi}{\partial x}\frac{\partial}{\partial y} - \frac{\partial \psi}{\partial y}\frac{\partial}{\partial x}\right)(\Delta\psi - F\psi + \beta_0 y) = \frac{1}{Re}\Delta^2\psi - \frac{r}{2}\Delta\psi + f(x, y, t),$$

where F is Froude number $F \approx O(1)$, Re is Reynolds number ($Re \geq 10^2$), β_0 is Rossby parameter ($\beta_0 \approx O(10^{-1})$), r is Ekman dissipative constant ($r \approx O(1)$), and $f(x,y,t) = -\dfrac{\mathrm{d}W}{\mathrm{d}t}$ (the definition of W will be given later) is Gaussian random field, which is a white noise in time, and satisfies some assumptions given later.

Let $A = -\dfrac{1}{Re}\Delta$, $A : L^2(D) \to L^2(D)$ with the domain $D(A) = H^2(D) \cap H_0^1(D)$ where $L^2(D)$ is Lebesgue space, $H^2(D)$, and $H_0^1(D)$ are Sobolev spaces. $|\cdot|_p$ is the norm of space $L^p(D)(1 \leq p \leq +\infty)$, and $\|\cdot\|$ is the norm of space $H_0^1(D)$. A is positive and self-adjoint with compact inverse operator A^{-1}. We denote by $0 < \lambda_1 < \lambda_2 \leq \cdots$ the eigenvalues of A, and by $\boldsymbol{e}_1, \boldsymbol{e}_2, \cdots$ a corresponding complete orthonormal system of eigenvectors. We notice that for any $u \in H_0^1(D)$, $\dfrac{1}{Re}\|u\|^2 \geq \lambda_1 |u|_2^2$. We denote by $\mathrm{e}^{t(-A)}$, $t \geq 0$ the semigroups generated by $-A$ in $L^2(D)$.

Here we assume that the stochastic process W be a two-sided in time Wiener process with the form

$$W(t) = \sum_{i=1}^{+\infty} \mu_i \omega_i(t) \boldsymbol{e}_i,$$

where $\omega_1, \omega_2, \cdots$ is a sequence of independent standard Brownian motions on a complete probability space $(\boldsymbol{\Omega}, \mathcal{F}, P)$ (with expectation denoted by E), and the coefficients μ_i satisfies

$$\sum_{j=1}^{+\infty} \frac{\mu_i^2}{\lambda_i^{\frac{1}{2}-2\beta_1}} < +\infty, \quad \text{for some } \beta_1 > 0.$$

We shall study the equation (4.1.1) with the following boundary conditions (boundary conditions with no penetration and free-slip, cf. [28] or [173, p. 34]),

$$\psi(x,y,t) = 0, \text{ on } \partial D,$$

$$\Delta\psi(x,y,t) = 0, \text{ on } \partial D.$$

For any $u \in L^2(D)$, by solving the elliptic equation with Dirichlet boundary condition

$$\begin{cases} F\psi - \Delta\psi = u, \\ \psi|_{\partial D} = 0, \end{cases}$$

we get $\psi = (FI - \Delta)^{-1}u = B(u)$. By the elliptic regularity theory (see, e.g., [82]), we know that $L^2(D) \to H_0^1(D) \cap H^2(D)$. Thus, the equation (4.1.1) can be rewritten as

$$u_t + J(\psi, u) - \beta_0 \psi_x = \frac{1}{Re}\Delta u + \left(\frac{F}{Re} - \frac{r}{2}\right) u - F\left(\frac{F}{Re} - \frac{r}{2}\right)\psi + \frac{\mathrm{d}W}{\mathrm{d}t},$$

where $\psi = B(u)$, J is a Jacobian operator which is defined by $J(\psi, u) = \dfrac{\partial \psi}{\partial x}\dfrac{\partial u}{\partial y} - \dfrac{\partial \psi}{\partial y}\dfrac{\partial u}{\partial x}$. Define

$$G(u) = -J(\psi, u) + \beta_0 \psi_x + \left(\frac{F}{Re} - \frac{r}{2}\right) u - F\left(\frac{F}{Re} - \frac{r}{2}\right)\psi.$$

In the following, we consider

$$u_t - \frac{1}{Re}\Delta u = G(u) + \frac{\mathrm{d}W}{\mathrm{d}t}, \tag{4.1.3}$$

with boundary condition

$$u|_{\partial D} = 0 \tag{4.1.4}$$

and initial condition

$$u(x, y, 0) = u_0. \tag{4.1.5}$$

We remark that the boundary value problem (4.1.3) and (4.1.4) corresponds to the dynamical system (4.1.2).

4.1.2 *Global Existence and Uniqueness of Solutions*

Now we rewrite (4.1.3)-(4.1.5) as the abstract stochastic differential equation with the initial condition

$$\begin{cases} \mathrm{d}u = -Au\mathrm{d}t + G(u)\mathrm{d}t + \mathrm{d}W, \\ u(0) = u_0. \end{cases} \tag{4.1.6}$$

The solution of the linear problem

$$\begin{cases} \mathrm{d}u = -Au\mathrm{d}t + \mathrm{d}W, \\ u(0) = 0 \end{cases}$$

is unique and given by the so-called stochastic convolution

$$W_A(t) = \int_0^t \mathrm{e}^{(t-s)(-A)}\mathrm{d}W(s).$$

For P-a.e. $\omega \in \mathbf{\Omega}$, where P-a.e. denotes almost everywhere according to probability, $W_A(t)$ has a continuous version with values in $D(A^{\frac{1}{4}+\beta})$ ($\beta <$

β_1), especially in $C_0(D)$ (see [54, 55, 72]), where $C_0(D) := \{u;\ u \in C(D),$ supp u is a compact subset in $D\}$.

We set

$$v(t) = u(t) - W_A(t),\ t \geq 0,$$

then u satisfies (4.1.6) if and only if v is a solution of the following problem

$$\begin{cases} \dfrac{\partial v}{\partial t} + Av = G(v + W_A(t)), \\ v(0) = u_0. \end{cases} \tag{4.1.7}$$

From now on, we shall study equation (4.1.7) for P-a.e. $\omega \in \boldsymbol{\Omega}$. Let us write (4.1.7) as an integral equation

$$v(t) = \mathrm{e}^{t(-A)}u_0 + \int_0^t \mathrm{e}^{(t-s)(-A)}G(v + W_A)\mathrm{d}s. \tag{4.1.8}$$

If v satisfies (4.1.8), we say that v is a mild solution to (4.1.7).

4.1.2.1 *Local Existence of Solutions to (4.1.7)*

We shall prove by Banach fixed point theorem that there exists $T > 0$, such that equation (4.1.8) has a solution in $C([0,T]\,;L^2(D))$. For any $m > 0$, define

$$\sum(m,T) = \{v \in C([0,T]\,;L^2(D));\, |v(t)|_2 \leq m, \forall t \in [0,T]\}.$$

Lemma 4.1.1. If $u_0 \in L^2(D)$ for P-a.e. $\omega \in \boldsymbol{\Omega}$ and $m > |u_0|_2$, then there exists $T > 0$, such that (4.1.8) has a unique solution in $\Sigma(m,T)$. Moreover, for P-a.e. $\omega \in \boldsymbol{\Omega}$, $v \in C((0,T]; H^\alpha(D))$ $(0 \leq \alpha < \frac{1}{2})$, where $H^\alpha(D)$ is a usual Sobolev space.

Proof. Firstly, we recall that $\mathrm{e}^{t(-A)}$ has the following properties (see, e.g., [170, pp. 69-75]):

$$\mathrm{e}^{t(-A)}A^\alpha = A^\alpha \mathrm{e}^{t(-A)},$$

$$|A^\alpha \mathrm{e}^{t(-A)}u|_2 \leq \frac{c}{t^\alpha}|u|_2,$$

$$|\mathrm{e}^{t(-A)}u|_2 \leq c|u|_2.$$

For the definition of the fractional differential operator A^α, one can refer to [170, pp. 69-75]). In the sequel $\omega \in \boldsymbol{\Omega}$ is fixed. Let

$$Mv(t) = \mathrm{e}^{t(-A)}u_0 + \int_0^t \mathrm{e}^{(t-s)(-A)}G(v(s) + W_A(s))\mathrm{d}s,$$

then

$$|Mv(t)|_2 = \sup_{\varphi\in L^2(D),|\varphi|_2=1} |\langle Mv,\varphi\rangle|,$$

where $\langle\cdot,\cdot\rangle$ is an inner product in $L^2(D)$, and

$$\langle Mv,\varphi\rangle = \langle \mathrm{e}^{t(-A)}u_0,\varphi\rangle + \int_0^t \langle \mathrm{e}^{(t-s)(-A)}G(v+W_A),\varphi\rangle \mathrm{d}s.$$

In the following we assume $\varphi\in C_0^\infty(D)$, $v+W_A\in H_0^1(D)$, where $C_0^\infty(D)$ is the space of infinitely differentiable functions on D with compact support strictly contained in D.

Let

$$\begin{aligned}
J =&\langle \mathrm{e}^{(t-s)(-A)}G(v+W_A),\varphi\rangle\\
=&-\int_D \mathrm{e}^{(t-s)(-A)}\frac{\partial\psi}{\partial x}\frac{\partial(v+W_A)}{\partial y}\varphi + \int_D \mathrm{e}^{(t-s)(-A)}\frac{\partial\psi}{\partial y}\frac{\partial(v+W_A)}{\partial x}\varphi\\
&+\int_D \mathrm{e}^{(t-s)(-A)}\left(\frac{F}{Re}-\frac{r}{2}\right)(v+W_A)\varphi - \int_D \mathrm{e}^{(t-s)(-A)}F\left(\frac{F}{Re}-\frac{r}{2}\right)\psi\varphi\\
&+\beta_0\int_D \mathrm{e}^{(t-s)(-A)}(B(v+W_A))_x\varphi\\
=&J_1+J_2+J_3+J_4+J_5,
\end{aligned}\tag{4.1.9}$$

where $\psi = B(v+W_A)$.

First, we estimate J_1. By integration by parts and Hölder Inequality, we have

$$\begin{aligned}
|J_1| =& \left|\int_D \mathrm{e}^{(t-s)(-A)}\frac{\partial\psi}{\partial x}\frac{\partial(v+W_A)}{\partial y}\varphi\right|\\
\le& \left(\int_D \left|(\mathrm{e}^{(t-s)(-A)}\frac{\partial\psi}{\partial x}\varphi)_y\right|^2\right)^{\frac{1}{2}} |v+W_A|_2\\
\le& \left(\int_D \left|(\mathrm{e}^{(t-s)(-A)}\varphi)_y\frac{\partial\psi}{\partial x} + \mathrm{e}^{(t-s)(-A)}\varphi\frac{\partial^2\psi}{\partial x\partial y}\right|^2\right)^{\frac{1}{2}} |v+W_A|_2\\
\le& c\left[\left(\int_D \left|(\mathrm{e}^{(t-s)(-A)}\varphi)_y\frac{\partial\psi}{\partial x}\right|^2\right)^{\frac{1}{2}} + |\mathrm{e}^{(t-s)(-A)}\varphi|_\infty|v+W_A|_2\right]|v+W_A|_2,
\end{aligned}$$

where we have used $\|\psi\|_{H^2}\le c|v+W_A|_2$ ($\|\cdot\|_{H^q}$, $-\infty<q<+\infty$, is the usual norm of space $H^q(D)$).

By Hölder Inequality, Sobolev Embedding Theorem, Galiardo-Nirenberg Inequality and Poincaré Inequality, we get

$$\left(\int_D \left|(\mathrm{e}^{(t-s)(-A)}\varphi)_y\frac{\partial\psi}{\partial x}\right|^2\right)^{\frac{1}{2}} \le \left(\int_D \left|(\mathrm{e}^{(t-s)(-A)}\varphi)_y\right|^4\right)^{\frac{1}{4}}\left(\int_D \left|\frac{\partial\psi}{\partial x}\right|^4\right)^{\frac{1}{4}}$$

$$\le c\|(\mathrm{e}^{(t-s)(-A)}\varphi)_y\|_{H^{\frac{1}{2}}}\|\psi_x\|_{H^1} \le c\|\mathrm{e}^{(t-s)(-A)}\varphi\|_{H^{\frac{3}{2}}}|v+W_A|_2$$
$$\le c|A^{\frac{3}{4}}\mathrm{e}^{(t-s)(-A)}\varphi|_2|v+W_A|_2 \le c(t-s)^{-\frac{3}{4}}|\varphi|_2|v+W_A|_2.$$

Similarly, we obtain

$$|\mathrm{e}^{(t-s)(-A)}\varphi|_\infty|v+W_A|_2^2 \le c\|\mathrm{e}^{(t-s)(-A)}\varphi\|_{H^{1+2\varepsilon_0}}|v+W_A|_2^2$$
$$\le c|A^{\frac{1}{2}+\varepsilon_0}\mathrm{e}^{(t-s)(-A)}\varphi|_2|v+W_A|_2^2 \le c(t-s)^{-(\frac{1}{2}+\varepsilon_0)}|\varphi|_2|v+W_A|_2^2,$$

where ε_0 is an arbitrary positive constant. Thus, we get

$$|J_1| \le c(t-s)^{-\frac{3}{4}}|\varphi|_2|v+W_A|_2^2 + c(t-s)^{-(\frac{1}{2}+\varepsilon_0)}|\varphi|_2|v+W_A|_2^2.$$

In the same way, we have

$$|J_2| \le c(t-s)^{-\frac{3}{4}}|\varphi|_2|v+W_A|_2^2 + c(t-s)^{-\frac{1}{2}+\varepsilon_0}|\varphi|_2|v+W_A|_2^2,$$
$$|J_3| \le c|\mathrm{e}^{(t-s)(-A)}(v+W_A)|_2|\varphi|_2 \le c|v+W_A|_2|\varphi|_2,$$
$$|J_4| = \left|F\left(\frac{F}{Re}-\frac{r}{2}\right)\right|\left|\int_D \mathrm{e}^{(t-s)(-A)}\psi\varphi\right| \le c|v+W_A|_2|\varphi|_2,$$
$$|J_5| = \beta_0\left|\int_D \mathrm{e}^{(t-s)(-A)}(B(v+W_A))_x\varphi\right| \le c|v+W_A|_2|\varphi|_2.$$

Therefore, we obtain

$$|Mv(t)|_2 \le |u_0|_2 + c(t^{\frac{1}{4}} + t^{\frac{1}{2}-\varepsilon_0})|v+W_A|_2^2 + c\cdot t|v+W_A|_2. \tag{4.1.10}$$

It is clear that for any $m > |u_0|_2$, there exists $T_1 > 0$ such that $Mv \in \sum(m, T_1)$.

For any v_1, $v_2 \in \sum(m, T_1)$,

$$Mv_1 - Mv_2 = \int_0^t \mathrm{e}^{(t-s)(-A)}(G(v_1+W_A(s)) - G(v_2+W_A(s)))\mathrm{d}s.$$

Similar to (4.1.10), we have

$$|Mv_1 - Mv_2|_2 \le c(t^{\frac{1}{4}} + t^{\frac{1}{2}-\varepsilon_0})\sup_{0\le s\le t}(|v_1(s)+W_A(s)|_2 + |v_2(s)+W_A(s)|_2)$$
$$\cdot \sup_{0\le s\le t}|v_1(s)-v_2(s)|_2 + c\cdot t\sup_{0\le s\le t}|v_1(s)-v_2(s)|_2.$$

So we choose $T_2 > 0$ such that M is a contraction mapping.

Let $T = \min\{T_1, T_2\}$. Applying the Banach fixed point theorem, we know that (4.1.8) has a unique solution v in $\sum(m, T)$. Moreover, using the same way to get (4.1.10), we know $v \in C((0,T]; H^\alpha(D))$, $0 \le \alpha < \frac{1}{2}$. In fact, we prove

$$|A^{\frac{\alpha}{2}}Mv|_2 \le t^{\frac{\alpha}{2}}|u_0|_2 + c(t^{\frac{1}{4}-\frac{\alpha}{2}} + t^{\frac{1}{2}-\varepsilon_0-\frac{\alpha}{2}})|v+W_A|_2^2 + ct^{1-\frac{\alpha}{2}}|v+W_A|_2.$$

4.1.2.2 *The Global Existence of the Solution of (4.1.7)*

To prove the global existence of the solution of (4.1.7), we must make *a priori* estimates about L^2-norm of the local solution $v(t)$ obtained in Lemma 4.1.1.

Lemma 4.1.2. If $v \in C([0,T]\,;L^2(D))$ for P-a.e. $\omega \in \mathbf{\Omega}$ satisfies equation (4.1.8), then

$$|v(t)|_2^2 \le (|u_0|_2^2 + c\nu_\infty^4 + c\nu_\infty^2)\mathrm{e}^{c(\nu_\infty^2+1)t},$$

where $\nu_\infty = \sup\limits_{0\le t\le T} |W_A(t)|_\infty$.

Proof. Let $\omega \in \mathbf{\Omega}$ be fixed. Let $\{u_0^n\} \subset C_0^\infty(D)$ such that $u_0^n \to u_0$ in $L^2(D)$, and $\{W_A^n\}$ be a stochastic process regular enough such that

$$W_A^n(t) = \int_0^t \mathrm{e}^{(t-s)(-A)}\mathrm{d}W^n(s) \to W_A(t) \text{ in } C([0,T_0]\times D) \text{ a.s. for } \omega \in \mathbf{\Omega},$$

where a.s. denotes almost sure. Let v^n be a mild solution of equation (4.1.7) given by Lemma 4.1.1. From the proof of Lemma 4.1.1, we know $v^n \in C([0,T^n]\,;L^2(D))$, such that $T^n \to T$ (we can select appropriate sequence $\{u_0^n\}$ and $\{W^n\}$ such that $T^n \ge T$), and $v^n \to v$ strongly in $C([0,T]\,;L^2(D))$. Moreover, v^n is regular enough and satisfies

$$\begin{cases} \dfrac{\partial v^n}{\partial t} + Av^n = G(v^n + W_A^n(t)), \\ v^n(0) = u_0^n. \end{cases} \tag{4.1.11}$$

Selecting v^n as a test function in (4.1.11), we get

$$\frac{1}{2}\frac{\mathrm{d}|v^n|_2}{\mathrm{d}t} + \frac{1}{Re}\|v^n\|^2 = \int_D G(v^n + W_A^n(t))v^n.$$

Since $\int_D J(B(v^n),v^n)v^n = 0$, $\int_D J(B(W_A^n),v^n)v^n = 0$, we have

$$\begin{aligned}
&\left|\int_D G\left(v^n + W_A^n\left(t\right)\right)v^n\right| \\
=&\left|\int_D \left(-J\left(B\left(v^n + W_A^n\left(t\right)\right), v^n + W_A^n\right) + \beta_0\left(B\left(v^n + W_A^n\right)\right)_x\right.\right. \\
&\left.+\left(\frac{F}{Re} - \frac{r}{2}\right)\cdot\left(v^n + W_A^n\right) - F\left(\frac{F}{Re} - \frac{r}{2}\right)B\left(v^n + W_A^n\right)\cdot v^n\right| \\
\le&\left|\int_D J\left(B\left(W_A^n\right), W_A^n\right)v^n\right| + \left|\int_D J\left(B\left(v^n\right), W_A^n\right)v^n\right| \\
&+\left|\int_D \left(\frac{F}{Re} - \frac{r}{2}\right)\left(v^n + W_A^n\right)v^n\right| + \left|\int_D F\left(\frac{F}{Re} - \frac{r}{2}\right)B\left(v^n + W_A^n\right)v^n\right|
\end{aligned}$$

$$
\begin{aligned}
&+ \beta_0 \left| \int_D \left(B\left(v^n + W_A^n\right)\right)_x v^n \right| \\
=&I_1 + I_2 + I_3 + I_4 + I_5.
\end{aligned}
$$

Let us estimate I_i $(1 \leq i \leq 5)$. By integration by parts, Hölder Inequality and Young Inequality, we get

$$
\begin{aligned}
I_1 &= \left| \int_D J\left(B\left(W_A^n\right), W_A^n\right) v^n \right| \\
&\leq \left| \int_D \frac{\partial B\left(W_A^n\right)}{\partial x} W_A^n \frac{\partial v^n}{\partial y} \right| + \left| \int_D \frac{\partial B\left(W_A^n\right)}{\partial x} W_A^n \frac{\partial v^n}{\partial x} \right| \\
&\leq |W_A^n|_\infty \left| \frac{\partial B\left(W_A^n\right)}{\partial x} \right|_2 \left| \frac{\partial v^n}{\partial y} \right|_2 + |W_A^n|_\infty \left| \frac{\partial B\left(W_A^n\right)}{\partial y} \right|_2 \left| \frac{\partial v^n}{\partial x} \right|_2 \\
&\leq c|W_A^n|_\infty^2 |W_A^n|_2^2 + \varepsilon \|v^n\|^2 \leq c|W_A^n|_\infty^4 + \varepsilon \|v^n\|^2,
\end{aligned}
$$

where ε is a positive constant.

In a similar way, we obtain

$$
\begin{aligned}
I_2 =& \left| \int_D J\left(B\left(v^n\right), W_A^n\right) v^n \right| \leq c|W_A^n|_\infty^2 |v^n|_2^2 + \varepsilon \|v^n\|^2, \\
I_3 =& \left| \int_D \left(\frac{F}{Re} - \frac{r}{2}\right) \left(v^n + W_A^n\right) v^n \right| \leq c|v^n + W_A^n|_2 |v^n|_2 \leq c|v^n|_2^2 + |W_A^n|_\infty^2, \\
I_4 =& \left| \int_D F \left(\frac{F}{Re} - \frac{r}{2}\right) B\left(v^n + W_A^n\right) v^n \right| \leq c|B\left(v^n + W_A^n\right)|_2 |v^n|_2 \\
\leq& c|v^n|_2^2 + |W_A^n|_\infty^2, \\
I_5 =& \beta_0 \left| \int_D \left(B\left(v^n + W_A^n\right)\right)_x v^n \right| \leq c|\left(B\left(v^n + W_A^n\right)\right)_x|_2 |v^n|_2 \\
\leq& c|v^n|_2^2 + |W_A^n|_\infty^2.
\end{aligned}
$$

Thus, choosing ε small enough, we get

$$
\frac{\mathrm{d}|v^n|_2^2}{\mathrm{d}t} + \frac{1}{Re}\|v^n\|^2 \leq c(|W_A^n|_\infty^2 + 1)|v^n|_2^2 + c|W_A^n|_\infty^4 + 3|W_A^n|_\infty^2.
$$

By Gronwall Inequality, we have

$$
|v^n|_2^2 \leq |u_0^n|_2^2 \mathrm{e}^{\int_0^t (c|W_A^n|_\infty^2 + c)\mathrm{d}s} + \int_0^t (c|W_A^n|_\infty^4 + 3|W_A^n|_\infty^2) \mathrm{e}^{\int_s^t (c|W_A^n|_\infty^2 + c)\mathrm{d}\tau} \mathrm{d}s.
$$

Letting $n \to \infty$ in the above equality, we obtain

$$
\begin{aligned}
&|v(t)|_2^2 \leq |u_0|_2^2 \mathrm{e}^{\int_0^t (c|W_A|_\infty^2 + c)\mathrm{d}s} + \int_0^t (c|W_A|_\infty^4 + 3|W_A|_\infty^2) \mathrm{e}^{\int_s^t (c|W_A|_\infty^2 + c)\mathrm{d}\tau} \mathrm{d}s \\
&\leq |u_0|_2^2 \mathrm{e}^{(c\nu_\infty^2 + c)t} + c(c\nu_\infty^4 + 3\nu_\infty^2)\mathrm{e}^{(c\nu_\infty^2 + c)t} \leq (|u_0|_2^2 + c\nu_\infty^4 + c\nu_\infty^2)\mathrm{e}^{(c\nu_\infty^2 + c)t}.
\end{aligned}
$$

According to Lemma 4.1.1 and Lemma 4.1.2, we obtain the following theorem.

Theorem 4.1.3. Let u_0 be given, $u_0 \in L^2(D)$, for P-a.e. $\omega \in \Omega$. The equation (4.1.7) has a unique global solution $v(x, y, t)$. Moreover, $v \in C((0, T]; H^\alpha(D))$ $(0 \le \alpha < \frac{1}{2})$ for P-a.e. $\omega \in \Omega$.

4.1.3 *Preliminaries of Random Attractors*

Before proving the existence of random attractors of the two-dimensional quasi-geostrophic dynamical system, let's recall the definition and some preliminaries of random attractors, see, e.g., [4, 51, 52].

Let (X, d) be a complete separable metric space (a Polish space), $(\boldsymbol{\Omega}, \mathcal{F}, P)$ be a complete probability space, and $\{\vartheta_t : \boldsymbol{\Omega} \to \boldsymbol{\Omega}, t \in \mathbb{R}\}$ be a family of measure preserving transform such that $\vartheta_0 = \mathrm{id}_{\boldsymbol{\Omega}}$ and $\vartheta_{t+s} = \vartheta_t \circ \vartheta_s$ for any t, $s \in \mathbb{R}$. $\{\vartheta_t\}$ is called a metric dynamical system in $\boldsymbol{\Omega}$, which represents a random dynamical system driven by white noise. Here we assume ϑ_t is ergodic under P.

Definition 4.1.4 (Random dynamical system). A measurable mapping $\psi : \mathbb{R}^+ \times \boldsymbol{\Omega} \times X \to X, (t, \omega, U) \mapsto \psi(t, \omega)U$ is called a random dynamic system, if ψ satisfies the cocycle property: for any t, $s \in \mathbb{R}^+$, P-a.s. $\omega \in \boldsymbol{\Omega}$, $\psi(0, \omega) = \mathrm{id}_X$, $\psi(t + s, \omega) = \psi(t, \vartheta_s\omega)\psi(s, \omega)$. If $\psi(t, \omega) : X \longrightarrow X$ is continuous, then ψ is called a continuous random dynamical system.

Random dynamical systems with continuous time are generated by infinite-dimensional stochastic evolutional equations effected by an additive white noise, which has a unique global solution, as well as by differential equations with random coefficients or stochastic differential equations.

Definition 4.1.5 (Random compact set). Let $K : \boldsymbol{\Omega} \to 2^X$, 2^X be the set of all the subsets of X. K is called a random compact set, if $K(\omega)$ is compact P-a.s., and for any $U \in X$, the mapping $\omega \to d(U, K(\omega))$ is measurable, where $d(U, K(\omega)) = \inf\limits_{U_1 \in K(\omega)} d(U, U_1)$.

Definition 4.1.6. Let $A(\omega)$, $B(\omega)$ be two random sets,

(i) $A(\omega)$ attracts $B(\omega)$, if $\lim\limits_{t \to +\infty} d(\psi(t, \vartheta_{-t}\omega)B(\vartheta_{-t}\omega), A(\omega)) = 0$, P-a.s.

(ii) $A(\omega)$ absorbs $B(\omega)$, if there exists $t_B(\omega)$ such that for any $t \ge t_B(\omega)$, $\psi(t, \vartheta_{-t}\omega)\ B(\vartheta_{-t}\omega) \subset A(\omega)$, P-a.s..

Definition 4.1.7 (Random attractor). A random set $\mathcal{A}(\omega)$ is called a random attractor of the random dynamical system ψ, if the following conditions are satisfied P-a.s.,

(i) $\mathcal{A}(\omega)$ is a random compact set;
(ii) $\mathcal{A}(\omega)$ is invariant, that is, $\forall t \geq 0$, $\psi(t,\omega)\mathcal{A}(\omega) = \mathcal{A}(\vartheta_t\omega)$;
(iii) $\mathcal{A}(\omega)$ attracts all deterministic bounded sets $B \subset X$, that is

$$\lim_{t\to+\infty} d(\psi(t,\vartheta_{-t}\omega)B, A(\omega)) = 0, \ P\text{-a.s.}$$

Remark 4.1.8. $\psi(t,\vartheta_{-t}\omega)U$ can be interpreted as the position at $t = 0$ of the trajectory whose position is U at initial time $-t$, that is, when t is moving, $\psi(t,\vartheta_{-t}\omega)U$ is always at the position at time $t = 0$. Thus, the random attractor is also called a random pull-back attractor.

Theorem 4.1.9 (cf. [51, 52]). If the continuous random dynamical system ψ has a random compact set $K(\omega)$ absorbing all the non-random bounded set $B(B \subset X)$, then ψ possesses a random pull-back attractor $\mathcal{A}(\omega)$, where

$$\mathcal{A}(\omega) = \overline{\cup_{B\subset X} \Lambda_B(\omega)}, \quad \Lambda_B(\omega) = \cap_{s\geq 0}\overline{\cup_{t\geq s}\psi(t,\vartheta_{-t}\omega)B},$$

moreover, $\mathcal{A}(\omega) \subset K(\omega)$, and $\mathcal{A}(\omega)$ is unique.

4.1.4 *Existence of Random Attractors*

For any $\alpha > 0$, let

$$z(t) = W_A^\alpha(t) = \int_{-\infty}^t e^{(t-s)(-A-\alpha)}dW(s),$$

where $W_A^\alpha(t)$ is the mild solution of the following initial value problem

$$\begin{cases} dz = (-Az - \alpha z)dt + dW(t), \\ z(0) = \displaystyle\int_{-\infty}^0 e^{-s(-A-\alpha)}dW(s). \end{cases}$$

Suppose that u is the mild solution of the following initial boundary value problem

$$(P_1)\begin{cases} du - \dfrac{1}{Re}\Delta u dt = G(u)dt + dW(t), \\ u(s,w) = u_s, \\ u|_{\partial D} = 0. \end{cases}$$

Let $v(t) = u(t) - z(t)$. Then $v(t)$ is a solution of the following problem

$$(P_2)\begin{cases} dv = -Av + G(v+z) + \alpha z, \\ v(s,\omega) = u_s - z(s), \\ v|_{\partial D} = 0. \end{cases} \tag{4.1.12}$$

4.1.4.1 *Existence, Uniqueness and Regularity of the Problem* (P_2)

In order to study the asymptotic behavior of solutions to problems (P_1), we have to obtain more regularity of v than that in Theorem 4.1.3. Hence, we prove the following regularity result.

Theorem 4.1.10.

(i) Let $u_s \in L^2(D)$ for P-a.e. $\omega \in \mathbf{\Omega}$. Then, for any $T > 0$, problem (P_2) has a unique solution $v \in C(s,T;L^2(D)) \cap L^2(s,T;H_0^1(D))$ in weak sense for P-a.e. $\omega \in \mathbf{\Omega}$, and the solution of problem (P_1) $u = v + z$ satisfies $u \in C(s,T;L^2(D)) \cap L^2(s,T;D(A^{\min\{\frac{1}{4}+\beta,\frac{1}{2}\}}))$ a.s. for any $\beta < \beta_1$, where β_1 is the constant appeared in the definition of stochastic process $W(t)$ in subsection 4.1.2.

(ii) If $u_s \in D(A^\theta)$ a.s. for some $\theta \in (0, 2\beta_1) \cap (0, \frac{1}{2}]$, then for P-a.e. $\omega \in \mathbf{\Omega}$, $v \in C(s,T;D(A^\theta)) \cap L^2(s,T;D(A^{\frac{1}{2}+\theta}))$.

Theorem 4.1.10 is proved by Faedo-Galerkin method (see, e.g., [137]). Because this method is standard, we only give *a priori* estimates.

Energy estimates of v. In the sequel, $\omega \in \mathbf{\Omega}$ is fixed. By choosing v as a test function in (4.1.12), we get

$$\frac{1}{2}\frac{\mathrm{d}|v|_2^2}{\mathrm{d}t} + \frac{1}{Re}\|v\|^2 = \int_D G(v+z)v + \alpha\int_D zv,$$

where

$$\int_D G(v+z)v = -\int_D [J(B(v+z), v+z)v + \beta_0(B(v+z))_x v] + \int_D \left[\left(\frac{F}{Re} - \frac{r}{2}\right)(v+z)v - F\left(\frac{F}{Re} - \frac{r}{2}\right)B(v+z)v\right].$$

By integration by parts, Hölder Inequality and Young Inequality, we have

$$-\int_D J(B(v+z), v+z)v = \int_D J(B(v+z), v+z)z$$
$$\le |(B(v+z))_x|_4|v_y|_2|z|_4 + |(B(v+z))_y|_4|v_x|_2|z|_4$$
$$\le c|v+z|_2|v_y|_2|z|_4 + c|v+z|_2|v_x|_2|z|_4 \le c|v|_2^2|z|_4^4 + c|z|_4^4 + 2\varepsilon\|v\|^2,$$

$$|\beta_0\int_D (B(z))_x v| \le c|(B(z))_x|_2|v|_2 \le c|z|_2|v|_2 \le c|z|_2^2 + \varepsilon|v|_2^2.$$

Similarly,

$$\left|\left(\frac{F}{Re} - \frac{r}{2}\right)\int_D zv\right| \le c|z|_2^2 + \varepsilon|v|_2^2,$$

$$\left|-F\left(\frac{F}{Re}-\frac{r}{2}\right)\int_D vB(z)\right| \le c|z|_2^2+\varepsilon|v|_2^2.$$

According to the definition of operator B and the assumptions of constants β_0, F, Re, r, we get

$$\beta_0\int_D (B(v))_x v+\left(\frac{F}{Re}-\frac{r}{2}\right)\int_D v^2-F\left(\frac{F}{Re}-\frac{r}{2}\right)\int_D vB(v)\le 0.$$

Thus, choosing $\varepsilon>0$ small enough, we obtain

$$\frac{\mathrm{d}|v|_2^2}{\mathrm{d}t}+\frac{1}{Re}\|v\|^2\le c|z|_4^4|v|_2^2+c|z|_2^2+c|z|_4^4. \tag{4.1.13}$$

By Poincaré Inequality $\frac{1}{Re}\|v\|^2\ge\lambda_1|v|_2^2$, we have

$$\frac{\mathrm{d}|v|_2^2}{\mathrm{d}t}\le\left(c|z|_4^4-\frac{\lambda_1}{Re}\right)|v|_2^2+c|z|_2^2+c|z|_4^4.$$

So, for any $t\in[s,T]$, by Gronwall Inequality, we get

$$|v(t)|_2^2\le \mathrm{e}^{\int_s^t(-\lambda_1+c|z(\tau)|_4^4)\mathrm{d}\tau}|v(s)|_2^2+\int_s^t \mathrm{e}^{\int_\sigma^t(-\lambda_1+c|z(\tau)|_4^4)\mathrm{d}\tau}(c|z|_2^2+c|z|_4^4)\mathrm{d}\sigma. \tag{4.1.14}$$

Integrating (4.1.13) from t_1 to $t_2([t_1,t_2]\subset[s,T])$, we have

$$\frac{1}{Re}\int_{t_1}^{t_2}\|v(\tau)\|^2\mathrm{d}\tau\le|v(t_1)|_2^2+\int_{t_1}^{t_2}(c|z|_4^4|v|_2^2+c|z|_2^2+c|z|_4^4)\mathrm{d}\tau. \tag{4.1.15}$$

For the energy estimates of $A^\theta v$, we need the following lemma.

Lemma 4.1.11 (cf. [72]). For any two real-valued functions $f,\ g\in H^{\frac{1}{2}+\theta}(D)$, $0<\theta<\frac{1}{2}$,

$$\|fg\|_{H^{2\theta}}\le\|f\|_{H^{\frac{1}{2}+\theta}}\|g\|_{H^{\frac{1}{2}+\theta}}.$$

Energy estimates of $A^\theta v$. As above, $\omega\in\mathbf{\Omega}$ is fixed. Suppose that v is a solution of problem (P_2), and $v\in C(s,T;\ D(A^\theta))\cap L^2(s,T;D(A^{\frac{1}{2}+\theta}))$. Because the proof for the case of $\theta=1/2$ is classical (see [201]), here we only prove the case $0<\theta<1/2$.

Firstly, we have

$$\frac{1}{2}\frac{\mathrm{d}|A^\theta v|_2^2}{\mathrm{d}t}+\frac{1}{Re}|A^{\frac{1}{2}+\theta}v|_2^2=\int_D A^\theta G(v+z)A^\theta v+\alpha\int_D A^\theta zA^\theta v.$$

According to the definition of operator A^θ, the interpolation inequalities and Lemma 4.1.11, we have

$$-\int_D A^\theta J(B(v+z),v+z)A^\theta v=-\int_D J(B(v+z),v+z)A^{2\theta}v$$

$$
\begin{aligned}
\leq & c\left\|\frac{\partial A^{2\theta}v}{\partial y}\right\|_{H^{-2\theta}}\left\|\frac{\partial B(v+z)}{\partial x}(v+z)\right\|_{H^{2\theta}} \\
& + c\left\|\frac{\partial A^{2\theta}v}{\partial x}\right\|_{H^{-2\theta}}\left\|\frac{\partial B(v+z)}{\partial y}(v+z)\right\|_{H^{2\theta}} \\
\leq & c\|A^{2\theta}v\|_{H^{1-2\theta}}\left(\left\|\frac{\partial B(v+z)}{\partial x}\right\|_{H^{\frac{1}{2}+\theta}}+\left\|\frac{\partial B(v+z)}{\partial y}\right\|_{H^{\frac{1}{2}+\theta}}\right)\|v+z\|_{H^{\frac{1}{2}+\theta}} \\
\leq & \varepsilon|A^{\frac{1}{2}+\theta}v|_2^2+c|A^{\frac{1}{2}\theta}(v+z)|_2^2|A^{\frac{1}{4}+\frac{\theta}{2}}(v+z)|_2^2 \\
\leq & \varepsilon|A^{\frac{1}{2}+\theta}v|_2^2+c(|A^{\theta}v|_2|v|_2+|A^{\frac{1}{2}\theta}z|_2^2)(|A^{\frac{1}{4}+\frac{\theta}{2}}v|_2^2+|A^{\frac{1}{4}+\frac{\theta}{2}}z|_2^2) \\
\leq & \varepsilon|A^{\frac{1}{2}+\theta}v|_2^2+c|A^{\theta}v|_2^2|v|_2^2+c|A^{\frac{1}{4}+\frac{\theta}{2}}v|_2^4+c|A^{\frac{1}{4}+\frac{\theta}{2}}z|_2^4. \qquad (4.1.16)
\end{aligned}
$$

By the interpolation inequalities and Young Inequality, we get

$$
|A^{\frac{1}{4}+\frac{\theta}{2}}v|_2^4 \leq c|A^{\frac{1}{4}}v|_2^2|A^{\frac{1}{4}+\theta}v|_2^2 \leq \varepsilon|A^{\frac{1}{2}+\theta}v|_2^2+c|v|_2^2|A^{\frac{1}{2}}v|_2^2|A^{\theta}v|_2^2.
$$

We derive from the above two inequalities that

$$
\begin{aligned}
& -\int_D A^{\theta}J(B(v+z),v+z)A^{\theta}v \\
\leq & 2\varepsilon|A^{\frac{1}{2}+\theta}v|_2^2+c|v|_2^2|A^{\theta}v|_2^2+c|v|_2^2|A^{\frac{1}{2}}v|_2^2|A^{\theta}v|_2^2+c|A^{\frac{1}{4}+\frac{\theta}{2}}z|_2^4.
\end{aligned}
$$

By Hölder Inequality and Cauchy-Schwarz Inequality, we obtain

$$
\beta_0\int_D A^{\theta}(B(v+z))_x A^{\theta}v \leq c|A^{\theta}(B(v+z))_x|_2|A^{\theta}v|_2 \leq c|A^{\theta}v|_2^2+c|z|_2^2.
$$

In a similar way to get (4.1.16), we have

$$
\left(\frac{F}{Re}-\frac{r}{2}\right)\left[\int_D(A^{\theta}(v+z)-FA^{\theta})A^{\theta}v\right] \leq c|A^{\theta}v|_2^2+\varepsilon|A^{\frac{1}{2}+\theta}v|_2^2+c|z|_2^2.
$$

Thus, choosing $\varepsilon>0$ small enough, we get

$$
\begin{aligned}
\frac{\mathrm{d}|A^{\theta}v|_2^2}{\mathrm{d}t}+\frac{1}{Re}|A^{\frac{1}{2}+\theta}v|_2^2 \leq\ & c|v|_2^2|A^{\frac{1}{2}}v|_2^2|A^{\theta}v|_2^2+c|A^{\theta}v|_2^2+c|v|_2^2|A^{\theta}v|_2^2 \\
& +c|A^{\frac{1}{4}+\frac{\theta}{2}}z|_2^4+c|z|_2^2.
\end{aligned}
$$

By Gronwall Inequality, for any $t\in[s,T]$, we have

$$
\begin{aligned}
|A^{\theta}v(t)|_2^2 \leq & \mathrm{e}^{\int_s^t(c|v|_2^2|A^{\frac{1}{2}}v|_2^2+c|v|_2^2+c)\mathrm{d}\tau}|A^{\theta}v(s)|_2^2 \\
& +\int_s^t \mathrm{e}^{\int_\sigma^t(c|v|_2^2|A^{\frac{1}{2}}v|_2^2+c|v|_2^2+c)\mathrm{d}\tau}(c|z|_2^2+c|A^{\frac{1}{4}+\frac{\theta}{2}}z|_2^2)\mathrm{d}\sigma. \qquad (4.1.17)
\end{aligned}
$$

Moreover, we obtain

$$
\begin{aligned}
\frac{1}{Re}\int_{t_1}^{t_2}|A^{\frac{1}{2}+\theta}v|_2^2\mathrm{d}\tau \leq |A^{\theta}v(t_1)|_2^2+\int_{t_1}^{t_2} & c(|v|_2^2|A^{\frac{1}{2}}v|_2^2|A^{\theta}v|_2^2 \\
& +|v|_2^2|A^{\theta}v|_2^2+|A^{\theta}v|_2^2+|A^{\frac{1}{4}+\frac{\theta}{2}}z|_2^4+|z|_2^2)\mathrm{d}\tau.
\end{aligned}
$$

4.1.4.2 *Dissipativity Property of Dynamic System (4.1.2) in $L^2(D)$*

Here, by studying the dissipativity property of dynamical system (4.1.2), we obtain the existence of the random attractors of this system.

Theorem 4.1.12. The boundary value problem of the two-dimensional quasi-geostrophic with a general white noise

$$\mathrm{d}u - \frac{1}{Re}\Delta u \mathrm{d}t = G(u)\mathrm{d}t + \mathrm{d}W(t), \ \ u|_{\partial D} = 0,$$

has a random attractor in $L^2(D)$.

Proof. According to the ergodicity of process z with values in $D(A^{\frac{1}{4}+\beta})$ ($\beta < \beta_1$), by $D(A^{\frac{1}{4}+\beta}) \hookrightarrow L^4$, we have (see [55])

$$\lim_{s_0 \to -\infty} \frac{1}{-s_0} \int_{s_0}^{0} |z(\tau)|_4^4 \mathrm{d}\tau = E(|z(0)|_4^4).$$

According to the results in [72], we know that for a α sufficiently large, $E(|z(0)|_4^4)$ is arbitrary small. Thus, by choosing a α large enough, we have

$$\lim_{s \to -\infty} \frac{1}{-s} \int_{s}^{0} (-\lambda_1 + c|z|_4^4)\mathrm{d}\tau = -\lambda_1 + cE(|z(0)|_4^4) \le -\frac{\lambda_1}{2}.$$

The above equality implies that there exists $S_0(\omega)$ such that $s < S_0(\omega)$,

$$\int_{s}^{0} (-\lambda_1 + c|z|_4^4)\mathrm{d}\tau \le -\frac{\lambda_1}{4}(-s). \tag{4.1.18}$$

Since $|z(s)|_2^2$ and $|z(s)|_4^4$ have at most polynomial growth as $s \to -\infty$ (see [72]), (4.1.14) and (4.1.18) imply that there exists an a.s. finite random variable $r_1(\omega)$ such that a.s.

$$|v(t)|_2^2 \le r_1(\omega), \ \ \text{for any} \ \ t \in [-1, 0], \tag{4.1.19}$$

when u_s belongs to a bounded set of $L^2(D)$.

Let $t_1 = -1$, $t_2 = 0$. According to (4.1.15) and (4.1.19), there exists a a.s. finite random variable $r_2(\omega)$, such that a.s.

$$\int_{-1}^{0} \|v(\tau)\|^2 \mathrm{d}\tau \le r_2(\omega). \tag{4.1.20}$$

According to (4.1.17), letting $t = 0$, $s \in [-1, 0]$, we have

$$\begin{aligned} |A^{\theta}v(0)|_2^2 \le & e^{\int_s^0 (c|v|_2^2 |A^{\frac{1}{2}}v|_2^2 + c)\mathrm{d}\tau} |A^{\theta}v(s)|_2^2 \\ & + \int_s^0 e^{\int_\sigma^0 (c|v|_2^2 |A^{\frac{1}{2}}v|_2^2 + c)\mathrm{d}\tau} (c|z|_2^2 + c|A^{\frac{1}{4}+\frac{\theta}{2}}z|_2^2)\mathrm{d}\sigma. \end{aligned} \tag{4.1.21}$$

Integrating (4.1.21) about s from -1 to 0, we obtain

$$|A^\theta v(0)|_2^2 \leq e^{\int_{-1}^0 (c|v|_2^2|A^{\frac{1}{2}}v|_2^2+c)\mathrm{d}\tau} \int_{-1}^0 |A^\theta(s)|_2^2 \mathrm{d}s + \int_{-1}^0 e^{\int_\sigma^0 (c|v|_2^2|A^{\frac{1}{2}}v|_2^2+c)\mathrm{d}\tau}(c|z|_2^2 + c|A^{\frac{1}{4}+\frac{\theta}{2}}z|_2^2)\mathrm{d}\sigma. \quad (4.1.22)$$

Combining (4.1.19) with (4.1.20), according to (4.1.22), we know that there exists an a.s. finite random variable $r_3(\omega)$ such that a.s.

$$|A^\theta v(0)|_2^2 \leq r_3(\omega).$$

Without loss of generality, we suppose that $\boldsymbol{\Omega} = \{\omega : \omega \in C(\mathbb{R}, D(A^{-\frac{1}{4}+\beta_1})),\ \omega(0) = 0\}$, P is Wiener measure, and $W(t,\omega) = \omega(t)$, $t \in \mathbb{R}$, $\omega \in \boldsymbol{\Omega}$. Here, we define a family of measure preserving and ergodic transform $(\theta_t)_{t\in\mathbb{R}}$ on $\boldsymbol{\Omega}$ by $\theta_s\omega(t) = \omega(t+s)-\omega(s)$, t, $s \in \mathbb{R}$. Define $S(t,s,\omega)u_s = u(t,\omega)$, and $\psi(t,\omega)u_0 = S(t,0,\omega)u_0$. Let $K(\omega)$ be a ball with radius of $r_3(\omega) + |A^\theta z(0,\omega)|_2^2$ in $D(A^\theta)$. $K(\omega)$ is a compact attracting set at time 0 (because the imbedding $H^{2\theta}(D) \hookrightarrow L^2(D)$ is compact). Applying Theorem 4.1.9, we prove Theorem 4.1.12.

4.1.5 *Stochastic Quasi-Geostrophic Equations on Unbounded Domains*

In this subsection, we discuss the existence of a continuous random dynamical system for the stochastic Q-G equation defined on unbounded domain. Let E be a bounded domain in $\mathbb{R}$ and $D = E \times \mathbb{R}$. Consider the stochastic Q-G equation defined on D:

$$\mathrm{d}(\Delta\psi - F\psi + \beta_0 y) + \left(\frac{\partial\psi}{\partial x}\frac{\partial}{\partial y} - \frac{\partial\psi}{\partial y}\frac{\partial}{\partial x}\right)(\Delta\psi - F\psi + \beta_0 y)\mathrm{d}t = \left(\frac{1}{Re}\Delta^2\psi - \frac{r}{2}\Delta\psi\right)\mathrm{d}t - \Phi\mathrm{d}w, \quad (4.1.23)$$

with the boundary condition

$$\psi(x,y,t) = \Delta\psi(x,y,t) = 0, \qquad (x,y) \in \partial D. \quad (4.1.24)$$

Next, in order to be convenient for the research, we simplify equation (4.1.23).

For any $u \in L^2(D)$, by the following elliptic equation with Dirichlet boundary condition

$$\begin{cases} F\psi - \Delta\psi = u, \\ \psi|_{\partial D} = 0, \end{cases} \quad (4.1.25)$$

we obtain $\psi = (FI - \Delta)^{-1}u = P(u)$. Applying the elliptic regular theory, we know that $P : L^2(D) \to H_0^1(D) \cap H^2(D)$. Therefore equation (4.1.23) can be rewritten as

$$\mathrm{d}u - \frac{1}{Re}\Delta u\mathrm{d}t = G(u)\mathrm{d}t + \Phi\mathrm{d}w,$$

where $\psi = P(u)$, $G(u) = -J(\psi, u) + \beta_0\psi_x + (\frac{F}{Re} - \frac{r}{2})u - F(\frac{F}{Re} - \frac{r}{2})\psi$, J is the Jacobian operator, and $J(\psi, u) = \frac{\partial\psi}{\partial x}\frac{\partial u}{\partial y} - \frac{\partial\psi}{\partial y}\frac{\partial u}{\partial x}$.

Here, we shall consider the following problem

$$\mathrm{d}u - \frac{1}{Re}\Delta u\mathrm{d}t = G(u)\mathrm{d}t + \Phi\mathrm{d}w, \quad (x, y) \in D, \ t > 0, \tag{4.1.26}$$

with the boundary condition

$$u|_{\partial D} = 0, \qquad t > 0, \tag{4.1.27}$$

and the initial condition

$$u(x, y, 0) = u_0(x, y), \qquad (x, y) \in D. \tag{4.1.28}$$

In the sequel, we consider the probability space $(\boldsymbol{\Omega}, \mathcal{F}, \mathbb{P})$ where

$$\boldsymbol{\Omega} = \{\omega \in C(\mathbb{R}, \mathbb{R}) : \omega(0) = 0\},$$

$\mathcal{F}$ is the Borel σ-algebra induced by the compact-open topology of $\boldsymbol{\Omega}$, and $\mathbb{P}$ the corresponding Wiener measure on $(\boldsymbol{\Omega}, \mathcal{F})$. Defined the time shift by

$$\theta_t\omega(\cdot) = \omega(\cdot + t) - \omega(t), \quad \omega \in \boldsymbol{\Omega}, \quad t \in \mathbb{R}.$$

Then $(\boldsymbol{\Omega}, \mathcal{F}, (\theta_t)_{t\in\mathbb{R}})$ is a metric dynamical system. Here, we need to convert the stochastic equation (4.1.23) with a random term into a deterministic one with a random parameter. To this end, we consider the stationary process

$$\eta(\theta_t\omega) = -\alpha\int_{-\infty}^0 \mathrm{e}^{\alpha\tau}(\theta_t\omega)(\tau)\mathrm{d}\tau, \quad t \in \mathbb{R},$$

which satisfies the following one-dimensional equation

$$\mathrm{d}\eta + \alpha\eta\mathrm{d}t = \mathrm{d}w(t), \tag{4.1.29}$$

where α is a positive constant.

It is known that there exists a θ_t-invariant set $\tilde{\boldsymbol{\Omega}} \subseteq \boldsymbol{\Omega}$ of full $\mathbb{P}$ measure such that $y(\theta_t\omega)$ is continuous in t for every $\omega \in \tilde{\boldsymbol{\Omega}}$, and the random variable $|\eta(\omega)|$ is tempered (see [4, 51, 73]). Therefore, it follows from Proposition 4.3.3 in [4] that there exists a tempered function $r(\omega) > 0$ such that

$$|\eta(\omega)|^2 \leq r(\omega), \tag{4.1.30}$$

where $r(\omega)$ satisfies, for P-a.e. $\omega \in \boldsymbol{\Omega}$,

$$r(\theta_t\omega) \le e^{\frac{\alpha}{2}|t|} r(\omega), \qquad t \in \mathbb{R}. \tag{4.1.31}$$

Then it follows from (4.1.30) and (4.1.31) that, for P-a.e. $\omega \in \boldsymbol{\Omega}$,

$$|\eta(\theta_t\omega)|^2 \le e^{\frac{\alpha}{2}|t|} r(\omega), \qquad t \in \mathbb{R}. \tag{4.1.32}$$

Put $z(\theta_t\omega) = (I - \Delta)^{-1}\Phi\eta(\theta_t\omega)$ where Δ is the Laplacian with domain $H_0^1(D) \cap H^2(D)$. By (4.1.29) we find that

$$\mathrm{d}z = \mathrm{d}(\Delta z) - \alpha(z - \Delta z)\mathrm{d}t + \Phi \mathrm{d}w.$$

Let $v(t,\omega) = u(t,\omega) - z(\theta_t\omega)$, where $u(t,\omega)$ satisfies (4.1.26)–(4.1.28). Then for $v(t,\omega)$ we infer that

$$v_t - \frac{1}{Re}\Delta v = G(v + z(\theta_t\omega)) + \alpha z(\theta_t\omega) + \Big(\frac{1}{Re} - \alpha\Big)\Delta z(\theta_t\omega), \tag{4.1.33}$$

with the boundary condition

$$v|_{\partial D} = 0, \tag{4.1.34}$$

and the initial condition

$$v(0,\omega) = v_0(\omega). \tag{4.1.35}$$

Referring to the Galerkin method, it can be proved that, for P-a.e. $\omega \in \boldsymbol{\Omega}$ and for all $v_0 \in L^2(D)$, problem (4.1.33)–(4.1.35) has a unique solution $v(\cdot,\omega,v_0) \in C([0,\infty), L^2(D))$ with $v(0,\omega,v_0) = v_0$. Then the solution $v(t,\omega,v_0)$ is continuous with respect to v_0 in $L^2(D)$ for all $t > 0$. Throughout this subsection, we always write

$$u(t,\omega,u_0) = v(t,\omega,v_0) + z(\theta_t\omega), \quad \text{with} \quad v_0 = u_0 - z(\omega). \tag{4.1.36}$$

Then u is a solution of problem (4.1.33)–(4.1.35) in some sense. We now define a mapping $\Psi : \mathbb{R}^+ \times \boldsymbol{\Omega} \times L^2(D) \to L^2(D)$ by

$$\Psi(t,\omega,u_0) = u(t,\omega,u_0), \quad \forall\ (t,\omega,u_0) \in \mathbb{R}^+ \times \boldsymbol{\Omega} \times L^2(D). \tag{4.1.37}$$

Note that Ψ satisfies conditions in Definition 4.1. Therefore, Ψ is a continuous random dynamical system associated with the stochastic Q-G equation on D. In the following, we shall establish uniform estimates for Ψ and prove that Ψ has a $\mathcal{D}$-random attractor in $L^2(D)$, where $\mathcal{D}$ is a collection of random subsets of $L^2(D)$ given by

$$\begin{aligned} \mathcal{D} = \{\ B : B = \{B(\omega)\}_{\omega\in\Omega}, B(\omega) \subseteq L^2(D) \\ \text{and } e^{-\delta t}\mathrm{d}(B(\theta_{-t}\omega)) \to 0 \text{ as } t \to -\infty\}, \end{aligned} \tag{4.1.38}$$

where δ is the positive constant in (4.1.47) and

$$\mathrm{d}(B(\theta_{-t}\omega)) = \sup_{u \in B(\theta_{-t}\omega)} \|u\|_{L^2(D)}.$$

Notice that $\mathcal{D}$ contains all tempered random sets, especially all bounded deterministic subsets of $L^2(D)$.

4.1.5.1 *Uniform estimates of solutions*

To prove the existence of bounded absorbing sets and the asymptotic compactness of the random dynamical system, here, we shall derive uniform estimates on the solutions of the stochastic Q-G equation defined on D when $t \to \infty$, which include the uniform estimates on the tails of solutions as both (x, y) and t approach infinity.

From now on, we always suppose that $\mathcal{D}$ is the collection of random subsets of $L^2(D)$ given by (4.1.38). Firstly, we derive the following uniform estimates on v in $L^2(D)$.

Lemma 4.1.13. Suppose that $\Phi \in L^2(D)$. Let $B = \{B(\omega)\}_{\omega\in\boldsymbol{\Omega}} \in \mathcal{D}$ and $v_0(\omega) \in B(\omega)$. Then for P-a.e. $\omega \in \boldsymbol{\Omega}$, there is $T = T(B, \omega) > 0$ such that for all $t \geq T$,

$$\|v(t, \theta_{-t}\omega, v_0(\theta_{-t}\omega))\| \leq r_1(\omega),$$

where $r_1(\omega)$ is a positive random function satisfying

$$\mathrm{e}^{-\delta t} r_1(\theta_{-t}\omega) \to 0 \quad \text{as} \ \ t \to \infty. \tag{4.1.39}$$

Proof. Taking the inner product of (4.1.33) with v in $L^2(D)$, we obtain

$$\begin{aligned}
&\frac{1}{2}\frac{\mathrm{d}}{\mathrm{d}t}\|v\|^2 + \frac{1}{Re}\|\boldsymbol{\nabla} v\|^2 \\
&= \int_D G(v + z(\theta_t\omega))v + \left(\alpha z(\theta_t\omega) + \Big(\frac{1}{Re} - \alpha\Big)\Delta z(\theta_t\omega), v\right),
\end{aligned} \tag{4.1.40}$$

where

$$\begin{aligned}
&\int_D G(v + z(\theta_t\omega))v \\
&= \int_D -J\left(P(v + z(\theta_t\omega)), v + z(\theta_t\omega)\right) v + \int_D \beta_0 \left(P(v + z(\theta_t\omega))\right)_x v \\
&+ \int_D \Big(\frac{F}{Re} - \frac{r}{2}\Big)(v + z(\theta_t\omega))v - \int_D F\Big(\frac{F}{Re} - \frac{r}{2}\Big)P(v + z(\theta_t\omega))v.
\end{aligned} \tag{4.1.41}$$

Now we estimate every term on the right-hand side of (4.1.41). Note that

$$\int_D J(P(v), v)v = 0, \qquad \int_D J(P(z(\theta_t\omega)), v)v = 0.$$

Then, for the first term on the right-hand side of (4.1.41), we have

$$\begin{aligned}
&\int_D -J\left(P(v + z(\theta_t\omega)), v + z(\theta_t\omega)\right) v \\
&\leq \left|\int_D J(P(z(\theta_t\omega)), z(\theta_t\omega))v\right| + \left|\int_D J(P(v), z(\theta_t\omega))v\right|.
\end{aligned}$$

By integration by parts and Hölder Inequality,

$$\begin{aligned}
&\left|\int_D J(P(z(\theta_t\omega)), z(\theta_t\omega))v\right| \\
&= \left|\int_D \frac{\partial P(z(\theta_t\omega))}{\partial x}\frac{\partial z(\theta_t\omega)}{\partial y}v - \frac{\partial P(z(\theta_t\omega))}{\partial y}\frac{\partial z(\theta_t\omega)}{\partial x}v\right| \\
&\leq \left|\int_D \frac{\partial P(z(\theta_t\omega))}{\partial x}z(\theta_t\omega)\frac{\partial v}{\partial y}\right| + \left|\int_D \frac{\partial P(z(\theta_t\omega))}{\partial y}z(\theta_t\omega)\frac{\partial v}{\partial x}\right| \\
&\leq \|z(\theta_t\omega)\|_\infty\left(\left\|\frac{\partial P(z(\theta_t\omega))}{\partial x}\right\|\left\|\frac{\partial v}{\partial y}\right\| + \left\|\frac{\partial P(z(\theta_t\omega))}{\partial y}\right\|\left\|\frac{\partial v}{\partial x}\right\|\right) \\
&\leq 2\|z(\theta_t\omega)\|_\infty\|\nabla P(z(\theta_t\omega))\|\|\nabla v\|. \qquad (4.1.42)
\end{aligned}$$

By (4.1.25), we deduce that $P(z(\theta_t\omega)) = \psi = (FI - \Delta)^{-1}z(\theta_t\omega)$. Then $F\psi - \Delta\psi = z(\theta_t\omega)$ holds. Taking the inner product of the above equality with ψ in $L^2(D)$, then applying integration by parts and Young Inequality, we infer that

$$F\|\psi\|^2 + \|\nabla\psi\|^2 = (z(\theta_t\omega), \psi) \leq \frac{F}{2}\|\psi\|^2 + \frac{1}{2F}\|z(\theta_t\omega)\|^2.$$

Then we have

$$\|\nabla P(z(\theta_t\omega))\| = \|\nabla\psi\| \leq \|z(\theta_t\omega)\|.$$

Substituting the above inequality into (4.1.42), we obtain that

$$\begin{aligned}
\left|\int_D J(P(z(\theta_t\omega)), z(\theta_t\omega))v\right| &\leq 2\|z(\theta_t\omega)\|_\infty\|z(\theta_t\omega)\|\|\nabla v\| \\
&\leq 4Re\|z(\theta_t\omega)\|_\infty^2\|z(\theta_t\omega)\|^2 + \frac{1}{4Re}\|\nabla v\|^2 \\
&\leq 4Re\beta_1^2\|z(\theta_t\omega)\|_{H^1}^2\|z(\theta_t\omega)\|^2 + \frac{1}{4Re}\|\nabla v\|^2 \\
&\leq 4Re\beta_1^2 C_0|\eta(\theta_t\omega)|^4 + \frac{1}{4Re}\|\nabla v\|^2. \qquad (4.1.43)
\end{aligned}$$

Similarly to (4.1.43), we have

$$\begin{aligned}
\left|\int_D J(P(v), z(\theta_t\omega))v\right| &\leq 2\|z(\theta_t\omega)\|_\infty\|\nabla P(v)\|\|\nabla v\| \\
&\leq 2\beta_1\|z(\theta_t\omega)\|_{H^1}\|v\|\|\nabla v\| \\
&\leq 4Re\beta_1^2\|\Phi\|^2|\eta(\theta_t\omega)|^2\|v\|^2 + \frac{1}{4Re}\|\nabla v\|^2.
\end{aligned}$$

Then we obtain

$$\begin{aligned}
&\int_D -J\left(P(v + z(\theta_t\omega)), v + z(\theta_t\omega)\right)v \\
&\leq \frac{1}{2Re}\|\nabla v\|^2 + 4Re\beta_1^2\|\Phi\|^2|\eta(\theta_t\omega)|^2\|v\|^2 + 4Re\beta_1^2 C_0|\eta(\theta_t\omega)|^4. \qquad (4.1.44)
\end{aligned}$$

Next, we estimate the three terms on the right-hand side of (4.1.41). The three terms on the right-hand side of (4.1.41) can be rewritten as

$$\begin{aligned}
&\int_D \beta_0 \left(P(v+z(\theta_t\omega))\right)_x v + \int_D \Big(\frac{F}{Re}-\frac{r}{2}\Big)(v+z(\theta_t\omega))v \\
&- \int_D F\Big(\frac{F}{Re}-\frac{r}{2}\Big)P(v+z(\theta_t\omega))v \\
= {}& \beta_0 \int_D (P(z(\theta_t\omega)))_x v + \int_D \Big(\frac{F}{Re}-\frac{r}{2}\Big)z(\theta_t\omega)v \\
&-F\Big(\frac{F}{Re}-\frac{r}{2}\Big)\int_D vP(z(\theta_t\omega)) \\
&+\left(\beta_0\int_D (P(v))_x v + \Big(\frac{F}{Re}-\frac{r}{2}\Big)\int_D v^2 - F\Big(\frac{F}{Re}-\frac{r}{2}\Big)\int_D vP(v)\right).
\end{aligned}$$

According to the definition of the operator P, we have

$$\begin{aligned}
&\beta_0\int_D (P(v))_x v + \Big(\frac{F}{Re}-\frac{r}{2}\Big)\int_D v^2 - F\Big(\frac{F}{Re}-\frac{r}{2}\Big)\int_D vP(v) \\
={}& \frac{\beta_0}{F}\int_D v(v_x+\Delta\psi_x) + \Big(\frac{F}{Re}-\frac{r}{2}\Big)\int_D v^2 - \Big(\frac{F}{Re}-\frac{r}{2}\Big)\int_D v(v+\Delta\psi) \\
={}& \frac{\beta_0}{F}\int_D (F\psi-\Delta\psi)\Delta\psi_x - \Big(\frac{F}{Re}-\frac{r}{2}\Big)\int_D (F\psi-\Delta\psi)\Delta\psi \\
={}& F\Big(\frac{F}{Re}-\frac{r}{2}\Big)\|\Delta\psi\|^2 + \Big(\frac{F}{Re}-\frac{r}{2}\Big)\|\Delta\psi\|^2.
\end{aligned}$$

By the assumption of the parameter β_0, F, Re, r, we obtain that

$$\beta_0\int_D (P(v))_x v + \Big(\frac{F}{Re}-\frac{r}{2}\Big)\int_D v^2 - F\Big(\frac{F}{Re}-\frac{r}{2}\Big)\int_D vP(v) \le 0.$$

So the three terms on the right-hand side of (4.1.41) are dominated by

$$\begin{aligned}
&\int_D \beta_0 \left(P(v+z(\theta_t\omega))\right)_x v + \int_D \Big(\frac{F}{Re}-\frac{r}{2}\Big)(v+z(\theta_t\omega))v \\
&- \int_E F\Big(\frac{F}{Re}-\frac{r}{2}\Big)P(v+z(\theta_t\omega))v \\
\le{}& \beta_0\left|\int_D (P(z(\theta_t\omega)))_x v\right| + \left|\int_D \Big(\frac{F}{Re}-\frac{r}{2}\Big)z(\theta_t\omega)v\right| \\
&+\left|F\Big(\frac{F}{Re}-\frac{r}{2}\Big)\int_D vP(z(\theta_t\omega))\right| \\
\le{}& \beta_0\|\nabla P(z(\theta_t\omega))\|\|v\| + \left|\frac{F}{Re}-\frac{r}{2}\right|\|z(\theta_t\omega)\|\|v\| \\
&+\left|F\Big(\frac{F}{Re}-\frac{r}{2}\Big)\right|\|z(\theta_t\omega)\|\|v\|
\end{aligned}$$

$$\leq \frac{3\lambda}{16Re}\|v\|^2 + C_1|\eta(\theta_t\omega)|^2. \tag{4.1.45}$$

Substituting (4.1.44) and (4.1.45) into (4.1.41), we obtain that

$$\int_D G(v + z(\theta_t\omega))v \leq \frac{1}{2Re}\|\nabla v\|^2 + \left(4Re\beta_1^2\|\Phi\|^2|\eta(\theta_t\omega)|^2\right)\|v\|^2 + \frac{3\lambda}{16Re}\|v\|^2 + C_2(|\eta(\theta_t\omega)|^2 + |\eta(\theta_t\omega)|^4).$$

By using the Young Inequality, the second term on the right-hand side of (4.1.40) is bounded by

$$\left|\left(\alpha z(\theta_t\omega) + \left(\frac{1}{Re} - \alpha\right)\Delta z(\theta_t\omega), v\right)\right| \leq \frac{\lambda}{16Re}\|v\|^2 + C_3|\eta(\theta_t\omega)|^2.$$

Therefore, we obtain that

$$\frac{\mathrm{d}}{\mathrm{d}t}\|v\|^2 + \frac{1}{Re}\|\nabla v\|^2 \leq \left(8Re\beta_1^2\|\Phi\|^2|\eta(\theta_t\omega)|^2\right)\|v\|^2 + \frac{\lambda}{2Re}\|v\|^2 + C_4(|\eta(\theta_t\omega)|^2 + |\eta(\theta_t\omega)|^4).$$

By Poincaré Inequality, the above inequality can be rewritten as

$$\frac{\mathrm{d}}{\mathrm{d}t}\|v\|^2 + \left(\frac{\lambda}{2Re} - 8Re\beta_1^2\|\Phi\|^2|\eta(\theta_t\omega)|^2\right)\|v\|^2 \leq C_4(|\eta(\theta_t\omega)|^2 + |\eta(\theta_t\omega)|^4). \tag{4.1.46}$$

Denote by

$$\delta = \frac{1}{2Re}\lambda \quad \text{and} \quad \beta = 8Re\beta_1^2\|\Phi\|^2. \tag{4.1.47}$$

Multiplying (4.1.46) by $e^{\int_0^t(\delta-\beta|\eta(\theta_\tau\omega)|^2)\mathrm{d}\tau}$ and then integrating over $(0, s)$ with $s \geq 0$, we infer that

$$\|v(s,\omega,v_0(\omega))\|^2 \leq e^{-\delta s+\beta\int_0^s|\eta(\theta_\tau\omega)|^2\mathrm{d}\tau}\|v_0(\omega)\|^2 + C_4\int_0^s(|\eta(\theta_\sigma\omega)|^2 + |\eta(\theta_\sigma\omega)|^4)e^{\delta(\sigma-s)+\beta\int_\sigma^s|\eta(\theta_\tau\omega)|^2\mathrm{d}\tau}\mathrm{d}\sigma.$$

Now we replace ω by $\theta_{-t}\omega$ with $t \geq 0$ in the above inequality to deduce that, for any $t \geq 0$ and $s \geq 0$,

$$\begin{aligned}
&\|v(s,\theta_{-t}\omega,v_0(\theta_{-t}\omega))\|^2 \\
\leq\ & e^{-\delta s+\beta\int_0^s|y(\theta_{\tau-t}\omega)|^2\mathrm{d}\tau}\|v_0(\theta_{-t}\omega)\|^2 \\
&+C_4\int_0^s(|\eta(\theta_{\sigma-t}\omega)|^2 + |\eta(\theta_{\sigma-t}\omega)|^4)e^{\delta(\sigma-s)+\beta\int_\sigma^s|\eta(\theta_{\tau-t}\omega)|^2\mathrm{d}\tau}\mathrm{d}\sigma \\
=\ & e^{-\delta s+\beta\int_{-t}^{s-t}|\eta(\theta_\tau\omega)|^2\mathrm{d}\tau}\|v_0(\theta_{-t}\omega)\|^2
\end{aligned}$$

$$+C_4 \int_{-t}^{s-t} (|\eta(\theta_\sigma \omega)|^2 + |\eta(\theta_\sigma \omega)|^4) e^{\delta(\sigma - s + t) + \beta \int_\sigma^{s-t} |\eta(\theta_\tau \omega)|^2 d\tau} d\sigma.$$

Then we obtain that, for any $t \geq 0$,

$$\|v(t, \theta_{-t}\omega, v_0(\theta_{-t}\omega))\|^2 \leq e^{-\delta t + \beta \int_{-t}^0 |\eta(\theta_\tau \omega)|^2 d\tau} \|v_0(\theta_{-t}\omega)\|^2$$
$$+C_4 \int_{-t}^{0} (|\eta(\theta_\sigma \omega)|^2 + |\eta(\theta_\sigma \omega)|^4) e^{\delta \sigma + \beta \int_\sigma^0 |\eta(\theta_\tau \omega)|^2 d\tau} d\sigma.$$

Note that $|\eta(\theta_t \omega)|^2 \leq e^{\frac{\alpha}{2} t} r(\omega)$, $t \geq 0$. Then we deduce that

$$\beta \int_{-t}^0 |\eta(\theta_\tau \omega)|^2 d\tau \leq \beta \int_{-t}^0 e^{\frac{\alpha}{2}\tau} r(\omega) d\tau \leq \beta r(\omega) \frac{2}{\alpha}(1 - e^{-\frac{\alpha}{2}t}) \leq \frac{2\beta}{\alpha} r(\omega).$$

So we obtain that, for any $t \geq T_0(\omega)$,

$$\|v(t, \theta_{-t}\omega, v_0(\theta_{-t}\omega))\|^2$$
$$\leq e^{-\delta t + \frac{2\beta}{\alpha} r(\omega)} \|v_0(\theta_{-t}\omega)\|^2$$
$$+C_4 \int_{-t}^{0} (|\eta(\theta_\sigma \omega)|^2 + |\eta(\theta_\sigma \omega)|^4) e^{\delta \sigma + \beta \int_\sigma^0 |\eta(\theta_\tau \omega)|^2 d\tau} d\sigma.$$

Note that $|\eta(\theta_\sigma \omega)|$ is tempered, and hence by (4.1.32), the integrand of the second term on the right-hand side of the above inequality converges to zero exponentially as $\sigma \to -\infty$. This shows that the following integral is convergent:

$$r_0(w) = C_4 \int_{-\infty}^{0} (|\eta(\theta_\sigma \omega)|^2 + |\eta(\theta_\sigma \omega)|^4) e^{\delta \sigma + \beta \int_\sigma^0 |\eta(\theta_\tau \omega)|^2 d\tau} d\sigma. \quad (4.1.48)$$

It follows from the above two relationships that, for all $t \geq T_0(\omega)$,

$$\|v(t, \theta_{-t}\omega, v_0(\theta_{-t}\omega))\|^2 \leq e^{-\delta t + \frac{2\beta}{\alpha} r(\omega)} \|v_0(\theta_{-t}\omega)\|^2 + r_0(w). \quad (4.1.49)$$

In addition, by assumption $B = \{B(\omega)\}_{\omega \in \Omega} \in \mathcal{D}$ and hence we have

$$e^{-\delta t} \|v_0(\theta_{-t}\omega)\|^2 \to 0 \qquad \text{as } t \to \infty,$$

from which and (4.1.48) we find that there is $T = T(B, \omega) > 0$ such that for all $t \geq T$,

$$\|v(t, \theta_{-t}\omega, v_0(\theta_{-t}\omega))\|^2 \leq 2 r_0(\omega).$$

Let $r_1(\omega) = \sqrt{2 r_0(\omega)}$. Then we obtain that, for all $t \geq T$,

$$\|v(t, \theta_{-t}\omega, v_0(\theta_{-t}\omega))\| \leq r_1(w). \quad (4.1.50)$$

In what follows, we prove $r_1(\omega)$ satisfies (4.1.39). Replacing ω by $\theta_{-t}\omega$ in (4.1.48) we obtain that

$$r_0(\theta_{-t}\omega) = C_4 \int_{-\infty}^{0} (|\eta(\theta_{\sigma - t}\omega)|^2 + |\eta(\theta_{\sigma - t}\omega)|^4) e^{\delta \sigma + \beta \int_\sigma^0 |\eta(\theta_{\tau - t}\omega)|^2 d\tau} d\sigma$$

$$
\begin{aligned}
&= C_4 \int_{-\infty}^{-t} (|\eta(\theta_\sigma\omega)|^2 + |\eta(\theta_\sigma\omega)|^4) e^{\delta(\sigma+t)+\beta\int_\sigma^{-t}|\eta(\theta_\tau\omega)|^2 d\tau} d\sigma \\
&\le C_4 \int_{-\infty}^{-t} (|\eta(\theta_\sigma\omega)|^2 + |\eta(\theta_\sigma\omega)|^4) e^{\frac{3}{2}\delta(\sigma+t)+\beta\int_\sigma^{-t}|\eta(\theta_\tau\omega)|^2 d\tau} d\sigma \\
&\le C_4 e^{\frac{3}{2}\delta t} \int_{-\infty}^{0} (|\eta(\theta_\sigma\omega)|^2 + |\eta(\theta_\sigma\omega)|^4) e^{\frac{3}{2}\delta\sigma+\beta\int_\sigma^{0}|\eta(\theta_\tau\omega)|^2 d\tau} d\sigma.
\end{aligned} \tag{4.1.51}
$$

The integrand converges to zero exponentially as $\sigma \to -\infty$, hence the last integral of the above formula is indeed convergent. Then we deduce that

$$
\begin{aligned}
& e^{-\sigma t} r_1(\theta_{-t}\omega) \\
= \;& e^{-\sigma t}\sqrt{2r_0(\theta_{-t}\omega)} \\
\le \;& \sqrt{2C_4} e^{-\frac{1}{4}\delta t} \left(\int_{-\infty}^{0} (|\eta(\theta_\sigma\omega)|^2 + |\eta(\theta_\sigma\omega)|^4) e^{\frac{3}{2}\delta\sigma+\beta\int_\sigma^{0}|\eta(\theta_\tau\omega)|^2 d\tau} d\sigma \right)^{\frac{1}{2}} \\
& \to 0 \quad \text{as } t \to \infty,
\end{aligned}
$$

which along with (4.1.50) completes the proof.

Lemma 4.1.14. Suppose that $\Phi \in H_0^1(D)$. Let $B = \{B(\omega)\}_{\omega\in\mathbf{\Omega}} \in \mathcal{D}$ and $v_0(\omega) \in B(\omega)$. Then for P-a.e. $\omega \in \mathbf{\Omega}$, there is $T_1 = T_1(B,\omega) > 0$ such that for all $t \ge T_1$,

$$
\begin{aligned}
& \|\nabla v(t, \theta_{-t}\omega, v_0(\theta_{-t}\omega))\|^2 \\
& \le C\left(1 + r_0(\omega) + r(\omega) + r_0^2(\omega) + r^2(\omega) + r_0^5(\omega) + r^5(\omega)\right).
\end{aligned} \tag{4.1.52}
$$

Proof. Taking the inner product of (4.1.33) with Δv in $L^2(D)$, we deduce that

$$
\begin{aligned}
& \frac{1}{2}\frac{d}{dt}\|\nabla v\|^2 + \frac{1}{Re}\|\Delta v\|^2 \\
= & -\int_D G(v + z(\theta_t\omega))\Delta v - \left(\alpha z(\theta_t\omega) + \left(\frac{1}{Re} - \alpha\right)\Delta z(\theta_t\omega), \Delta v\right),
\end{aligned} \tag{4.1.53}
$$

where

$$
\begin{aligned}
& -\int_D G(v + z(\theta_t\omega))\Delta v \\
= & \int_D J(P(v + z(\theta_t\omega)), v + z(\theta_t\omega))\Delta v - \int_D \left(\frac{F}{Re} - \frac{r}{2}\right)(v + z(\theta_t\omega))\Delta v \\
& -\beta_0 \int_D (P(v + z(\theta_t\omega)))_x \Delta v + \int_D F\left(\frac{F}{Re} - \frac{r}{2}\right) P(v + z(\theta_t\omega))\Delta v.
\end{aligned} \tag{4.1.54}
$$

Next, we estimate every term on the right-hand side of (4.1.54). Firstly, we consider $\|P(v+z(\theta_t\omega))\|$ and $\|\Delta P(v+z(\theta_t\omega))\|$. By (4.1.25), we have $P(v+z(\theta_t\omega)) = \psi = (FI-\Delta)^{-1}(v+z(\theta_t\omega))$. Then

$$F\psi - \Delta\psi = v + z(\theta_t\omega).$$

Taking the inner product of the above equality with ψ in $L^2(D)$, then applying integration by parts and Young Inequality, we obtain that

$$F\|\psi\|^2 + \|\nabla\psi\|^2 = (v+z(\theta_t\omega),\psi) \le \frac{F}{2}\|\psi\|^2 + \frac{1}{2F}\|v+z(\theta_t\omega)\|^2.$$

Then we have

$$\|P(v+z(\theta_t\omega))\| = \|\psi\| \le \frac{1}{F}\|v+z(\theta_t\omega)\|,$$
$$\|\nabla P(v+z(\theta_t\omega))\| = \|\nabla\psi\| \le \|v+z(\theta_t\omega)\|.$$

Moreover, we obtain that

$$F\|\nabla\psi\|^2 + \|\Delta\psi\|^2 = -(v+z(\theta_t\omega),\Delta\psi) \le \frac{1}{2}\|\Delta\psi\|^2 + \frac{1}{2}\|v+z(\theta_t\omega)\|^2.$$

Then we have

$$\|\Delta P(v+z(\theta_t\omega))\| = \|\Delta\psi\| \le \|v+z(\theta_t\omega)\|.$$

By applying Hölder Inequality, Gagliardo-Nirenberg Inequality, Young Inequality, we have

$$\begin{aligned}
&\left|\int_D J(P(v+z(\theta_t\omega)), v+z(\theta_t\omega))\Delta v\right| \\
&= \left|\int_D \frac{\partial P(v+z(\theta_t\omega))}{\partial x}\frac{\partial(v+z(\theta_t\omega))}{\partial y}\Delta v \right.\\
&\quad \left. - \frac{\partial P(v+z(\theta_t\omega))}{\partial y}\frac{\partial(v+z(\theta_t\omega))}{\partial x}\Delta v\right| \\
&\le 2\|\Delta v\|\|\nabla P(v+z(\theta_t\omega))\|_4\|\nabla(v+z(\theta_t\omega))\|_4 \\
&\le C_5\|\Delta v\|\|\Delta P(v+z(\theta_t\omega))\|^{\frac{3}{4}}\|\Delta(v+z(\theta_t\omega))\|^{\frac{3}{4}}\|v+z(\theta_t\omega)\|^{\frac{1}{4}} \\
&\le C_6\|\Delta v\|\|v+z(\theta_t\omega)\|^{\frac{5}{4}}\|\Delta(v+z(\theta_t\omega))\|^{\frac{3}{4}} \\
&\le \frac{1}{8Re}\|\Delta v\|^2 + C_7\|v+z(\theta_t\omega)\|^{\frac{5}{2}}\|\Delta(v+z(\theta_t\omega))\|^{\frac{3}{2}} \\
&\le \frac{1}{4Re}\|\Delta v\|^2 + C_8(\|\Delta z(\theta_t\omega)\|^2 + \|v+z(\theta_t\omega)\|^{10}).
\end{aligned}$$

Similarly, the terms on the right-hand side of (4.1.54) are bounded by

$$\left|\int_D \left(-\beta_0(P(v+z(\theta_t\omega)))_x - \left(\frac{F}{Re} - \frac{r}{2}\right)(v+z(\theta_t\omega))\right.\right.$$

$$
\begin{aligned}
&+F\Big(\frac{F}{Re}-\frac{r}{2}\Big)P(v+z(\theta_t\omega))\Big)\,\Delta v\Big| \\
&\leq \beta_0\|\nabla P(v+z(\theta_t\omega))\|\|\Delta v\| + \Big|\frac{F}{Re}-\frac{r}{2}\Big|\,\|v+z(\theta_t\omega)\|\|\Delta v\| \\
&\quad+\Big|F\Big(\frac{F}{Re}-\frac{r}{2}\Big)\Big|\,\|\nabla P(v+z(\theta_t\omega))\|\|\nabla v\| \\
&\leq \Big(\beta_0+\Big|\frac{F}{Re}-\frac{r}{2}\Big|\Big)\,\|v+z(\theta_t\omega)\|\|\Delta v\| + \Big|F\Big(\frac{F}{Re}-\frac{r}{2}\Big)\Big|\,\|v \\
&\quad+z(\theta_t\omega)\|\|\nabla v\| \\
&\leq \frac{1}{4Re}\|\Delta v\|^2+\frac{1}{8Re}\|\nabla v\|^2+C_9\|v+z(\theta_t\omega)\|^2.
\end{aligned}
$$

Then, we infer that

$$
\begin{aligned}
-\int_D G(v+z(\theta_t\omega))\Delta v \leq{}& \frac{1}{2Re}\|\Delta v\|^2+\frac{1}{8Re}\|\nabla v\|^2+C_9\|v+z(\theta_t\omega)\|^2 \\
&+C_8(\|\Delta z(\theta_t\omega)\|^2+\|v+z(\theta_t\omega)\|^{10}).
\end{aligned}
$$

For the second term on the right-hand side of (4.1.53), we deduce that

$$
\begin{aligned}
&\Big(\alpha z(\theta_t\omega)+\Big(\frac{1}{Re}-\alpha\Big)\Delta z(\theta_t\omega),\Delta v\Big) \\
&\leq \frac{1}{4Re}\|\Delta v\|^2+C_{10}(\|z(\theta_t\omega)\|^2+\|\Delta z(\theta_t\omega)\|^2).
\end{aligned}
$$

Therefore, we obtain

$$
\begin{aligned}
&\frac{\mathrm{d}}{\mathrm{d}t}\|\nabla v\|^2+\frac{1}{2Re}\|\Delta v\|^2 \\
&\leq \frac{1}{4Re}\|\nabla v\|^2+C_{11}(1+\|\Delta z(\theta_t\omega)\|^2+\|v+z(\theta_t\omega)\|^{10}+\|z(\theta_t\omega)\|^2).
\end{aligned}
$$

Applying integration by parts and Young Inequality, we deduce that

$$
\|\nabla v\|^2 \leq \|\Delta v\|\|v\| \leq \|\Delta v\|^2+\frac{1}{4}\|v\|^2,
$$

namely,

$$
-\frac{1}{4}\|v\|^2+\|\nabla v\|^2 \leq \|\Delta v\|^2.
$$

Thus we have

$$
\begin{aligned}
&\frac{\mathrm{d}}{\mathrm{d}t}\|\nabla v\|^2+\frac{1}{4Re}\|\nabla v\|^2 \\
&\leq \frac{1}{8Re}\|v\|^2+C_{12}(1+\|\Delta z(\theta_t\omega)\|^2+\|v+z(\theta_t\omega)\|^{10}+\|z(\theta_t\omega)\|^2) \\
&\leq C_{13}(\|v\|^2+\|v\|^{10})+C_{14}(1+|\eta(\theta_t\omega)|^2+|\eta(\theta_t\omega)|^{10}),
\end{aligned}
$$

which implies that

$$\begin{aligned}
&\frac{\mathrm{d}}{\mathrm{d}t}\|\boldsymbol{\nabla} v\|^2\\
&\le \frac{1}{8Re}\|v\|^2 + C_{12}(1 + \|\Delta z(\theta_t\omega)\|^2 + \|v + z(\theta_t\omega)\|^{10} + \|z(\theta_t\omega)\|^2)\\
&\le C_{13}(\|v\|^2 + \|v\|^{10}) + C_{14}(1 + |\eta(\theta_t\omega)|^2 + |\eta(\theta_t\omega)|^{10}).
\end{aligned}$$

Then there exists a positive constant $T_1(B)$, take $t \ge T_1(B)$ and $s \in (t, t+1)$. Integrating the above inequality over $(s, t+1)$, we infer that

$$\begin{aligned}
&\|\boldsymbol{\nabla} v(t+1, \omega, v_0(\omega))\|^2\\
&\le C_{14}\int_s^{t+1}(1 + |\eta(\theta_\tau\omega)|^2 + |\eta(\theta_\tau\omega)|^{10})\mathrm{d}\tau + \|\boldsymbol{\nabla} v(s, \omega, v_0(\omega))\|^2\\
&\quad + C_{13}\int_s^{t+1}(\|v(\tau, \omega, v_0(\omega))\|^2 + \|v(\tau, \omega, v_0(\omega))\|^{10})\mathrm{d}\tau\\
&\le \|\boldsymbol{\nabla} v(s, \omega, v_0(\omega))\|^2 + C_{14}\int_t^{t+1}(1 + |\eta(\theta_\tau\omega)|^2 + |\eta(\theta_\tau\omega)|^{10})\mathrm{d}\tau\\
&\quad + C_{13}\int_t^{t+1}(\|v(\tau, \omega, v_0(\omega))\|^2 + \|v(\tau, \omega, v_0(\omega))\|^{10})\mathrm{d}\tau.
\end{aligned}$$

Now integrating the above with respect to s over $(t, t+1)$, we infer that

$$\begin{aligned}
\|\boldsymbol{\nabla} v(t+1, \omega, v_0(\omega))\|^2 \le &\int_t^{t+1}\|\boldsymbol{\nabla} v(s, \omega, v_0(\omega))\|^2\mathrm{d}s\\
&+ C_{14}\int_t^{t+1}(1 + |\eta(\theta_\tau\omega)|^2 + |\eta(\theta_\tau\omega)|^{10})\mathrm{d}\tau\\
&+ C_{13}\int_t^{t+1}(\|v(\tau, \omega, v_0(\omega))\|^2 + \|v(\tau, \omega, v_0(\omega))\|^{10})\mathrm{d}\tau.
\end{aligned}$$

Replacing ω by $\theta_{-t-1}\omega$, we obtain that

$$\begin{aligned}
&\|\boldsymbol{\nabla} v(t+1, \theta_{-t-1}\omega, v_0(\theta_{-t-1}\omega))\|^2 \qquad (4.1.55)\\
&\le \int_t^{t+1}\|\boldsymbol{\nabla} v(s, \theta_{-t-1}\omega, v_0(\theta_{-t-1}\omega))\|^2\mathrm{d}s\\
&\quad + C_{14}\int_t^{t+1}(1 + |\eta(\theta_{\tau-t-1}\omega)|^2 + |\eta(\theta_{\tau-t-1}\omega)|^{10})\mathrm{d}\tau\\
&\quad + C_{13}\int_t^{t+1}(\|v(\tau, \theta_{-t-1}\omega, v_0(\theta_{-t-1}\omega))\|^2 + \|v(\tau, \theta_{-t-1}\omega, v_0(\theta_{-t-1}\omega))\|^{10})\mathrm{d}\tau.
\end{aligned}$$

By (4.1.46), we deduce that

$$\|v(t+1, \omega, v_0(\omega))\|^2 - \|v(t, \omega, v_0(\omega))\|^2 + \frac{1}{Re}\int_t^{t+1}\|\boldsymbol{\nabla} v(s, \omega, v_0(\omega))\|^2\mathrm{d}s$$

$$\leq \int_t^{t+1} \left(\left(\beta|\eta(\theta_\tau\omega)|^2 + \frac{\lambda}{2Re}\right)\|v(\tau,\omega,v_0(\omega))\|^2 + C_4(|\eta(\theta_\tau\omega)|^2 + |\eta(\theta_\tau\omega)|^4) \right) \mathrm{d}\tau.$$

Replacing ω by $\theta_{-t-1}\omega$, we have

$$\frac{1}{Re}\int_t^{t+1} \|\nabla v(s,\theta_{-t-1}\omega,v_0(\theta_{-t-1}\omega))\|^2 \mathrm{d}s$$
$$\leq \|v(t,\theta_{-t-1}\omega,v_0(\theta_{-t-1}\omega))\|^2 + C_4\int_t^{t+1} (|\eta(\theta_{\tau-t-1}\omega)|^2 + |\eta(\theta_{\tau-t-1}\omega)|^4)\mathrm{d}\tau$$
$$+ \int_t^{t+1} \left(\beta|\eta(\theta_{\tau-t-1}\omega)|^2 + \frac{\lambda}{2Re}\right)\|v(\tau,\theta_{-t-1}\omega,v_0(\theta_{-t-1}\omega))\|^2 \mathrm{d}\tau. \quad (4.1.56)$$

Applying Lemma 4.1.13 and (4.1.56), we deduce that

$$\int_t^{t+1} \|\nabla v(s,\theta_{-t-1}\omega,v_0(\theta_{-t-1}\omega))\|^2 \mathrm{d}s$$
$$\leq C\left(r_0(\omega) + r(\omega) + r_0^2(\omega) + r^2(\omega)\right). \quad (4.1.57)$$

By (4.1.55) and (4.1.57), we obtain that

$$\|\nabla v(t+1,\theta_{-t-1}\omega,v_0(\theta_{-t-1}\omega))\|^2$$
$$\leq C\left(1 + r_0(\omega) + r(\omega) + r_0^2(\omega) + r^2(\omega) + r_0^5(\omega) + r^5(\omega)\right), \quad (4.1.58)$$

which completes the proof.

To prove the asymptotic compactness of the random dynamical system, we will derive the uniform estimates on the tails of solutions when (x,y) and t approach to infinite. To this end, for every $(x,y) \in D = E \times \mathbb{R}$, where $x \in D$, and $y \in \mathbb{R}$. Given $k > 0$, denote by $D_k = \{(x,y) \in D : |y| < k\}$, and D/D_k the complement of D_k.

Lemma 4.1.15. Suppose that $\Phi \in L^2(D)$. Let $B = \{B(\omega)\}_{\omega\in\mathbf{\Omega}} \in \mathcal{D}$ and $v_0(\omega) \in B(\omega)$. Then for every $\varepsilon > 0$, and P-a.e. $\omega \in \mathbf{\Omega}$, there exist $T_2 = T_2(B,\omega,\varepsilon) > 0$ and $k_0 = k_0(\omega,\varepsilon) > 0$ such that for all $t \geq T_2$,

$$\int_{D/D_{k_0}} |v(t,\theta_{-t}\omega,v_0(\theta_{-t}\omega))|^2 \leq \varepsilon. \quad (4.1.59)$$

Proof. Take a smooth χ such that $0 \leq \chi \leq 1$ for all $s \in \mathbb{R}$ and

$$\chi(s) = \begin{cases} 0, & \text{if } |s| < 1, \\ 1, & \text{if } |s| > 2. \end{cases}$$

Then there exists a positive constant c such that $|\chi'(s)| + |\chi''(s)| \leq c$ for all $s \in \mathbb{R}$. Multiplying (4.1.33) by $\chi^2(\frac{y^2}{k^2})v$ and then integrating with respect to x, y respectively on D, we obtain that

$$\int_D \chi^2(\frac{y^2}{k^2})vv_t - \frac{1}{Re}\int_D \chi^2(\frac{y^2}{k^2})v\Delta v$$

$$= \int_D \chi^2(\frac{y^2}{k^2}) vG(v + z(\theta_t\omega)) + \int_D \chi^2(\frac{y^2}{k^2}) v(\alpha z(\theta_t\omega) + (\frac{1}{Re} - \alpha)\Delta z(\theta_t\omega)). \tag{4.1.60}$$

Note the first term on the left-hand side of the above, and we have

$$\int_D \chi^2(\frac{y^2}{k^2}) vv_t = \frac{1}{2}\frac{\mathrm{d}}{\mathrm{d}t}\int_D \chi^2(\frac{y^2}{k^2})|v|^2.$$

Applying integration by part, we deduce that

$$-\frac{1}{Re}\int_D \chi^2(\frac{y^2}{k^2}) v\Delta v = \frac{1}{Re}\int_D \chi^2(\frac{y^2}{k^2})|\nabla v|^2 + \frac{1}{Re}\int_D v\left(\nabla v \nabla \chi^2(\frac{y^2}{k^2})\right).$$

Substituting the above equations into (4.1.60), we obtain that

$$\begin{aligned} &\frac{1}{2}\frac{\mathrm{d}}{\mathrm{d}t}\int_D \chi^2(\frac{y^2}{k^2})|v|^2 + \frac{1}{Re}\int_D \chi^2(\frac{y^2}{k^2})|\nabla v|^2 \\ =& -\frac{1}{Re}\int_D v\left(\nabla v\nabla\chi^2(\frac{y^2}{k^2})\right) + \int_D \chi^2(\frac{y^2}{k^2}) vG(v + z(\theta_t\omega)) \\ &+ \int_D \chi^2(\frac{y^2}{k^2}) v\left(\alpha z(\theta_t\omega) + (\frac{1}{Re} - \alpha)\Delta z(\theta_t\omega)\right). \end{aligned} \tag{4.1.61}$$

Next, we estimate every term on the right-hand side of (4.1.61). For the first term we infer that

$$\begin{aligned} \left|-\frac{1}{Re}\int_D v\left(\nabla v\nabla\chi^2(\frac{y^2}{k^2})\right)\right| &\leq \frac{1}{Re}\int_D |v||\nabla v||2\chi\chi'(\frac{y^2}{k^2})|\frac{2|y|}{k^2} \\ &\leq \frac{1}{Re}\int_{k\leq|y|\leq\sqrt{2}k} |v||\nabla v||2\chi\chi'(\frac{y^2}{k^2})|\frac{2|y|}{k^2} \\ &\leq \frac{c}{k}\int_{k\leq|y|\leq\sqrt{2}k} |v||\nabla v| \\ &\leq \frac{c}{k}\|v\|\|\nabla v\| \leq \frac{c}{k}(\|v\|^2 + \|\nabla v\|^2). \end{aligned}$$

For the second term on the right-hand side of (4.1.61) we obtain that

$$\begin{aligned} &\int_D \chi^2(\frac{y^2}{k^2}) vG(v + z(\theta_t\omega)) \\ =& -\int_D \chi^2(\frac{y^2}{k^2}) vJ\left(P(v + z(\theta_t\omega), (v + z(\theta_t\omega))\right)) \\ &+\beta_0\int_D \chi^2(\frac{y^2}{k^2}) vP(v + z(\theta_t\omega))_x \\ &+\left(\frac{F}{Re} - \frac{r}{2}\right)\int_D \chi^2(\frac{y^2}{k^2}) v(v + z(\theta_t\omega)) \end{aligned}$$

$$-F\left(\frac{F}{Re}-\frac{r}{2}\right)\int_D \chi^2(\frac{y^2}{k^2})vP(v+z(\theta_t\omega)).$$

Now, we estimate the terms on the right-hand side of the above equality. Since

$$\begin{aligned}
&-\int_D \chi^2(\frac{y^2}{k^2})vJ\left(P(v+z(\theta_t\omega)),v+z(\theta_t\omega)\right)\\
&=\int_D \chi^2(\frac{y^2}{k^2})v\frac{\partial P(v+z(\theta_t\omega))}{\partial y}\frac{\partial(v+z(\theta_t\omega))}{\partial x}\\
&\quad-\int_D \chi^2(\frac{y^2}{k^2})v\frac{\partial P(v+z(\theta_t\omega))}{\partial x}\frac{\partial(v+z(\theta_t\omega))}{\partial y}\\
&=-\int_D\left(\frac{\partial P(v+z(\theta_t\omega))}{\partial y}\chi^2(\frac{y^2}{k^2})\frac{\partial v}{\partial x}\right.\\
&\quad\left.+\frac{\partial P(v+z(\theta_t\omega))}{\partial x}(2\chi\chi'(\frac{y^2}{k^2})\frac{2y}{k^2}v+\chi^2(\frac{y^2}{k^2})\frac{\partial v}{\partial y})\right)(v+z(\theta_t\omega))\\
&\le 2\|\nabla P(v+z(\theta_t\omega))\|_4\|v+z(\theta_t\omega)\|_4\|\nabla v\|\\
&\quad+\frac{c}{k}\int_{k\le|y|\le\sqrt{2}k}|\nabla P(v+z(\theta_t\omega))|\cdot|v|\cdot|v+z(\theta_t\omega)|\\
&\le C\left(\|v+z(\theta_t\omega)\|\|v+z(\theta_t\omega)\|_4\|\nabla v\|+\|\nabla P(v+z(\theta_t\omega))\|_4\|v\|_4\|v\right.\\
&\quad\left.+z(\theta_t\omega)\|\right)\\
&\le C\left(\|v+z(\theta_t\omega)\|^{\frac{3}{2}}\|\nabla(v+z(\theta_t\omega))\|^{\frac{1}{2}}\|\nabla v\|+\|v+z(\theta_t\omega)\|^2\|\nabla v\|^{\frac{1}{2}}\|v\|^{\frac{1}{2}}\right)\\
&\le C(\|\nabla(v+z(\theta_t\omega))\|+\|\nabla v\|^2\|v+z(\theta_t\omega)\|^3+\|\nabla v\|+\|v+z(\theta_t\omega)\|^4\|v\|)\\
&\le C(\|\nabla v\|+\|\nabla v\|^4+\|\nabla z(\theta_t\omega)\|+\|v\|^2+\|v+z(\theta_t\omega)\|^6+\|v+z(\theta_t\omega)\|^8),
\end{aligned}$$

by

$$\begin{aligned}
&\beta_0\int_D \chi^2(\frac{y^2}{k^2})v(P(v+z(\theta_t\omega)))_x\\
&\le \beta_0\|\nabla P(v+z(\theta_t\omega))\|\|v\|\\
&\le \beta_0\|v+z(\theta_t\omega)\|\|v\|\le C(\|z(\theta_t\omega)\|^2+\|v\|^2),
\end{aligned}$$

$$\begin{aligned}
&\left(\frac{F}{Re}-\frac{r}{2}\right)\int_D \chi^2(\frac{y^2}{k^2})v(v+z(\theta_t\omega))\\
&\quad-F\left(\frac{F}{Re}-\frac{r}{2}\right)\int_D \chi^2(\frac{y^2}{k^2})vP(v+z(\theta_t\omega))\\
&\le C(\|v\|^2+\|z(\theta_t\omega)\|^2),
\end{aligned}$$

we have

$$\int_D \chi^2(\frac{y^2}{k^2})vG(v+z(\theta_t\omega)) \leq C(\|v\|^2 + \|\nabla v\| + \|\nabla v\|^4 + \|z(\theta_t\omega)\|^2 + \|\nabla z(\theta_t\omega)\| + \|v+z(\theta_t\omega)\|^6 + \|v+z(\theta_t\omega)\|^8). \tag{4.1.62}$$

By using the Young Inequality, the last term on the right-hand side of (4.1.61) is bounded by

$$\left|\int_D \chi^2(\frac{y^2}{k^2})v\left(\alpha z(\theta_t\omega) + (\frac{1}{Re}-\alpha)\Delta z(\theta_t\omega)\right)\right|$$
$$\leq \frac{1}{4Re}\lambda\int_D \chi^2(\frac{y^2}{k^2})|v|^2 + C(\|z(\theta_t\omega)\|^2 + \|\Delta z(\theta_t\omega)\|^2).$$

So (4.1.61) can be rewritten as

$$\frac{1}{2}\frac{\mathrm{d}}{\mathrm{d}t}\int_D \chi^2(\frac{y^2}{k^2})|v|^2 + \frac{1}{Re}\int_D \chi^2(\frac{y^2}{k^2})|\nabla v|^2$$
$$\leq \frac{1}{4Re}\lambda\int_D \chi^2(\frac{y^2}{k^2})|v|^2 + C(\|v\|^2 + \|\nabla v\| + \|\nabla v\|^2 + \|\nabla v\|^4 + \|z(\theta_t\omega)\|^2 + \|\Delta z(\theta_t\omega)\|^2 + \|\nabla z(\theta_t\omega)\| + \|v+z(\theta_t\omega)\|^6 + \|v+z(\theta_t\omega)\|^8).$$

Note that

$$\begin{aligned}\int_D \left|\nabla(\chi(\frac{y^2}{k^2})v)\right|^2 &= \int_D \left|v\nabla\chi(\frac{y^2}{k^2}) + \chi(\frac{y^2}{k^2})\nabla v\right|^2\\
&\leq 2\int_D |v|^2\left|\nabla\chi(\frac{y^2}{k^2})\right|^2 + 2\int_D\left|\chi(\frac{y^2}{k^2})\right|^2|\nabla v|^2\\
&\leq 2\int_{k\leq|y|\leq\sqrt{2}k}|v|^2\left|\chi'(\frac{y^2}{k^2})\right|^2\frac{|2y|^2}{k^4} + 2\int_D \chi^2(\frac{y^2}{k^2})|\nabla v|^2\\
&\leq \frac{c}{k^2}\int_{k\leq|y|\leq\sqrt{2}k}|v|^2 + 2\int_D \chi^2(\frac{y^2}{k^2})|\nabla v|^2\\
&\leq \frac{c}{k^2}\|v\|^2 + 2\int_D \chi^2(\frac{y^2}{k^2})|\nabla v|^2.\end{aligned}$$

Applying Poincaré Inequality, we infer that

$$\int_D \chi^2(\frac{y^2}{k^2})|v|^2 \leq \frac{1}{\lambda}\int_D\left|\nabla(\chi(\frac{y^2}{k^2})v)\right|^2 \leq \frac{c}{k^2\lambda}\|v\|^2 + \frac{2}{\lambda}\int_D \chi^2(\frac{y^2}{k^2})|\nabla v|^2,$$

namely,

$$\int_D \chi^2(\frac{y^2}{k^2})|\nabla v|^2 \geq \frac{1}{2}\lambda\int_D \chi^2(\frac{y^2}{k^2})|v|^2 - \frac{c}{2k^2}\|v\|^2.$$

Then, we obtain that

$$\begin{aligned}&\frac{\mathrm{d}}{\mathrm{d}t}\int_D \chi^2(\frac{y^2}{k^2})|v|^2 + \frac{1}{2Re}\lambda\int_D \chi^2(\frac{y^2}{k^2})|v|^2\\&\leq \frac{C}{k^2}(\|v\|^2 + \|\nabla v\| + \|\nabla v\|^2 + \|\nabla v\|^4 + \|z(\theta_t\omega)\|^2\\&+ \|\Delta z(\theta_t\omega)\|^2 + \|\nabla z(\theta_t\omega)\| + \|v + z(\theta_t\omega)\|^6 + \|v + z(\theta_t\omega)\|^8).\end{aligned}$$

Applying Gronwall Inequality over $(0, t)$, we obtain that, for all $t \geq 0$,

$$\begin{aligned}&\int_D \chi^2(\frac{y^2}{k^2})|v(t,\omega,v_0(\omega))|^2\\&\leq \mathrm{e}^{-\frac{1}{2Re}\lambda t}\int_D \chi^2(\frac{y^2}{k^2})|v_0(\omega)|^2 + \frac{C}{k^2}\int_0^t \mathrm{e}^{\frac{1}{2Re}\lambda(s-t)}(\|\nabla z(\theta_s\omega)\|\\&+\|v(s,\omega,v_0(\omega))\|^2 + \|\nabla v(s,\omega,v_0(\omega))\| + \|z(\theta_s\omega)\|^2\\&+\|\nabla v(s,\omega,v_0(\omega))\|^2 + \|\nabla v(s,\omega,v_0(\omega))\|^4 + \|\Delta z(\theta_s\omega)\|^2\\&+\|v(s,\omega,v_0(\omega)) + z(\theta_s\omega)\|^6 + \|v(s,\omega,v_0(\omega)) + z(\theta_s\omega)\|^8)\mathrm{d}s.\end{aligned}$$

Replacing ω by $\theta_{-t}\omega$ in the above, we find that, for all $t \geq 0$,

$$\begin{aligned}&\int_D \chi^2(\frac{y^2}{k^2})|v(t,\theta_{-t}\omega,v_0(\theta_{-t}\omega))|^2\\&\leq \mathrm{e}^{-\frac{1}{2Re}\lambda t}\int_D \chi^2(\frac{y^2}{k^2})|v_0(\theta_{-t}\omega)|^2 + \frac{C}{k^2}\int_0^t \mathrm{e}^{\frac{1}{2Re}\lambda(s-t)}(\|z(\theta_{s-t}\omega)\|^2\\&+ \|\nabla z(\theta_{s-t}\omega)\| + \|v(s,\theta_{-t}\omega,v_0(\theta_{-t}\omega))\|^2 + \|\nabla v(s,\theta_{-t}\omega,v_0(v\omega))\|\\&+ \|\nabla v(s,\theta_{-t}\omega,v_0(\theta_{-t}\omega))\|^2 + \|v(s,\theta_{-t}\omega,v_0(\theta_{-t}\omega)) + z(\theta_{s-t}\omega)\|^8\\&+ \|\Delta z(\theta_{s-t}\omega)\|^2 + \|\nabla v(s,\theta_{-t}\omega,v_0(\theta_{-t}\omega))\|^4\\&+ \|v(s,\theta_{-t}\omega,v_0(\theta_{-t}\omega)) + z(\theta_{s-t}\omega)\|^6)\mathrm{d}s. \qquad (4.1.63)\end{aligned}$$

In the following, we estimate every term on the right-hand side of (4.1.63). For the first term, we infer that

$$\begin{aligned}&\mathrm{e}^{-\frac{1}{2Re}\lambda t}\int_D \chi^2(\frac{y^2}{k^2})|v_0(\theta_{-t}\omega)|^2\\&\leq \mathrm{e}^{-\frac{1}{2Re}\lambda t}\|v_0(\theta_{-t}\omega)\|^2 \quad \text{for all } t \geq T_0(\omega). \qquad (4.1.64)\end{aligned}$$

Since $v_0(\theta_{-t}\omega) \in B(\theta_{-t}\omega)$ and $B = \{B(\omega)\}_{\omega\in\Omega} \in \mathcal{D}$, the right-hand side of (4.1.64) tends to zero as $t \to \infty$. Therefore, given a $T' = T'(B,\omega,\varepsilon) > 0$ such that for all $t \geq T'$,

$$\mathrm{e}^{-\frac{1}{2Re}\lambda t}\int_D \chi^2(\frac{y^2}{k^2})|v_0(\theta_{-t}\omega)|^2 \leq \varepsilon. \qquad (4.1.65)$$

Note that

$$
\begin{aligned}
&\int_0^t \mathrm{e}^{\frac{1}{2Re}\lambda(s-t)}(\|v(s,\theta_{-t}\omega,v_0(\theta_{-t}\omega))+z(\theta_{s-t}\omega)\|^6 \\
&+\|v(s,\theta_{-t}\omega,v_0(\theta_{-t}\omega))+z(\theta_{s-t}\omega)\|^8)\mathrm{d}s \\
&\le C\int_0^t \mathrm{e}^{\frac{1}{2Re}\lambda(s-t)}(\|v(s,\theta_{-t}\omega,v_0(\theta_{-t}\omega))\|^6+\|v(s,\theta_{-t}\omega,v_0(\theta_{-t}\omega))\|^8)\mathrm{d}s \\
&\quad +C\int_0^t \mathrm{e}^{\frac{1}{2Re}\lambda(s-t)}(\|z(\theta_{s-t}\omega)\|^6+\|z(\theta_{s-t}\omega)\|^8)\mathrm{d}s \\
&\le C\int_0^t \mathrm{e}^{\frac{1}{2Re}\lambda(s-t)}(\|v(s,\theta_{-t}\omega,v_0(\theta_{-t}\omega))\|^6+\|v(s,\theta_{-t}\omega,v_0(\theta_{-t}\omega))\|^8)\mathrm{d}s \\
&\quad +C\int_{-t}^0 \mathrm{e}^{-\frac{1}{2Re}\lambda s}(\|z(\theta_s\omega)\|^6+\|z(\theta_s\omega)\|^8)\mathrm{d}s \\
&\le \varepsilon(1+r(\omega)) \quad \text{as} \quad t\to\infty.
\end{aligned} \tag{4.1.66}
$$

Through a simple computation, we obtain that

$$
\begin{aligned}
&\frac{C}{k^2}\int_0^t \mathrm{e}^{\frac{1}{2Re}\lambda(s-t)}(\|z(\theta_{s-t}\omega)\|^2+\|\Delta z(\theta_{s-t}\omega)\|^2+\|\nabla z(\theta_{s-t}\omega)\|^2)\mathrm{d}s \\
&=\frac{C}{k^2}\int_{-t}^0 \mathrm{e}^{\frac{1}{2Re}\lambda s}(\|z(\theta_s\omega)\|^2+\|\Delta z(\theta_s\omega)\|^2+\|\nabla z(\theta_s\omega)\|^2)\mathrm{d}s \\
&\le \frac{C}{k^2}\int_{-t}^0 \mathrm{e}^{\frac{1}{2Re}\lambda s}|\eta(\theta_s\omega)|^2\mathrm{d}s\le\varepsilon(r(\omega)),
\end{aligned} \tag{4.1.67}
$$

and

$$
\begin{aligned}
&\frac{C}{k^2}\int_0^t \mathrm{e}^{\frac{1}{2Re}\lambda(s-t)}(\|v(s,\theta_{-t}\omega,v_0(\theta_{-t}\omega))\|^2+\|\nabla v(s,\theta_{-t}\omega,v_0(v\omega))\| \\
&+\|\nabla v(s,\theta_{-t}\omega,v_0(\theta_{-t}\omega))\|^2+\|\nabla v(s,\theta_{-t}\omega,v_0(\theta_{-t}\omega))\|^4)\mathrm{d}s \\
&\le \varepsilon(r(\omega),r_0(\omega)).
\end{aligned} \tag{4.1.68}
$$

According to (4.1.63)–(4.1.68), we infer that there exist positive random function $r_2(\omega)$, $T_2=T_2(B,\omega)$ and $k_0(\varepsilon,\omega)$ such that for all $t\ge T_2$,

$$
\int_D \chi^2\Big(\frac{y^2}{k^2}\Big)|v(t,\theta_{-t}\omega,v_0(\theta_{-t}\omega))|^2\le\varepsilon(1+r_2(\omega)),
$$

where $r_2(\omega)$ is a positive random function. So we deduce that

$$
\begin{aligned}
\int_{|y|\ge k_0}|v(t,\theta_{-t}\omega,v_0(\theta_{-t}\omega))|^2 &\le \int_D \chi^2\Big(\frac{y^2}{k^2}\Big)|v(t,\theta_{-t}\omega,v_0(\theta_{-t}\omega))|^2 \\
&\le \varepsilon(1+r_2(\omega)),
\end{aligned} \tag{4.1.69}
$$

which completes the proof.

In the following, the asymptotic compactness of the solution of the problem (4.1.33)–(4.1.35) is proved.

Lemma 4.1.16. Suppose that $\Phi \in L^2(D)$. Let $B = \{B(\omega)\}_{\omega\in\boldsymbol{\Omega}} \in \mathcal{D}$, $t_n \to \infty$ and $v_{0,n}(\omega) \in B(\theta_{-t_n}\omega)$. Then for P-a.e. $\omega \in \boldsymbol{\Omega}$, the sequence $\{v(t_n, \theta_{-t_n}\omega, v_{0,n})\}_{n=1}^{\infty}$ has a convergent subsequence in $L^2(D)$.

Proof. According to Lemma 4.1.13, we infer that, for P-a.e. $\omega \in \boldsymbol{\Omega}$,

$$\{v(t_n, \theta_{-t_n}\omega, v_{0,n}(\theta_{-t_n}\omega))\}_{n=1}^{\infty} \text{ is bounded in } L^2(D).$$

Hence, there exists $\tilde{v} \in L^2(D)$ such that up to a subsequence,

$$v(t_n, \theta_{-t_n}\omega, v_{0,n}(\theta_{-t_n}\omega)) \rightharpoonup \tilde{v} \quad \text{weakly in } L^2(D). \tag{4.1.70}$$

In the following, we prove the weak convergence of (4.1.70) is actually strong convergence. Applying Lemma 4.1.15, we obtain that, given $\varepsilon > 0$, for P-a.e. $\omega \in \boldsymbol{\Omega}$, there exist $T' = T'(B, \omega, \varepsilon)$ and $k_1 = k_1(\omega, \varepsilon)$ such that for all $t \geq T'$,

$$\int_{D/D_{k_1}} |v(t, \theta_{-t}\omega, v_0(\theta_{-t}\omega))|^2 \leq \varepsilon.$$

Due to $t_n \to \infty$, there is $N_1 = N_1(B, \omega, \varepsilon)$ such that $t_n \geq T'$ for every $n \geq N_1$. Hence the above inequality implies that

$$\int_{D/D_{k_1}} |v(t_n, \theta_{-t_n}\omega, v_{0,n}(\theta_{-t_n}\omega))|^2 \leq \varepsilon, \qquad \forall\, n \geq N_1. \tag{4.1.71}$$

Furthermore, by Lemmas 4.1.13 and 4.1.14, there exists $T'' = T''(B, \omega)$ such that for all $t \geq T''$,

$$\|v(t, \theta_{-t}\omega, v_0(\theta_{-t}\omega))\|^2_{H^1(D)} \leq C(1 + r'(\omega)).$$

Let $N_2 = N_2(B, \omega)$ be large enough such that $t_n \geq T''$ for $n \geq N_2$. Then by the above inequality we infer that

$$\|v(t_n, \theta_{-t_n}\omega, v_{0,n}(\theta_{-t_n}\omega))\|^2_{H^1(D)} \leq C(1 + r'(\omega)) \qquad \forall\, n \geq N_2. \tag{4.1.72}$$

By the compactness of embedding $H^1(D_{k_1}) \hookrightarrow L^2(D_{k_1})$, it follows from (4.1.72) that, up to a subsequence,

$$v(t_n, \theta_{-t_n}\omega, v_{0,n}(\theta_{-t_n}\omega)) \to \tilde{v} \quad \text{strongly in } L^2(D_{k_1}),$$

which shows that for the given $\varepsilon > 0$, there exists $N_3 = N_3(B, \omega, \varepsilon)$ such that for all $n \geq N_3$,

$$\|v(t_n, \theta_{-t_n}\omega, v_{0,n}(\theta_{-t_n}\omega)) - \tilde{v}\|^2_{L^2(D_{k_1})} \leq \varepsilon. \tag{4.1.73}$$

Note that $\tilde{v} \in L^2(D)$. Therefore there exists $k_2 = k_2(\varepsilon)$ such that

$$\int_{D/D_{k_2}} |\tilde{v}|^2 \leq \varepsilon. \tag{4.1.74}$$

Let $k_3 = \max\{k_1, k_2\}$ and $N_4 = \max\{N_1, N_3\}$. By (4.1.71), (4.1.73) and (4.1.74), we obtain that for all $n \geq N_4$,

$$\begin{aligned}
&\|v(t_n, \theta_{-t_n}\omega, v_{0,n}(\theta_{-t_n}\omega)) - \tilde{v}\|^2_{L^2(D)} \\
&\leq \int_{D_{k_3}} |v(t_n, \theta_{-t_n}\omega, v_{0,n}(\theta_{-t_n}\omega)) - \tilde{v}|^2 \\
&\quad + \int_{D/D_{k_3}} |v(t_n, \theta_{-t_n}\omega, v_{0,n}(\theta_{-t_n}\omega)) - \tilde{v}|^2 \leq 5\varepsilon,
\end{aligned}$$

which shows that

$$v(t_n, \theta_{-t_n}\omega, v_{0,n}(\theta_{-t_n}\omega)) \to \tilde{v} \quad \text{strongly in } L^2(D).$$

4.1.5.2 *The Existence of Random Attractors to (4.1.23)*

In this part, we prove the existence of a $\mathcal{D}$-random attractor for the random dynamical system Ψ associated with the stochastic quasi-geostrophic equation on the unbounded domain D. By (4.1.36) and (4.1.37), Ψ satisfies

$$\begin{aligned}
\Psi(t, \theta_{-t}\omega, u_0(\theta_{-t}\omega)) &= u(t, \theta_{-t}\omega, u_0(\theta_{-t}\omega)) \\
&= v(t, \theta_{-t}\omega, v_0(\theta_{-t}\omega)) + z(\omega), \qquad (4.1.75)
\end{aligned}$$

where $v_0(\theta_{-t}\omega) = u_0(\theta_{-t}\omega) - z(\theta_{-t}\omega)$. Let $B = \{B(\omega)\}_{\omega\in\boldsymbol{\Omega}} \in \mathcal{D}$ and define

$$\tilde{B}(\omega) = \left\{v \in L^2(D): \ \ \|v\| \leq \|u(\omega)\| + \|z(\omega)\|, \ \ u(\omega) \in B(\omega)\right\}. \quad (4.1.76)$$

We claim that $\tilde{B} = \{\tilde{B}(\omega)\}_{\omega\in\boldsymbol{\Omega}} \in \mathcal{D}$ provided $B = \{B(\omega)\}_{\omega\in\boldsymbol{\Omega}} \in \mathcal{D}$. Note that $B = \{B(\omega)\}_{\omega\in\boldsymbol{\Omega}} \in \mathcal{D}$ implies that

$$\lim_{t\to\infty} \mathrm{e}^{-\delta t} d(B(\theta_{-t}\omega)) = 0. \qquad (4.1.77)$$

Since $z(\omega)$ is tempered, by (4.1.76)–(4.1.77) we deduce that

$$\lim_{t\to\infty} \mathrm{e}^{-\delta t} d(\tilde{B}(\theta_{-t}\omega)) \leq \lim_{t\to\infty} \mathrm{e}^{-\delta t} d(B(\theta_{-t}\omega)) + \lim_{t\to\infty} \mathrm{e}^{-\delta t}\|z(\theta_{-t}\omega)\| = 0,$$

which shows $\tilde{B} = \{\tilde{B}(\omega)\}_{\omega\in\boldsymbol{\Omega}} \in \mathcal{D}$. Then by Lemma 4.1.13, for P-a.e. $\omega \in \boldsymbol{\Omega}$, if $v_0(\omega) \in \tilde{B}(\omega)$, there is $T_1 = T_1(\tilde{B}, \omega)$ such that for all $t \geq T_1$,

$$\|v(t, \theta_{-t}\omega, v_0(\theta_{-t}\omega))\| \leq r(\omega), \qquad (4.1.78)$$

where $r(\omega)$ is a positive random function satisfies

$$\mathrm{e}^{-\delta t} r(\theta_{-t}\omega) \to 0 \quad \text{as} \quad t \to \infty. \qquad (4.1.79)$$

Denote by

$$K(\omega) = \{u \in L^2(D): \ \ \|u\| \leq r(\omega) + \|z(\omega)\|\}. \qquad (4.1.80)$$

Then we have the following result.

Lemma 4.1.17. Suppose that $\Phi \in L^2(D)$. Let $K = \{K(\omega)\}_{\omega\in\boldsymbol{\Omega}}$ be given by (4.1.80). Then $K = \{K(\omega)\}_{\omega\in\boldsymbol{\Omega}} \in \mathcal{D}$ is a closed absorbing set of Ψ in $\mathcal{D}$.

Proof. According to (4.1.79) we deduce that

$$\lim_{t\to\infty} \mathrm{e}^{-\delta t} d(K(\theta_{-t}\omega)) \le \lim_{t\to\infty} \mathrm{e}^{-\delta t} r(\theta_{-t}\omega) + \lim_{t\to\infty} \mathrm{e}^{-\delta t}\|z(\theta_{-t}\omega)\| = 0,$$

which implies that $K = \{K(\omega)\}_{\omega\in\boldsymbol{\Omega}} \in \mathcal{D}$. We now show that K is also an absorbing set of Ψ in $\mathcal{D}$. Given $B = \{B(\omega)\}_{\omega\in\boldsymbol{\Omega}} \in \mathcal{D}$ and $u_0(\omega) \in B(\omega)$, by (4.1.75) and (4.1.78) we obtain that, for all $t \ge T_1$,

$$u(t, \theta_{-t}\omega, v_0(\theta_{-t}\omega)) \le \|v(t, \theta_{-t}\omega, v_0(\theta_{-t}\omega))\| + \|z(\omega)\| \le r(\omega) + \|z(\omega)\|,$$

which along with (4.1.75) and (4.1.80) implies that

$$\Psi(t, \theta_{-t}\omega, B(\theta_{-t}\omega)) \subseteq K(\omega), \quad \forall\, t \ge T_1, \tag{4.1.81}$$

and hence $K = \{K(\omega)\}_{\omega\in\boldsymbol{\Omega}} \in \mathcal{D}$ is a closed absorbing set of Ψ in $\mathcal{D}$.

The $\mathcal{D}$-pullback asymptotic compactness of Ψ is given by the following lemma.

Lemma 4.1.18. Suppose that $\Phi \in L^2(D)$. Then the random dynamical system Ψ is $\mathcal{D}$-pullback asymptotically compact in $L^2(D)$; that is, for P-a.e. $\omega \in \boldsymbol{\Omega}$, the sequence $\{\Psi(t_n, \theta_{-t_n}\omega, u_{0,n})\}$ has a convergent subsequence in $L^2(D)$ provided $t_n \to \infty$, $B = \{B(\omega)\}_{\omega\in\boldsymbol{\Omega}} \in \mathcal{D}$ and $u_{0,n} \in B(\theta_{-t_n}\omega)$.

Proof. Since $B = \{B(\omega)\}_{\omega\in\boldsymbol{\Omega}}$ belongs to $\mathcal{D}$, so does $\tilde{B} = \{\tilde{B}(\omega)\}_{\omega\in\boldsymbol{\Omega}}$ which is given by (4.1.76). Then according to Lemma 4.1.16, it follows that, for P-a.e. $\omega \in \boldsymbol{\Omega}$, up to a subsequence $\{v(t_n, \theta_{-t_n}\omega, v_{0,n})\}$ is convergent in $L^2(D)$, where $v_{0,n} = u_{0,n} - z(\theta_{-t_n}\omega) \in \tilde{B}(\theta_{-t_n}\omega)$. This along with (4.1.75) shows that, up to a subsequence $\{\Psi(t_n, \theta_{-t_n}\omega, u_{0,n})\}$ is convergent in $L^2(D)$.

Applying Lemma 4.1.17, Lemma 4.1.18 and the following Proposition 4.1.19, we directly establish the existence of a $\mathcal{D}$-random attractor for Ψ.

Prosition 4.1.19. (cf. [73]) Let $\mathcal{D}$ be an inclusion-closed collection of random subsets of z and ϕ a continuous RDS on z over $(\boldsymbol{\Omega}, F, P, (\theta_t)_{t\in R})$. Suppose that $\{K(\omega)\}_{\omega\in\boldsymbol{\Omega}} \in \mathcal{D}$ is a closed absorbing set of ϕ and ϕ is $\mathcal{D}$-pullback asymptotically compact in z. Then ϕ has a unique $\mathcal{D}$-random attractor which is given by $\{\mathcal{A}(\omega)\}_{\omega\in\boldsymbol{\Omega}}$, $A(\omega) = \bigcap_{k\ge 0}\bigcup_{t\ge k}\phi(t, \theta_{-t}\omega, k(\theta_{-t}\omega))$.

Theorem 4.1.20. Suppose that $\Phi \in H_0^1(D)$. Let $\mathcal{D}$ be the collection of random sets given by (4.1.38). Then the random dynamical system Ψ has a unique $\mathcal{D}$-random attractor in $L^2(D)$.

4.2 Global Well-Posedness and Attractors of 3D Stochastic Primitive Equations of the Large-Scale Oceans

In this section, we consider the global well-posedness and the long-time dynamics of the three-dimensional viscous primitive equations with a random force, which is an additive white noise in time. Firstly, we prove the global existence and uniqueness of strong solutions of the initial boundary value problem of the stochastic primitive equations. Next, we obtain the existence of random attractors of the corresponding random dynamical system.

The arrangement of this section is as follows. In subsection 4.2.1, we introduce three-dimensional stochastic primitive equations of the large-scale oceans, and introduce the main results of this section. The new formulation of the initial boundary value problem of the stochastic primitive equations will be given in subsection 4.2.2. In subsection 4.2.3, we prove the local existence of strong solutions of the initial boundary value problem of the stochastic primitive equations, and make *a priori* estimates of the local strong solutions; we shall prove the main results of this section in the last two subsection.

4.2.1 *Three-Dimensional Stochastic Equations of the Large-Scale Oceans*

In the rectangular coordinates frame, the three-dimensional viscous primitive equations of large-scale oceans under a random force are written as

$$\frac{\partial \boldsymbol{v}}{\partial t} + (\boldsymbol{v} \cdot \boldsymbol{\nabla})\boldsymbol{v} + \theta \frac{\partial \boldsymbol{v}}{\partial z} + f\boldsymbol{k} \times \boldsymbol{v} + \boldsymbol{\nabla} p - \frac{1}{Re_1}\Delta \boldsymbol{v} - \frac{1}{Re_2}\frac{\partial^2 \boldsymbol{v}}{\partial z^2} = \Psi(t,x,y,z), \tag{4.2.1}$$

$$\frac{\partial p}{\partial z} + T = 0, \tag{4.2.2}$$

$$\boldsymbol{\nabla} \cdot \boldsymbol{v} + \frac{\partial \theta}{\partial z} = 0, \tag{4.2.3}$$

$$\frac{\partial T}{\partial t} + \boldsymbol{v} \cdot \boldsymbol{\nabla} T + \theta \frac{\partial T}{\partial z} - \frac{1}{Rt_1}\Delta T - \frac{1}{Rt_2}\frac{\partial^2 T}{\partial z^2} = Q, \tag{4.2.4}$$

where the unknown functions are $\boldsymbol{v},\ \theta,\ p,\ T$, $\boldsymbol{v} = (v^{(1)}, v^{(2)})$ is the horizontal velocity, θ is the vertical velocity, p is pressure, T is temperature, $f = f_0 + \beta y$ is Coriolis parameter, $\boldsymbol{k}$ is the vertical unit vector, $Re_1,\ Re_2$ are Reynolds numbers, $Rt_1,\ Rt_2$ are horizontal and vertical heat diffusivity respectively, $Q(x,y,z)$ is the given heat source, $\boldsymbol{\nabla} = (\partial_x,\ \partial_y)$, and $\Delta = \partial_x^2 + \partial_y^2$. The random force $\Psi(t,x,y,z)$ will be given in subsection 4.2.2.

The space domain of equations (4.2.1)-(4.2.4) is

$$\Omega = \{(x, y, z) : \ (x, y) \in M \text{ and } z \in (-g(x, y), 0)\},$$

where M is a smooth bounded domain in $\mathbb{R}^2$, g is a sufficiently smooth function. Without loss of generality, we assume $g = 1$, that is, $\Omega = M \times (-1, 0)$. The boundary conditions of (4.2.1)-(4.2.4) are given by

$$M \times \{0\} = \Gamma_u : \quad \frac{\partial \boldsymbol{v}}{\partial z} = 0, \ \theta = 0, \ \frac{\partial T}{\partial z} = -\alpha_u T, \tag{4.2.5}$$

$$M \times \{-1\} = \Gamma_b : \quad \frac{\partial \boldsymbol{v}}{\partial z} = 0, \ \theta = 0, \ \frac{\partial T}{\partial z} = 0, \tag{4.2.6}$$

$$\partial M \times [-1, 0] = \Gamma_l : \quad \boldsymbol{v} \cdot \boldsymbol{n} = 0, \ \frac{\partial \boldsymbol{v}}{\partial \boldsymbol{n}} \times \boldsymbol{n} = 0, \ \frac{\partial T}{\partial \boldsymbol{n}} = 0, \tag{4.2.7}$$

where α_u is a positive constant, and $\boldsymbol{n}$ is a normal vector of Γ_l.

Remark 4.2.1. For simplicity, we omit the salinity equation. When we take the salinity equation into account, and replace the boundary conditions $\frac{\partial \boldsymbol{v}}{\partial z}|_{\Gamma_u} = 0, \frac{\partial T}{\partial z}|_{\Gamma_u} = -\alpha_u T$ by $\frac{\partial \boldsymbol{v}}{\partial z}|_{\Gamma_u} = \tau, \frac{\partial T}{\partial z}|_{\Gamma_u} = -\alpha_u(T - T^*)$ for smooth enough τ, T^*, satisfying compatibility conditions $\tau \cdot \boldsymbol{n}|_{\partial M} = 0$, $\frac{\partial \tau}{\partial \boldsymbol{n}} \times \boldsymbol{n}|_{\partial M} = 0$, $\frac{\partial T^*}{\partial \boldsymbol{n}}|_{\partial M} = 0$, the results of this section are still valid.

In this section, we assume that constants Re_1, Re_2, Rt_1, Rt_2, are all equal to 1. Integrating (4.2.3) with respect to z from -1 to z, and applying (4.2.5) and (4.2.6), we have

$$\theta(t, x, y, z) = \Phi(v)(t, x, y, z) = -\int_{-1}^{z} \nabla \cdot \boldsymbol{v}(t, x, y, z') \, \mathrm{d}z', \tag{4.2.8}$$

moreover, $\int_{-1}^{0} \nabla \cdot \boldsymbol{v} \, \mathrm{d}z = 0$. Supposing that p_b is a certain unknown function defined in Γ_b, integrating (4.2.2) with respect to z from -1 to z, we rewrite (4.2.1)-(4.2.4) as

$$\frac{\partial \boldsymbol{v}}{\partial t} + (\boldsymbol{v} \cdot \nabla)\boldsymbol{v} + \Phi(\boldsymbol{v})\frac{\partial \boldsymbol{v}}{\partial z} + f\boldsymbol{k} \times \boldsymbol{v} + \nabla p_b - \int_{-1}^{z} \nabla T \mathrm{d}z' - \Delta \boldsymbol{v} - \frac{\partial^2 \boldsymbol{v}}{\partial z^2} = \Psi, \tag{4.2.9}$$

$$\frac{\partial T}{\partial t} + (\boldsymbol{v} \cdot \nabla)T + \Phi(\boldsymbol{v})\frac{\partial T}{\partial z} - \Delta T - \frac{\partial^2 T}{\partial z^2} = Q, \tag{4.2.10}$$

$$\int_{-1}^{0} \nabla \cdot \boldsymbol{v} \, \mathrm{d}z = 0. \tag{4.2.11}$$

The boundary conditions of equations (4.2.9)-(4.2.11) are given by

$$\Gamma_u : \quad \frac{\partial \boldsymbol{v}}{\partial z} = 0, \ \frac{\partial T}{\partial z} = -\alpha_u T, \tag{4.2.12}$$

$$\Gamma_b: \quad \frac{\partial \boldsymbol{v}}{\partial z} = 0, \quad \frac{\partial T}{\partial z} = 0, \tag{4.2.13}$$

$$\Gamma_l: \quad \boldsymbol{v} \cdot \boldsymbol{n} = 0, \quad \frac{\partial \boldsymbol{v}}{\partial \boldsymbol{n}} \times \boldsymbol{n} = 0, \quad \frac{\partial T}{\partial \boldsymbol{n}} = 0. \tag{4.2.14}$$

The initial condition is

$$U|_{t=t_0} = (\boldsymbol{v}|_{t=t_0}, T|_{t=t_0}) = U_{t_0} = (\boldsymbol{v}_{t_0}, T_{t_0}). \tag{4.2.15}$$

We call (4.2.9)-(4.2.15) as the initial boundary value problem of the 3D viscous stochastic primitive equations, denoted by IBVP.

Next, we shall state main results of this section.

Theorem 4.2.2 (Global well-posedness of IBVP). If $Q \in H^1(\Omega)$, $U_{t_0} = (\boldsymbol{v}_{t_0}, T_{t_0}) \in V$, then, for any given $\mathcal{T} > t_0$, the initial boundary value problem (4.2.9)-(4.2.15) has a unique strong solution U on $[t_0, \mathcal{T}]$. Moreover, U depends continuously on the initial data.

The definitions of space V and strong solutions of system (4.2.9)-(4.2.15) and assumptions of the random force will be given in subsection 4.2.2.

Theorem 4.2.3 [Existence of attractors of random dynamical system (4.2.9)-(4.2.14)]. The random dynamical system (4.2.9)-(4.2.14) possesses a unique random pull-back attractor $\mathcal{A}(\omega)$, which captures all the trajectories started at time $-\infty$ and evolved, under the action of the shift ϑ_t, from $-\infty$ to the present time $t = 0$. The attractor $\mathcal{A}(\omega)$ has the following properties:

(i) (weak compactness) $\mathcal{A}(\omega)$ is bounded and weakly closed in V;

(ii) (invariant) For any $t \geq 0$, $\psi(t, \omega)\mathcal{A}(\omega) = \mathcal{A}(\vartheta_t \omega)$;

(iii) (attracting) For any deterministic bounded set B in V, the sets $\psi(t, \vartheta_{-t}\omega)B$ converge to $\mathcal{A}(\omega)$ with respect to the weak topology of V as $t \to \infty$, that is,

$$\lim_{t \to +\infty} \mathrm{d}_V^w(\psi(t, \vartheta_{-t}\omega)B, \mathcal{A}(\vartheta_t\omega)) = 0, \ \ P\text{-a.s.}$$

where the definitions of $\mathcal{A}(\omega)$, $\vartheta_t, t \in \mathbb{R}$, and $\psi(t, \omega), t \geq 0$ will be given in subsection 4.2.5, and the distance d_V^w is induced by the weak topology of V.

4.2.2 *New Formulation of IBVP*

Before giving the new formulation of initial-boundary value problem of the stochastic primitive equations, let's define some function spaces and an auxilliary Ornstein-Uhlenbeck process.

4.2.2.1 *Some Function Spaces*

$L^p(\Omega)$ is the usual Lebesgue space with the norm $|\cdot|_p$, $1 \le p \le \infty$. $H^m(\Omega)$ is the usual Sobolev space with the norm

$$\|h\|_m = \left[\int_\Omega \left(\sum_{1\le k\le m} \sum_{i_j=1,2,3;j=1,\cdots,k} |\boldsymbol{\nabla}_{i_1}\cdots \boldsymbol{\nabla}_{i_k} h|^2 + |h|^2 \right)\right]^{\frac{1}{2}},$$

where m is a positive integer, $\boldsymbol{\nabla}_1 = \dfrac{\partial}{\partial x}, \boldsymbol{\nabla}_2 = \dfrac{\partial}{\partial y}, \boldsymbol{\nabla}_3 = \dfrac{\partial}{\partial z}$. Denote $\int_\Omega \cdot \mathrm{d}\Omega$ and $\int_M \cdot \mathrm{d}M$ by $\int_\Omega \cdot$, $\int_M \cdot$ respectively.

Next, let's define the work space of IBVP. Let

$$\mathcal{V}_1 := \Big\{ \boldsymbol{v} \in (C^\infty(\Omega))^2; \frac{\partial \boldsymbol{v}}{\partial z}\Big|_{\Gamma_u,\Gamma_b} = 0, \boldsymbol{v}\cdot\boldsymbol{n}\Big|_{\Gamma_l} = 0, \frac{\partial \boldsymbol{v}}{\partial \boldsymbol{n}} \times \boldsymbol{n}|_{\Gamma_l} = 0,$$
$$\int_{-1}^0 \boldsymbol{\nabla}\cdot \boldsymbol{v}\mathrm{d}z = 0\Big\},$$
$$\mathcal{V}_2 := \left\{ T \in C^\infty(\Omega); \frac{\partial T}{\partial z}|_{\Gamma_u} = -\alpha_u T, \frac{\partial T}{\partial z}|_{\Gamma_b} = 0, \frac{\partial T}{\partial \boldsymbol{n}}|_{\Gamma_l} = 0 \right\},$$
$$V_1 = \text{the closure of } \widetilde{\mathcal{V}_1} \text{ with respect to norm } \|\cdot\|_1,$$
$$V_2 = \text{the closure of } \widetilde{\mathcal{V}_2} \text{ with respect to norm } \|\cdot\|_1,$$
$$H_1 = \text{the closure of } \widetilde{\mathcal{V}_1} \text{ with respect to norm } |\cdot|_2,$$
$$V = V_1 \times V_2, \qquad H = H_1 \times L^2(\Omega).$$

The inner products and norms of spaces V, H are given by

$$(\boldsymbol{U},\boldsymbol{U}_1)_V = (\boldsymbol{v},\boldsymbol{v}_1)_{V_1} + (T,T_1)_{V_2},$$

$$(\boldsymbol{U},\boldsymbol{U}_1) = (v^{(1)},(v_1)^{(1)}) + (v^{(2)},(v_1)^{(2)}) + (T,T_1),$$

$$\|\boldsymbol{U}\| = (\boldsymbol{U},\boldsymbol{U})_V^{\frac{1}{2}} = (\boldsymbol{v},\boldsymbol{v})_{V_1}^{\frac{1}{2}} + (T,T)_{V_2}^{\frac{1}{2}} = \|\boldsymbol{v}\| + \|T\|,\ |\boldsymbol{U}|_2 = (\boldsymbol{U},\boldsymbol{U})^{\frac{1}{2}},$$

where $\boldsymbol{U} = (\boldsymbol{v},T)$, $\boldsymbol{U}_1 = (\boldsymbol{v}_1,T_1) \in V$, and $(\cdot,\cdot)$ denotes the inner product in $L^2(\Omega)$.

4.2.2.2 *White Noise and Ornstein-Uhlenbeck Process*

Before defining white noise, we define functionals: $a : V \times V \to \mathbb{R}$, $a_1 : V_1 \times V_1 \to \mathbb{R}$, $a_2 : V_2 \times V_2 \to \mathbb{R}$, and their corresponding linear operators

$A : V \to V'$, $A_1 : V_1 \to V_1'$, $A_2 : V_2 \to V_2'$ by $a(\boldsymbol{U}, \boldsymbol{U}_1) = (A\boldsymbol{U}, \boldsymbol{U}_1) = a_1(\boldsymbol{v}, \boldsymbol{v}_1) + a_2(T, T_1)$, where

$$a_1(\boldsymbol{v}, \boldsymbol{v}_1) = (A_1\boldsymbol{v}, \boldsymbol{v}_1) = \int_\Omega \left(\nabla \boldsymbol{v} \cdot \nabla \boldsymbol{v}_1 + \frac{\partial \boldsymbol{v}}{\partial z} \cdot \frac{\partial \boldsymbol{v}_1}{\partial z} \right),$$

$$a_2(T, T_1) = (A_2 T, T_1) = \int_\Omega \left(\nabla T \cdot \nabla T_1 + \frac{\partial T}{\partial z} \frac{\partial T_1}{\partial z} \right) + \alpha_u \int_{\Gamma_u} T T_1.$$

Lemma 4.2.4.

(i) a is coercive and continuous, and $A : V \to V'$ is isomorphic. Moreover

$$c\|\boldsymbol{U}\|^2 \le a(\boldsymbol{U}, \boldsymbol{U}_1) \le c\|\boldsymbol{U}\| \|\boldsymbol{U}_1\|.$$

(ii) $A : V \to V'$ can be extended to be a self-adjoint unbounded operator on H with a compact inverse operator $A^{-1} : H \to H$, and with the domain of definition of the operator $D(A) = V \cap \left[(H^2(\Omega))^2 \times H^2(\Omega) \right]$.

Proof of Lemma 4.2.4. By $\|\boldsymbol{v}\|_{L^2}^2 \le C_M \|\nabla \boldsymbol{v}\|_{L^2}^2$ (see [76, p. 55]), we prove the first part of (i). Since the operator A is similar to the usual positive symmetrical Laplacian operator on H_0^1, the other part of Lemma 4.2.4 can be proved by the usual argument. Here, we omit the details of the proof. For more details, the reader can refer to [141, Lemma 2.4].

Denote $0 < \lambda_1 < \lambda_2 \le \cdots$ by the eigenvalues of A_1, by $\boldsymbol{e}_1, \boldsymbol{e}_2, \cdots$ the corresponding complete orthonormal system of eigenvectors. For any $\boldsymbol{v} \in V_1$, $\|\boldsymbol{v}\|^2 \ge \lambda_1 |\boldsymbol{v}|_2^2$. Let $\mathrm{e}^{t(-A_1)}$, $t \ge 0$, be the semigroup on H_1 generated by $-A_1$.

In this section, we assume that the random force $\Psi(t, x, y, z)$ is an additive white noise in time, with the form

$$\Psi(t, x, y, z) = G\frac{\partial W}{\partial t}, \tag{4.2.16}$$

where the derivative is in the Itô integral sense. The stochastic process W is a two-sided in time cylindrical Wiener process in H_1 with the form $W(t) = \sum_{i=1}^{+\infty} \omega_i(t, \omega)\boldsymbol{e}_i$, and G is a Hilbert-Schmidt operator from H_1 to $H^{1+2\gamma_0}(\Omega) \times H^{1+2\gamma_0}(\Omega) (\gamma_0 > 0)$, that is, $\sum_{i=1}^{+\infty} \|G\boldsymbol{e}_i\|_{1+2\gamma_0}^2 < +\infty$. Here $\omega_1, \omega_2, \cdots$ is a sequence of independent standard one-dimensional Brownian motions on complete probability space $(\boldsymbol{\Omega}, \mathcal{F}, P)$ with expectation denoted by E, and $H^{1+2\gamma_0}(\Omega)$ is the usual non-integer order Sobolev space.

Now, we define an auxiliary Ornstein-Uhlenbeck process. For any $\alpha \geq 0$, let

$$Z_\alpha(t) = W_{A_1}^\alpha(t) = \int_{-\infty}^t \mathrm{e}^{(t-s)(-A_1-\alpha)} G \mathrm{d}W(s) \tag{4.2.17}$$

be a solution of the stochastic Stokes equation

$$\mathrm{d}z = (-A_1 Z - \alpha Z)\mathrm{d}t + G\mathrm{d}W(t),$$

with $Z(0) = \int_{-\infty}^0 \mathrm{e}^{s(A_1+\alpha)} G\mathrm{d}W(s)$. In the following, for simplicity, denote $Z_\alpha(t)$ by $Z(t)$. The damp term αZ is not necessary in the proof of the global well-posedness of problem IBVP, but it is very important in studying the long-time behavior of strong solutions of problem IBVP. So we introduce αZ from the beginning.

Lemma 4.2.5 (cf. [54, 55]). If $Z(t)$ is the Ornstein-Uhlenbeck process defined above, then it is a stationary ergodic solution with continuous trajectories, taking values in $D(A_1^{1+\gamma})$, for any $\gamma < \gamma_0$.

Remark 4.2.6. An example for Ψ is $\Psi = \dfrac{\partial W}{\partial t}$, where W is a two-sided in time finite-dimensional Brownian motion with the form

$$W = \sum_{i=1}^m \delta_i \omega_i(t, \omega) \mathrm{e}_i.$$

In the above formula, $\omega_1, \cdots, \omega_m$ are independent standard one-dimensional Brownian motions on complete probability space $(\boldsymbol{\Omega}, \mathcal{F}, P)$ (with expectation denoted by E), and δ_i are real coefficients. For this example, Z is a stationary ergodic solution with continuous trajectories which take values in $D(A_1^k)$, for any $k \in \mathbb{N}$.

Remark 4.2.7. Another example of Ψ is $\Psi = \dfrac{\partial W}{\partial t}$, where W is a two-sided in time infinite-dimensional Brownian motion with the form

$$W(t) = \sum_{i=1}^{+\infty} \mu_i \omega_i(t, \omega) \boldsymbol{e}_i.$$

Here $\omega_1, \omega_2, \cdots$ is a sequence of independent standard one-dimensional Brownian motions on a complete probability space $(\boldsymbol{\Omega}, \mathcal{F}, P)$ (with expectation denoted by E), and the coefficients μ_i satisfy $\sum_{i=1}^{+\infty} \lambda_i^{1+2\gamma_0} \mu_i^2 <$

$+\infty$, for some $\gamma_0 > 0$. In this case,

$$\begin{aligned}E|A_1^{1+\gamma}\boldsymbol{Z}(t)|_2^2 =& E\sum_{i=1}^{+\infty}\left|\int_{-\infty}^{t}\lambda_i^{1+\gamma}\mathrm{e}^{(t-s)(-\lambda_i-\alpha)}\mu_i\mathrm{d}\omega_i\right|^2\\ =& \sum_{i=1}^{+\infty}\left|\int_{-\infty}^{t}\lambda_i^{2+2\gamma}\mathrm{e}^{2(t-s)(-\lambda_i-\alpha)}\mu_i^2\mathrm{d}s\right|\\ =& \sum_{i=1}^{+\infty}\frac{\lambda_i^{2+2\gamma}\mu_i^2}{2(\lambda_i+\alpha)} < +\infty,\ \forall\gamma<\gamma_0.\end{aligned}$$

Remark 4.2.8. If the random force $\Psi(t,x,y,z)$ is independent of variable z, then the process $\boldsymbol{Z}$ is also independent of z. In order to prove the global well-posedness of IBVP, we only need lower regular G such that $\boldsymbol{Z} \in L^\infty(\mathbb{R};(H^1(M))^2)\cap L^2(\mathbb{R};(H^2(M))^2)$. But, in the process of proving the existence of random attractors for the 3D stochastic primitive equations, we need $\boldsymbol{Z} \in L^\infty(\mathbb{R};(H^2(M))^2)$.

4.2.2.3 *New Formulation of the Initial Boundary Value Problem (4.2.9)-(4.2.15)*

Definition 4.2.9. For any $\mathcal{T} > t_0$, a stochastic process $\boldsymbol{U}(t,\omega) = (\boldsymbol{v},T)$ is called a **strong solution (weak solution)** of initial boundary value problem (4.2.9)-(4.2.15) in $[t_0,\mathcal{T}]$, if U satisfies for P-a.e. $\omega \in \boldsymbol{\Omega}$

$$\begin{aligned}&\int_\Omega \boldsymbol{v}(t)\cdot\varphi_1 - \int_{t_0}^{\mathcal{T}}\int_\Omega\Big\{[(\boldsymbol{v}\cdot\boldsymbol{\nabla})\varphi_1+\Phi(\boldsymbol{v})\partial_z\varphi_1]\cdot\boldsymbol{v}-\Big[(f\boldsymbol{k}\times\boldsymbol{v})\cdot\varphi_1\\ &+\left(\int_{-1}^{z}T\mathrm{d}z'\right)\boldsymbol{\nabla}\cdot\varphi_1\Big]\Big\}+\int_{t_0}^{\mathcal{T}}\int_\Omega\boldsymbol{v}\cdot A_1\varphi_1\\ =&\int_\Omega\boldsymbol{v}_{t_0}\cdot\varphi_1+\int_\Omega[GW(t,\omega)-GW(t_0,\omega)]\cdot\varphi_1,\\ &\int_\Omega T(t)\varphi_2-\int_{t_0}^{\mathcal{T}}\int_\Omega\Big\{[(\boldsymbol{v}\cdot\boldsymbol{\nabla})\varphi_2+\Phi(\boldsymbol{v})\partial_z\varphi_2]T-TA_2\varphi_2\Big\}\\ =&\int_\Omega T_{t_0}\varphi_2+\int_{t_0}^{\mathcal{T}}\int_\Omega Q\varphi_2,\end{aligned}$$

for any $t \in [t_0,\mathcal{T}]$ and $\varphi = (\varphi_1,\varphi_2) \in D(A_1)\times D(A_2)$, moreover $\boldsymbol{U} \in L^\infty(t_0,\mathcal{T};V)\cap L^2(t_0,\mathcal{T};(H^2(\Omega))^3)$ ($\boldsymbol{U} \in L^\infty(t_0,\mathcal{T};H)\cap L^2(t_0,\mathcal{T};V)$), and $\boldsymbol{U}$ is progressively measurable in these topologies.

Assume that $\boldsymbol{Y}$ is a solution of the following initial boundary value problem,

$$\begin{cases} \dfrac{\partial \boldsymbol{Y}}{\partial t} + \nabla p_{b_1} - \Delta \boldsymbol{Y} - \dfrac{\partial^2 \boldsymbol{Y}}{\partial z^2} = 0, \\ \dfrac{\partial \boldsymbol{Y}}{\partial z}|_{\Gamma_u, \Gamma_b} = 0, \ \boldsymbol{Y} \cdot \boldsymbol{n}|_{\Gamma_l} = 0, \ \dfrac{\partial \boldsymbol{Y}}{\partial \boldsymbol{n}} \times \boldsymbol{n}|_{\Gamma_l} = 0, \\ \displaystyle\int_{-1}^{0} \nabla \cdot \boldsymbol{Y} \mathrm{d}z = 0, \\ \boldsymbol{Y}(t_0, \omega) = \boldsymbol{v}_{t_0} - \boldsymbol{Z}_{t_0}. \end{cases}$$

If $\boldsymbol{v}_{t_0} \in V_1$, then for any $\mathcal{T} > t_0$ and P-a.e. $\omega \in \boldsymbol{\Omega}$,

$$\boldsymbol{Y} \in L^\infty(t_0, \mathcal{T}; V_1) \cap L^2(t_0, \mathcal{T}; (H^2(\Omega))^2), \tag{4.2.18}$$

see [87].

Let $\boldsymbol{u}(t) = \boldsymbol{v}(t) - \boldsymbol{Z}(t) - \boldsymbol{Y}$. A stochastic process $\boldsymbol{U}(t, \omega) = (\boldsymbol{v}, T)$ is a strong solution of (4.2.9)-(4.2.15) on $[t_0, \mathcal{T}]$, if and only if $(\boldsymbol{u}, T)$ is a strong solution of the following problem on $[t_0, \mathcal{T}]$,

$$\begin{aligned} &\frac{\partial \boldsymbol{u}}{\partial t} + [(\boldsymbol{u} + \boldsymbol{Z} + \boldsymbol{Y}) \cdot \nabla](\boldsymbol{u} + \boldsymbol{Z} + \boldsymbol{Y}) + \Phi(\boldsymbol{u} + \boldsymbol{Z} + \boldsymbol{Y}) \frac{\partial (\boldsymbol{u} + \boldsymbol{Z} + \boldsymbol{Y})}{\partial z} \\ &+ f \boldsymbol{k} \times (\boldsymbol{u} + \boldsymbol{Z} + \boldsymbol{Y}) - \alpha \boldsymbol{Z} + \nabla p_{b_2} - \int_{-1}^{z} \nabla T \mathrm{d}z' - \Delta \boldsymbol{u} - \frac{\partial^2 \boldsymbol{u}}{\partial z^2} = 0, \end{aligned} \tag{4.2.19}$$

$$\frac{\partial T}{\partial t} + [(\boldsymbol{u} + \boldsymbol{Z} + \boldsymbol{Y}) \cdot \nabla] T + \Phi(\boldsymbol{u} + \boldsymbol{Z} + \boldsymbol{Y}) \frac{\partial T}{\partial z} - \Delta T - \frac{\partial^2 T}{\partial z^2} = Q, \tag{4.2.20}$$

$$\int_{-1}^{0} \nabla \cdot \boldsymbol{u} \ \mathrm{d}z = 0, \tag{4.2.21}$$

$(\boldsymbol{u}, T)$ satisfies boundary conditions (4.2.12)-(4.2.14), (4.2.22)

$$(\boldsymbol{u}|_{t=t_0}, T|_{t=t_0}) = (0, T_{t_0}). \tag{4.2.23}$$

Definition 4.2.10. *$\boldsymbol{Z}$, $\boldsymbol{Y}$ are defined as before, $T_{t_0} \in V_2$, and $\mathcal{T}$ is a given positive time. For P-a.e. $\omega \in \boldsymbol{\Omega}$, $(\boldsymbol{u}, T)$ is called a strong solution of the initial boundary value problem (4.2.19)-(4.2.23) in $[t_0, \mathcal{T}]$, if $(\boldsymbol{u}, T)$ satisfies (4.2.19)-(4.2.23) in the weak sense, and satisfies*

$$\begin{aligned} &\boldsymbol{u} \in L^\infty(t_0, \mathcal{T}; V_1) \cap L^2(t_0, \mathcal{T}; (H^2(\Omega))^2), \\ &T \in L^\infty(t_0, \mathcal{T}; V_2) \cap L^2(0, \mathcal{T}; H^2(\Omega)), \\ &\frac{\partial \boldsymbol{u}}{\partial t} \in L^1(0, \mathcal{T}; (L^2(\Omega))^2), \ \frac{\partial T}{\partial t} \in L^1(0, \mathcal{T}; L^2(\Omega)). \end{aligned}$$

Remark 4.2.11.

(i) In order to consider the local well-posedness of IBVP, we need to study the initial boundary value problem (4.2.19)-(4.2.23). However, only by studying the global well-posedness of IBVP and the long-time behavior of strong solutions to IBVP, instead of (4.2.19)-(4.2.23), we need to consider (4.2.32)-(4.2.36).

(ii) For almost all the given paths of process $\boldsymbol{Z}(t)$, we can study (4.2.19)-(4.2.23) and the coming (4.2.32)-(4.2.36) in a similar way to study the deterministic equations.

4.2.3 *Existence and A Priori Estimate of Local Solutions*

4.2.3.1 *Existence of Local Solutions*

It's known from Remark 4.2.11 that we obtain the local existence of strong solutions of IBVP by proving that of the initial boundary value problem (4.2.19)-(4.2.23). At first, let's recall some interpolation inequalities used frequently later, see [3, 76].

(1) $\forall h_1 \in H^1(M)$,

$$\|h_1\|_{L^4} \le c\|h_1\|_{L^2}^{\frac{1}{2}}\|h_1\|_{H^1}^{\frac{1}{2}}, \tag{4.2.24}$$

$$\|h_1\|_{L^5} \le c\|h_1\|_{L^3}^{\frac{3}{5}}\|h_1\|_{H^1}^{\frac{2}{5}}, \tag{4.2.25}$$

$$\|h_1\|_{L^6} \le c\|h_1\|_{L^4}^{\frac{2}{3}}\|h_1\|_{H^1}^{\frac{1}{3}}. \tag{4.2.26}$$

(2)

$$|h_2|_4 \le c|h_2|_2^{\frac{1}{4}}\|h_2\|_1^{\frac{3}{4}}, \qquad \forall h_2 \in H^1(\Omega). \tag{4.2.27}$$

Proposition 4.2.12 (Existence of local strong solutions of (4.2.19)-(4.2.23)). If $Q \in H^1(\Omega)$, $T_{t_0} \in V_2$, then, for P-a.e. $\omega \in \boldsymbol{\Omega}$, there exists $\mathcal{T}^* > t_0$ such that $(\boldsymbol{u}, T)$ is a local strong solution of (4.2.19)-(4.2.23) on the interval $[t_0, \mathcal{T}^*)$.

Before proving Proposition 4.2.12, we present a Lemma used frequently later.

Lemma 4.2.13. If $\boldsymbol{v}_1 \in V_1$, $\boldsymbol{v}_2 \in V_1$ or V_2, $\boldsymbol{v}_3 \in L^2(\Omega)\times L^2(\Omega)$ or $L^2(\Omega)$, then,

(i) $\left|\int_\Omega \boldsymbol{v}_3 \cdot (\boldsymbol{v}_1 \cdot \boldsymbol{\nabla})\boldsymbol{v}_2\right|$

$$\leq c(|\boldsymbol{v}_1|_4^2 + |\boldsymbol{v}_1|_4^8)|\nabla \boldsymbol{v}_2|_2^2 + \varepsilon\Big[|\boldsymbol{v}_3|_2^2 + \int_\Omega (|\nabla \boldsymbol{v}_{2z}|^2 + |\Delta \boldsymbol{v}_2|^2)\Big],$$

(ii) $$\left|\int_\Omega \Phi(\boldsymbol{v}_1)\boldsymbol{v}_{2z}\cdot \boldsymbol{v}_3\right| \leq c|\nabla \boldsymbol{v}_1|_2^{\frac{1}{2}}(|\nabla \boldsymbol{v}_1|_2^2 + |\Delta \boldsymbol{v}_1|_2^2)^{\frac{1}{4}}|\boldsymbol{v}_{2z}|_2^{\frac{1}{2}}|\nabla \boldsymbol{v}_{2z}|_2^{\frac{1}{2}}|\boldsymbol{v}_3|_2.$$

Proof of Lemma 4.2.13. By Hölder Inequality, (4.2.27) and Young Inequality, we get

$$\int_\Omega |\boldsymbol{v}_1||\nabla \boldsymbol{v}_2||\boldsymbol{v}_3| \leq c|\boldsymbol{v}_1|_4^2\left(\int_\Omega |\nabla \boldsymbol{v}_2|^4\right)^{\frac{1}{2}} + \varepsilon|\boldsymbol{v}_3|_2^2$$
$$\leq c(|\boldsymbol{v}_1|_4^2 + |\boldsymbol{v}_1|_4^8)|\nabla \boldsymbol{v}_2|_2^2 + \varepsilon\left[|\boldsymbol{v}_3|_2^2 + \int_\Omega (|\nabla \boldsymbol{v}_{2z}|^2 + |\Delta \boldsymbol{v}_2|^2)\right].$$

Using Hölder Inequality, Minkowski Inequality and (4.2.24), we have

$$\int_M \left(\int_{-1}^0 |\nabla \boldsymbol{v}_1|\mathrm{d}z \int_{-1}^0 |\boldsymbol{v}_{2z}||\boldsymbol{v}_3|\mathrm{d}z\right)$$
$$\leq \int_M \left[\left(\int_{-1}^0 |\nabla \boldsymbol{v}_1|^2\mathrm{d}z\right)^{\frac{1}{2}}\left(\int_{-1}^0 |\boldsymbol{v}_{2z}|^2\mathrm{d}z\right)^{\frac{1}{2}}\left(\int_{-1}^0 |\boldsymbol{v}_3|^2\mathrm{d}z\right)^{\frac{1}{2}}\right]$$
$$\leq c\left[\int_{-1}^0 (\|\nabla \boldsymbol{v}_1\|_{L^2}\|\nabla \boldsymbol{v}_1\|_{H^1})\int_{-1}^0 (\|\boldsymbol{v}_{2z}\|_{L^2}\|\boldsymbol{v}_{2z}\|_{H^1})\right]^{\frac{1}{2}}|\boldsymbol{v}_3|_2.$$

Proof of Proposition 4.2.12. Let $\omega \in \boldsymbol{\Omega}$ be fixed. We shall prove Proposition 4.2.12 with Faedo-Galerkin method. Since the procedure is similar to that of the proof of the local existence of strong solutions of the three-dimensional incompressible Navier-Stokes equations (see [202]), we only give *a priori* estimates of the approximate solutions. Let $(\boldsymbol{u}_m, T_m)$ be approximate solutions of (4.2.19)-(4.2.23), where $(\boldsymbol{u}_m, T_m) = \sum_{i=1}^m \alpha_{i,m}(t)\phi_i(x)$, and $\{\phi_m\}$ is a complete orthogonal basis of the space V. Let $\boldsymbol{e} = \boldsymbol{Z} + \boldsymbol{Y}$. Then $(\boldsymbol{u}_m, T_m)$ satisfies

$$\int_\Omega \boldsymbol{h}_m \cdot \frac{\partial \boldsymbol{u}_m}{\partial t} + \int_\Omega \boldsymbol{h}_m \cdot \left\{[(\boldsymbol{u}_m + \boldsymbol{e})\cdot\nabla](\boldsymbol{u}_m + \boldsymbol{e}) + \Phi(\boldsymbol{u}_m + \boldsymbol{e})\frac{\partial(\boldsymbol{u}_m + \boldsymbol{e})}{\partial z}\right\}$$
$$-\int_\Omega \boldsymbol{h}_m \cdot \alpha\boldsymbol{Z} + \int_\Omega \boldsymbol{h}_m \cdot [f\boldsymbol{k}\times(\boldsymbol{u}_m + \boldsymbol{e})] - \int_\Omega \boldsymbol{h}_m \cdot \int_{-1}^z \nabla T_m \mathrm{d}z'$$
$$+\int_\Omega \boldsymbol{h}_m \cdot A_1\boldsymbol{u}_m = 0, \tag{4.2.28}$$

$$\int_\Omega q_m\frac{\partial T_m}{\partial t} + \int_\Omega q_m\left\{[(\boldsymbol{u}_m + \boldsymbol{e})\cdot\nabla]T_m + \Phi(\boldsymbol{u}_m + \boldsymbol{e})\frac{\partial T_m}{\partial z}\right\} + \int_\Omega q_m A_2 T_m$$
$$= \int_\Omega q_m Q, \tag{4.2.29}$$

$\boldsymbol{u}_m(t_0) = 0, T_m(t_0) = T_{0m} \to T_{t_0} \quad \text{in } V_2,$

where $\boldsymbol{h}_m \in V_{1m}$, $q_m \in V_{2m}$, and $V_{1m} \times V_{2m} = \text{span}\{\phi_1, \cdots, \phi_m\}$.

L^2 estimates of T_m, $\boldsymbol{u}_m$. Letting $q_m = T_m$ in (4.2.29), by integration by parts and Lemma 4.2.4, we have

$$\frac{\mathrm{d}|T_m|_2^2}{\mathrm{d}t} + c\|T_m\|^2 \le c|Q|_2^2,$$

which implies that T_m is uniformly m bounded in $L^\infty(t_0, \mathcal{T}; L^2(\Omega)) \cap L^2(t_0, \mathcal{T}; V_2)$ for any $\mathcal{T} > t_0$.

In the same way of the proof of Lemma 4.2.13, according to (4.2.27) and (4.2.24), we get

$$\begin{aligned}
&-\int_\Omega \boldsymbol{u}_m \cdot [(\boldsymbol{e} \cdot \boldsymbol{\nabla})\boldsymbol{u}_m + (\boldsymbol{u}_m \cdot \boldsymbol{\nabla})\boldsymbol{e} + (\boldsymbol{e} \cdot \boldsymbol{\nabla})\boldsymbol{e}] \\
\le&\varepsilon\|\boldsymbol{u}_m\|^2 + c(|\boldsymbol{e}|_4^8 + \|\boldsymbol{e}\|^4)|\boldsymbol{u}_m|_2^2 + c\|\boldsymbol{e}\|^2, \\
&-\int_\Omega \boldsymbol{u}_m \cdot \left[\Phi(\boldsymbol{u}_m)\frac{\partial \boldsymbol{e}}{\partial z} + \Phi(\boldsymbol{e})\frac{\partial \boldsymbol{u}_m}{\partial z} + \Phi(\boldsymbol{e})\frac{\partial \boldsymbol{e}}{\partial z}\right] \\
\le&\varepsilon\|\boldsymbol{u}_m\|^2 + c\|\boldsymbol{e}\|^2\|\boldsymbol{e}\|_2^2|\boldsymbol{u}_m|_2^2 + c\|\boldsymbol{e}\|^2.
\end{aligned}$$

Letting $\boldsymbol{h}_m = \boldsymbol{u}_m$ in (4.2.28), and noticing that

$$\int_\Omega \left[(\boldsymbol{u}_m \cdot \boldsymbol{\nabla})\boldsymbol{u}_m + \Phi(\boldsymbol{u}_m)\frac{\partial \boldsymbol{u}_m}{\partial z}\right] \cdot \boldsymbol{u}_m = 0,$$

by Lemma 4.2.4, Hölder Inequality and Young Inequality, we obtain

$$\frac{\mathrm{d}|\boldsymbol{u}_m|_2^2}{\mathrm{d}t} + c\|\boldsymbol{u}_m\|^2 \le c(|\boldsymbol{e}|_4^8 + \|\boldsymbol{e}\|^4 + \|\boldsymbol{e}\|^2\|\boldsymbol{e}\|_2^2)|\boldsymbol{u}_m|_2^2 + c(\|\boldsymbol{e}\|^2 + |T_m|_2^2 + |\boldsymbol{Z}|_2^2).$$

By Lemma 4.2.5 and (4.2.18), we know that for any $\mathcal{T} > t_0$, $\boldsymbol{u}_m$ is uniformly m bounded in $L^\infty(t_0, \mathcal{T}; H_1) \cap L^2(t_0, \mathcal{T}; V_1)$.

H^1 estimates of $\boldsymbol{u}_m$, T_m. In the same way as proving Lemma 4.2.13, we get by (4.2.27), (4.2.24) and Young Inequality

$$\begin{aligned}
&\left|\int_\Omega A_1\boldsymbol{u}_m \cdot [(\boldsymbol{u}_m + \boldsymbol{e}) \cdot \boldsymbol{\nabla}]\,(\boldsymbol{u}_m + \boldsymbol{e})\right| \\
\le&c(|\boldsymbol{u}_m|_2^2\|\boldsymbol{u}_m\|^6 + |\boldsymbol{e}|_2^2\|\boldsymbol{e}\|^6 + \|\boldsymbol{e}\|^{\frac{1}{2}}\|\boldsymbol{e}\|_2^{\frac{3}{2}})\|\boldsymbol{u}_m\|^2 + c|\boldsymbol{e}|_2^{\frac{1}{2}}\|\boldsymbol{e}\|^2\|\boldsymbol{e}\|_2^{\frac{3}{2}} + \varepsilon|A_1\boldsymbol{u}_m|_2^2, \\
&\left|\int_\Omega A_1\boldsymbol{u}_m \cdot \Phi(\boldsymbol{u}_m + \boldsymbol{e})\frac{\partial(\boldsymbol{u}_m + \boldsymbol{e})}{\partial z}\right| \\
\le&c\|\boldsymbol{e}\|^2\|\boldsymbol{e}\|_2^2(1 + \|\boldsymbol{u}_m\|^2) + c|A_1\boldsymbol{u}_m|_2^2\|\boldsymbol{u}_m\| + \varepsilon|A_1\boldsymbol{u}_m|_2^2.
\end{aligned}$$

Letting $h_m = A_1\boldsymbol{u}_m$ in (4.2.28), by the above two inequalities, Hölder Inequality and Young Inequality, we have

$$\frac{\mathrm{d}(A_1\boldsymbol{u}_m, \boldsymbol{u}_m)}{\mathrm{d}t} + \frac{1}{2}|A_1\boldsymbol{u}_m|_2^2 \le c|A_1\boldsymbol{u}_m|_2^2\|\boldsymbol{u}_m\| + c\|\boldsymbol{e}\|^2\|\boldsymbol{e}\|_2^2 + c\|T_m\|^2$$

$$+ c(|\boldsymbol{e}|_2^2 + |\boldsymbol{Z}|_2^2) + c(|\boldsymbol{u}_m|_2^2\|\boldsymbol{u}_m\|^6 + \|\boldsymbol{e}\|^2\|\boldsymbol{e}\|_2^2 + |\boldsymbol{e}|_2^2\|\boldsymbol{e}\|^6$$
$$+ \|\boldsymbol{e}\|^{\frac{1}{2}}\|\boldsymbol{e}\|_2^{\frac{3}{2}})\|\boldsymbol{u}_m\|^2. \tag{4.2.30}$$

Similarly,

$$\frac{\mathrm{d}\|T_m\|^2}{\mathrm{d}t} + \frac{1}{2}|A_2T_m|_2^2 \leq c(|\boldsymbol{u}_m|_2^2\|\boldsymbol{u}_m\|^6 + |\boldsymbol{e}|_2^2\|\boldsymbol{e}\|^6)\|T_m\|^2$$
$$+ c(\|\boldsymbol{e}\|^2\|\boldsymbol{e}\|_2^2 + \|\boldsymbol{u}_m\|^2\|\boldsymbol{u}_m\|_2^2)\|T_m\|^2 + |Q|_2^2. \tag{4.2.31}$$

Due to $\boldsymbol{u}_m(t_0) = 0$ and (4.2.30), by Lemma 4.2.4, there exists $\mathcal{T}_1 > t_0$ independent of m such that for any $t \in [t_0, \mathcal{T}_1]$, $\|\boldsymbol{u}_m(t)\|$ is small enough. According to (4.2.30), (4.2.31), using Lemmas 4.2.4 and 4.2.5, by a standard argument used in proving the local well-posedness of the three-dimensional incompressible Navier-Stokes equations, we prove Proposition 4.2.12.

4.2.3.2 *A Priori Estimates of the Local Strong Solutions of (4.2.9)-(4.2.15)*

According to Proposition 4.2.12, we get the local existence of strong solutions of (4.2.9)-(4.2.15). Suppose that $(\boldsymbol{v}, T)$ is a strong solution of (4.2.9)-(4.2.15) on $[t_0, \mathcal{T}^*)$. In order to obtain the global existence of the strong solutions of (4.2.9)-(4.2.15), we must make *a priori* estimates of local strong solution $(\boldsymbol{v}, T)$. We shall prove that if $\mathcal{T}^* < +\infty$, then $\limsup_{t\to\mathcal{T}^{*-}}(\|\boldsymbol{v}\| + \|T\|) < +\infty$. In order to do that, we need to introduce a barotropic flow $\bar{\boldsymbol{w}}$ and baroclinic flow $\tilde{\boldsymbol{w}}$, and consider some of their properties. Here $\boldsymbol{w} = \boldsymbol{v} - \boldsymbol{Z}$, and $(\boldsymbol{w}, T)$ satisfies the following initial boundary value problem

$$\frac{\partial \boldsymbol{w}}{\partial t} + [(\boldsymbol{w} + \boldsymbol{Z}) \cdot \boldsymbol{\nabla}](\boldsymbol{w} + \boldsymbol{Z}) + \Phi(\boldsymbol{w} + \boldsymbol{Z})\frac{\partial(\boldsymbol{w} + \boldsymbol{Z})}{\partial z} + f\boldsymbol{k} \times (\boldsymbol{w} + \boldsymbol{Z})$$
$$- \alpha\boldsymbol{Z} + \boldsymbol{\nabla}p_{b_3} - \int_{-1}^{z} \boldsymbol{\nabla}T\mathrm{d}z' - \Delta\boldsymbol{w} - \frac{\partial^2\boldsymbol{w}}{\partial z^2} = 0, \tag{4.2.32}$$

$$\frac{\partial T}{\partial t} + [(\boldsymbol{w} + \boldsymbol{Z}) \cdot \boldsymbol{\nabla}]T + \Phi(\boldsymbol{w} + \boldsymbol{Z})\frac{\partial T}{\partial z} - \Delta T - \frac{\partial^2 T}{\partial z^2} = Q, \tag{4.2.33}$$

$$\int_{-1}^{0} \boldsymbol{\nabla} \cdot \boldsymbol{w}\mathrm{d}z = 0 \tag{4.2.34}$$

$$(\boldsymbol{w}, T) \text{ satisfies boundary conditions (4.2.12)-(4.2.14)}, \tag{4.2.35}$$

$$(\boldsymbol{w}|_{t=t_0}, T|_{t=t_0}) = (\boldsymbol{v}_{t_0} - \boldsymbol{Z}_{t_0}, T_{t_0}). \tag{4.2.36}$$

We define the strong solutions of the initial boundary value problem (4.2.32)-(4.2.36) similar to that in the definition 4.2.10. If $(\boldsymbol{v}, T)$ is a strong solution of (4.2.9)-(4.2.15) on $[t_0, \mathcal{T}^*)$, according to the definition of $(\boldsymbol{w}, T)$, $(\boldsymbol{w}, T)$ is the strong solution of (4.2.32)-(4.2.36) on $[t_0, \mathcal{T}^*)$.

We define the barotropic flow by $\bar{\boldsymbol{w}} = \int_{-1}^{0} \boldsymbol{w} \mathrm{d}z$, and the baroclinic flow by $\tilde{\boldsymbol{w}} = \boldsymbol{w} - \bar{\boldsymbol{w}}$. We notice that

$$\bar{\tilde{\boldsymbol{w}}} = \int_{-1}^{0} \tilde{\boldsymbol{w}} \mathrm{d}z = 0, \quad \boldsymbol{\nabla} \cdot \bar{\boldsymbol{w}} = 0. \tag{4.2.37}$$

By integration by parts and (4.2.37), we get

$$\int_{-1}^{0} \Phi(\boldsymbol{w}) \frac{\partial \boldsymbol{w}}{\partial z} \mathrm{d}z = \int_{-1}^{0} \boldsymbol{w} \boldsymbol{\nabla} \cdot \boldsymbol{w} \mathrm{d}z = \int_{-1}^{0} \tilde{\boldsymbol{w}} \boldsymbol{\nabla} \cdot \tilde{\boldsymbol{w}} \mathrm{d}z,$$

$$\int_{-1}^{0} (\boldsymbol{w} \cdot \boldsymbol{\nabla}) \boldsymbol{w} \mathrm{d}z = \int_{-1}^{0} (\tilde{\boldsymbol{w}} \cdot \boldsymbol{\nabla}) \tilde{\boldsymbol{w}} \mathrm{d}z + (\bar{\boldsymbol{w}} \cdot \boldsymbol{\nabla}) \bar{\boldsymbol{w}}.$$

Integrating (4.2.32) with respect to z from -1 to 0, and applying (4.2.35) and the above two equalities, we obtain

$$\frac{\partial \bar{\boldsymbol{w}}}{\partial t} - \Delta \bar{\boldsymbol{w}} + \boldsymbol{\nabla} p_{b_3} + \overline{(\tilde{\boldsymbol{w}} + \tilde{\boldsymbol{Z}}) \boldsymbol{\nabla} \cdot (\tilde{\boldsymbol{w}} + \tilde{\boldsymbol{Z}}) + \left[(\tilde{\boldsymbol{w}} + \tilde{\boldsymbol{Z}}) \cdot \boldsymbol{\nabla}\right] (\tilde{\boldsymbol{w}} + \tilde{\boldsymbol{Z}})} + [(\bar{\boldsymbol{w}}$$
$$+ \bar{\boldsymbol{Z}}) \cdot \boldsymbol{\nabla}](\bar{\boldsymbol{w}} + \bar{\boldsymbol{Z}}) + f \boldsymbol{k} \times (\bar{\boldsymbol{w}} + \bar{\boldsymbol{Z}}) - \alpha \bar{\boldsymbol{Z}} - \int_{-1}^{0} \int_{-1}^{z} \boldsymbol{\nabla} T \mathrm{d}z' \mathrm{d}z = 0 \text{ in } M, \tag{4.2.38}$$

$$\boldsymbol{\nabla} \cdot \bar{\boldsymbol{w}} = 0 \qquad \text{in } M, \tag{4.2.39}$$

$$\bar{\boldsymbol{w}} \cdot \boldsymbol{n} = 0, \ \frac{\partial \bar{\boldsymbol{w}}}{\partial \boldsymbol{n}} \times \boldsymbol{n} = 0 \quad \text{on } \partial M. \tag{4.2.40}$$

Subtracting (4.2.38) from (4.2.32), we know that $\tilde{\boldsymbol{w}}$ satisfies

$$\frac{\partial \tilde{\boldsymbol{w}}}{\partial t} - \Delta \tilde{\boldsymbol{w}} - \frac{\partial^2 \tilde{\boldsymbol{w}}}{\partial z^2} + \left[(\tilde{\boldsymbol{w}} + \tilde{\boldsymbol{Z}}) \cdot \boldsymbol{\nabla}\right] (\tilde{\boldsymbol{w}} + \tilde{\boldsymbol{Z}}) + \Phi(\tilde{\boldsymbol{w}} + \tilde{\boldsymbol{Z}}) \frac{\partial (\tilde{\boldsymbol{w}} + \tilde{\boldsymbol{Z}})}{\partial z}$$
$$+ \left[(\tilde{\boldsymbol{w}} + \tilde{\boldsymbol{Z}}) \cdot \boldsymbol{\nabla}\right] (\bar{\boldsymbol{w}} + \bar{\boldsymbol{Z}}) + \left[(\bar{\boldsymbol{w}} + \bar{\boldsymbol{Z}}) \cdot \boldsymbol{\nabla}\right] (\tilde{\boldsymbol{w}} + \tilde{\boldsymbol{Z}}) + f \boldsymbol{k} \times (\tilde{\boldsymbol{w}} + \tilde{\boldsymbol{Z}})$$
$$- \overline{(\tilde{\boldsymbol{w}} + \tilde{\boldsymbol{Z}}) \boldsymbol{\nabla} \cdot (\tilde{\boldsymbol{w}} + \tilde{\boldsymbol{Z}}) + \left[(\tilde{\boldsymbol{w}} + \tilde{\boldsymbol{Z}}) \cdot \boldsymbol{\nabla}\right] (\tilde{\boldsymbol{w}} + \tilde{\boldsymbol{Z}})} - \alpha \tilde{\boldsymbol{Z}} - \int_{-1}^{z} \boldsymbol{\nabla} T \mathrm{d}z'$$
$$+ \int_{-1}^{0} \int_{-1}^{z} \boldsymbol{\nabla} T \mathrm{d}z' \mathrm{d}z = 0 \ \text{ in } \Omega, \tag{4.2.41}$$

$$\frac{\partial \tilde{\boldsymbol{w}}}{\partial z} = 0 \text{ on } \Gamma_u, \ \frac{\partial \tilde{\boldsymbol{w}}}{\partial z} = 0 \text{ on } \Gamma_b, \ \tilde{\boldsymbol{w}} \cdot \boldsymbol{n} = 0, \ \frac{\partial \tilde{\boldsymbol{w}}}{\partial \boldsymbol{n}} \times \boldsymbol{n} = 0 \text{ on } \Gamma_l. \tag{4.2.42}$$

L^2 estimates of T, $\boldsymbol{w}$. Taking $L^2(\Omega)$ inner product of equation (4.2.33) with T, by integration by parts, $T(t,x,y,z) = -\int_z^0 \frac{\partial T}{\partial z'}\mathrm{d}z' + T|_{z=0}$, Hölder Inequality and Cauchy-Schwarz Inequality, we have

$$\frac{\mathrm{d}|T|_2^2}{\mathrm{d}t} + \int_\Omega |\nabla T|^2 + \int_\Omega |\frac{\partial T}{\partial z}|^2 + \alpha_u |T|_{z=0}|_2^2 \le c|Q|_2^2.$$

By Gronwall Inequality, we derive from the above inequality

$$|T(t)|_2^2 \le \mathrm{e}^{-ct}|T_{t_0}|_2^2 + c|Q|_2^2, \tag{4.2.43}$$

where $t \ge t_0$. From the above two inequalities, for $\mathcal{T} > t_0$ given, there exists a positive constant $C_1(\mathcal{T}, \|U_{t_0}\|, \|Q\|_1)$ such that

$$\int_{t_0}^{\mathcal{T}} \left[\int_\Omega \left(|\nabla T|^2 + |\frac{\partial T}{\partial z}|^2 + |T|^2\right) + |T|_{z=0}|_2^2\right] + |T(t)|_2^2 \le C_1, \tag{4.2.44}$$

where $t \in [t_0, \mathcal{T})$, and $\int_{t_0}^{\mathcal{T}} \cdot \mathrm{d}s$ is denoted by $\int_{t_0}^{\mathcal{T}} \cdot$. In this book, $C_i(\cdot,\cdot)$ denote the positive constants dependent on the quantities appearing in parentheses, $i \in \mathbb{N}$.

Taking $L^2(\Omega) \times L^2(\Omega)$ inner product of equation (4.2.32) with $\boldsymbol{w}$, we get

$$\begin{aligned}&\frac{\mathrm{d}|\boldsymbol{w}|_2^2}{\mathrm{d}t} + \int_\Omega |\nabla \boldsymbol{w}|^2 + \int_\Omega \left|\frac{\partial \boldsymbol{w}}{\partial z}\right| \\ \le & c(|\boldsymbol{Z}|_4^8 + \|\boldsymbol{Z}\|^4 + \|\boldsymbol{Z}\|^2\|\boldsymbol{Z}\|_2^2)|\boldsymbol{w}|_2^2 + c(\|\boldsymbol{Z}\|^2 + |T|_2^2 + |\boldsymbol{Z}|_2^2).\end{aligned}$$

By $\lambda_1|\boldsymbol{w}|_2^2 \le \|\boldsymbol{w}\|^2$ and Gronwall Inequality, for $t \ge t_0$, we derive from the above inequality that

$$\begin{aligned}|\boldsymbol{w}(t)|_2^2 \le & \mathrm{e}^{\int_{t_0}^t [-\lambda_1 + c(|\boldsymbol{Z}|_4^8 + \|\boldsymbol{Z}\|^4 + \|\boldsymbol{Z}\|^2\|\boldsymbol{Z}\|_2^2)]\mathrm{d}\tau}|\boldsymbol{w}(t_0)|_2^2 \\ & + c\int_{t_0}^t \mathrm{e}^{\int_\sigma^t [-\lambda_1 + c(|\boldsymbol{Z}|_4^8 + \|\boldsymbol{Z}\|^4 + \|\boldsymbol{Z}\|^2\|\boldsymbol{Z}\|_2^2)]\mathrm{d}\tau}(\|\boldsymbol{Z}\|^2 + |T|_2^2 + |\boldsymbol{Z}|_2^2)\mathrm{d}\sigma.\end{aligned} \tag{4.2.45}$$

For $\mathcal{T} > t_0$ given, according to Lemma 4.2.5 and (4.2.45), there exists $C_2(\mathcal{T}, \|U_{t_0}\|, \|Q\|_1, \|\boldsymbol{Z}_{t_0}\|_2)$ such that

$$\int_{t_0}^{\mathcal{T}} \int_\Omega \left(|\nabla \boldsymbol{w}|^2 + |\frac{\partial \boldsymbol{w}}{\partial z}|^2\right) + |\boldsymbol{w}(t)|_2^2 \le C_2, \text{ for any } t \in [t_0, \mathcal{T}). \tag{4.2.46}$$

By Minkowski Inequality and Hölder Inequality, we derive from (4.2.36)

$$\int_{t_0}^{\mathcal{T}} \int_M (|\nabla \bar{\boldsymbol{w}}|^2 + |\bar{\boldsymbol{w}}|^2) + \|\bar{\boldsymbol{w}}(t)\|_{L^2}^2 \le C_2, \forall t \in [t_0, \mathcal{T}). \tag{4.2.47}$$

L^4 estimates of T. Taking $L^2(\Omega)$ inner product of equation (4.2.33) with $|T|^2T$, we have

$$\frac{1}{4}\frac{\mathrm{d}|T|_4^4}{\mathrm{d}t}+3\int_\Omega|\nabla T|^2|T|^2+3\int_\Omega\left|\frac{\partial T}{\partial z}\right|^2|T|^2+\alpha_u\int_M|T|_{z=0}|^4$$
$$=\int_\Omega Q|T|^2T-\int_\Omega\left\{[(\boldsymbol{w}+\boldsymbol{Z})\cdot\nabla]T+\Phi(\boldsymbol{w}+\boldsymbol{Z})\frac{\partial T}{\partial z}\right\}|T|^2T.$$

Since $T^4(t,x,y,z)=-\int_z^0\frac{\partial T^4}{\partial z'}\mathrm{d}z'+T^4|_{z=0}$, by Hölder Inequality and Cauchy-Schwarz Inequality, we get

$$|T|_4^4\le c\left(\int_\Omega|T|^2\left|\frac{\partial T}{\partial z}\right|^2\right)+\frac{1}{2}\int_\Omega T^4+|T|_{z=0}|_4^4.$$

By integration by parts, Hölder Inequality and Young Inequality, we derive from the above two relationships

$$\frac{\mathrm{d}|T|_4^4}{\mathrm{d}t}+3\int_\Omega|\nabla T|^2|T|^2+3\int_\Omega|\frac{\partial T}{\partial z}|^2|T|^2+\alpha_u\int_M|T|_{z=0}|^4\le c|Q|_4^4.$$

By Gronwall Inequality, we obtain

$$|T(t)|_4^4\le \mathrm{e}^{-ct}|T_{t_0}|_4^4+c|Q|_4^4\le C_3, \tag{4.2.48}$$

where $t\ge t_0$, and C_3 is a positive constant.

L^3 estimates of $\tilde{\boldsymbol{w}}$. By integration by parts, we get

$$-\int_\Omega\left[(\tilde{\boldsymbol{w}}\cdot\nabla)\tilde{\boldsymbol{w}}-(\int_{-1}^z\nabla\cdot\tilde{\boldsymbol{w}}\mathrm{d}z')\frac{\partial\tilde{\boldsymbol{w}}}{\partial z}\right]\cdot|\tilde{\boldsymbol{w}}|\tilde{\boldsymbol{w}}=0,$$
$$-\int_\Omega\{[(\bar{\boldsymbol{w}}+\bar{\boldsymbol{Z}})\cdot\nabla]\,\tilde{\boldsymbol{w}}\}\cdot|\tilde{\boldsymbol{w}}|\tilde{\boldsymbol{w}}=\frac{1}{3}\int_\Omega|\tilde{\boldsymbol{w}}|^3\nabla\cdot(\bar{\boldsymbol{w}}+\bar{\boldsymbol{Z}})=0,$$
$$-\int_\Omega\{\left[(\tilde{\boldsymbol{w}}+\tilde{\boldsymbol{Z}})\cdot\nabla\right](\bar{\boldsymbol{w}}+\bar{\boldsymbol{Z}})\}\cdot|\tilde{\boldsymbol{w}}|\tilde{\boldsymbol{w}}$$
$$=\int_\Omega(\bar{\boldsymbol{w}}+\bar{\boldsymbol{Z}})\cdot\left[(\tilde{\boldsymbol{w}}+\tilde{\boldsymbol{Z}})\cdot\nabla\right]|\tilde{\boldsymbol{w}}|\tilde{\boldsymbol{w}}+\int_\Omega|\tilde{\boldsymbol{w}}|\tilde{\boldsymbol{w}}\cdot(\bar{\boldsymbol{w}}+\bar{\boldsymbol{Z}})\nabla\cdot(\tilde{\boldsymbol{w}}+\tilde{\boldsymbol{Z}}),$$
$$\int_\Omega\overline{(\tilde{\boldsymbol{w}}+\tilde{\boldsymbol{Z}})\nabla\cdot(\tilde{\boldsymbol{w}}+\tilde{\boldsymbol{Z}})+\left[(\tilde{\boldsymbol{w}}+\tilde{\boldsymbol{Z}})\cdot\nabla\right](\tilde{\boldsymbol{w}}+\tilde{\boldsymbol{Z}})}\cdot|\tilde{\boldsymbol{w}}|\tilde{\boldsymbol{w}}$$
$$=\int_\Omega\int_{-1}^0(\tilde{w}+\tilde{Z})^{(1)}(\tilde{\boldsymbol{w}}+\tilde{\boldsymbol{Z}})\mathrm{d}z\cdot\partial_x(|\tilde{\boldsymbol{w}}|\tilde{\boldsymbol{w}})$$
$$+\int_\Omega\int_{-1}^0(\tilde{w}+\tilde{Z})^{(2)}(\tilde{\boldsymbol{w}}+\tilde{\boldsymbol{Z}})\mathrm{d}z\cdot\partial_y(|\tilde{\boldsymbol{w}}|\tilde{\boldsymbol{w}}),$$

where $\tilde{\boldsymbol{w}}+\tilde{\boldsymbol{Z}}=((\tilde{w}+\tilde{Z})^{(1)},(\tilde{w}+\tilde{Z})^{(2)})$. Taking $L^2(\Omega)\times L^2(\Omega)$ inner product of equation (4.2.41) with $|\tilde{\boldsymbol{w}}|\tilde{\boldsymbol{w}}$, and applying the above equality, we have

$$
\begin{aligned}
&\frac{1}{3}\frac{\mathrm{d}|\tilde{\boldsymbol{w}}|_3^3}{\mathrm{d}t}+\int_\Omega\left(|\nabla\tilde{\boldsymbol{w}}|^2|\tilde{\boldsymbol{w}}|+\frac{4}{9}|\nabla|\tilde{\boldsymbol{w}}|^{\frac{3}{2}}|^2\right)+\int_\Omega(|\partial_z\tilde{\boldsymbol{w}}|^2|\tilde{\boldsymbol{w}}|+\frac{4}{9}|\partial_z|\tilde{\boldsymbol{w}}|^{\frac{3}{2}}|^2)\\
=&\int_\Omega(\bar{\boldsymbol{w}}+\bar{\boldsymbol{Z}})\cdot\Big[(\tilde{\boldsymbol{w}}+\tilde{\boldsymbol{Z}})\cdot\nabla\Big]|\tilde{\boldsymbol{w}}|\tilde{\boldsymbol{w}}+\int_\Omega|\tilde{\boldsymbol{w}}|\tilde{\boldsymbol{w}}\cdot(\bar{\boldsymbol{w}}+\bar{\boldsymbol{Z}})\nabla\cdot(\tilde{\boldsymbol{w}}+\tilde{\boldsymbol{Z}})\\
&-\int_\Omega\Big[(\tilde{\boldsymbol{Z}}\cdot\nabla)\tilde{\boldsymbol{w}}+(\tilde{\boldsymbol{w}}\cdot\nabla)\tilde{\boldsymbol{Z}}+(\tilde{\boldsymbol{Z}}\cdot\nabla)\tilde{\boldsymbol{Z}}\Big]\cdot|\tilde{\boldsymbol{w}}|\tilde{\boldsymbol{w}}\\
&-\int_\Omega\left[\Phi(\tilde{\boldsymbol{Z}})\frac{\partial\tilde{\boldsymbol{w}}}{\partial z}+\Phi(\tilde{\boldsymbol{w}})\frac{\partial\tilde{\boldsymbol{Z}}}{\partial z}+\Phi(\tilde{\boldsymbol{Z}})\frac{\partial\tilde{\boldsymbol{Z}}}{\partial z}\right]\cdot|\tilde{\boldsymbol{w}}|\tilde{\boldsymbol{w}}\\
&-\int_\Omega\{[(\bar{\boldsymbol{w}}+\bar{\boldsymbol{Z}})\cdot\nabla]\,\tilde{\boldsymbol{Z}}\}\cdot|\tilde{\boldsymbol{w}}|\tilde{\boldsymbol{w}}+\int_\Omega\int_{-1}^0(\tilde{w}+\tilde{Z})^{(1)}(\tilde{\boldsymbol{w}}+\tilde{\boldsymbol{Z}})\mathrm{d}z\cdot\partial_x(|\tilde{\boldsymbol{w}}|\tilde{\boldsymbol{w}})\\
&+\int_\Omega\int_{-1}^0(\tilde{w}+\tilde{Z})^{(2)}(\tilde{\boldsymbol{w}}+\tilde{\boldsymbol{Z}})\mathrm{d}z\cdot\partial_y(|\tilde{\boldsymbol{w}}|\tilde{\boldsymbol{w}})-\int_\Omega(f\boldsymbol{k}\times\tilde{\boldsymbol{Z}})\cdot|\tilde{\boldsymbol{w}}|\tilde{\boldsymbol{w}}\\
&-\int_\Omega\left(\int_{-1}^z T\mathrm{d}z'-\int_{-1}^0\int_{-1}^z T\mathrm{d}z'\mathrm{d}z\right)\nabla\cdot|\tilde{\boldsymbol{w}}|\tilde{\boldsymbol{w}}+\int_\Omega\alpha\tilde{\boldsymbol{Z}}\cdot|\tilde{\boldsymbol{w}}|\tilde{\boldsymbol{w}}.
\end{aligned}
$$

By Hölder Inequality, we derive from the above equality

$$
\begin{aligned}
&\frac{1}{3}\frac{\mathrm{d}|\tilde{\boldsymbol{w}}|_3^3}{\mathrm{d}t}+\int_\Omega\left(|\nabla\tilde{\boldsymbol{w}}|^2|\tilde{\boldsymbol{w}}|+\frac{4}{9}|\nabla|\tilde{\boldsymbol{w}}|^{\frac{3}{2}}|^2\right)+\int_\Omega\left(|\partial_z\tilde{\boldsymbol{w}}|^2|\tilde{\boldsymbol{w}}|+\frac{4}{9}|\partial_z|\tilde{\boldsymbol{w}}|^{\frac{3}{2}}|^2\right)\\
\leq& c\left(\|\bar{\boldsymbol{w}}\|_{L^4}+\|\bar{\boldsymbol{Z}}\|_{L^4}\right)\left(\int_\Omega|\nabla\tilde{\boldsymbol{w}}|^2|\tilde{\boldsymbol{w}}|\right)^{\frac{1}{2}}\left\{\left[\int_M\left(\int_{-1}^0|\tilde{\boldsymbol{w}}|^3\mathrm{d}z\right)^2\right]^{\frac{1}{4}}\right.\\
&\left.+\left[\int_M\left(\int_{-1}^0|\tilde{\boldsymbol{w}}|^2\mathrm{d}z\right)^2\right]^{\frac{1}{8}}\left[\int_M\left(\int_{-1}^0|\tilde{\boldsymbol{Z}}|^4\mathrm{d}z\right)^2\right]^{\frac{1}{8}}\right\}\\
&+c\left(\int_\Omega|\nabla\tilde{\boldsymbol{w}}|^2|\tilde{\boldsymbol{w}}|\right)^{\frac{1}{2}}|\tilde{\boldsymbol{w}}|_2^{\frac{1}{2}}\left\{\left[\int_M\left(\int_{-1}^0|\tilde{\boldsymbol{Z}}|^2\mathrm{d}z\right)^4\right]^{\frac{1}{4}}+\||\overline{T}|\|_{L^4}\right\}\\
&+c\left(\int_\Omega|\nabla\tilde{\boldsymbol{w}}|^2|\tilde{\boldsymbol{w}}|\right)^{\frac{1}{2}}\left[\int_M\left(\int_{-1}^0|\tilde{\boldsymbol{w}}|^2\mathrm{d}z\right)^{\frac{5}{2}}\right]^{\frac{1}{2}}+I_1+I_2+I_3,
\end{aligned}
$$

where

$$
I_1=\int_\Omega\Big\{-[(\bar{\boldsymbol{w}}+\bar{\boldsymbol{Z}})\cdot\nabla]\,\tilde{\boldsymbol{Z}}+(\bar{\boldsymbol{w}}+\bar{\boldsymbol{Z}})\,\nabla\cdot\tilde{\boldsymbol{Z}}-\Big[\Big(\tilde{\boldsymbol{w}}+\tilde{\boldsymbol{Z}}\Big)\cdot\nabla\Big]\,\tilde{\boldsymbol{Z}}
$$

$$-\Phi\left(\tilde{Z}\right)\frac{\partial \tilde{Z}}{\partial z}\Big\}\cdot|\tilde{w}|\tilde{w},$$

$$I_2=-\int_\Omega\left[\left(\tilde{Z}\cdot\nabla\right)\tilde{w}+\Phi\left(\tilde{Z}\right)\frac{\partial\tilde{w}}{\partial z}+\Phi\left(\tilde{w}\right)\frac{\partial\tilde{Z}}{\partial z}\right]\cdot|\tilde{w}|\tilde{w},$$

$$I_3=-\int_\Omega\left(fk\times\tilde{Z}-\alpha\tilde{Z}\right)\cdot|\tilde{w}|\tilde{w}.$$

Applying Minkowski Inequality, (4.2.24) and Hölder Inequality, we get

$$\left[\int_M\left(\int_{-1}^0|\tilde{w}|^3\mathrm{d}z\right)^2\right]^{\frac{1}{2}}\le c|\tilde{w}|_3^{\frac{3}{2}}\left[\int_{-1}^0\left(\|\nabla|\tilde{w}|^{\frac{3}{2}}\|_{L^2}^2+\||\tilde{w}|^{\frac{3}{2}}\|_{L^2}^2\right)\mathrm{d}z\right]^{\frac{1}{2}}.$$

By Minkowski Inequality and Hölder Inequality, (4.2.25), we have

$$\int_M\left(\int_{-1}^0|\tilde{w}|^2\mathrm{d}z\right)^{\frac{5}{2}}\le\left(\int_{-1}^0\|\tilde{w}\|_{L^3}^{\frac{6}{5}}\|\tilde{w}\|_{H^1}^{\frac{4}{5}}\mathrm{d}z\right)^{\frac{5}{2}}\le c\|\tilde{w}\|^2|\tilde{w}|_3^3.$$

Applying Hölder Inequality and Minkowski Inequality, (4.2.24), $|\tilde{w}|_4\le|\tilde{w}|_3^{\frac{1}{2}}\|\tilde{w}\|^{\frac{1}{2}}$, $|\tilde{w}|_6^6=\int_\Omega|\tilde{w}|^{\frac{3}{2}\cdot4}\le|\tilde{w}|_3^{\frac{3}{2}}\||\tilde{w}|^{\frac{3}{2}}\|^3$ and Young Inequality, we obtain

$$\begin{aligned}I_1\le&c(\|\bar{w}\|_{L^4}+\|\bar{Z}\|_{L^4})\left[\int_{-1}^0\left(\int_M|\nabla\tilde{Z}|^4\right)^{\frac{1}{2}}\mathrm{d}z\right]^{\frac{1}{2}}|\tilde{w}|_4^2+c|\nabla\tilde{Z}|_2|\tilde{w}|_6^3\\&+c|\tilde{Z}|_6|\nabla\tilde{Z}|_2|\tilde{w}|_6^2\\&+c\left[\int_{-1}^0\left(\int_M|\nabla\tilde{Z}|^4\right)^{\frac{1}{2}}\mathrm{d}z\right]^{\frac{1}{2}}\left[\int_{-1}^0\left(\int_M|\frac{\partial\tilde{Z}}{\partial z}|^4\right)^{\frac{1}{2}}\mathrm{d}z\right]^{\frac{1}{2}}|\tilde{w}|_4^2\\\le&\varepsilon\int_\Omega(|\nabla|\tilde{w}|^{\frac{3}{2}}|^2+|\partial_z|\tilde{w}|^{\frac{3}{2}}|^2)+c(1+|w|_2^{\frac{3}{2}}\|w\|^{\frac{3}{2}}\|Z\|^{\frac{3}{2}}\|Z\|_2^{\frac{3}{2}}\\&+|Z|_2^{\frac{3}{2}}\|Z\|^3\|Z\|_2^{\frac{3}{2}}+\|Z\|^4+|Z|_6^6+\|w\|^2)|\tilde{w}|_3^3+c\|w\|^{\frac{3}{2}}+c\|Z\|^3\\&+c\|Z\|^2\|Z\|_2^2+c\|w\|^2.\end{aligned}$$

By integration by parts and Hölder Inequality, we get

$$\begin{aligned}I_2\le&\int_\Omega\left[\left(\int_{-1}^z|\tilde{w}|\right)|\frac{\partial\tilde{Z}}{\partial z}||\nabla\tilde{w}||\tilde{w}|+|\nabla\tilde{Z}||\tilde{w}|^3+\left(\int_{-1}^z|\tilde{w}|\right)|\nabla\tilde{Z}||\frac{\partial\tilde{w}}{\partial z}||\tilde{w}|\right]\\&+\int_\Omega\left[|\tilde{Z}||\nabla\tilde{w}||\tilde{w}|^2+\left(\int_{-1}^z|\nabla\tilde{Z}|\right)|\frac{\partial\tilde{w}}{\partial z}||\tilde{w}|^2\right]\\\le&\varepsilon\int_\Omega(|\nabla\tilde{w}|^2|\tilde{w}|+|\partial_z\tilde{w}|^2|\tilde{w}|+|\nabla|\tilde{w}|^{\frac{3}{2}}|^2+|\partial_z|\tilde{w}|^{\frac{3}{2}}|^2)\end{aligned}$$

$$+ (1 + |\boldsymbol{Z}|_4^8 + \|\boldsymbol{Z}\|^4 + \|\boldsymbol{Z}\|\|\boldsymbol{Z}\|_2^3 + \|\boldsymbol{Z}\|^2\|\boldsymbol{Z}\|_2^2)|\tilde{\boldsymbol{w}}|_3^3.$$

Thus, by $I_3 \le c|\boldsymbol{Z}|_3^3 + c|\tilde{\boldsymbol{w}}\|_3^3$, we obtain

$$\frac{\mathrm{d}|\tilde{\boldsymbol{w}}|_3^3}{\mathrm{d}t} + \int_\Omega \left(|\nabla \tilde{\boldsymbol{w}}|^2|\tilde{\boldsymbol{w}}| + \frac{4}{9}|\nabla|\tilde{\boldsymbol{w}}|^{\frac{3}{2}}|^2 \right) + \int_\Omega \left(|\partial_z \tilde{\boldsymbol{w}}|^2|\tilde{\boldsymbol{w}}| + \frac{4}{9}|\partial_z|\tilde{\boldsymbol{w}}|^{\frac{3}{2}}|^2 \right)$$

$$\begin{aligned}\le & c(1 + \|\bar{\boldsymbol{w}}\|_{L^2}^2\|\bar{\boldsymbol{w}}\|_{H^1}^2 + \|\tilde{\boldsymbol{w}}\|^2 + |\boldsymbol{w}|_2^{\frac{3}{2}}\|\boldsymbol{w}\|^{\frac{3}{2}}\|\boldsymbol{Z}\|^{\frac{3}{2}}\|\boldsymbol{Z}\|_2^{\frac{3}{2}} + |\boldsymbol{Z}|_2^{\frac{3}{2}}\|\boldsymbol{Z}\|^3\|\boldsymbol{Z}\|_2^{\frac{3}{2}} \\ & + \|\boldsymbol{Z}\|^2 + |\boldsymbol{Z}|_4^8 + \|\boldsymbol{Z}\|^4 + \|\boldsymbol{Z}\|\|\boldsymbol{Z}\|_2^3 + \|\boldsymbol{Z}\|^2\|\boldsymbol{Z}\|_2^2 + |\boldsymbol{Z}|_6^6)|\tilde{\boldsymbol{w}}|_3^3 + c\|\boldsymbol{w}\|^{\frac{3}{2}} \\ & + c|\boldsymbol{w}|_2^2\|\boldsymbol{w}\|^2 + c\|\boldsymbol{w}\|^2 + c\|\boldsymbol{Z}\|^3 + c\|\boldsymbol{Z}\|^8 + c|T|_4^4 + c\|\boldsymbol{Z}\|^2\|\boldsymbol{Z}\|_2^2.\end{aligned}$$

By Gronwall Inequality, Lemma 4.2.5, (4.2.46)-(4.3.48), for any $\mathcal{T} > t_0$, there exists $C_4(\mathcal{T}, \|\boldsymbol{U}_{t_0}\|, \|Q\|_1, \|\boldsymbol{Z}_{t_0}\|_2)$, such that

$$|\tilde{\boldsymbol{w}}(t)|_3^3 \le C_4, \quad \forall t \in [t_0, \mathcal{T}). \tag{4.2.49}$$

Remark 4.2.14. In estimating $\int_\Omega \Phi(\tilde{\boldsymbol{w}})\frac{\partial \tilde{\boldsymbol{Z}}}{\partial z} \cdot |\tilde{\boldsymbol{w}}|\tilde{\boldsymbol{w}}$ in I_2, we can't reduce the index 3 of $\|\boldsymbol{Z}\|\|\boldsymbol{Z}\|_2^3$. Thus, when the random force only satisfies $\boldsymbol{Z} \in L^\infty(\mathbb{R}; V_1) \cap L^2(\mathbb{R}; (H^2(\Omega))^2)$, we can't prove the global well-posedness of IBVP.

L^4 estimates of $\tilde{\boldsymbol{w}}$. Taking $L^2(\Omega) \times L^2(\Omega)$ inner product of equation (4.2.41) with $|\tilde{\boldsymbol{w}}|^2\tilde{\boldsymbol{w}}$, similarly to L^3 estimates of $\tilde{\boldsymbol{w}}$, we get

$$\frac{1}{4}\frac{\mathrm{d}|\tilde{\boldsymbol{w}}|_4^4}{\mathrm{d}t} + \int_\Omega (|\nabla \tilde{\boldsymbol{w}}|^2|\tilde{\boldsymbol{w}}|^2 + \frac{1}{2}|\nabla|\tilde{\boldsymbol{w}}|^2|^2) + \int_\Omega (|\partial_z \tilde{\boldsymbol{w}}|^2|\tilde{\boldsymbol{w}}|^2 + \frac{1}{2}|\partial_z|\tilde{\boldsymbol{w}}|^2|^2)$$

$$\begin{aligned}\le & c(\|\bar{\boldsymbol{w}}\|_{L^4} + \|\bar{\boldsymbol{Z}}\|_{L^4}) \left(\int_\Omega |\nabla \tilde{\boldsymbol{w}}|^2|\tilde{\boldsymbol{w}}|^2 \right)^{\frac{1}{2}} \left\{ \left[\int_M (\int_{-1}^0 |\tilde{\boldsymbol{w}}|^4 \mathrm{d}z)^2 \right]^{\frac{1}{4}} \right. \\ & + \left[\int_M \left(\int_{-1}^0 |\tilde{\boldsymbol{w}}|^4 \mathrm{d}z \right)^2 \right]^{\frac{1}{8}} \left[\int_M \left(\int_{-1}^0 |\tilde{\boldsymbol{Z}}|^4 \mathrm{d}z \right)^2 \right]^{\frac{1}{8}} \Bigg\} \\ & + c \left(\int_\Omega |\nabla \tilde{\boldsymbol{w}}|^2|\tilde{\boldsymbol{w}}|^2 \right)^{\frac{1}{2}} \left[\int_M \left(\int_{-1}^0 |\tilde{\boldsymbol{w}}|^2 \mathrm{d}z \right)^2 \right]^{\frac{1}{4}} \left\{ \left[\int_M \left(\int_{-1}^0 |\tilde{\boldsymbol{Z}}|^2 \mathrm{d}z \right)^4 \right]^{\frac{1}{4}} \right. \\ & \left. + c\||\overline{T}|\|_{L^4} \right\} + c \left(\int_\Omega |\nabla \tilde{\boldsymbol{w}}|^2|\tilde{\boldsymbol{w}}|^2 \right)^{\frac{1}{2}} \left[\int_M \left(\int_{-1}^0 |\tilde{\boldsymbol{w}}|^2 \mathrm{d}z \right)^3 \right]^{\frac{1}{2}} + J_1 + J_2 + J_3,\end{aligned}$$

where

$$J_1 = -\int_\Omega \Big\{ [(\bar{\boldsymbol{w}} + \bar{\boldsymbol{Z}}) \cdot \nabla] \, \tilde{\boldsymbol{Z}} - (\bar{\boldsymbol{w}} + \bar{\boldsymbol{Z}}) \nabla \cdot \tilde{\boldsymbol{Z}} + \Big[(\tilde{\boldsymbol{w}} + \tilde{\boldsymbol{Z}}) \cdot \nabla \Big] \, \tilde{\boldsymbol{Z}}$$

$$+\Phi(\tilde{\boldsymbol{Z}})\frac{\partial \tilde{\boldsymbol{Z}}}{\partial z}\Big\}\cdot|\tilde{\boldsymbol{w}}|^2\tilde{\boldsymbol{w}},$$

$$J_2 = -\int_\Omega \left[(\tilde{\boldsymbol{Z}}\cdot\boldsymbol{\nabla})\tilde{\boldsymbol{w}} + \Phi(\tilde{\boldsymbol{Z}})\frac{\partial \tilde{\boldsymbol{w}}}{\partial z} + \Phi(\tilde{\boldsymbol{w}})\frac{\partial \tilde{\boldsymbol{Z}}}{\partial z}\right]\cdot|\tilde{\boldsymbol{w}}|^2\tilde{\boldsymbol{w}},$$

$$J_3 = -\int_\Omega (f\boldsymbol{k}\times\tilde{\boldsymbol{Z}} - \alpha\tilde{\boldsymbol{Z}})\cdot|\tilde{\boldsymbol{w}}|^2\tilde{\boldsymbol{w}}.$$

By Minkowski Inequality, (4.2.24) and Hölder Inequality, we have

$$\left[\int_M\left(\int_{-1}^0|\tilde{\boldsymbol{w}}|^4\mathrm{d}z\right)^2\right]^{\frac{1}{2}} \le c|\tilde{\boldsymbol{w}}|_4^2\left[\int_{-1}^0(\|\boldsymbol{\nabla}|\tilde{\boldsymbol{w}}|^2\|_{L^2(M)}^2 + \||\tilde{\boldsymbol{w}}|^2\|_{L^2(M)}^2)\mathrm{d}z\right]^{\frac{1}{2}}.$$

Similarly, by (4.2.26), we obtain

$$\int_M\left(\int_{-1}^0|\tilde{\boldsymbol{w}}|^2\mathrm{d}z\right)^3 \le \left[\int_{-1}^0\left(\int_M|\tilde{\boldsymbol{w}}|^6\right)^{\frac{1}{3}}\mathrm{d}z\right]^3$$
$$\le c\|\tilde{\boldsymbol{w}}\|^2|\tilde{\boldsymbol{w}}|_4^4.$$

By Hölder Inequality, Minkowski Inequality, (4.2.24), (4.2.27), $|\tilde{\boldsymbol{w}}|_6^6 \le |\tilde{\boldsymbol{w}}|_4^3\||\tilde{\boldsymbol{w}}|^2\|^{\frac{3}{2}}$, $|\tilde{\boldsymbol{w}}|_8^8 = |\tilde{\boldsymbol{w}}|_4^2\||\tilde{w}|^2\|^3$ and Young Inequality, we get

$$J_1 \le \varepsilon\int_\Omega(|\boldsymbol{\nabla}|\tilde{\boldsymbol{w}}|^2|^2 + |\partial_z|\tilde{\boldsymbol{w}}|^2|^2) + c|\boldsymbol{w}|_2^2\|\boldsymbol{w}\|^2 + c|\boldsymbol{Z}|_2^2\|\boldsymbol{Z}\|^2 + \|\boldsymbol{Z}\|^{\frac{8}{5}}\|\boldsymbol{Z}\|_2^{\frac{8}{5}}$$

$$+ c|\boldsymbol{Z}|_4^4 + c(1 + \|\boldsymbol{Z}\|^{\frac{4}{3}}\|\boldsymbol{Z}\|_2^{\frac{4}{3}} + \|\boldsymbol{Z}\|^4 + \|\boldsymbol{Z}\|^{\frac{2}{3}}\|\boldsymbol{Z}\|_2^2 + \|\boldsymbol{Z}\|^{\frac{8}{5}}\|\boldsymbol{Z}\|_2^{\frac{8}{5}})|\tilde{\boldsymbol{w}}|_4^4.$$

By integration by parts and Hölder Inequality, we have

$$J_2 \le \varepsilon\int_\Omega(|\boldsymbol{\nabla}\tilde{\boldsymbol{w}}|^2|\tilde{\boldsymbol{w}}|^2 + |\partial_z\tilde{\boldsymbol{w}}|^2|\tilde{\boldsymbol{w}}|^2 + |\boldsymbol{\nabla}|\tilde{\boldsymbol{w}}|^2|^2 + |\partial_z|\tilde{\boldsymbol{w}}|^2|^2)$$

$$+(1 + |\boldsymbol{Z}|_4^8 + \|\boldsymbol{Z}\|^4 + \|\boldsymbol{Z}\|\|\boldsymbol{Z}\|_2^3 + \|\boldsymbol{Z}\|^2\|\boldsymbol{Z}\|_2^2)|\tilde{\boldsymbol{w}}|_4^4.$$

Thus, we obtain

$$\frac{\mathrm{d}|\tilde{\boldsymbol{w}}|_4^4}{\mathrm{d}t} + \int_\Omega(|\boldsymbol{\nabla}\tilde{\boldsymbol{w}}|^2|\tilde{\boldsymbol{w}}|^2 + \frac{1}{2}|\boldsymbol{\nabla}|\tilde{\boldsymbol{w}}|^2|^2) + \int_\Omega\left(|\partial_z\tilde{\boldsymbol{w}}|^2|\tilde{\boldsymbol{w}}|^2 + \frac{1}{2}|\partial_z|\tilde{\boldsymbol{w}}|^2|^2\right)$$
$$\le c(1 + |\boldsymbol{w}|_2^2\|\boldsymbol{w}\|^2 + |\boldsymbol{Z}|_2^2\|\boldsymbol{Z}\|^2 + |\boldsymbol{Z}|_8^4 + \|\boldsymbol{w}\|^2 + \|\boldsymbol{Z}\|^4 + \|\boldsymbol{Z}\|^{\frac{4}{3}}\|\boldsymbol{Z}\|_2^{\frac{4}{3}}$$
$$+ \|\boldsymbol{Z}\|^{\frac{2}{3}}\|\boldsymbol{Z}\|_2^2 + \|\boldsymbol{Z}\|^{\frac{8}{5}}\|\boldsymbol{Z}\|_2^{\frac{8}{5}} + \|\boldsymbol{Z}\|\|\boldsymbol{Z}\|_2^3 + \|\boldsymbol{Z}\|^2\|\boldsymbol{Z}\|_2^2)|\tilde{\boldsymbol{w}}|_4^4$$
$$+ c|\boldsymbol{w}|_2^2\|\boldsymbol{w}\|^2 + c|\boldsymbol{Z}|_2^2\|\boldsymbol{Z}\|^2 + c|\boldsymbol{Z}|_4^4 + c|T|_4^4 + \|\boldsymbol{Z}\|^{\frac{8}{5}}\|\boldsymbol{Z}\|_2^{\frac{8}{5}}.$$

By Gronwall Inequality, Lemma 4.2.5, (4.2.46)-(4.2.48), for any $\mathcal{T} > t_0$, we know that there exists $C_5(\mathcal{T}, \|U_{t_0}\|, \|Q\|_1, \|\boldsymbol{Z}_{t_0}\|_2)$ such that

$$\int_{t_0}^{\mathcal{T}} \int_{\Omega} \left[\left(|\boldsymbol{\nabla} \tilde{\boldsymbol{w}}|^2 |\tilde{\boldsymbol{w}}|^2 + \frac{1}{2} |\boldsymbol{\nabla} |\tilde{\boldsymbol{w}}|^2|^2 \right) + \left(|\partial_z \tilde{\boldsymbol{w}}|^2 |\tilde{\boldsymbol{w}}|^2 + \frac{1}{2} |\partial_z |\tilde{\boldsymbol{w}}|^2|^2 \right) \right] + |\tilde{\boldsymbol{w}}(t)|_4^4 \leq C_5, \tag{4.2.50}$$

where $t_0 \leq t < \mathcal{T}$.

L^2 **estimates of** $\boldsymbol{\nabla} \bar{\boldsymbol{w}}$**.** According to Lemma 4.2.13, we get

$$\begin{aligned}
&\left| \int_M \left[(\bar{\boldsymbol{w}} + \bar{\boldsymbol{Z}}) \cdot \boldsymbol{\nabla} \right] (\bar{\boldsymbol{w}} + \bar{\boldsymbol{Z}}) \cdot \Delta \bar{\boldsymbol{w}} \right| \\
\leq& (\|\bar{\boldsymbol{w}}\|_{L^2} \|\bar{\boldsymbol{w}}\|_{H^1} + \|\bar{\boldsymbol{Z}}\|_{L^4}^2) \|\boldsymbol{\nabla} \bar{\boldsymbol{Z}}\|_{L^2} \|\boldsymbol{\nabla} \bar{\boldsymbol{Z}}\|_{H^1} \\
&+ c(\|\bar{\boldsymbol{w}}\|_{H^1}^2 + \|\bar{\boldsymbol{w}}\|_{L^2}^2 \|\bar{\boldsymbol{w}}\|_{H^1}^2 + \|\bar{\boldsymbol{Z}}\|_{H^1}^2 + \|\bar{\boldsymbol{Z}}\|_{H^1}^4) \|\boldsymbol{\nabla} \bar{\boldsymbol{w}}\|_{L^2}^2 + \varepsilon \|\Delta \bar{\boldsymbol{w}}\|_{L^2}^2.
\end{aligned}$$

By Hölder Inequality and Minkowski Inequality, we have

$$\left| \int_M \int_{-1}^0 [\tilde{\boldsymbol{w}} \boldsymbol{\nabla} \cdot \tilde{\boldsymbol{w}} + (\tilde{\boldsymbol{w}} \cdot \boldsymbol{\nabla}) \tilde{\boldsymbol{w}}] \, \mathrm{d}z \cdot \Delta \bar{\boldsymbol{w}} \right| \leq c \int_{\Omega} |\boldsymbol{\nabla} \tilde{\boldsymbol{w}}|^2 |\tilde{\boldsymbol{w}}|^2 + \varepsilon \|\Delta \bar{\boldsymbol{w}}\|_{L^2}^2.$$

By Hölder Inequality, $|\boldsymbol{Z}|_\infty \leq c\|\boldsymbol{Z}\|_2$, Minkowski Inequality and (4.2.24), we get

$$\begin{aligned}
&\left| \int_M \int_{-1}^0 [\tilde{\boldsymbol{Z}} \boldsymbol{\nabla} \cdot \tilde{\boldsymbol{w}} + (\tilde{\boldsymbol{Z}} \cdot \boldsymbol{\nabla}) \tilde{\boldsymbol{w}} + \tilde{\boldsymbol{w}} \boldsymbol{\nabla} \cdot \tilde{\boldsymbol{Z}} + (\tilde{\boldsymbol{w}} \cdot \boldsymbol{\nabla}) \tilde{\boldsymbol{Z}} + \tilde{\boldsymbol{Z}} \boldsymbol{\nabla} \cdot \tilde{\boldsymbol{Z}} \right. \\
&\left. + (\tilde{\boldsymbol{Z}} \cdot \boldsymbol{\nabla}) \tilde{\boldsymbol{Z}}] \mathrm{d}z \cdot \Delta \bar{\boldsymbol{w}} \right| \\
\leq& c\|\boldsymbol{Z}\|_2^2 \|\boldsymbol{w}\|^2 + \|\boldsymbol{Z}\| \|\boldsymbol{Z}\|_2 |\boldsymbol{w}|_2 \|\boldsymbol{w}\| + |\boldsymbol{Z}|_2 \|\boldsymbol{Z}\|^2 \|\boldsymbol{Z}\|_2 + \varepsilon \|\Delta \bar{\boldsymbol{w}}\|_{L^2}^2.
\end{aligned}$$

By integration by parts, we have

$$\int_M \boldsymbol{\nabla} p_{b_3} \cdot \Delta \bar{\boldsymbol{w}} = 0, \quad -\int_M \int_{-1}^0 \int_{-1}^z \boldsymbol{\nabla} T \mathrm{d}z' \mathrm{d}z \cdot \Delta \bar{\boldsymbol{w}} = 0.$$

Taking $L^2(M) \times L^2(M)$ inner product of equation (4.2.38) with $-\Delta \bar{\boldsymbol{w}}$, according to the above relationships, applying Hölder Inequality and Young Inequality, choosing ε small enough, we obtain

$$\begin{aligned}
&\frac{\mathrm{d}\|\boldsymbol{\nabla} \bar{\boldsymbol{w}}\|_{L^2}^2}{\mathrm{d}t} + \|\Delta \bar{\boldsymbol{w}}\|_{L^2}^2 \\
\leq& c(\|\bar{\boldsymbol{w}}\|_{H^1}^2 + \|\bar{\boldsymbol{w}}\|_{L^2}^2 \|\bar{\boldsymbol{w}}\|_{H^1}^2 + \|\bar{\boldsymbol{Z}}\|_{H^1}^2 + \|\bar{\boldsymbol{Z}}\|_{H^1}^4) \|\boldsymbol{\nabla} \bar{\boldsymbol{w}}\|_{L^2}^2 + c \int_{\Omega} |\boldsymbol{\nabla} \tilde{\boldsymbol{w}}|^2 |\tilde{\boldsymbol{w}}|^2 \\
&+ c\|\boldsymbol{Z}\|_2^2 \|\boldsymbol{w}\|^2 + \|\boldsymbol{Z}\| \|\boldsymbol{Z}\|_2 |\boldsymbol{w}|_2 \|\boldsymbol{w}\| + |\boldsymbol{Z}|_2 \|\boldsymbol{Z}\|^2 \|\boldsymbol{Z}\|_2^2 + c|\boldsymbol{Z}|_2^2.
\end{aligned}$$

Applying to Gronwall Inequality, Lemma 4.2.5, (4.2.46), (4.2.47) and (4.2.50), we know that for any $\mathcal{T} > t_0$, there exists $C_6(\mathcal{T}, \|U_{t_0}\|, \|Q\|_1, \|\boldsymbol{Z}_{t_0}\|_2)$ such that

$$\|\boldsymbol{\nabla}\bar{\boldsymbol{w}}(t)\|_{L^2}^2 \leq C_6, \quad \forall\, t \in [t_0, \mathcal{T}). \tag{4.2.51}$$

L^2 **estimates of** $\partial_z \boldsymbol{w}$**.** Taking the derivative of equation (4.2.32) with respect to z, we get

$$\begin{aligned}
&\frac{\partial \boldsymbol{w}_z}{\partial t} - \Delta \boldsymbol{w}_z - \frac{\partial^2 \boldsymbol{w}_z}{\partial z^2} + [(\boldsymbol{w}+\boldsymbol{Z})\cdot\boldsymbol{\nabla}](\boldsymbol{w}_z+\boldsymbol{Z}_z) + \Phi(\boldsymbol{w}+\boldsymbol{Z})\frac{\partial(\boldsymbol{w}_z+\boldsymbol{Z}_z)}{\partial z}\\
&+ [(\boldsymbol{w}_z+\boldsymbol{Z}_z)\cdot\boldsymbol{\nabla}](\boldsymbol{w}+\boldsymbol{Z}) - (\boldsymbol{w}_z+\boldsymbol{Z}_z)\boldsymbol{\nabla}\cdot(\boldsymbol{w}+\boldsymbol{Z}) + f\boldsymbol{k}\times(\boldsymbol{w}_z+\boldsymbol{Z}_z)\\
&- \boldsymbol{\nabla}T - \alpha \boldsymbol{Z}_z = 0.
\end{aligned} \tag{4.2.52}$$

By integration by parts, Hölder Inequality, Minkowski Inequality, Sobolev Inequality and (4.2.27), we have

$$\begin{aligned}
&-\int_\Omega \left\{[(\boldsymbol{w}+\boldsymbol{Z})\cdot\boldsymbol{\nabla}]\boldsymbol{Z}_z + \Phi(\boldsymbol{w}+\boldsymbol{Z})\frac{\partial \boldsymbol{Z}_z}{\partial z}\right\}\cdot \boldsymbol{w}_z\\
\leq&|\boldsymbol{w}+\boldsymbol{Z}|_4|\boldsymbol{\nabla}\boldsymbol{Z}_z|_2|\boldsymbol{w}_z|_4 + \int_\Omega \left|\int_{-1}^z (\boldsymbol{w}+\boldsymbol{Z})\mathrm{d}z'\right| (|\boldsymbol{Z}_{zz}||\boldsymbol{\nabla}\boldsymbol{w}_z| + |\boldsymbol{\nabla}\boldsymbol{Z}_z||\boldsymbol{w}_{zz}|)\\
\leq&\varepsilon\|\boldsymbol{w}_z\|^2 + c\|\boldsymbol{Z}\|_2^2 + c|\boldsymbol{w}+\boldsymbol{Z}|_4^8|\boldsymbol{w}_z|_2^2 + c\|\boldsymbol{w}+\boldsymbol{Z}\|^2\|\boldsymbol{Z}\|_{2+2\gamma}^2,
\end{aligned}$$

where $0 < \gamma < \gamma_0$, and γ_0 is in the definition of Ψ. By integration by parts, Hölder Inequality, Young Inequality and (4.2.27), we get

$$\begin{aligned}
&-\int_\Omega \{[(\boldsymbol{w}_z+\boldsymbol{Z}_z)\cdot\boldsymbol{\nabla}](\boldsymbol{w}+\boldsymbol{Z}) - (\boldsymbol{w}_z+\boldsymbol{Z}_z)\boldsymbol{\nabla}\cdot(\boldsymbol{w}+\boldsymbol{Z})\}\cdot\boldsymbol{w}_z\\
&\leq \varepsilon\|\boldsymbol{w}_z\|^2 + c\|\boldsymbol{Z}\|_2^2 + c|\boldsymbol{w}+\boldsymbol{Z}|_4^8|\boldsymbol{Z}_z|_2^2 + c|\boldsymbol{w}+\boldsymbol{Z}|_4^8|\boldsymbol{w}_z|_2^2.
\end{aligned}$$

Taking $L^2(\Omega)\times L^2(\Omega)$ inner product of equation (4.2.52) with $\boldsymbol{w}_z$, applying the above inequalities, Hölder Inequality and Poincaré Inequality, choosing ε small enough, we obtain

$$\begin{aligned}
\frac{\mathrm{d}|\boldsymbol{w}_z|_2^2}{\mathrm{d}t} + \int_\Omega |\boldsymbol{\nabla}\boldsymbol{w}_z|^2 + \int_\Omega |\frac{\partial \boldsymbol{w}_z}{\partial z}|^2 \leq& c(|\bar{\boldsymbol{w}}|_{H^1}^8 + |\tilde{\boldsymbol{w}}|_4^8 + |\boldsymbol{Z}|_4^8)(|\boldsymbol{w}_z|_2^2 + |\boldsymbol{Z}_z|_2^2)\\
&+ c\|\boldsymbol{Z}\|_2^2 + c\|\boldsymbol{w}+\boldsymbol{Z}\|^2\|\boldsymbol{Z}\|_{2+2\gamma}^2 + c|T|_2^2.
\end{aligned}$$

By Gronwall Inequality, Lemma 4.2.5, (4.2.46), (4.2.47), (4.2.50) and (4.2.51), we know that for any $\mathcal{T} > t_0$, there exists $C_7(\mathcal{T}, \|U_{t_0}\|, \|Q\|_1, \|\boldsymbol{Z}_{t_0}\|_2)$ such that

$$\int_{t_0}^{\mathcal{T}} \|\boldsymbol{w}_z\|^2 + |\boldsymbol{w}_z(t)|_2^2 \leq C_7, \quad \forall\, t \in [t_0, \mathcal{T}). \tag{4.2.53}$$

L^2 **estimates of** $\nabla \boldsymbol{w}$**.** Similarly to Lemma 4.2.13, by (4.2.27), we get

$$\begin{aligned}
&\left|\int_\Omega \{[(\boldsymbol{w}+\boldsymbol{Z})\cdot\nabla](\boldsymbol{w}+\boldsymbol{Z})\}\cdot\Delta\boldsymbol{w}\right|\\
\le&\varepsilon(|\Delta\boldsymbol{w}|_2^2+|\nabla\boldsymbol{w}_z|_2^2)+c(|\boldsymbol{w}|_4^8+|\boldsymbol{Z}|_4^8)(|\nabla\boldsymbol{w}|_2^2+|\nabla\boldsymbol{Z}|_2^2)+c\|\boldsymbol{Z}\|_2^2,\\
&\left|\int_\Omega [\Phi(\boldsymbol{w}+\boldsymbol{Z})(\boldsymbol{w}_z+\boldsymbol{Z}_z)]\cdot\Delta\boldsymbol{w}\right|\\
\le&\varepsilon|\Delta\boldsymbol{w}|_2^2+c\|\boldsymbol{Z}\|_2^2+c(|\boldsymbol{w}_z|_2^2+|\nabla\boldsymbol{w}_z|_2^2+|\boldsymbol{w}_z|_2^2|\nabla\boldsymbol{w}_z|_2^2)(|\nabla\boldsymbol{w}|_2^2+|\nabla\boldsymbol{Z}|_2^2)\\
&+c(|\boldsymbol{Z}_z|_2^2+|\nabla\boldsymbol{Z}_z|_2^2+|\boldsymbol{Z}_z|_2^2|\nabla\boldsymbol{Z}_z|_2^2)(|\nabla\boldsymbol{w}|_2^2+|\nabla\boldsymbol{Z}|_2^2).
\end{aligned}$$

Taking $L^2(\Omega)\times L^2(\Omega)$ inner product of equation (4.2.32) with $-\Delta\boldsymbol{w}$, using the above two inequalities, Hölder Inequality, and choosing ε small enough, we have

$$\begin{aligned}
&\frac{\mathrm{d}|\nabla\boldsymbol{w}|_2^2}{\mathrm{d}t}+\int_\Omega|\Delta\boldsymbol{w}|^2+\int_\Omega|\nabla\boldsymbol{w}_z|^2\\
\le&c(|\boldsymbol{w}|_4^8+|\boldsymbol{Z}|_4^8+|\boldsymbol{w}_z|_2^2+|\nabla\boldsymbol{w}_z|_2^2+|\boldsymbol{w}_z|_2^2|\nabla\boldsymbol{w}_z|_2^2+|\boldsymbol{Z}_z|_2^2+|\nabla\boldsymbol{Z}_z|_2^2\\
&+|\boldsymbol{Z}_z|_2^2|\nabla\boldsymbol{Z}_z|_2^2)|\nabla\boldsymbol{w}|_2^2+c|\nabla T|_2^2+c(|\boldsymbol{w}|_4^8+|\boldsymbol{Z}|_4^8+|\boldsymbol{w}_z|_2^2+|\nabla\boldsymbol{w}_z|_2^2\\
&+|\boldsymbol{w}_z|_2^2|\nabla\boldsymbol{w}_z|_2^2+|\boldsymbol{Z}_z|_2^2+|\nabla\boldsymbol{Z}_z|_2^2+|\boldsymbol{Z}_z|_2^2|\nabla\boldsymbol{Z}_z|_2^2)|\nabla\boldsymbol{Z}|_2^2+c\|\boldsymbol{Z}\|_2^2.
\end{aligned}$$

By Gronwall Inequality, Lemma 4.2.5, (4.2.44), (4.2.50), (4.2.51), and (4.2.53), we know that for any $\mathcal{T}>t_0$, there exists $C_8(\mathcal{T},\ \|U_{t_0}\|,\ \|Q\|_1,\ \|\boldsymbol{Z}_{t_0}\|_2)$ such that

$$\int_{t_0}^{\mathcal{T}}|\Delta\boldsymbol{w}|_2^2+|\nabla\boldsymbol{w}(t)|_2^2\le C_8,\quad\text{for any}\quad t\in[t_0,\mathcal{T}).\tag{4.2.54}$$

L^2 **estimates of** ∇T, T_z**.** Similarly to Lemma 4.2.13, by (4.2.27), we have

$$\begin{aligned}
&\left|\int_\Omega [(\boldsymbol{w}+\boldsymbol{Z})\cdot\nabla]T(\Delta T+T_{zz})\right|\\
\le&c(|\boldsymbol{w}|_4^8+|\boldsymbol{Z}|_4^8)|\nabla T|_2^2+\varepsilon(|\Delta T|_2^2+|\nabla T_z|_2^2+|T_{zz}|_2^2),\\
&\left|\int_\Omega \Phi(\boldsymbol{w}+\boldsymbol{Z})T_z(\Delta T+T_{zz})\right|\\
\le&\varepsilon(|\Delta T|_2^2+|T_{zz}|_2^2+|\nabla T_z|_2^2)+c[(|\nabla\boldsymbol{w}|_2^4+|\nabla\boldsymbol{w}|_2^2|\Delta\boldsymbol{w}|_2^2)\\
&+c(|\nabla\boldsymbol{Z}|_2^4+|\nabla\boldsymbol{Z}|_2^2|\Delta\boldsymbol{Z}|_2^2)]|T_z|_2^2.
\end{aligned}$$

Taking $L^2(\Omega)$ inner product of equation (4.2.33) with $-(\Delta T+T_{zz})$, using the above two inequalities, applying Hölder and Young Inequalities, and choosing ε small enough, we get

$$\frac{\mathrm{d}(|\nabla T|_2^2+|T_z|_2^2+\alpha_u|T|_{z=0}|_2^2)}{\mathrm{d}t}+|\Delta T|_2^2+|T_{zz}|_2^2+(|\nabla T_z|_2^2+\alpha_u|\nabla T|_{z=0}|_2^2)$$

$$\leq c(|\boldsymbol{w}|_4^8 + |\boldsymbol{Z}|_4^8)|\nabla T|_2^2 + c\left[(|\nabla \boldsymbol{w}|_2^4 + |\nabla \boldsymbol{w}|_2^2|\Delta \boldsymbol{w}|_2^2)\right.$$
$$\left. + c(|\nabla \boldsymbol{Z}|_2^4 + |\nabla \boldsymbol{Z}|_2^2|\Delta \boldsymbol{Z}|_2^2)\right] |T_z|_2^2 + c|Q|_2^2.$$

By Gronwall Inequality, Lemma 4.2.5, (4.2.44), (4.2.50), (4.2.51), (4.2.53) and (4.2.54), we know that for any $\mathcal{T} > t_0$, there exists $C_9(\mathcal{T}, \|U_{t_0}\|, \|Q\|_1, \|\boldsymbol{Z}_{t_0}\|_2)$ such that

$$|\nabla T(t)|_2^2 + |T_z(t)|_2^2 \leq C_9, \quad \text{for any } t \in [t_0, \mathcal{T}). \tag{4.2.55}$$

4.2.4 *The Global Well-Posedness of IBVP*

Proof of Theorem 4.2.2. Step one, **the global existence of strong solutions.** According to Proposition 4.2.12, we prove Theorem 4.2.2 by the method of contradiction. Indeed, let $(\boldsymbol{u}, T)$ be a strong solution of the initial boundary value problem (4.2.19)-(4.2.23) on the maximal interval $[0, \mathcal{T}^*)$, i.e., $(\boldsymbol{w}, T)$ be a strong solution of the initial boundary value problem (4.2.32)-(4.2.36) on the maximal interval $[0, \mathcal{T}^*)$. If $\mathcal{T}^* < +\infty$, then $\limsup_{t\to\mathcal{T}^{*-}}(\|\boldsymbol{w}\| + \|T\|) = +\infty$, which contradicts with (4.2.44), (4.2.46), (4.2.53), (4.2.54) and (4.2.55).

Step two, **the uniqueness of global strong solutions.** Let $(\boldsymbol{w}_1, T_1)$ and $(\boldsymbol{w}_2, T_2)$ be two strong solutions of (4.2.32)-(4.2.36) on $[t_0, \mathcal{T}]$ with p'_{b_3}, p''_{b_3} and initial data $((\boldsymbol{w}_{t_0})_1, (T_{t_0})_1)$, $((\boldsymbol{w}_{t_0})_2, (T_{t_0})_2)$ respectively.

Define $\boldsymbol{w} = \boldsymbol{w}_1 - \boldsymbol{w}_2$, $T = T_1 - T_2$, $p_{b_3} = p'_{b_3} - p''_{b_3}$. Then $\boldsymbol{w}$, T, p_{b_3} satisfy

$$\frac{\partial \boldsymbol{w}}{\partial t} - \Delta \boldsymbol{w} - \frac{\partial^2 \boldsymbol{w}}{\partial z^2} + [(\boldsymbol{w}_1 + \boldsymbol{Z})\cdot\nabla]\,\boldsymbol{w} + (\boldsymbol{w}\cdot\nabla)(\boldsymbol{w}_2 + \boldsymbol{Z}) + \Phi(\boldsymbol{w}_1 + \boldsymbol{Z})\frac{\partial \boldsymbol{w}}{\partial z}$$
$$+ \Phi(\boldsymbol{w})\frac{\partial(\boldsymbol{w}_2 + \boldsymbol{Z})}{\partial z} + f\boldsymbol{k}\times\boldsymbol{w} + \nabla p_{b_3} - \int_{-1}^{z}\nabla T \mathrm{d}z' = 0, \tag{4.2.56}$$

$$\frac{\partial T}{\partial t} - \Delta T - \frac{\partial^2 T}{\partial z^2} + [(\boldsymbol{w}_1 + \boldsymbol{Z})\cdot\nabla]T + (\boldsymbol{w}\cdot\nabla)T_2 + \Phi(\boldsymbol{w}_1 + \boldsymbol{Z})\frac{\partial T}{\partial z}$$
$$+ \Phi(\boldsymbol{w})\frac{\partial T_2}{\partial z} = 0, \tag{4.2.57}$$

$$\int_{-1}^{0}\nabla\cdot\boldsymbol{w}\mathrm{d}z = 0,$$

$$\boldsymbol{w}|_{t=t_0} = (\boldsymbol{w}_{t_0})_1 - (\boldsymbol{w}_{t_0})_2, \ T|_{t=t_0} = (T_{t_0})_1 - (T_{t_0})_2,$$

$(\boldsymbol{w}, T)$ satisfies boundary condition (4.2.12)-(4.2.14).

Taking $L^2(\Omega)\times L^2(\Omega)$ inner product of equation (4.2.56) with $\boldsymbol{w}$, we get

$$\frac{1}{2}\frac{\mathrm{d}|\boldsymbol{w}|_2^2}{\mathrm{d}t} + \int_\Omega |\nabla \boldsymbol{w}|^2 + \int_\Omega |\boldsymbol{w}_z|^2$$

$$= -\int_\Omega \{[(\boldsymbol{w}_1 + \boldsymbol{Z}) \cdot \boldsymbol{\nabla}]\, \boldsymbol{w} + \Phi(\boldsymbol{w}_1 + \boldsymbol{Z})\frac{\partial \boldsymbol{w}}{\partial z}\} \cdot \boldsymbol{w} - \int_\Omega (f\boldsymbol{k} \times \boldsymbol{w} + \boldsymbol{\nabla} p_{b_3}) \cdot \boldsymbol{w}$$
$$- \int_\Omega \left[(\boldsymbol{w} \cdot \boldsymbol{\nabla})(\boldsymbol{w}_2 + \boldsymbol{Z}) + \Phi(\boldsymbol{w})\frac{\partial(\boldsymbol{w}_2 + \boldsymbol{Z})}{\partial z}\right] \cdot \boldsymbol{w} + \int_\Omega (\int_{-1}^{z} \boldsymbol{\nabla} T \mathrm{d}z') \cdot \boldsymbol{w}.$$

By integration by parts, and Lemma 4.2.13, we have

$$\left|\int_\Omega [(\boldsymbol{w} \cdot \boldsymbol{\nabla})(\boldsymbol{w}_2 + \boldsymbol{Z})] \cdot \boldsymbol{w}\right|$$
$$\leq \varepsilon \int_\Omega (|\boldsymbol{\nabla} \boldsymbol{w}|^2 + |\boldsymbol{w}_z|^2) + c(|\boldsymbol{Z}|_4^8 + |\boldsymbol{w}_2|_4^8)|\boldsymbol{w}|_2^2,$$
$$\left|\int_\Omega \Phi(\boldsymbol{w})\frac{\partial(\boldsymbol{w}_2 + \boldsymbol{Z})}{\partial z} \cdot \boldsymbol{w}\right|$$
$$\leq \varepsilon \int_\Omega |\boldsymbol{\nabla} \boldsymbol{w}|^2 + c(|\boldsymbol{w}_{2z}|_2^2|\boldsymbol{\nabla} \boldsymbol{w}_{2z}|_2^2 + |\boldsymbol{w}_{2z}|_2^4 + |\boldsymbol{Z}_z|_2^2|\boldsymbol{\nabla} \boldsymbol{Z}_z|_2^2 + |\boldsymbol{Z}_z|_2^4)|\boldsymbol{w}|_2^2.$$

Thus, by integration by parts, Hölder Inequality and Young Inequality, we obtain

$$\frac{1}{2}\frac{\mathrm{d}|\boldsymbol{w}|_2^2}{\mathrm{d}t} + \int_\Omega |\boldsymbol{\nabla} \boldsymbol{w}|^2 + \int_\Omega |\boldsymbol{w}_z|^2$$
$$\leq 2\varepsilon \int_\Omega (|\boldsymbol{\nabla} \boldsymbol{w}|^2 + |\boldsymbol{w}_z|^2) + \varepsilon|\boldsymbol{\nabla} T|_2^2$$
$$+ c(1 + |\boldsymbol{Z}|_4^8 + |\boldsymbol{w}_2|_4^8 + |\boldsymbol{w}_{2z}|_2^2|\boldsymbol{\nabla} \boldsymbol{w}_{2z}|_2^2 + |\boldsymbol{w}_{2z}|_2^4 + |\boldsymbol{Z}_z|_2^2|\boldsymbol{\nabla} \boldsymbol{Z}_z|_2^2 + |\boldsymbol{Z}_z|_2^4)|\boldsymbol{w}|_2^2.$$

In the same way, we have

$$\frac{1}{2}\frac{\mathrm{d}|T|_2^2}{\mathrm{d}t} + \int_\Omega |\boldsymbol{\nabla} T|^2 + \int_\Omega |T_z|^2 + \alpha_u |T|_{z=0}|_2^2$$
$$\leq \varepsilon \int_\Omega (|\boldsymbol{\nabla} \boldsymbol{w}|^2 + |\boldsymbol{w}_z|^2) + \varepsilon \int_\Omega (|\boldsymbol{\nabla} T|^2 + |T_z|^2)$$
$$+ c(|T_2|_4^8 + |T_{2z}|_2^2|\boldsymbol{\nabla} T_{2z}|_2^2 + |T_{2z}|_2^4)(|\boldsymbol{w}|_2^2 + |T|_2^2).$$

According to the above two inequalities, and choosing ε small enough, we get

$$\frac{\mathrm{d}(|\boldsymbol{w}|_2^2 + |T|_2^2)}{\mathrm{d}t} + \int_\Omega (|\boldsymbol{\nabla} \boldsymbol{w}|^2 + |\boldsymbol{w}_z|^2) + \int_\Omega (|\boldsymbol{\nabla} T|^2 + |T_z|^2) + \alpha_u |T|_{z=0}|_2^2$$
$$\leq c(1 + |T_2|_4^8 + |\boldsymbol{Z}|_4^8 + |\boldsymbol{w}_2|_4^8 + |\boldsymbol{w}_{2z}|_2^2|\boldsymbol{\nabla} \boldsymbol{v}_{2z}|_2^2 + |\boldsymbol{w}_{2z}|_2^4 + |\boldsymbol{Z}_z|_2^2|\boldsymbol{\nabla} \boldsymbol{Z}_z|_2^2$$
$$+ |\boldsymbol{Z}_z|_2^4)|\boldsymbol{w}|_2^2 + c\left[|T_2|_4^8 + (|T_{2z}|_2^2 + 1)|\boldsymbol{\nabla} T_{2z}|_2^2 + |T_{2z}|_2^4 + |T_{2z}|_2^2\right] |T|_2^2. \tag{4.2.58}$$

By Gronwall Inequality, the result of step one and (4.2.58), we prove the uniqueness of global strong solutions.

4.2.5 *Existence of Random Attractors*

We shall use Theorem 4.1.9 in subsection 4.1.3 to prove the existence of the random attractors of the 3D stochastic primitive equations of large-scale oceans. Firstly, we construct a random dynamical system corresponding to the initial boundary value problem (4.2.9)-(4.2.14). Let

$$\boldsymbol{\Omega} = \{\omega;\ \omega \in C(\mathbb{R}, l^2),\ \omega(0) = 0\},$$

$\mathcal{F}$ be a Borel σ-algebra induced by the compact open topology of $\boldsymbol{\Omega}$, and P be Wiener measure on $(\boldsymbol{\Omega}, \mathcal{F})$. Write

$$(\omega_1(t,\omega), \omega_2(t,\omega), \omega_3(t,\omega), \ldots) = \omega(t).$$

Define

$$\vartheta_t \omega(s) = \omega(t+s) - \omega(t). \tag{4.2.59}$$

Then ϑ_t satisfies $\vartheta_{t+s} = \vartheta_t \circ \vartheta_s$, for any $t,\ s \in \mathbb{R}$, and ϑ_t is ergodic under P. According to Theorem 4.2.2, let

$$U(t,\omega) = S(t,s;\omega)U_s,\ (\boldsymbol{w}(t,\omega), T(t,\omega)) = \phi(t,s;\omega)(\boldsymbol{w}_s, T_s),$$

where $\boldsymbol{U}(t,\omega) = (\boldsymbol{v}(t,\omega), T(t,\omega))$ is a strong solution of (4.2.9)-(4.2.15) with the initial data $\boldsymbol{U}(s) = \boldsymbol{U}_s$ on $[s,t]$ and $(\boldsymbol{w}, T)$ is a strong solution to (4.2.32)-(4.2.36) on $[s,t]$ with initial data $[\boldsymbol{w}_s, T_s]$. Thus, for $s \le r \le t$, $S(t,s;\omega) = S(t,r;\omega)S(r,s;\omega)$. Due to (4.2.59), for any $s,t \in \mathbb{R}^+$, $U_0 \in V$, we get P-a.s.

$$S(t+s,0;\omega)\boldsymbol{U}_0 = S(t,0;\vartheta_s\omega)S(s,0;\omega)\boldsymbol{U}_0.$$

Define $\psi : \mathbb{R}^+ \times \boldsymbol{\Omega} \times V \to V$, $\psi(t,\omega)\boldsymbol{U}_0 = S(t,0;\omega)\boldsymbol{U}_0$. According to Theorem 4.2.2 and the following Proposition 4.2.16, we know that ψ is a continuous random dynamical system on V with weak topology over $(\boldsymbol{\Omega}, \mathcal{F}, P, \{\vartheta_t\}_{t\in\mathbb{R}})$, which is corresponding to the random dynamical system generated by (4.2.9)-(4.2.14).

In the process of proving Theorem 4.2.3 by Theorem 4.1.9 in subsection 4.1.3, the key step is to prove the existence of a bounded absorbing set which is compact in V with weak topology. We use the following proposition to prove this conclusion.

Proposition 4.2.15 [the existence of bounded absorbing sets of the random dynamical system (4.2.9)-(4.2.14)]. If $Q \in H^1(\Omega)$, and $B_\rho = \{\boldsymbol{U}; \|\boldsymbol{U}\| \le \rho, U \in V\}$, then there exist $r_0(\omega, \|Q\|_1)$ and $t(\omega,\rho) \le -1$ such that for any $t_0 \le t(\omega,\rho)$, $\boldsymbol{U}_{t_0} \in B_\rho$,

$$\|S(0,t_0;\omega)\boldsymbol{U}_{t_0}\| \le r_0(\omega),$$

that is, for any bounded set $B \subset V$, there exists $-t_0(B) > 0$ big enough, such that

$$\psi(-s, \vartheta_s\omega)B = S(0, s; \omega)B \subset B_{r_0(\omega)}, \quad \forall\ s \le t_0.$$

Proof of Proposition 4.2.15. According to the ergodicity of process $\boldsymbol{Z}$ in $D(A_1^{1+\gamma})$, we get

$$\begin{aligned}&\lim_{s\to-\infty}\frac{1}{-s}\int_s^0(|\boldsymbol{Z}(\tau)|_4^8+\|\boldsymbol{Z}(\tau)\|^4+\|\boldsymbol{Z}(\tau)\|^2\|\boldsymbol{Z}(\tau)\|_2^2)\mathrm{d}\tau\\ =&E(|\boldsymbol{Z}(0)|_4^8+\|\boldsymbol{Z}(0)\|^4+\|\boldsymbol{Z}(0)\|^2\|\boldsymbol{Z}(0)\|_2^2),\end{aligned}$$

cf. [55]. By (4.2.17), we know that $E(|\boldsymbol{Z}(0)|_4^8+\|\boldsymbol{Z}(0)\|^4+\|\boldsymbol{Z}(0)\|^2\|\boldsymbol{Z}(0)\|_2^2) \to 0$ as $\alpha \to +\infty$. Thus, choosing α large enough, we have

$$\lim_{s\to-\infty}\frac{1}{-s}\int_s^0\left[-\lambda_1+c(|\boldsymbol{Z}(\tau)|_4^8+\|\boldsymbol{Z}(\tau)\|^4+\|\boldsymbol{Z}(\tau)\|^2\|\boldsymbol{Z}(\tau)\|_2^2)\right]\mathrm{d}\tau \le -\frac{\lambda_1}{2},$$

which implies that there exists $s_0(\omega)$ such that for $s < s_0(\omega)$,

$$\int_s^0\left[-\lambda_1+c(|\boldsymbol{Z}(\tau)|_4^8+\|\boldsymbol{Z}(\tau)\|^4+\|\boldsymbol{Z}(\tau)\|^2\|\boldsymbol{Z}(\tau)\|_2^2)\right]\mathrm{d}\tau \le -\frac{\lambda_1}{4}(-s).$$

By (4.2.16) and (4.2.44), with a similar argument as in [69], we know that $\|\boldsymbol{Z}(\tau)\|^2 + |T(\tau)|_2^2 + |\boldsymbol{Z}(\tau)|_2^2$ has at most polynomial growth as $\tau \to -\infty$. Thus, (4.2.45) and the above inequality imply that there exist an a.s. finite random variable $R_0(\omega)$ and a $t_0(\rho, \omega)$, such that a.s.

$$|\boldsymbol{w}(t)|_2^2 \le R_0(\omega), \ \forall\ s \le t \le 0, \tag{4.2.60}$$

where $s \le t_0$, $(\boldsymbol{w}, T)$ is a strong solution of (4.2.32)-(4.2.36) with initial data $(\boldsymbol{w}_s, T_s) = (\boldsymbol{v}_s - \boldsymbol{Z}_s, T_s)$ and $(\boldsymbol{v}_s, T_s) \in B_\rho$.

According to (4.2.60), we know that there exists an a.s. finite random variable $R_1(\omega)$ satisfying

$$c_2\int_t^{t+1}\left[\int_\Omega\left(|\nabla\boldsymbol{w}|^2+|\frac{\partial\boldsymbol{w}}{\partial z}|^2+|\boldsymbol{w}|^2\right)\right] \le R_1(\omega). \tag{4.2.61}$$

By the uniform Gronwall Lemma (Lemma 3.2.13), $|\tilde{\boldsymbol{w}}|_3^3 \le |\tilde{\boldsymbol{w}}|_2^{\frac{3}{2}}\|\tilde{\boldsymbol{w}}\|_2^{\frac{3}{2}}$, (4.2.48), (4.2.60), (4.2.61) and Lemma 4.2.5, we get

$$|\tilde{\boldsymbol{w}}(t+1)|_3^3 \le R_2(\omega), \tag{4.2.62}$$

where $R_2(\omega)$ is an a.s. finite random variable, and $t \in [s, -1]$.

Similarly, according to the above inequality and $|\tilde{\boldsymbol{w}}|_4^4 \le |\tilde{\boldsymbol{w}}|_3^2\|\tilde{\boldsymbol{w}}\|^2$, we obtain

$$|\tilde{\boldsymbol{w}}(t+2)|_4^4 \le R_3(\omega), \tag{4.2.63}$$

where $R_3(\omega)$ is an a.s. finite random variable and $t \in [s, -2]$. For $t \in [s, -2]$, we derive from (4.2.62)

$$\int_{t+2}^{t+3} \left[\int_{\Omega} \left(|\nabla \tilde{w}|^2 |\tilde{w}|^2 + \frac{1}{2} |\nabla |\tilde{w}|^2|^2 \right) + \int_{\Omega} \left(|\partial_z \tilde{w}|^2 |\tilde{w}|^2 + \frac{1}{2} |\partial_z |\tilde{w}|^2|^2 \right) \right]$$
$$\leq R_3(\omega)^2 + R_3(\omega) = R_4(\omega).$$

Thus,

$$\|\nabla \tilde{w}(t+3)\|_{L^2}^2 \leq R_5(\omega),$$

where $R_5(\omega)$ is an a.s. finite random variable and $t \in [s, -3]$.

Similarly, there exist a.s. finite limited random variables $R_6(\omega)$, $R_7(\omega)$, $R_8(\omega)$ such that

$$|w_z(t+4)|_2^2 \leq R_6(\omega), \quad t \in [s, -4], \tag{4.2.64}$$
$$|\nabla w(t+5)|_2^2 \leq R_7(\omega), \quad t \in [s, -5], \tag{4.2.65}$$
$$|\nabla T(t+6)|_2^2 + |T_z(t+6)|_2^2 \leq R_8(\omega), \quad t \in [s, -6]. \tag{4.2.66}$$

According to (4.2.43), (4.2.60), (4.2.64)-(4.2.66), we know that there exist $r_0(\omega)$ and $t(\omega, \rho) \leq -1$, such that for any $t_0 \leq t(\omega, \rho)$, $U_{t_0} \in B_\rho$,

$$\|S(0, t_0; \omega) U_{t_0}\| \leq r_0(\omega).$$

In order to prove Theorem 4.2.3, we need the following property of $\{S(t, s; \omega)\}_{t \geq s}$.

Proposition 4.2.16. For any $t \geq s$, the mapping $S(t, s; \omega)$ is weakly continuous on V.

Proof of Proposition 4.2.16. Let $\{U_n\}$ be a sequence in V, and U_n weakly converges to U in V. Then $\{U_n\}$ is bounded in V. According to the *a priori* estimates in subsection 4.2.3 and the proof of Proposition 4.2.15, we know that for any $t \geq s$, $\{S(t, s; \omega) U_n\}$ is bounded in V. So we extract a subsequence $\{S(t, s; \omega) U_{n_k}\}$ such that $S(t, s; \omega) U_{n_k}$ weakly converging to $\mathcal{U}$ in V. Because the imbedding $V \hookrightarrow L^2(\Omega) \times L^2(\Omega) \times L^2(\Omega)$ is compact, U_{n_k} strongly converges to U in $L^2(\Omega) \times L^2(\Omega) \times L^2(\Omega)$. According to (4.2.58), we know that $S(t, s; \omega) U$ strongly converges to $S(t, s; \omega) U$ in $L^2(\Omega) \times L^2(\Omega) \times L^2(\Omega)$. Then, $\mathcal{U} = S(t, s; \omega) U$. Thus, the sequence $\{S(t, s; \omega) U_n\}$ has a subsequence $\{S(t, s; \omega) U_{n_k}\}$ such that $S(t, s; \omega) U_{n_k}$ weakly converges to $S(t, s; \omega) U$ in V. The proposition is proved.

Proof of Theorem 4.2.3. According to Propositions 4.2.15 and 4.2.16, and applying Theorem 4.1.9, we prove Theorem 4.2.3.

4.3 The Primitive Equations of Large-Scale Oceans with Random Boundary

In this section, we study the global well-posedness and the existence of attractor of the three-dimensional viscous primitive equations with random boundary. We prove the global well-posedness of the initial boundary value problem of the stochastic primitive equations. Moreover, by studying the long-time behavior of the strong solutions, we obtain the existence of random attractors of the corresponding random dynamical system. These results will be given in Theorem 4.3.8 and Theorem 4.3.9.

The arrangement of this section is as follows. In subsection 4.3.1, we introduce the random boundary value problem of the three-dimensional primitive equations. The new formulation of the boundary value problem of the primitive equations will be given in subsection 4.3.2, and the main results of this section will also be given here. In subsection 4.3.3, we prove the global existence and uniqueness of the weakly strong solutions of initial boundary value problem of the primitive equations. In the last subsection, we prove the existence of the random attractors of the random dynamical system corresponding to the random boundary value problem of the three-dimensional primitive equations.

4.3.1 *Model*

The non-dimensional form of the three-dimensional primitive equations (including the salinity equation) in a Cartesian coordinate frame is written as

$$\frac{\partial \boldsymbol{v}}{\partial t} + (\boldsymbol{v}\cdot\boldsymbol{\nabla})\boldsymbol{v} + \theta\frac{\partial \boldsymbol{v}}{\partial z} + f\boldsymbol{k}\times\boldsymbol{v} + \boldsymbol{\nabla} p - \frac{1}{Re_1}\Delta\boldsymbol{v} - \frac{1}{Re_2}\frac{\partial^2 \boldsymbol{v}}{\partial z^2} = 0, \tag{4.3.1}$$

$$\frac{\partial p}{\partial z} + \beta_1\rho = 0, \tag{4.3.2}$$

$$\boldsymbol{\nabla}\cdot\boldsymbol{v} + \frac{\partial\theta}{\partial z} = 0, \tag{4.3.3}$$

$$\frac{\partial T}{\partial t} + (\boldsymbol{v}\cdot\boldsymbol{\nabla})T + \theta\frac{\partial T}{\partial z} - \frac{1}{Rt_1}\Delta T - \frac{1}{Rt_2}\frac{\partial^2 T}{\partial z^2} = Q_1, \tag{4.3.4}$$

$$\frac{\partial S}{\partial t} + (\boldsymbol{v}\cdot\boldsymbol{\nabla})S + \theta\frac{\partial S}{\partial z} - \frac{1}{Rs_1}\Delta S - \frac{1}{Rs_2}\frac{\partial^2 S}{\partial z^2} = Q_2, \tag{4.3.5}$$

$$\rho = 1 - \beta_2(T-1) + \beta_3(S-1), \tag{4.3.6}$$

where the unknown functions are $\boldsymbol{v}$, θ, p, T, S, ρ, $\boldsymbol{v} = (v^{(1)}, v^{(2)})$ is the horizontal velocity, θ is the vertical velocity, p is pressure, T is temperature, S is salinity, ρ is density, f is Coriolis parameter, $\beta_1, \beta_2, \beta_3$ are positive constants, k is the vertical unit vector, Re_1, Re_2 are Reynold numbers, Rt_1, Rt_2 are respectively horizontal and vertical heat dissipative coefficients, Rs_1, Rs_2 are respectively horizontal and vertical salinity dissipative coefficients, $Q_1(x,y,z)$, $Q_2(x,y,z)$ are given functions, $\nabla = (\partial_x,\ \partial_y)$, and $\Delta = \partial_x^2 + \partial_y^2$.

The space domain of equations (4.3.1)-(4.3.6) is

$$\Omega = \{(x,y,z):\ (x,y) \in M,\ z \in (-g(x,y),0)\},$$

where M is a smooth bounded domain in $\mathbb{R}^2$, and g is a sufficiently smooth function. Let's assume $g = 1$, then $\Omega = M \times (-1,0)$. The boundary conditions of equations (4.3.1)-(4.3.6) are given by

$$\frac{\partial \boldsymbol{v}}{\partial z} = \tau,\ \theta = 0,\ \frac{\partial T}{\partial z} = -\alpha_u T,\ \frac{\partial S}{\partial z} = 0 \qquad \text{on } M \times \{0\} = \Gamma_u, \tag{4.3.7}$$

$$\frac{\partial \boldsymbol{v}}{\partial z} = 0,\ \theta = 0,\ \frac{\partial T}{\partial z} = 0,\ \frac{\partial S}{\partial z} = 0 \qquad \text{on } M \times \{-1\} = \Gamma_b, \tag{4.3.8}$$

$$\boldsymbol{v} \cdot \boldsymbol{n} = 0,\ \frac{\partial \boldsymbol{v}}{\partial \boldsymbol{n}} \times \boldsymbol{n} = 0,\ \frac{\partial T}{\partial \boldsymbol{n}} = 0,\ \frac{\partial S}{\partial \boldsymbol{n}} = 0 \quad \text{on } \partial M \times [-1,0] = \Gamma_l, \tag{4.3.9}$$

where α_u is a positive constant, and $\boldsymbol{n}$ is the normal vector of Γ_l. **In this section, we assume that the wind stress $\boldsymbol{\tau}$ is random.** The form of $\boldsymbol{\tau}$ is given in the next subsection.

Remark 4.3.1. If the boundary condition $\left.\frac{\partial T}{\partial z}\right|_{\Gamma_u} = -\alpha_u T$ is replaced by $\left.\frac{\partial T}{\partial z}\right|_{\Gamma_u} = -\alpha_u(T - T^*)$, where T^* is sufficiently smooth, and satisfies the compatible condition $\left.\frac{\partial T^*}{\partial \boldsymbol{n}}\right|_{\partial M} = 0$, then the results of this section are still valid.

Let p_b be an unknown function in Γ_b. Since

$$\theta(t,x,y,z) = \Phi(\boldsymbol{v})(t,x,y,z) = -\int_{-1}^{z} \nabla \cdot \boldsymbol{v}(t,x,y,z')\,\mathrm{d}z',$$

we rewrite (4.3.1)-(4.3.6) as

$$\frac{\partial \boldsymbol{v}}{\partial t} + (\boldsymbol{v}\cdot\nabla)\boldsymbol{v} + \Phi(\boldsymbol{v})\frac{\partial \boldsymbol{v}}{\partial z} + f\boldsymbol{k}\times\boldsymbol{v} + \nabla p_b + \int_{-1}^{z} \nabla(\nu_1 T - \nu_2 S)\mathrm{d}z'$$
$$-\frac{1}{Re_1}\Delta\boldsymbol{v} - \frac{1}{Re_2}\frac{\partial^2 \boldsymbol{v}}{\partial z^2} = 0, \tag{4.3.10}$$

$$\frac{\partial T}{\partial t} + (\boldsymbol{v} \cdot \boldsymbol{\nabla})T + \Phi(\boldsymbol{v})\frac{\partial T}{\partial z} - \frac{1}{Rt_1}\Delta T - \frac{1}{Rt_2}\frac{\partial^2 T}{\partial z^2} = Q_1, \tag{4.3.11}$$

$$\frac{\partial S}{\partial t} + (\boldsymbol{v} \cdot \boldsymbol{\nabla})S + \Phi(\boldsymbol{v})\frac{\partial S}{\partial z} - \frac{1}{Rs_1}\Delta S - \frac{1}{Rs_2}\frac{\partial^2 S}{\partial z^2} = Q_2, \tag{4.3.12}$$

$$\int_{-1}^{0} \boldsymbol{\nabla} \cdot \boldsymbol{v} \, \mathrm{d}z = 0, \tag{4.3.13}$$

where $\nu_1 = \beta_1\beta_2$, $\nu_2 = \beta_1\beta_3$. The boundary conditions of equations (4.3.10)-(4.3.13) are given by

$$\frac{\partial \boldsymbol{v}}{\partial z} = \tau, \quad \frac{\partial T}{\partial z} = -\alpha_u T, \quad \frac{\partial S}{\partial z} = 0 \qquad \text{on } \Gamma_u, \tag{4.3.14}$$

$$\frac{\partial \boldsymbol{v}}{\partial z} = 0, \quad \frac{\partial T}{\partial z} = 0, \quad \frac{\partial S}{\partial z} = 0 \qquad \text{on } \Gamma_b, \tag{4.3.15}$$

$$\boldsymbol{v} \cdot \boldsymbol{n} = 0, \quad \frac{\partial \boldsymbol{v}}{\partial \boldsymbol{n}} \times \boldsymbol{n} = 0, \quad \frac{\partial T}{\partial \boldsymbol{n}} = 0, \quad \frac{\partial S}{\partial \boldsymbol{n}} = 0 \qquad \text{on } \Gamma_l, \tag{4.3.16}$$

and the initial condition is

$$(\boldsymbol{v}|_{t=t_0}, T|_{t=t_0}, S|_{t=t_0}) = (\boldsymbol{v}_{t_0}, T_{t_0}, S_{t_0}). \tag{4.3.17}$$

4.3.2 *The New Formulation of the Initial-Boundary Value Problem (4.3.10)-(4.3.17)*

Before giving the new formulation of the initial-boundary value problem (4.3.10)-(4.3.17), we define some function spaces and an auxilliary Ornstein-Uhlenbeck process.

4.3.2.1 *Some Function Spaces*

Let $L^p(\Omega)$ be the usual Lebesgue space with the norm $|\cdot|_p$, $1 \leq p \leq \infty$, and $H^m(\Omega)$ be a Sobolev space with the norm $\|h\|_m = [\int_\Omega (\sum_{1\leq k\leq m} \sum_{i_j=1,2,3;j=1,\cdots,k} |\boldsymbol{\nabla}_{i_1}\cdots\boldsymbol{\nabla}_{i_k}h|^2 + |h|^2)]^{\frac{1}{2}}$, where m is a positive integer, $\boldsymbol{\nabla}_1 = \frac{\partial}{\partial x}$, $\boldsymbol{\nabla}_2 = \frac{\partial}{\partial y}$, $\boldsymbol{\nabla}_3 = \frac{\partial}{\partial z}$. $\int_\Omega \cdot \mathrm{d}\Omega$ and $\int_M \cdot \mathrm{d}M$ are denoted by $\int_\Omega \cdot$, and $\int_M \cdot$ respectively. Let

$$\mathcal{V}_1 := \Big\{ \boldsymbol{w} \in (C^\infty(\Omega))^2; \frac{\partial \boldsymbol{w}}{\partial z}\Big|_{\Gamma_u,\Gamma_b} = 0, \boldsymbol{w} \cdot \boldsymbol{n}|_{\Gamma_l} = 0, \frac{\partial \boldsymbol{w}}{\partial \boldsymbol{n}} \times \boldsymbol{n}\Big|_{\Gamma_l} = 0,$$
$$\int_{-1}^{0} \boldsymbol{\nabla} \cdot \boldsymbol{w} \mathrm{d}z = 0 \Big\},$$

$$\widetilde{\mathcal{V}_2} := \left\{ T \in C^\infty(\Omega);\ \frac{\partial T}{\partial z}\bigg|_{\Gamma_u} = -\alpha_u T,\ \frac{\partial T}{\partial z}\bigg|_{\Gamma_b} = 0,\ \frac{\partial T}{\partial \boldsymbol{n}}\bigg|_{\Gamma_l} = 0 \right\},$$

$$\widetilde{\mathcal{V}_3} := \left\{ S \in C^\infty(\Omega);\ \frac{\partial S}{\partial z}\bigg|_{\Gamma_u} = 0,\ \frac{\partial S}{\partial z}\bigg|_{\Gamma_b} = 0,\ \frac{\partial S}{\partial \boldsymbol{n}}\bigg|_{\Gamma_l} = 0,\ \int_\Omega S = 0 \right\},$$

V_1=the closure of $\widetilde{\mathcal{V}_1}$ with respect to norm $\|\cdot\|_1$,
V_2=the closure of $\widetilde{\mathcal{V}_2}$ with respect to norm $\|\cdot\|_1$,
V_3=the closure of $\widetilde{\mathcal{V}_3}$ with respect to norm $\|\cdot\|_1$,
H_1=the closure of $\widetilde{\mathcal{V}_1}$ with respect to norm $|\cdot|_2$,
$V = V_1 \times V_2 \times V_3$, $H = H_1 \times L^2(\Omega) \times \dot{L}^2(\Omega)$,
where $\dot{L}^2(\Omega) = \{S; S \in L^2(\Omega),\ \int_\Omega S = 0\}$. The inner products and norms of spaces V, H are given by

$$(\mathcal{U},\mathcal{U}_1)_V = (\boldsymbol{w},\boldsymbol{w}_1)_{V_1} + (T,T_1)_{V_2} + (S,S_1)_{V_3},$$

$$(\mathcal{U},\mathcal{U}_1) = (w^{(1)},(w_1)^{(1)}) + (w^{(2)},(w_1)^{(2)}) + (T,T_1) + (S,S_1),$$

$$\|\mathcal{U}\| = (\boldsymbol{w},\boldsymbol{w})_{V_1}^{\frac{1}{2}} + (T,T)_{V_2}^{\frac{1}{2}} + (S,S)_{V_3}^{\frac{1}{2}} = \|\boldsymbol{w}\| + \|T\| + \|S\|,\ |\mathcal{U}|_2 = (\mathcal{U},\mathcal{U})^{\frac{1}{2}},$$

where $\mathcal{U} = (\boldsymbol{w},T,S)$, $\mathcal{U}_1 = (\boldsymbol{w}_1,T_1,S_1) \in V$, and $(\cdot,\cdot)$ is the inner product in $L^2(\Omega)$.

4.3.2.2 *An Auxilliary Ornstein-Uhlenbeck Process*

Firstly, we define the functionals $a : V \times V \to \mathbb{R}$, $a_1 : V_1 \times V_1 \to \mathbb{R}$, $a_2 : V_2 \times V_2 \to \mathbb{R}$, $a_3 : V_3 \times V_3 \to \mathbb{R}$, and their corresponding linear operators $A : V \to V'$, $A_1 : V_1 \to V_1'$, $A_2 : V_2 \to V_2'$, $A_3 : V_3 \to V_3'$ by

$$a(\mathcal{U},\mathcal{U}_1) = (A\mathcal{U},\mathcal{U}_1) = a_1(\boldsymbol{w},\boldsymbol{w}_1) + a_2(T,T_1) + a_3(S,S_1),$$

where

$$a_1(\boldsymbol{w},\boldsymbol{w}_1) = (A_1\boldsymbol{w},\boldsymbol{w}_1) = \int_\Omega \left(\frac{1}{Re_1}\nabla\boldsymbol{w}\cdot\nabla\boldsymbol{w}_1 + \frac{1}{Re_2}\frac{\partial \boldsymbol{w}}{\partial z}\cdot\frac{\partial \boldsymbol{w}_1}{\partial z} \right),$$

$$a_2(T,T_1) = (A_2T,T_1) = \int_\Omega \left(\frac{1}{Rt_1}\nabla T\cdot\nabla T_1 + \frac{1}{Rt_2}\frac{\partial T}{\partial z}\frac{\partial T_1}{\partial z} \right) + \frac{\alpha_u}{Rt_2}\int_{\Gamma_u} TT_1,$$

$$a_3(S,S_1) = (A_3S,S_1) = \int_\Omega \left(\frac{1}{Rs_1}\nabla S\cdot\nabla S_1 + \frac{1}{Rs_2}\frac{\partial S}{\partial z}\frac{\partial S_1}{\partial z} \right).$$

Lemma 4.3.2.

(i) a is coercive and continuous, and $A : V \to V'$ is isomorphism. Moreover,

$$a(\mathcal{U},\mathcal{U}_1) \leq c\|\boldsymbol{w}\|\|\boldsymbol{w}_1\| + c\|T\|\|T_1\| + c\|S\|\|S_1\| \leq c\|U\|\|U_1\|,$$
$$a(\mathcal{U},\mathcal{U} \geq c\|\boldsymbol{w}\|^2 + c\|T\|^2 + c\|S\|^2 \geq c\|U\|^2. \tag{4.3.18}$$

(ii) $A : V \to V'$ can be extended to a self-adjoint unbounded linear operator on H with a compact inverse operator $A^{-1} : H \to H$ and with $D(A) = V \cap [(H^2(\Omega))^2 \times H^2(\Omega)\times\, H^2(\Omega)]$.

Proof of Lemma 4.3.2. By $\|\boldsymbol{w}\|_{L^2}^2 \leq C_M\|\boldsymbol{\nabla w}\|_{L^2}^2$ (cf. [76, p. 55]) and Poincaré Inequality, we prove (4.3.18). Since the operator A is similar to the usual positive symmetric Laplacian operator in H_0^1, the other part of Lemma 4.3.2 can be proven with the usual argument. Here, we omit the details of the proof. For more details, the reader can refer to [140, Lemma 2.4].

Denote by $0 < \lambda_1 < \lambda_2 \leq \cdots$ the eigenvalues of A_1, and by $\boldsymbol{e}_1, \boldsymbol{e}_2, \cdots$ corresponding complete orthogonal system of eigenvectors. For any $\boldsymbol{w} \in V_1$, $\|\boldsymbol{w}\|^2 \geq \lambda_1|\boldsymbol{w}|_2^2$. Denote by $\mathrm{e}^{t(-A_1)}$, $t \geq 0$ the semigroup on H_1 generated by $-A_1$.

In this section, we suppose that the random wind stress $\tau(t, x, y)$ is an additive white noise in time with the form

$$\boldsymbol{\tau}(t, x, y) = G^{\frac{1}{2}}\frac{\partial \boldsymbol{W}}{\partial t}, \tag{4.3.19}$$

where the derivative is in the Itô integration sense, the random process W is a two-sided time cylindrical Wiener process in $\{\boldsymbol{u}; \boldsymbol{u} \in L^2(M), \boldsymbol{u} \cdot \boldsymbol{n}|_{\partial M} = 0, \dfrac{\partial \boldsymbol{u}}{\partial \boldsymbol{n}} \times \boldsymbol{n}|_{\partial M} = 0\} = \mathrm{Span}\{\boldsymbol{f}_i\}$ with the form $\boldsymbol{W}(t) = \sum_{i=1}^{+\infty} \omega_i(t, \omega)\boldsymbol{f}_i$, and G is the non-negative bounded operator on $L^2(M)$, with $\sum_{i=1}^{+\infty} \lambda_i^{2+2\gamma_0+1}|G^{\frac{1}{2}}\mathcal{D}^*\mathrm{e}_i|_2^2 < +\infty$. The definition of Wiener process appears in [54]. Here $\omega_1, \omega_2, \cdots$ is a sequence of independent Brownian motions on complete probability space $(\boldsymbol{\Omega}, \mathcal{F}, P)$ with expectation E, and $\mathcal{D}^*$ is the dual operator of $\mathcal{D}$ given later.

For any $\boldsymbol{u} \in L^2(M)$, $\boldsymbol{u} \cdot \boldsymbol{n}|_{\partial M} = 0$, $\dfrac{\partial \boldsymbol{u}}{\partial \boldsymbol{n}} \times \boldsymbol{n}|_{\partial M} = 0$, let μ be a real number such that the elliptic boundary value problem

$$\begin{cases} -\frac{1}{Re_1}\Delta \boldsymbol{Y}_1 - \frac{1}{Re_2}\frac{\partial^2 \boldsymbol{Y}_1}{\partial z^2} + \boldsymbol{\nabla} p_b = \mu \boldsymbol{Y}_1, \\ \frac{\partial \boldsymbol{Y}_1}{\partial z}|_{\Gamma_u} = \boldsymbol{u}, \ \frac{\partial \boldsymbol{Y}_1}{\partial z}|_{\Gamma_b} = 0, \ \boldsymbol{Y}_1 \cdot \boldsymbol{n}|_{\Gamma_l} = 0, \ \frac{\partial \boldsymbol{Y}_1}{\partial \boldsymbol{n}} \times \boldsymbol{n}|_{\Gamma_l} = 0, \\ \int_{-1}^{0} \boldsymbol{\nabla} \cdot \boldsymbol{Y}_1 \mathrm{d}z = 0, \end{cases}$$

has a unique solution $\boldsymbol{Y}_1 = \mathcal{D}\boldsymbol{u}$. Now let's define an auxilliary Ornstein-Uhlenbeck process. Using similar arguments as in [55] or [56], we know that the process

$$\boldsymbol{Z}(t) = (-\mu + A_1) \int_{-\infty}^{t} \mathrm{e}^{(t-s)(-A_1)} \mathcal{D} G^{\frac{1}{2}} \mathrm{d}W(s) \tag{4.3.20}$$

is the unique Markovian generalized solution of the following problem,

$$\begin{cases} \frac{\partial \boldsymbol{Z}}{\partial t} = \frac{1}{Re_1}\Delta \boldsymbol{Z} + \frac{1}{Re_2}\frac{\partial^2 \boldsymbol{Z}}{\partial z^2} - \boldsymbol{\nabla} p_{b_1}, \\ \frac{\partial \boldsymbol{Z}}{\partial z}|_{\Gamma_u} = G^{\frac{1}{2}}\frac{\partial W}{\partial t}, \ \frac{\partial \boldsymbol{Z}}{\partial z}|_{\Gamma_b} = 0, \ \boldsymbol{Z} \cdot \boldsymbol{n}|_{\Gamma_l} = 0, \ \frac{\partial \boldsymbol{Z}}{\partial \boldsymbol{n}} \times \boldsymbol{n}|_{\Gamma_l} = 0, \\ \int_{-1}^{0} \boldsymbol{\nabla} \cdot \boldsymbol{Z} \mathrm{d}z = 0, \\ \boldsymbol{Z}(0) = (-\mu + A_1) \int_{-\infty}^{0} \mathrm{e}^{-s(-A_1)} \mathcal{D} G^{\frac{1}{2}} \mathrm{d}W(s). \end{cases}$$

According to the above definition, we have the following lemma about the properties of process $\boldsymbol{Z}(t)$.

Lemma 4.3.3 (cf. [53, Proposition 13.2.4]). If $\boldsymbol{Z}(t)$ is an Ornstein-Uhlenbeck process defined as above, then $\boldsymbol{Z}(t)$ is a stationary ergodic process with continuous trajectories, $\boldsymbol{Z}(t) \in H^{2+2\gamma}(\Omega) \times H^{2+2\gamma}(\Omega)$, for any $\gamma < \gamma_0$.

In order to obtain the existence of random attractors of the dynamical system corresponding to the random boundary value problem (4.3.21)-(4.3.27), we suppose that $\boldsymbol{Z}$ satisfies the following condition

$$-\lambda_1 + cE(|\boldsymbol{Z}(0)|_4^8 + \|\boldsymbol{Z}(0)\|^4 + \|\boldsymbol{Z}(0)\|^2\|\boldsymbol{Z}(0)\|_2^2) < 0. \tag{C}$$

In the following two examples, if the coefficients δ_i and μ_i are small enough, then $\boldsymbol{Z}$ satisfies the condition (C).

Now we give two examples of $\boldsymbol{\tau}$.

Remark 4.3.4. One example of wind stress is $\boldsymbol{\tau} = \frac{\partial \boldsymbol{W}}{\partial t}$, where $\boldsymbol{W}$ is a two-sided finite-dimensional Brownian motion with the form

$$\boldsymbol{W} = \sum_{i=1}^{m} \delta_i \omega_i(t,\omega) \boldsymbol{f}_i.$$

In the above formula, $\omega_1, \cdots, \omega_m$ are independent standard one-dimensional Brownian motion on a complete probability space $(\boldsymbol{\Omega}, \mathcal{F}, P)$ with expectation denoted by E, and δ_i are real coefficients. For this example, $\boldsymbol{Z}$ has a stationary ergodic process with continuous trajectories, which takes values in $H^k(\Omega)$ for any $k \in \mathbb{N}$.

Remark 4.3.5. Another example of $\boldsymbol{\tau}$ is $\boldsymbol{\tau} = \frac{\partial \boldsymbol{W}}{\partial t}$, where $\boldsymbol{W}$ is a two-sided infinite-dimensional Brownian motion with the form

$$\boldsymbol{W}(t) = \sum_{i=1}^{+\infty} \mu_i \omega_i(t,\omega) \boldsymbol{f}_i.$$

Here $\omega_1, \omega_2, \cdots$ is a sequence of independent standard one-dimensional Brownian motions on a complete probability space $(\boldsymbol{\Omega}, \mathcal{F}, P)$ (with expectation denoted by E), and the coefficients μ_i satisfy $\sum_{i=1}^{+\infty} \lambda_i^{3+2\gamma_0} \mu_i^2 |\mathcal{D}^* \mathbf{e}_i|_2^2 < +\infty$, for some $\gamma_0 > 0$. For this example,

$$\begin{aligned}
E\|\boldsymbol{Z}(t)\|_{2+2\gamma}^2 &\le CE \sum_{i=1}^{+\infty} \left| \int_{-\infty}^{t} \lambda_i^{2+\gamma} \mathrm{e}^{(t-s)(-\lambda_i)} \mu_i \mathrm{d}\omega_i \right|^2 |\mathcal{D}^* \boldsymbol{e}_i|_2^2 \\
&= \sum_{i=1}^{+\infty} \int_{-\infty}^{t} \lambda_i^{4+2\gamma} \mathrm{e}^{2(t-s)(-\lambda_i)} \mu_i^2 \mathrm{d}s |\mathcal{D}^* \boldsymbol{e}_i|_2^2 \\
&= \sum_{i=1}^{+\infty} \frac{\lambda_i^{3+2\gamma} \mu_i^2}{2} |\mathcal{D}^* \boldsymbol{e}_i|_2^2 < +\infty, \quad \text{for any} \quad \gamma < \gamma_0,
\end{aligned}$$

where $C > 0$.

4.3.2.3 *The New Formulation of (4.3.10)-(4.3.17)*

Let $\boldsymbol{w}(t) = \boldsymbol{v}(t) - \boldsymbol{Z}(t)$. By the definition of the process $\boldsymbol{Z}$ and (4.3.10)-(4.3.17), we get

$$\begin{aligned}
&\frac{\partial \boldsymbol{w}}{\partial t} + [(\boldsymbol{w}+\boldsymbol{Z}) \cdot \boldsymbol{\nabla}](\boldsymbol{w}+\boldsymbol{Z}) + \Phi(\boldsymbol{w}+\boldsymbol{Z}) \frac{\partial(\boldsymbol{w}+\boldsymbol{Z})}{\partial z} + f\boldsymbol{k} \times (\boldsymbol{w}+\boldsymbol{Z}) \\
&+ \boldsymbol{\nabla} p_{b_2} + \int_{-1}^{z} \boldsymbol{\nabla}(\nu_1 T - \nu_2 S) \mathrm{d}z' - \frac{1}{Re_1} \Delta \boldsymbol{w} - \frac{1}{Re_2} \frac{\partial^2 \boldsymbol{w}}{\partial z^2} = 0,
\end{aligned} \tag{4.3.21}$$

$$\frac{\partial T}{\partial t} + [(\boldsymbol{w} + \boldsymbol{Z}) \cdot \boldsymbol{\nabla}] T + \Phi(\boldsymbol{w} + \boldsymbol{Z}) \frac{\partial T}{\partial z} - \frac{1}{Rt_1} \Delta T - \frac{1}{Rt_2} \frac{\partial^2 T}{\partial z^2} = Q_1, \tag{4.3.22}$$

$$\frac{\partial S}{\partial t} + [(\boldsymbol{w} + \boldsymbol{Z}) \cdot \boldsymbol{\nabla}] S + \Phi(\boldsymbol{w} + \boldsymbol{Z}) \frac{\partial S}{\partial z} - \frac{1}{Rs_1} \Delta S - \frac{1}{Rs_2} \frac{\partial^2 S}{\partial z^2} = Q_2, \tag{4.3.23}$$

$$\int_{-1}^{0} \boldsymbol{\nabla} \cdot \boldsymbol{w} \, \mathrm{d}z = 0, \tag{4.3.24}$$

$$\frac{\partial \boldsymbol{w}}{\partial z} = 0, \ \frac{\partial T}{\partial z} = -\alpha_u T, \ \frac{\partial S}{\partial z} = 0 \qquad \text{on } \Gamma_u, \tag{4.3.25}$$

$$\frac{\partial \boldsymbol{w}}{\partial z} = 0, \ \frac{\partial T}{\partial z} = 0, \ \frac{\partial S}{\partial z} = 0 \qquad \text{on } \Gamma_b, \tag{4.3.26}$$

$$\boldsymbol{w} \cdot \boldsymbol{n} = 0, \ \frac{\partial \boldsymbol{w}}{\partial \boldsymbol{n}} \times \boldsymbol{n} = 0, \ \frac{\partial T}{\partial \boldsymbol{n}} = 0, \ \frac{\partial S}{\partial \boldsymbol{n}} = 0 \qquad \text{on } \Gamma_l, \tag{4.3.27}$$

$$(\boldsymbol{w}|_{t=t_0}, T|_{t=t_0}, S|_{t=t_0}) = (\boldsymbol{w}_{t_0}, T_{t_0}, S_{t_0}) = (\boldsymbol{v}_{t_0} - \boldsymbol{Z}_{t_0}, T_{t_0}, S_{t_0}). \tag{4.3.28}$$

Before defining the weakly strong solution of (4.3.21)-(4.3.28), we define baroclinic flow $\tilde{\boldsymbol{w}}$, and find the equation satisfied by $\tilde{\boldsymbol{w}}$. For any $\boldsymbol{w} \in V_1$, we denote baroclinic flow by $\tilde{\boldsymbol{w}} = \boldsymbol{w} - \bar{\boldsymbol{w}}$, where $\bar{\boldsymbol{w}} = \int_{-1}^{0} \boldsymbol{w} \mathrm{d}z$. Then $\tilde{\boldsymbol{w}}$ satisfies

$$\begin{aligned} & \frac{\partial \tilde{\boldsymbol{w}}}{\partial t} - \frac{1}{Re_1} \Delta \tilde{\boldsymbol{w}} - \frac{1}{Re_2} \frac{\partial^2 \tilde{\boldsymbol{w}}}{\partial z^2} + \left[(\tilde{\boldsymbol{w}} + \tilde{\boldsymbol{Z}}) \cdot \boldsymbol{\nabla}\right] (\tilde{\boldsymbol{w}} + \tilde{\boldsymbol{Z}}) \\ & + \Phi(\tilde{\boldsymbol{w}} + \tilde{\boldsymbol{Z}}) \frac{\partial (\tilde{\boldsymbol{w}} + \tilde{\boldsymbol{Z}})}{\partial z} + \left[(\tilde{\boldsymbol{w}} + \tilde{\boldsymbol{Z}}) \cdot \boldsymbol{\nabla}\right] (\bar{\boldsymbol{w}} + \bar{\boldsymbol{Z}}) + \left[(\bar{\boldsymbol{w}} + \bar{\boldsymbol{Z}}) \cdot \boldsymbol{\nabla}\right] (\tilde{\boldsymbol{w}} + \tilde{\boldsymbol{Z}}) \\ & + f\boldsymbol{k} \times (\tilde{\boldsymbol{w}} + \tilde{\boldsymbol{Z}}) - \overline{(\tilde{\boldsymbol{w}} + \tilde{\boldsymbol{Z}}) \boldsymbol{\nabla} \cdot (\tilde{\boldsymbol{w}} + \tilde{\boldsymbol{Z}}) + \left[(\tilde{\boldsymbol{w}} + \tilde{\boldsymbol{Z}}) \cdot \boldsymbol{\nabla}\right] (\tilde{\boldsymbol{w}} + \tilde{\boldsymbol{Z}})} \\ & + \int_{-1}^{z} \boldsymbol{\nabla}(\nu_1 T - \nu_2 S) \mathrm{d}z' - \int_{-1}^{0} \int_{-1}^{z} \boldsymbol{\nabla}(\nu_1 T - \nu_2 S) \mathrm{d}z' \mathrm{d}z = 0 \ \text{ in } \Omega, \end{aligned} \tag{4.3.29}$$

$$\frac{\partial \tilde{\boldsymbol{w}}}{\partial z} = 0 \text{ on } \Gamma_u, \ \frac{\partial \tilde{\boldsymbol{w}}}{\partial z} = 0 \text{ on } \Gamma_b, \ \tilde{\boldsymbol{w}} \cdot \boldsymbol{n} = 0, \ \frac{\partial \tilde{\boldsymbol{w}}}{\partial \boldsymbol{n}} \times \boldsymbol{n} = 0 \text{ on } \Gamma_l. \tag{4.3.30}$$

Definition 4.3.6. Let $\boldsymbol{Z}$ be the process defined as above, $\boldsymbol{\mathcal{U}}_{t_0} = (\boldsymbol{w}_{t_0}, T_{t_0}, S_{t_0})$ satisfies $\boldsymbol{\mathcal{U}}_{t_0} \in H$, $\tilde{\boldsymbol{w}}_{t_0} \in (L^4(\Omega))^2$, $T_{t_0}, S_{t_0} \in L^4(\Omega)$, $\partial_z \boldsymbol{w}_{t_0} \in (L^2(\Omega))^2$, $\partial_z T_{t_0}, \partial_z S_{t_0} \in L^2(\Omega)$, $\mathcal{T} > t_0$ is the fixed time. For P-a.e. $\omega \in \boldsymbol{\Omega}$, $(\boldsymbol{w}, T, S)$ is called a weakly strong solution of (4.3.21)-(4.3.28) on $[t_0, \mathcal{T}]$, if it satisfies (4.3.21)-(4.3.24) in weak sense such that

$$\boldsymbol{w} \in L^2(t_0, \mathcal{T}; V_1) \cap L^\infty(t_0, \mathcal{T}; H_1), \quad \tilde{\boldsymbol{w}} \in L^\infty(t_0, \mathcal{T}; (L^4(\Omega))^2),$$

$\partial_z \boldsymbol{w} \in L^\infty(t_0, \mathcal{T}; (L^2(\Omega))^2) \cap L^2(t_0, \mathcal{T}; (H^1(\Omega))^2),$
$T \in L^\infty(t_0, \mathcal{T}; L^4(\Omega)) \cap L^2(0, \mathcal{T}; V_2), S \in L^\infty(t_0, \mathcal{T}; L^4(\Omega)) \cap L^2(t_0, \mathcal{T}; V_3),$
$\partial_z T \in L^\infty(t_0, \mathcal{T}; L^2(\Omega)) \cap L^2(t_0, \mathcal{T}; H^1(\Omega)),$
$\partial_z S \in L^\infty(t_0, \mathcal{T}; L^2(\Omega)) \cap L^2(t_0, \mathcal{T}; H^1(\Omega)),$
$\frac{\partial \boldsymbol{v}}{\partial t} \in L^2(t_0, V_1'), \ \frac{\partial T}{\partial t} \in L^2(t_0, \mathcal{T}; V_2'), \ \frac{\partial S}{\partial t} \in L^2(t_0, \mathcal{T}; V_3'),$
where V_i' is the dual space of V_i, $i = 1, 2, 3$.

Remark 4.3.7. For almost all the given paths of process $\boldsymbol{Z}(t)$, we can study (4.3.21)-(4.3.24) as deterministic evolution equations.

Theorem 4.3.8 [Global well-posedness of (4.3.21)-(4.3.28)]. If Q_1, $Q_2 \in H^1(\Omega)$, and $\boldsymbol{\mathcal{U}}_{t_0} = (\boldsymbol{w}_{t_0}, T_{t_0}, S_{t_0})$ satisfies: $\boldsymbol{\mathcal{U}}_{t_0} \in H$, $\tilde{\boldsymbol{w}}_{t_0} \in (L^4(\Omega))^2$, $T_{t_0}, S_{t_0} \in L^4(\Omega)$, $\partial_z \boldsymbol{w}_{t_0} \in (L^2(\Omega))^2$, $\partial_z T_{t_0}, \partial_z S_{t_0} \in L^2(\Omega)$, then, for any $\mathcal{T} > t_0$ given, the initial boundary value problem (4.3.21)-(4.3.28) has a unique weakly strong solution U on $[t_0, \mathcal{T}]$, and U is dependent continuously on the initial data.

Theorem 4.3.9 (Existence of random attractors). If the ancillary Ornstein-Uhlenbeck process Z satisfies condition (C), then the random dynamical system corresponding to (4.3.21)-(4.3.27) possesses a unique random attractor $\mathcal{A}(\omega)$, which captures all the trajectories started at time $-\infty$, and evolved, under the action of the shift ϑ_t, from $-\infty$ to the present time $t = 0$. The random attractor $\mathcal{A}(\omega)$ has the following properties:

(i) (weakly compact) $\mathcal{A}(\omega)$ is bounded and weak closed in V;

(ii) (invariant) for any $t \geq 0$, $\psi(t, \omega)\mathcal{A}(\omega) = \mathcal{A}(\vartheta_t \omega)$;

(iii) (attracting) for any deterministic bounded set B in V, the set $\psi(t, \vartheta_{-t}\omega)B$ converge to $\mathcal{A}(\vartheta_t \omega)$ with respect to the weak topology of V as $t \to +\infty$, that is,

$$\lim_{t \to +\infty} \mathrm{d}_V^{\boldsymbol{w}}(\psi(t, \vartheta_{-t}\omega)B, \mathcal{A}(\omega)) = 0, P\text{-a.s.}$$

where the definitions of random attractor, $\vartheta_t, t \in \mathbb{R}$, and $\psi(t, \omega), t \geq 0$ will be given in subsection 4.3.4, and distance $\mathrm{d}_V^{\boldsymbol{w}}$ is induced by the weak topology of V.

4.3.3 *The Global Well-Posedness of the Primitive Equations with Random Boundary*

4.3.3.1 *Global Existence of Weakly Strong Solutions*

We shall use Faedo-Galerkin method to prove the global existence of the weakly strong solutions of (4.3.21)-(4.3.28). Here we only give the *a priori* estimates.

Befor making *a priori* estimates of the weakly strong solutions of (4.3.21)-(4.3.28), we give a lemma used frequently later.

Lemma 4.3.10. If $\boldsymbol{v}_1 \in H^2(\Omega) \times H^2(\Omega)$, $\boldsymbol{v}_2 \in H^2(\Omega) \times H^2(\Omega)$ or $H^2(\Omega)$, and $\boldsymbol{v}_3 \in L^2(\Omega) \times L^2(\Omega)$ or $L^2(\Omega)$, then

$$\text{(i)} \qquad \left| \int_\Omega \boldsymbol{v}_3 \cdot (\boldsymbol{v}_1 \cdot \nabla) \boldsymbol{v}_2 \right|$$

$$\leq c(|\boldsymbol{v}_1|_4^2 + |\boldsymbol{v}_1|_4^8) |\nabla \boldsymbol{v}_2|_2^2 + \varepsilon \left[|\boldsymbol{v}_3|_2^2 + \int_\Omega (|\nabla \boldsymbol{v}_{2z}|^2 + |\Delta \boldsymbol{v}_2|^2) \right],$$

$$\text{(ii)} \qquad \left| \int_\Omega \Phi(\boldsymbol{v}_1) \boldsymbol{v}_{2z} \cdot \boldsymbol{v}_3 \right|$$

$$\leq c |\nabla \boldsymbol{v}_1|_2^{\frac{1}{2}} (|\nabla \boldsymbol{v}_1|_2^2 + |\Delta \boldsymbol{v}_1|_2^2)^{\frac{1}{4}} |\boldsymbol{v}_{2z}|_2^{\frac{1}{2}} |\nabla \boldsymbol{v}_{2z}|_2^{\frac{1}{2}} |\boldsymbol{v}_3|_2,$$

where ε is a small enough positive constant.

Proof. Appying Hölder Inequality, (4.2.27) and Young Inequality, we have

$$\int_\Omega |\boldsymbol{v}_1| |\nabla \boldsymbol{v}_2| |\boldsymbol{v}_3|$$

$$\leq c |\boldsymbol{v}_1|_4^2 \left(\int_\Omega |\nabla \boldsymbol{v}_2|^4 \right)^{\frac{1}{2}} + \varepsilon |\boldsymbol{v}_3|_2^2$$

$$\leq c(|\boldsymbol{v}_1|_4^2 + |\boldsymbol{v}_1|_4^8) |\nabla \boldsymbol{v}_2|_2^2 + \varepsilon \left[|\boldsymbol{v}_3|_2^2 + \int_\Omega (|\nabla \boldsymbol{v}_{2z}|^2 + |\Delta \boldsymbol{v}_2|^2) \right].$$

By Hölder Inequality, Minkowski Inequality and (4.2.24), we get

$$\int_M \left(\int_{-1}^0 |\nabla \boldsymbol{v}_1| \int_{-1}^0 |\boldsymbol{v}_{2z}| |\boldsymbol{v}_3| \right)$$

$$\leq \int_M \left[\left(\int_{-1}^0 |\nabla \boldsymbol{v}_1|^2 \right)^{\frac{1}{2}} \left(\int_{-1}^0 |\boldsymbol{v}_{2z}|^2 \right)^{\frac{1}{2}} \left(\int_{-1}^0 |\boldsymbol{v}_3|^2 \right)^{\frac{1}{2}} \right]$$

$$\leq c \left[\int_{-1}^0 (|\nabla \boldsymbol{v}_1|_{L^2} |\nabla \boldsymbol{v}_1|_{H^1}) \int_{-1}^0 (|\boldsymbol{v}_{2z}|_{L^2} |\boldsymbol{v}_{2z}|_{H^1}) \right]^{\frac{1}{2}} |\boldsymbol{v}_3|_2.$$

Now we make *a priori* estimates of weakly strong solutions of (4.3.21)-(4.3.28). Let $\omega \in \boldsymbol{\Omega}$ be fixed.

L^2 estimates of T, S, w. Taking $L^2(\Omega)$ inner product of equation (4.3.22) with T, then by integration by parts, $T(t,x,y,z) = -\int_z^0 \frac{\partial T}{\partial z'} \mathrm{d}z' + T|_{z=0}$, Hölder Inequality and Cauchy-Schwarz Inequality, we have

$$\frac{\mathrm{d}|T|_2^2}{\mathrm{d}t} + \frac{1}{Rt_1} \int_\Omega |\nabla T|^2 + \frac{1}{Rt_2} \int_\Omega \left| \frac{\partial T}{\partial z} \right|^2 + \frac{\alpha_u}{Rt_2} |T|_{z=0}|_2^2 \leq c |Q_1|_2^2.$$

By Gronwall Inequality, we derive from the above inequality

$$|T(t)|_2^2 \le \mathrm{e}^{-ct}|T_{t_0}|_2^2 + c|Q_1|_2^2, \tag{4.3.31}$$

where $t \ge t_0$. According to the above two inequalities, for any $\mathcal{T} > t_0$, there exists $C_1(\mathcal{T},\ |T_{t_0}|_2,\ |Q_1|_2) > 0$ such that

$$\int_{t_0}^{\mathcal{T}} \left[\int_\Omega \left(|\nabla T|^2 + \left|\frac{\partial T}{\partial z}\right|^2 + |T|^2\right) + |T|_{z=0}|_2^2\right] + |T(t)|_2^2 \le C_1, \tag{4.3.32}$$

where $t \in [t_0, \mathcal{T})$, and $\int_{t_0}^{\mathcal{T}} \cdot$ is denoted by $\int_{t_0}^{\mathcal{T}} \cdot \mathrm{d}s$.

Taking $L^2(\Omega)$ inner product of equation (4.3.23) with S, by integration by parts, Poincaré Inequality and Gronwall Inequality, we obtain

$$|S(t)|_2^2 \le \mathrm{e}^{-ct}|S_{t_0}|_2^2 + c|Q_2|_2^2, \tag{4.3.33}$$

where $t \ge t_0$. Moreover, for any $\mathcal{T} > t_0$ given, there exists $C_2(\mathcal{T}, |T_{t_0}|_2, |Q_2|_2) > 0$ such that

$$\int_{t_0}^{\mathcal{T}} \int_\Omega \left(|\nabla S|^2 + \left|\frac{\partial S}{\partial z}\right|^2 + |S|^2\right) + |S(t)|_2^2 \le C_2, \tag{4.3.34}$$

where $t \in [t_0, \mathcal{T})$.

Taking $L^2(\Omega) \times L^2(\Omega)$ inner product of equation (4.3.21) with $\boldsymbol{w}$, then by Lemma 4.3.10, Hölder Inequality and Young Inequality, we get

$$\begin{aligned} &\frac{\mathrm{d}|\boldsymbol{w}|_2^2}{\mathrm{d}t} + \frac{1}{Re_1}\int_\Omega |\nabla \boldsymbol{w}|^2 + \frac{1}{Re_2}\int_\Omega \left|\frac{\partial \boldsymbol{w}}{\partial z}\right|^2 \\ \le &c(|\boldsymbol{Z}|_4^8 + \|\boldsymbol{Z}\|^4 + \|\boldsymbol{Z}\|^2\|\boldsymbol{Z}\|_2^2)|\boldsymbol{w}|_2^2 + c(\|\boldsymbol{Z}\|^2 + |T|_2^2 + |S|_2^2). \end{aligned}$$

Using $\lambda_1|\boldsymbol{w}|_2^2 \le \|\boldsymbol{w}\|^2$ and Gronwall Inequality, for $t \ge t_0$, we derive from the above inequality

$$\begin{aligned} |\boldsymbol{w}(t)|_2^2 \le &\mathrm{e}^{\int_{t_0}^t [-\lambda_1 + c(|\boldsymbol{Z}|_4^8 + \|\boldsymbol{Z}\|^4 + \|\boldsymbol{Z}\|^2\|\boldsymbol{Z}\|_2^2)]\mathrm{d}\tau}|\boldsymbol{w}(t_0)|_2^2 \\ &+ c\int_{t_0}^t \mathrm{e}^{\int_\sigma^t [-\lambda_1 + c(|\boldsymbol{Z}|_4^8 + \|\boldsymbol{Z}\|^4 + \|\boldsymbol{Z}\|^2\|\boldsymbol{Z}\|_2^2)]\mathrm{d}\tau}(\|\boldsymbol{Z}\|^2 + |T|_2^2 + |S|_2^2)\mathrm{d}\sigma. \end{aligned} \tag{4.3.35}$$

By Lemma 4.3.3 and the above inequality, for any $\mathcal{T} > t_0$ given, there exists $C_3(\mathcal{T}, \boldsymbol{\mathcal{U}}_{t_0}, Q_1, Q_2, \boldsymbol{Z}_{t_0}) > 0$ such that

$$\int_{t_0}^{\mathcal{T}} \int_\Omega \left(|\nabla \boldsymbol{w}|^2 + \left|\frac{\partial \boldsymbol{w}}{\partial z}\right|^2\right) + |\boldsymbol{w}(t)|_2^2 \le C_3, \quad \forall\, t \in [t_0, \mathcal{T}). \tag{4.3.36}$$

By Minkowski Inequality and Hölder Inequality, we derive from (4.3.36) that

$$\int_{t_0}^{\mathcal{T}}\int_M (|\nabla\bar{\boldsymbol{w}}|^2+|\bar{\boldsymbol{w}}|^2)+\|\bar{\boldsymbol{w}}(t)\|_{L^2}^2\leq C_3, \forall t\in[t_0,\mathcal{T}), \tag{4.3.37}$$

which implies

$$\int_{t_0}^{\mathcal{T}}|\bar{\boldsymbol{v}}|_4^4=\int_{t_0}^{\mathcal{T}}\|\bar{\boldsymbol{v}}\|_{L^4}^4\leq\int_{t_0}^{\mathcal{T}}\|\bar{\boldsymbol{v}}\|_{L^2}^2\|\bar{\boldsymbol{v}}\|_{H^1}^2\leq C_3^2. \tag{4.3.38}$$

L^3, L^4 **estimates of** $\tilde{\boldsymbol{w}}$**.** Similar to the above inequality (4.2.49),

$$\begin{aligned}
&\frac{\mathrm{d}|\tilde{\boldsymbol{w}}|_3^3}{\mathrm{d}t}+\frac{1}{Re_1}\int_\Omega\left(|\nabla\tilde{\boldsymbol{w}}|^2|\tilde{\boldsymbol{w}}|+\frac{4}{9}|\nabla|\tilde{\boldsymbol{w}}|^{\frac{3}{2}}|^2\right)\\
&+\frac{1}{Re_2}\int_\Omega\left(|\partial_z\tilde{\boldsymbol{w}}|^2|\tilde{\boldsymbol{w}}|+\frac{4}{9}|\partial_z|\tilde{\boldsymbol{w}}|^{\frac{3}{2}}|^2\right)\\
\leq& c(1+\|\bar{\boldsymbol{w}}\|_{L^2}^2\|\bar{\boldsymbol{w}}\|_{H^1}^2+\|\tilde{\boldsymbol{w}}\|^2+|\boldsymbol{w}|_2^{\frac{3}{2}}\|\boldsymbol{w}\|^{\frac{3}{2}}\|\boldsymbol{Z}\|^{\frac{3}{2}}\|\boldsymbol{Z}\|_2^{\frac{3}{2}}+|\boldsymbol{Z}|_2^{\frac{3}{2}}\|\boldsymbol{Z}\|^3\|\boldsymbol{Z}\|_2^{\frac{3}{2}}\\
&+\|\boldsymbol{Z}\|^2+|\boldsymbol{Z}|_4^8+\|\boldsymbol{Z}\|^4+\|\boldsymbol{Z}\|\|\boldsymbol{Z}\|_2^3+\|\boldsymbol{Z}\|^2\|\boldsymbol{Z}\|_2^2+|\boldsymbol{Z}|_6^6)|\tilde{\boldsymbol{w}}|_3^3\\
&+c|\boldsymbol{w}|_2^2\|\boldsymbol{w}\|^2+c\|\boldsymbol{w}\|^{\frac{3}{2}}+c\|\boldsymbol{w}\|^2+c\|\boldsymbol{Z}\|^3+c\|\boldsymbol{Z}\|^8\\
&+c|T|_2^2\|T\|^2+c|S|_2^2\|S\|^2+c\|\boldsymbol{Z}\|^2\|\boldsymbol{Z}\|_2^2,
\end{aligned}$$

which implies that there exists $C_4(\mathcal{T},\boldsymbol{\mathcal{U}}_{t_0},Q_1,Q_2,\boldsymbol{Z}_{t_0})>0$ such that

$$|\tilde{\boldsymbol{w}}(t)|_3^3\leq C_4,\quad \text{for any}\ \ t\in[t_0,\mathcal{T}). \tag{4.3.39}$$

Similarly,

$$\begin{aligned}
&\frac{\mathrm{d}|\tilde{\boldsymbol{w}}|_4^4}{\mathrm{d}t}+\frac{1}{Re_1}\int_\Omega\left(|\nabla\tilde{\boldsymbol{w}}|^2|\tilde{\boldsymbol{w}}|^2+\frac{1}{2}|\nabla|\tilde{\boldsymbol{w}}|^2|^2\right)+\frac{1}{Re_2}\int_\Omega\Big(|\partial_z\tilde{\boldsymbol{w}}|^2|\tilde{\boldsymbol{w}}|^2\\
&+\frac{1}{2}|\partial_z|\tilde{\boldsymbol{w}}|^2|^2\Big)\\
\leq& c(1+|\boldsymbol{w}|_2^2\|\boldsymbol{w}\|^2+|\boldsymbol{Z}|_2^2\|\boldsymbol{Z}\|^2+|\boldsymbol{Z}|_8^4+\|\boldsymbol{w}\|^2+\|\boldsymbol{Z}\|^4+\|\boldsymbol{Z}\|^{\frac{4}{3}}\|\boldsymbol{Z}\|_2^{\frac{4}{3}}\\
&+\|\boldsymbol{Z}\|^{\frac{2}{3}}\|\boldsymbol{Z}\|_2^2+\|\boldsymbol{Z}\|^{\frac{8}{5}}\|\boldsymbol{Z}\|_2^{\frac{8}{5}}+\|\boldsymbol{Z}\|\|\boldsymbol{Z}\|_2^3+\|\boldsymbol{Z}\|^2\|\boldsymbol{Z}\|_2^2)|\tilde{\boldsymbol{w}}|_4^4+c|\boldsymbol{w}|_2^2\|\boldsymbol{w}\|^2\\
&+c|\boldsymbol{Z}|_2^2\|\boldsymbol{Z}\|^2+c|\boldsymbol{Z}|_4^4+c|T|_2^2\|T\|^2+c|S|_2^2\|S\|^2+\|\boldsymbol{Z}\|^{\frac{8}{5}}\|\boldsymbol{Z}\|_2^{\frac{8}{5}}.
\end{aligned}$$

According to the above inequality, we know that there exists $C_5(\mathcal{T},\boldsymbol{\mathcal{U}}_{t_0}$, Q_1, Q_2, $\boldsymbol{Z}_{t_0})>0$ such that

$$|\tilde{\boldsymbol{w}}(t)|_4^4\leq C_5, \tag{4.3.40}$$

where $t_0\leq t<\mathcal{T}$.

L^2 estimates of $\partial_z \boldsymbol{w}$. Taking the derivative of (4.3.21) with respective to z, we get

$$\begin{aligned}&\frac{\partial \boldsymbol{w}_z}{\partial t}-\frac{1}{Re_1}\Delta \boldsymbol{w}_z-\frac{1}{Re_2}\frac{\partial^2 \boldsymbol{w}_z}{\partial z^2}+\left[(\boldsymbol{w}+\boldsymbol{Z})\cdot\boldsymbol{\nabla}\right](\boldsymbol{w}_z+\boldsymbol{Z}_z)\\&+\Phi(\boldsymbol{w}+\boldsymbol{Z})\frac{\partial(\boldsymbol{w}_z+\boldsymbol{Z}_z)}{\partial z}+\left[(\boldsymbol{w}_z+\boldsymbol{Z}_z)\cdot\boldsymbol{\nabla}\right](\boldsymbol{w}+\boldsymbol{Z})\\&-(\boldsymbol{w}_z+\boldsymbol{Z}_z)\boldsymbol{\nabla}\cdot(\boldsymbol{w}+\boldsymbol{Z})+f\boldsymbol{k}\times(\boldsymbol{w}_z+\boldsymbol{Z}_z)+\boldsymbol{\nabla}(\nu_1 T-\nu_2 S)=0.\end{aligned}\tag{4.3.41}$$

Using integration by parts, Hölder Inequality, Minkowski Inequality, Sobolev Inequality and (4.2.27), we have

$$-\int_\Omega\left\{\left[(\boldsymbol{w}+\boldsymbol{Z})\cdot\boldsymbol{\nabla}\right]\boldsymbol{Z}_z+\Phi(\boldsymbol{w}+\boldsymbol{Z})\frac{\partial \boldsymbol{Z}_z}{\partial z}\right\}\cdot\boldsymbol{w}_z$$

$$\le\varepsilon\|\boldsymbol{w}_z\|^2+c\|\boldsymbol{Z}\|_2^2+c(|\tilde{\boldsymbol{w}}+\tilde{\boldsymbol{Z}}|_4^8+|\bar{\boldsymbol{w}}+\bar{\boldsymbol{Z}}|_{L^4}^4)|\boldsymbol{w}_z|_2^2+c\|\boldsymbol{w}+\boldsymbol{Z}\|^2\|\boldsymbol{Z}\|_{2+2\gamma}^2,$$

$$-\int_\Omega\{\left[(\boldsymbol{w}_z+\boldsymbol{Z}_z)\cdot\boldsymbol{\nabla}\right](\boldsymbol{w}+\boldsymbol{Z})-(\boldsymbol{w}_z+\boldsymbol{Z}_z)\boldsymbol{\nabla}\cdot(\boldsymbol{w}+\boldsymbol{Z})\}\cdot\boldsymbol{w}_z$$

$$\le\varepsilon\|\boldsymbol{w}_z\|^2+c\|\boldsymbol{Z}\|_2^2+c(|\tilde{\boldsymbol{w}}+\tilde{\boldsymbol{Z}}|_4^8+|\bar{\boldsymbol{w}}+\bar{\boldsymbol{Z}}|_{L^4}^4)|\boldsymbol{Z}_z|_2^2|\boldsymbol{w}_z|_2^2,$$

where $0<\gamma<\gamma_0$, and γ_0 is given in definition of τ. Taking $L^2(\Omega)\times L^2(\Omega)$ inner product of (4.3.41) with $\boldsymbol{w}_z$, then by integration by parts, and Hölder Inequality, Poincaré Inequality and the above two inequalities, choosing ε small enough, we obtain

$$\begin{aligned}&\frac{\mathrm{d}|\boldsymbol{w}_z|_2^2}{\mathrm{d}t}+\frac{1}{Re_1}\int_\Omega|\boldsymbol{\nabla}\boldsymbol{w}_z|^2+\frac{1}{Re_2}\int_\Omega\left|\frac{\partial \boldsymbol{w}_z}{\partial z}\right|^2\\\le&c(|\bar{\boldsymbol{w}}|_2^2\|\bar{\boldsymbol{w}}\|_2^2+|\boldsymbol{Z}|_4^4+|\tilde{\boldsymbol{w}}|_4^8+|\boldsymbol{Z}|_4^8)(|\boldsymbol{w}_z|_2^2+|\boldsymbol{Z}_z|_2^2)\\&+c\|\boldsymbol{Z}\|_2^2+c\|\boldsymbol{w}+\boldsymbol{Z}\|^2\|\boldsymbol{Z}\|_{2+2\gamma}^2+c|T|_2^2+c|S|_2^2.\end{aligned}$$

According to the above inequality, by Gronwall Inequality, Lemma 4.3.3, (4.3.32), (4.3.34), (4.3.37) and (4.3.40), we know that for any $\mathcal{T}>t_0$ given, there exists $C_6(\mathcal{T},\mathcal{U}_{t_0},Q_1,Q_2,\boldsymbol{Z}_{t_0})>0$ such that

$$\int_{t_0}^{\mathcal{T}}\|\boldsymbol{w}_z\|^2+|\boldsymbol{w}_z(t)|_2^2\le C_6,\quad\forall\, t\in[t_0,\mathcal{T}).\tag{4.3.42}$$

L^2 estimates of $\partial_z T$, $\partial_z S$. Taking the derivative of (4.3.22) with respect to z, we get

$$\frac{\partial T_z}{\partial t}-\frac{1}{Rt_1}\Delta T_z-\frac{1}{Rt_2}\frac{\partial^2 T_z}{\partial z^2}+\left[(\boldsymbol{w}+\boldsymbol{Z})\cdot\boldsymbol{\nabla}\right]T_z+\Phi(\boldsymbol{w}+\boldsymbol{Z})\frac{\partial T_z}{\partial z}$$

$$+\left[(\boldsymbol{w}_z+\boldsymbol{Z}_z)\cdot\boldsymbol{\nabla}\right]T-T_z\boldsymbol{\nabla}\cdot(\boldsymbol{w}+\boldsymbol{Z})=Q_{1z}. \tag{4.3.43}$$

By integration by parts, Hölder Inequality, Poincaré Inequality and Young Inequality, we obtain

$$\begin{aligned}
&\left|\int_\Omega \left[\left((\boldsymbol{w}_z+\boldsymbol{Z}_z)\cdot\boldsymbol{\nabla}\right)T-T_z\boldsymbol{\nabla}\cdot(\boldsymbol{w}+\boldsymbol{Z})\right]T_z\right| \\
\leq & c\int_\Omega \left[\left(|\boldsymbol{\nabla}\boldsymbol{w}_z|+|\boldsymbol{\nabla}\boldsymbol{Z}_z|\right)|T||T_z|+|\boldsymbol{w}_z+\boldsymbol{Z}_z||T||\boldsymbol{\nabla}T_z|\right. \\
&\left.+\left(|\tilde{\boldsymbol{w}}|+|\bar{\boldsymbol{w}}|+|\boldsymbol{Z}|\right)|\boldsymbol{\nabla}T_z||T_z|\right] \\
\leq & c\int_\Omega \left(|\boldsymbol{\nabla}\boldsymbol{w}_z|^2+|\boldsymbol{\nabla}\boldsymbol{Z}_z|^2\right)+\frac{\varepsilon}{2}|\boldsymbol{\nabla}T_z|_2^2+c\left(|T|_4^2+|\tilde{\boldsymbol{w}}|_4^2\right)|T_z|_4^2 \\
&+c\left(|\boldsymbol{w}_z|_4^2+|\boldsymbol{Z}_z|_4^2\right)|T|_4^2+c\|\bar{\boldsymbol{w}}\|_{L^4(M)}^2\int_{-1}^0\left(\int_M|T_z|^4\right)^{\frac{1}{2}}+|\boldsymbol{Z}|_4^2|T_z|_4^2 \\
\leq & \varepsilon\left(|T_{zz}|_2^2+|\boldsymbol{\nabla}T_z|_2^2\right)+c\left(|\boldsymbol{w}_{zz}|_2^2+\int_\Omega|\boldsymbol{\nabla}\boldsymbol{w}_z|^2\right)+c\|\boldsymbol{Z}\|_2^2+c\|\boldsymbol{Z}\|_2^2|T|_4^2 \\
&+c|T|_4^8|\boldsymbol{w}_z|_2^2+c\left(|T|_4^8+|\boldsymbol{Z}|_4^8+|\tilde{\boldsymbol{w}}|_4^8+\|\bar{\boldsymbol{w}}\|_{L^4(M)}^4\right)|T_z|_2^2.
\end{aligned}$$

According to (4.3.25), and taking the trace of equation (4.3.22) on $z=0$, we have

$$\begin{aligned}
&-\frac{1}{Rt_2}\int_M\left(T_z|_{z=0}T_{zz}|_{z=0}\right) \\
=&\alpha_u\int_M T|_{z=0}\left[\frac{\partial T|_{z=0}}{\partial t}+\left((\boldsymbol{w}+\boldsymbol{Z})\cdot\boldsymbol{\nabla}\right)T|_{z=0}-\frac{1}{Rt_1}\Delta T|_{z=0}-Q_1|_{z=0}\right] \\
=&\alpha_u\left(\frac{1}{2}\frac{\mathrm{d}|T(z=0)|_2^2}{\mathrm{d}t}+\frac{1}{Rt_1}|\boldsymbol{\nabla}T(z=0)|_2^2\right) \\
&+\alpha_u\int_M T|_{z=0}\left[\left((\boldsymbol{w}+\boldsymbol{Z})\cdot\boldsymbol{\nabla}\right)T|_{z=0}-Q_1|_{z=0}\right].
\end{aligned}$$

By integration by parts, we get

$$\begin{aligned}
&-\alpha_u\int_M T|_{z=0}\left[\left((\boldsymbol{w}+\boldsymbol{Z})\cdot\boldsymbol{\nabla}\right)T|_{z=0}-Q_1|_{z=0}\right] \\
=&\frac{\alpha_u}{2}\int_M T^2|_{z=0}\mathrm{div}\,(\boldsymbol{w}+\boldsymbol{Z})|_{z=0}+\alpha_u\int_M T|_{z=0}Q_1|_{z=0} \\
=&\frac{\alpha_u}{2}\int_M T^2|_{z=0}\left(\int_z^0\mathrm{div}\,(\boldsymbol{w}_z+\boldsymbol{Z}_z)\,\mathrm{d}z'+\mathrm{div}\,(\boldsymbol{w}+\boldsymbol{Z})\right) \\
&+\alpha_u\int_M T|_{z=0}Q_1|_{z=0} \\
\leq & c|T|_{z=0}|_4^4+c\|\boldsymbol{w}_z\|^2+c\|\boldsymbol{Z}\|_2^2+c\|\boldsymbol{w}\|^2+c|T|_{z=0}|_2^2+c|Q_1|_{z=0}|_2^2.
\end{aligned}$$

Taking $L^2(\Omega)$ inner product of (4.3.43) with T_z, then by integration by parts, Hölder Inequality and the above inequality, choosing ε small enough, we obtain

$$\frac{\mathrm{d}(|T_z|_2^2+\alpha_u|T|_{z=0}|_2^2)}{\mathrm{d}t}+\frac{1}{Rt_1}\int_\Omega|\nabla T_z|^2+\frac{1}{Rt_2}\int_\Omega|T_{zz}|^2+\frac{\alpha_u}{R_{t1}}|\nabla T(z=0)|_2^2$$

$$\leq c(1+|T|_4^8+|\boldsymbol{Z}|_4^8+|\tilde{\boldsymbol{w}}|_4^8+\|\bar{\boldsymbol{w}}\|_{L^4(M)}^4)|T_z|_2^2+c\|\boldsymbol{w}_z\|^2+c\|\boldsymbol{w}\|^2+c|T|_4^8|\boldsymbol{w}_z|_2^2$$

$$+c\|\boldsymbol{Z}\|_2^2+c\|\boldsymbol{Z}\|_2^2|T|_4^2+c|T|_{z=0}|_4^4+c|T|_{z=0}|_2^2+c|Q_1|_{z=0}|_2^2+c|Q_{1z}|_2^2.$$

According to the above inequality, we know that there exists $C_7(\mathcal{T},\boldsymbol{\mathcal{U}}_{t_0},Q_1,Q_2,\boldsymbol{Z}_{t_0})$ such that

$$\int_{t_0}^{\mathcal{T}}\|T_z\|^2+|T_z(t)|_2^2\leq C_7,\quad\forall\, t\in[t_0,\mathcal{T}).\tag{4.3.44}$$

In the process of proving (4.3.44), we have applied

$$|T(t)|_4^4+\int_t^{\mathcal{T}}|T|_{z=0}|_4^4\leq C,$$

which is obtained by taking inner product of equation (4.3.22) with $|T|^2T$ and making some estimates. Similar to (4.3.44),

$$\int_{t_0}^{\mathcal{T}}\|S_z\|^2+|S_z(t)|_2^2\leq C_8,\ \text{to any}\ t\in[t_0,\mathcal{T})\tag{4.3.45}$$

where $C_8(\mathcal{T},\boldsymbol{\mathcal{U}}_{t_0},Q_1,Q_2,\boldsymbol{Z}_{t_0})>0$. In the process of proving (4.3.45), we have used $|S(t)|_4^4\leq c$.

4.3.3.2 *Uniqueness of Weakly Strong Solution*

Let $(\boldsymbol{w}_1,T_1,S_1)$ and $(\boldsymbol{w}_2,T_2,S_2)$ be two weakly strong solutions of (4.3.21)-(4.3.27) corresponding to p'_{b_2}, p''_{b_2}, and initial data $((\boldsymbol{w}_{t_0})_1,(T_{t_0})_1,(S_{t_0})_1)$, $((\boldsymbol{w}_{t_0})_2,(T_{t_0})_2,(S_{t_0})_2)$, respectively.

Let $\boldsymbol{w}=\boldsymbol{w}_1-\boldsymbol{w}_2$, $T=T_1-T_2$, $S=S_1-S_2$, $p_{b_2}=p'_{b_2}-p''_{b_2}$. Then $\boldsymbol{w}$, T, S, p_{b_2} satisfy

$$\frac{\partial\boldsymbol{w}}{\partial t}-\frac{1}{Re_1}\Delta\boldsymbol{w}-\frac{1}{Re_2}\frac{\partial^2\boldsymbol{w}}{\partial z^2}+[(\boldsymbol{w}_1+\boldsymbol{Z})\cdot\nabla]\,\boldsymbol{w}+(\boldsymbol{w}\cdot\nabla)(\boldsymbol{w}_2+\boldsymbol{Z})$$

$$+\Phi(\boldsymbol{w}_1+\boldsymbol{Z})\frac{\partial\boldsymbol{w}}{\partial z}+\Phi(\boldsymbol{w})\frac{\partial(\boldsymbol{w}_2+\boldsymbol{Z})}{\partial z}+f\boldsymbol{k}\times\boldsymbol{w}+\nabla p_{b_2}$$

$$-\int_{-1}^{z}\nabla(\nu_1T-\nu_2S)\mathrm{d}z'=0,\tag{4.3.46}$$

$$\frac{\partial T}{\partial t}-\frac{1}{Rt_1}\Delta T-\frac{1}{Rt_2}\frac{\partial^2 T}{\partial z^2}+[(\boldsymbol{w}_1+\boldsymbol{Z})\cdot\boldsymbol{\nabla}]\,T+(\boldsymbol{w}\cdot\boldsymbol{\nabla})T_2+\Phi(\boldsymbol{w}_1+\boldsymbol{Z})\frac{\partial T}{\partial z}$$
$$+\Phi(\boldsymbol{w})\frac{\partial T_2}{\partial z}=0,$$
$$\frac{\partial S}{\partial t}-\frac{1}{Rs_1}\Delta S-\frac{1}{Rs_2}\frac{\partial^2 S}{\partial z^2}+[(\boldsymbol{w}_1+\boldsymbol{Z})\cdot\boldsymbol{\nabla}]\,S+(\boldsymbol{w}\cdot\boldsymbol{\nabla})S_2+\Phi(\boldsymbol{w}_1+\boldsymbol{Z})\frac{\partial S}{\partial z}$$
$$+\Phi(\boldsymbol{w})\frac{\partial S_2}{\partial z}=0,$$
$$\int_{-1}^{0}\boldsymbol{\nabla}\cdot\boldsymbol{w}\mathrm{d}z=0,$$
$$\boldsymbol{w}_{t_0}=(\boldsymbol{w}_{t_0})_1-(\boldsymbol{w}_{t_0})_2,\ T_{t_0}=(T_{t_0})_1-(T_{t_0})_2,\ S_{t_0}=(S_{t_0})_1-(S_{t_0})_2,$$
$(\boldsymbol{w},T,S)$ satisfies boundary conditions (4.3.25)-(4.3.27).

Taking $L^2(\Omega)\times L^2(\Omega)$ inner product of (4.3.46) with $\boldsymbol{w}$, we get

$$\frac{1}{2}\frac{\mathrm{d}|\boldsymbol{w}|_2^2}{\mathrm{d}t}+\frac{1}{Re_1}\int_\Omega|\boldsymbol{\nabla}\boldsymbol{w}|^2+\frac{1}{Re_2}\int_\Omega|\boldsymbol{w}_z|^2$$
$$=-\int_\Omega\left\{[(\boldsymbol{w}_1+\boldsymbol{Z})\cdot\boldsymbol{\nabla}]\,\boldsymbol{w}+\Phi(\boldsymbol{w}_1+\boldsymbol{Z})\frac{\partial\boldsymbol{w}}{\partial z}\right\}\cdot w-\int_\Omega(f\boldsymbol{k}\times\boldsymbol{w}+\boldsymbol{\nabla}p_{b_2})\cdot\boldsymbol{w}$$
$$-\int_\Omega\left[(\boldsymbol{w}\cdot\boldsymbol{\nabla})(\boldsymbol{w}_2+\boldsymbol{Z})+\Phi(\boldsymbol{w})\frac{\partial(\boldsymbol{w}_2+\boldsymbol{Z})}{\partial z}\right]\cdot\boldsymbol{w}$$
$$+\int_\Omega\left[\int_{-1}^{z}\boldsymbol{\nabla}(\nu_1T-\nu_2S)\mathrm{d}z'\right]\cdot\boldsymbol{w}.$$

By integration by parts and Lemma 4.3.10, we obtain

$$\left|\int_\Omega[(\boldsymbol{w}\cdot\boldsymbol{\nabla})(\boldsymbol{w}_2+\boldsymbol{Z})]\cdot\boldsymbol{w}\right|$$
$$\le\varepsilon\int_\Omega(|\boldsymbol{\nabla}\boldsymbol{w}|^2+|\boldsymbol{w}_z|^2)+c(|\bar{\boldsymbol{w}}|_2^2\|\bar{\boldsymbol{w}}\|^2+|\boldsymbol{Z}|_4^4+|\boldsymbol{Z}|_4^8+|\tilde{\boldsymbol{w}}_2|_4^8)|\boldsymbol{w}|_2^2,$$
$$\left|\int_\Omega\Phi(\boldsymbol{w})\frac{\partial(\boldsymbol{w}_2+\boldsymbol{Z})}{\partial z}\cdot\boldsymbol{w}\right|$$
$$\le\varepsilon\int_\Omega|\boldsymbol{\nabla}\boldsymbol{w}|^2+c(|\boldsymbol{w}_{2z}|_2^2|\boldsymbol{\nabla}\boldsymbol{w}_{2z}|_2^2+|\boldsymbol{w}_{2z}|_2^4+|\boldsymbol{Z}_z|_2^2|\boldsymbol{\nabla}\boldsymbol{Z}_z|_2^2+|\boldsymbol{Z}_z|_2^4)|\boldsymbol{w}|_2^2.$$

By integration by parts, Hölder Inequality and Young Inequality, we derive from the above three relationships that

$$\frac{1}{2}\frac{\mathrm{d}|\boldsymbol{w}|_2^2}{\mathrm{d}t}+\frac{1}{Re_1}\int_\Omega|\boldsymbol{\nabla}\boldsymbol{w}|^2+\frac{1}{Re_2}\int_\Omega|\boldsymbol{w}_z|^2$$

$$\leq 2\varepsilon \int_{\Omega} (|\nabla \boldsymbol{w}|^2 + |\boldsymbol{w}_z|^2) + \varepsilon |\nabla T|_2^2 + c(1 + |\bar{\boldsymbol{w}}|_2^2 \|\bar{\boldsymbol{w}}\|^2 + |\boldsymbol{Z}|_4^8 + |\tilde{\boldsymbol{w}}_2|_4^8$$
$$+ |\boldsymbol{w}_{2z}|_2^2 |\nabla \boldsymbol{w}_{2z}|_2^2 + |\boldsymbol{w}_{2z}|_2^4 + |\boldsymbol{Z}_z|_2^2 |\nabla \boldsymbol{Z}_z|_2^2 + |\boldsymbol{Z}_z|_2^4) |\boldsymbol{w}|_2^2. \tag{4.3.47}$$

Similar to (4.3.47),

$$\frac{1}{2}\frac{\mathrm{d}|T|_2^2}{\mathrm{d}t} + \frac{1}{Rt_1}\int_{\Omega} |\nabla T|^2 + \frac{1}{Rt_2}\int_{\Omega} |T_z|^2 + \frac{\alpha_u}{Rt_2} |T|_{z=0}|_2^2$$
$$\leq \varepsilon \int_{\Omega} (|\nabla \boldsymbol{w}|^2 + |\boldsymbol{w}_z|^2) + \varepsilon \int_{\Omega} (|\nabla T|^2 + |T_z|^2) + c(|T_2|_4^8 + |T_{2z}|_2^2 |\nabla T_{2z}|_2^2$$
$$+ |T_{2z}|_2^4)(|\boldsymbol{w}|_2^2 + |T|_2^2), \tag{4.3.48}$$

$$\frac{1}{2}\frac{\mathrm{d}|S|_2^2}{\mathrm{d}t} + \frac{1}{Rs_1}\int_{\Omega} |\nabla S|^2 + \frac{1}{Rs_2}\int_{\Omega} |S_z|^2$$
$$\leq \varepsilon \int_{\Omega} (|\nabla \boldsymbol{w}|^2 + |\boldsymbol{w}_z|^2) + \varepsilon \int_{\Omega} (|\nabla S|^2 + |S_z|^2)$$
$$+ c(|S_2|_4^8 + |S_{2z}|_2^2 |\nabla S_{2z}|_2^2 + |S_{2z}|_2^4)(|\boldsymbol{w}|_2^2 + |S|_2^2). \tag{4.3.49}$$

According to (4.3.47)-(4.3.49), choosing ε small enough, then by Gronwall Inequality and the results of *a prior* estimates in subsection 4.3.3.1, we have proved the uniqueness of weakly strong solutions of (4.3.21)-(4.3.27).

4.3.4 *Existence of Random Attractors*

In this subsection, we shall prove Theorem 4.3.9. Firstly, we give the definition of strong solutions of (4.3.21)-(4.3.28) and a result of the global existence of strong solutions.

Definition 4.3.11. Let $\boldsymbol{Z}$ be an Ornstein-Uhlenbeck process in subsection 4.3.2, $\boldsymbol{w}_{t_0} \in V_1$, $T_{t_0} \in V_2$, $S_{t_0} \in V_3$, and $\mathcal{T} > t_0$ is a given time. For P-a.e. $\omega \in \boldsymbol{\Omega}$, $(\boldsymbol{w}, T, S)$ is called a strong solution of initial boundary value problem (4.3.21)-(4.3.28) on $[t_0, \mathcal{T}]$, if it satisfies (4.3.21)-(4.3.24) in weak sense such that

$$\boldsymbol{w} \in L^\infty(t_0, \mathcal{T}; V_1) \cap L^2(t_0, \mathcal{T}; (H^2(\Omega))^2),$$
$$T \in L^\infty(t_0, \mathcal{T}; V_2) \cap L^2(t_0, \mathcal{T}; H^2(\Omega)),$$
$$S \in L^\infty(t_0, \mathcal{T}; V_3) \cap L^2(t_0, \mathcal{T}; H^2(\Omega)),$$
$$\frac{\partial \boldsymbol{w}}{\partial t} \in L^2(t_0, \mathcal{T}; (L^2(\Omega))^2), \quad \frac{\partial T}{\partial t}, \frac{\partial S}{\partial t} \in L^2(t_0, \mathcal{T}; L^2(\Omega)).$$

Theorem 4.3.12 (The global existence of strong solutions). If

$Q_1, Q_2 \in H^1(\Omega)$, $\mathcal{U}_{t_0} = (\boldsymbol{w}_{t_0}, T_{t_0}, S_{t_0}) \in V$, then for any $\mathcal{T} > t_0$ given, the initial boundary value problem (4.3.21)-(4.3.28) has a strong solution U on $[t_0, \mathcal{T}]$, and U continuously dependent on the initial data.

Proof. According to the *a prior* estimates results in subsection 4.3.3, in order to prove Theorem 4.3.12, we only have to make L^2 estimates of $\nabla \boldsymbol{w}$, ∇T, ∇S.

L^2 estimates of $\nabla \boldsymbol{w}$. Similarly to Lemma 4.3.10, we have

$$\begin{aligned}
&\left|\int_\Omega \{[(\boldsymbol{w}+\boldsymbol{Z})\cdot\nabla](\boldsymbol{w}+\boldsymbol{Z})\}\cdot\Delta\boldsymbol{w}\right| \\
\leq&\varepsilon(|\Delta\boldsymbol{w}|_2^2+|\nabla\boldsymbol{w}_z|_2^2)+c(|\bar{\boldsymbol{w}}|_2^2\|\bar{\boldsymbol{w}}\|_2^2+|\boldsymbol{Z}|_4^4+|\tilde{\boldsymbol{w}}|_4^8+|\boldsymbol{Z}|_4^8)(|\nabla\boldsymbol{w}|_2^2+|\nabla\boldsymbol{Z}|_2^2)\\
&+c\|\boldsymbol{Z}\|_2^2,
\end{aligned}$$

$$\begin{aligned}
&\left|\int_\Omega [\Phi(\boldsymbol{w}+\boldsymbol{Z})(\boldsymbol{w}_z+\boldsymbol{Z}_z)]\cdot\Delta\boldsymbol{w}\right| \\
\leq&c(|\boldsymbol{Z}_z|_2^2+|\nabla\boldsymbol{Z}_z|_2^2+|\boldsymbol{Z}_z|_2^2|\nabla\boldsymbol{Z}_z|_2^2)(|\nabla\boldsymbol{w}|_2^2+|\nabla\boldsymbol{Z}|_2^2)+\varepsilon|\Delta\boldsymbol{w}|_2^2+c\|\boldsymbol{Z}\|_2^2\\
&+c(|\boldsymbol{w}_z|_2^2+|\nabla\boldsymbol{w}_z|_2^2+|\boldsymbol{w}_z|_2^2|\nabla\boldsymbol{w}_z|_2^2)(|\nabla\boldsymbol{w}|_2^2+|\nabla\boldsymbol{Z}|_2^2).
\end{aligned}$$

Taking $L^2(\Omega)\times L^2(\Omega)$ inner product of (4.3.21) with $-\Delta\boldsymbol{w}$, using Hölder Inequality, $(f\boldsymbol{k}\times\boldsymbol{w})\cdot\Delta\boldsymbol{w}=0$, $\displaystyle\int_\Omega \nabla p_{b_2}\cdot\Delta\boldsymbol{w}=0$ and the above two inequalities, then choosing ε small enough, we get

$$\begin{aligned}
&\frac{\mathrm{d}|\nabla\boldsymbol{w}|_2^2}{\mathrm{d}t}+\frac{1}{Re_1}\int_\Omega|\Delta\boldsymbol{w}|^2+\frac{1}{Re_2}\int_\Omega|\nabla\boldsymbol{w}_z|^2\\
\leq&c(|\boldsymbol{w}|_4^8+|\boldsymbol{Z}|_4^8+|\boldsymbol{w}_z|_2^2+|\nabla\boldsymbol{w}_z|_2^2+|\boldsymbol{w}_z|_2^2|\nabla\boldsymbol{w}_z|_2^2+|\boldsymbol{Z}_z|_2^2+|\boldsymbol{Z}_z|_2^2|\nabla\boldsymbol{Z}_z|_2^2\\
&+|\nabla\boldsymbol{Z}_z|_2^2)|\nabla\boldsymbol{w}|_2^2+c|\nabla T|_2^2+c|\nabla S|_2^2+c(|\bar{\boldsymbol{w}}|_2^2\|\bar{\boldsymbol{w}}\|_2^2+|\boldsymbol{Z}|_4^4+|\tilde{\boldsymbol{w}}|_4^8+|\boldsymbol{Z}|_4^8\\
&+|\boldsymbol{w}_z|_2^2+|\nabla\boldsymbol{w}_z|_2^2+|\boldsymbol{w}_z|_2^2|\nabla\boldsymbol{w}_z|_2^2+|\boldsymbol{Z}_z|_2^2+|\nabla\boldsymbol{Z}_z|_2^2+|\boldsymbol{Z}_z|_2^2|\nabla\boldsymbol{Z}_z|_2^2)|\nabla\boldsymbol{Z}|_2^2\\
&+c\|\boldsymbol{Z}\|_2^2.
\end{aligned}$$

The above inequality implies that for any $\mathcal{T} > t_0$, there exists $C_9(\mathcal{T}, \mathcal{U}_{t_0}, Q_1, Q_2, \boldsymbol{Z}_{t_0}) > 0$ such that

$$\int_{t_0}^{\mathcal{T}} |\Delta\boldsymbol{w}|_2^2+|\nabla\boldsymbol{w}(t)|_2^2\leq C_9,\quad \forall\, t\in[t_0,\mathcal{T}). \tag{4.3.50}$$

L^2 estimates of ∇T, ∇S. In a similar way to get Lemma 4.3.10, have

$$\begin{aligned}
&\left|\int_\Omega [(\boldsymbol{w}+\boldsymbol{Z})\cdot\nabla]\,T(\Delta T)\right|\\
\leq&c(|\bar{\boldsymbol{w}}|_2^2\|\bar{\boldsymbol{w}}\|_2^2+|\boldsymbol{Z}|_4^4+|\tilde{\boldsymbol{w}}|_4^8+|\boldsymbol{Z}|_4^8)|\nabla T|_2^2+\varepsilon(|\Delta T|_2^2+|\nabla T_z|_2^2),
\end{aligned}$$

$$\begin{aligned}&\left|\int_{\Omega} \Phi(\boldsymbol{w}+\boldsymbol{Z})T_z(\Delta T)\right| \\ \leq & \varepsilon(|\Delta T|_2^2+|\nabla T_z|_2^2)+c[(|\nabla \boldsymbol{w}|_2^4+|\nabla \boldsymbol{w}|_2^2|\Delta \boldsymbol{w}|_2^2) \\ & +c(|\nabla \boldsymbol{Z}|_2^4+|\nabla \boldsymbol{Z}|_2^2|\Delta \boldsymbol{Z}|_2^2)]|T_z|_2^2.\end{aligned}$$

Taking $L^2(\Omega)$ inner product of (4.3.22) with $-\Delta T$, by Hölder Inequality and Young Inequality, and choosing ε small enough, we derive from the above two inequalities

$$\begin{aligned}&\frac{\mathrm{d}|\nabla T|_2^2}{\mathrm{d}t}+\frac{1}{Rt_1}|\Delta T|_2^2+\frac{1}{Rt_2}|\nabla T_z|_2^2+\frac{\alpha_u}{Rt_2}|\nabla T|_{z=0}|_2^2 \\ \leq\ & c(|\bar{\boldsymbol{w}}|_2^2\|\bar{\boldsymbol{w}}\|_2^2+|\boldsymbol{Z}|_4^4+|\tilde{\boldsymbol{w}}|_4^8+|\boldsymbol{Z}|_4^8)|\nabla T|_2^2+c\left[(|\nabla \boldsymbol{w}|_2^4+|\nabla \boldsymbol{w}|_2^2|\Delta \boldsymbol{w}|_2^2)\right. \\ & \left.+c(\|\boldsymbol{Z}\|^4+\|\boldsymbol{Z}\|_2^4)\right]|T_z|_2^2+c|Q_1|_2^2.\end{aligned}$$

Thus, for any $\mathcal{T}>t_0$ given, there exists $C_{10}(\mathcal{T}, \boldsymbol{U}_{t_0}, Q_1, Q_2, \boldsymbol{Z}_{t_0})>0$ such that

$$|\nabla T(t)|_2^2 \leq C_{10},\ \forall\ t\in[t_0,\mathcal{T}). \tag{4.3.51}$$

Similarly, there exists $C_{11}(\mathcal{T},\boldsymbol{\mathcal{U}}_{t_0},Q_1,Q_2,\boldsymbol{Z}_{t_0})>0$ such that

$$|\nabla S(t)|_2^2 \leq C_{11},\ \forall\ t\in[t_0,\mathcal{T}). \tag{4.3.52}$$

Now we use Theorem 4.1.9 in subsection 4.1.3 to prove the existence of random attractors of the primitive equations with random boundary, that is, Theorem 4.3.9. Firstly, we construct the random dynamical system corresponding to the random boundary value problem (4.3.21)-(4.3.27). Let

$$\boldsymbol{\Omega}=\{\omega;\ \omega\in C(\mathbb{R},l^2),\ \omega(0)=0\},$$

$\mathcal{F}$ be a Borel σ-algebra induced by the compact open topology of $\boldsymbol{\Omega}$, and P be a Wiener measure in $(\boldsymbol{\Omega},\mathcal{F})$. Write

$$(\omega_1(t,\omega),\omega_2(t,\omega),\omega_3(t,\omega),\cdots)=\omega(t).$$

Define

$$\vartheta_t\omega(s)=\omega(t+s)-\omega(t). \tag{4.3.53}$$

Then, ϑ_t satisfies $\vartheta_{t+s}=\vartheta_t\circ\vartheta_s$ for any $t,\ s\in\mathbb{R}$, and ϑ_t is ergodic under P. According to Theorem 4.3.12, Let

$$\boldsymbol{\mathcal{U}}(t,\omega)=\phi(t,s;\omega)\boldsymbol{\mathcal{U}}_s,$$

where $\boldsymbol{\mathcal{U}}(t,\omega)=(\boldsymbol{w}(t,\omega),T(t,\omega),S(t,\omega))$ is a strong solution of (4.3.21)-(4.3.27) with the initial value $\boldsymbol{\mathcal{U}}(s)=\boldsymbol{\mathcal{U}}_s=(\boldsymbol{w}_s,T_s,S_s)$. Thus, for $s\leq$

$r \leq t$, there exists $\phi(t,s;\omega) = \phi(t,r;\omega)\phi(r,s;\omega)$. Due to (4.3.53), for any $s,t \in \mathbb{R}^+$, $\boldsymbol{U}_0 \in V$, we have P-a.s.

$$\phi(t+s,0;\omega)\boldsymbol{\mathcal{U}}_0 = \phi(t,0;\vartheta_s\omega)\phi(s,0;\omega)\boldsymbol{\mathcal{U}}_0.$$

Define $\psi : \mathbb{R}^+ \times \boldsymbol{\Omega} \times V \to V$, $\psi(t,\omega)\boldsymbol{\mathcal{U}}_0 = \phi(t,0;\omega)\boldsymbol{\mathcal{U}}_0$. According to Theorem 4.3.12 and the following Proposition 4.3.14, we know that ψ is a continuous random dynamical system on V with weak topology over $(\Omega, \mathcal{F}, P, \{\vartheta_t\}_{t\in\mathbb{R}})$, which corresponds to the random dynamical system generated by the boundary value problem (4.3.21)-(4.3.27).

In the process of proving Theorem 4.3.9 with Theorem 4.1.9 in section 4.1.3, the key step is to prove the existence of bounded absorbing sets. Then we shall prove this conclusion.

Proposition 4.3.13 (Existence of bounded absorbing sets). If Q_1, $Q_2 \in H^1(\Omega)$, $B_\rho = \{\boldsymbol{\mathcal{U}};\ \|\boldsymbol{\mathcal{U}}\| \leq \rho,\ \boldsymbol{\mathcal{U}} \in V\}$, then, there exist $r_0(\omega, \|Q_1\|_1, \|Q_2\|_1)$ and $t(\omega,\rho) \leq -1$ such that for any $t_0 \leq t(\omega,\rho)$, $\boldsymbol{\mathcal{U}}_{t_0} \in B_\rho$.

$$\|\phi(0,t_0;\omega)\boldsymbol{\mathcal{U}}_{t_0}\| \leq r_0(\omega).$$

According to Remark 4.1.8, for any bounded set $B \subset V$, there exists $-t_0(B) > 0$ big enough, such that

$$\psi(-s,\vartheta_s\omega)B = \phi(0,s;\omega)B \subset B_{r_0(\omega)},\ \forall\ s \leq t_0.$$

Proof. According to the ergodicity of the process $\boldsymbol{Z}$ in $D(A_1^{1+\gamma})$, we have

$$\begin{aligned}&\lim_{s\to-\infty}\frac{1}{-s}\int_s^0 (|\boldsymbol{Z}(\tau)|_4^8 + \|\boldsymbol{Z}(\tau)\|^4 + \|\boldsymbol{Z}(\tau)\|^2\|\boldsymbol{Z}(\tau)\|_2^2)\mathrm{d}\tau\\ &= E(|\boldsymbol{Z}(0)|_4^8 + \|\boldsymbol{Z}(0)\|^4 + \|\boldsymbol{Z}(0)\|^2\|\boldsymbol{Z}(0)\|_2^2).\end{aligned}$$

According to condition (C) in subsection 4.3.2, we know that there exists a positive constant c_0 such that

$$\lim_{s\to-\infty}\frac{1}{-s}\int_s^0 \left[-\lambda_1 + c(|\boldsymbol{Z}(\tau)|_4^8 + \|\boldsymbol{Z}(\tau)\|^4 + \|\boldsymbol{Z}(\tau)\|^2\|\boldsymbol{Z}(\tau)\|_2^2)\right]\mathrm{d}\tau \leq -c_0,$$

which implies that there exists a $s_0(\omega)$ such that for $s < s_0(\omega)$,

$$\int_s^0 \left[-\lambda_1 + c(|\boldsymbol{Z}(\tau)|_4^8 + \|\boldsymbol{Z}(\tau)\|^4 + \|\boldsymbol{Z}(\tau)\|^2\|\boldsymbol{Z}(\tau)\|_2^2)\right]\mathrm{d}\tau \leq -\frac{c_0}{2}(-s). \tag{4.3.54}$$

According to the definition of process $\boldsymbol{Z}$ and (4.3.32), with the method in [72], we know that $\|\boldsymbol{Z}(\tau)\|^2 + |T(\tau)|_2^2 + |\boldsymbol{Z}(\tau)|_2^2$ has at most polynomial growth as $\tau \to -\infty$. Thus, (4.3.35) and (4.3.54) imply that there exist $t_0(\rho,\omega)$ and a finite random variable $R_0(\omega)$ a.s. such that a.s.

$$|\boldsymbol{w}(t)|_2^2 \leq R_0(\omega),\ \forall\ s \leq t \leq 0, \tag{4.3.55}$$

where and $s \le t_0$, and $(\boldsymbol{w}, T, S)$ is a strong solution of (4.3.21)-(4.3.27) with initial data $(\boldsymbol{w}_s, T_s, S_s)$ and $(\boldsymbol{w}_s, T_s, S_s) \in B_\rho$. When $s \le t \le -1$, integrating the inequality before (4.3.35) from t to $t+1$, according to (4.3.55), we know that there exists a finite random variable $R_1(\omega)$ such that a.s.

$$c_2 \int_t^{t+1} \left[\int_\Omega \left(|\nabla \boldsymbol{w}|^2 + \left| \frac{\partial \boldsymbol{w}}{\partial z} \right|^2 + |\boldsymbol{w}|^2 \right) \right] \le R_1(\omega). \tag{4.3.56}$$

Applying the uniform Gronwall Inequality, $|\tilde{\boldsymbol{w}}|_3^3 \le |\tilde{\boldsymbol{w}}|_2^{\frac{3}{2}} \|\tilde{\boldsymbol{w}}\|_2^{\frac{3}{2}}$, (4.3.55), (4.3.56) and Lemma 4.3.3, we derive from the inequality before (4.3.39)

$$|\tilde{\boldsymbol{w}}(t+1)|_3^3 \le R_2(\omega), \tag{4.3.57}$$

where $R_2(\omega)$ is a a.s. finite random variable, $t \in [s, -1]$.

Applying the uniform Gronwall Inequality, $|\tilde{\boldsymbol{w}}|_4^4 \le |\tilde{\boldsymbol{w}}|_3^2 \|\tilde{\boldsymbol{w}}\|^2$ (4.3.55)-(4.3.57) and Lemma 4.33, we derive from the inequality before (4.3.40),

$$|\tilde{\boldsymbol{w}}(t+2)|_4^4 \le R_3(\omega), \tag{4.3.58}$$

where $R_3(\omega)$ is an a.s. finite random variable, $t \in [s, -2]$.

Similarly, there exist a.s. finite random variables $R_4(\omega), R_5(\omega), R_6(\omega)$ and $R_7(\omega)$ such that

$$|\boldsymbol{w}_z(t+3)|_2^2 \le R_4(\omega), \ \forall\, t \in [s, -3], \tag{4.3.59}$$

$$|T_z(t+4)|_2^2 + |S_z(t+4)|_2^2 \le R_5(\omega), \ \forall\, t \in [s, -4], \tag{4.3.60}$$

$$|\nabla \boldsymbol{w}(t+5)|_2^2 \le R_6(\omega), \ \forall\, t \in [s, -5], \tag{4.3.61}$$

$$|\nabla T(t+6)|_2^2 + |\nabla S(t+6)|_2^2 \le R_7(\omega), \ \forall\, t \in [s, -6]. \tag{4.3.62}$$

According to (4.3.31), (4.3.33), (4.3.55) and (4.3.59)-(4.3.62), we know that there exist $r_0(\omega)$ and $t(\omega, \rho) \le -1$ such that for any $t_0 \le t(\omega, \rho)$, $\mathcal{U}_{t_0} \in B_\rho$

$$\|\phi(0, t_0; \omega)\mathcal{U}_{t_0}\| \le r_0(\omega).$$

In order to prove Theorem 4.3.9, we need the following properties about $\{\phi(t, s; \omega)\}_{t \ge s}$.

Proposition 4.3.14. For any $t \ge s$, the mapping $\phi(t, s; \omega)$ is weakly continuous from V to V.

Proof of Proposition 4.3.14. Let $\{\mathcal{U}_n\}$ be a sequence in V and $\mathcal{U}_n$ weakly converge to $\mathcal{U}^1$ in V. Then $\{\mathcal{U}_n\}$ is bounded in V. According to the *a priori* estimates of subsections 4.3.3 and 4.3.4 and the proof of Proposition 4.3.13, we know that for any $t \ge s$, $\{\phi(t, s; \omega)\mathcal{U}_n\}$ is bounded in

V. So we can extract a subsequence $\{\phi(t,s;\omega)\mathcal{U}_{n_k}\}$ such that $\phi(t,s;\omega)\mathcal{U}_{n_k}$ strongly converge to $\mathcal{U}^1$ in H. According to (4.3.47)-(4.3.49), we obtain $\phi(t,s;\omega)\mathcal{U}_{n_k}$ strongly converge to $\phi(t,s;\omega)\mathcal{U}^1$ in H. Then $\mathcal{U} = \phi(t,s;\omega)\mathcal{U}^1$. Thus, the sequence $\{S(t,s;\omega)U_n\}$ has a subsequence $\{\phi(t,s;\omega)\mathcal{U}_{n_k}\}$ such that $\phi(t,s;\omega)\mathcal{U}_{n_k}$ weakly converge to $\phi(t,s;\omega)\mathcal{U}^1$. Proposition 4.3.14 is proved.

Proof of Theorem 4.3.9. According to Proposition 4.3.13 and Proposition 4.3.14, applying Theorem 4.1.9, we prove Theorem 4.3.9.

Remark 4.3.15. Recently, there are some research works about the stochastic primitive equations with white noise, see [58, 78, 79, 80, 85].

Chapter 5

Stability and Instability Theory

In this chapter we recall some stability and instability theory of waves in the atmospheric and oceanic dynamics. We discuss neutral stability, linear stability, formal stability and nonlinear stability of waves in geophysics fluid dynamics. Here several methods are used in stability or instability of waves, such as normal mode approach, energy method, and variation method.

5.1 Stability and Instability of Gravity Waves

In this section, we mainly introduce some neutral stability and instability theory of internal gravity waves and internal inertial gravity waves through the normal mode approach, which appears in chapter 11 of [145]. Here, we give the definitions of gravity waves, internal inertial gravity waves and neutral stability. Internal gravity waves are generated by vertical disturbance under gravity in the atmosphere. Internal inertial gravity waves are internal gravity waves in the cases where Coriolis force is considered. A stationary solution u_e of the dynamical system $u_t = X(u)$ (that is $X(u_e) = 0$) is neutral stable, if the spectrums of the linear operator $DX(u_e)$ are all pure imaginary numbers. Neutral stability is a special example of spectral stability (if the spectrums of the linear operator $DX(u_e)$ have no strict positive real part, then the stationary solution u_e is spectral stable), and for Hamiltonian systems, they are all the same.

5.1.1 *Stability and Instability of Internal Gravity Waves in Stratified Flows*

In a two-layer fluid system with an interface and discontinuous density, when the upper and lower velocity is not the same, internal gravity waves generated at the interface are called interfacial waves. We suppose that each layer is inviscous, homogeneous and incompressible. Assume that the density of the lower layer ($0 \le z \le h_1$) is constant ρ_1, the basic flow is $\bar{u}_1$, the density of the upper layer ($h_1 \le z \le H = h_1 + h_2$) is constant ρ_2, and the basic flow is $\bar{u}_2$, where $\bar{u}_1$ and $\bar{u}_2$ are constants. Here, we assume

$$\rho_2 < \rho_1, \ \bar{u}_2 > \bar{u}_1.$$

In this way, the interface of the two fluids is

$$z = h_1(x, y, t)$$

and the total depth of the fluid is

$$H = h_1 + h_2.$$

For simplicity, we suppose that the motions only occur in (x, z) surface, and the disturbed interface satisfies

$$h_1 = H_1 + h_1', \quad h_2 = H_2 + h_2', \tag{5.1.1}$$

$$u_j = \bar{u}_j + u_j', \quad \omega_j = \omega_j', \quad p_j = \bar{p}_j + p_j', \ j = 1, 2, \tag{5.1.2}$$

where $j = 1$ stands for the lower layer, $j = 2$ stands for the upper layer, and H_1, H_2 are positive constants. Obviously, $\bar{p}_j$ satisfies the hydrostatic equilibrium equation

$$\frac{\partial \bar{p}_j}{\partial z} = -g\rho_j.$$

Suppose that the disturbance quantities are small. The linear equations (linearization of the two-dimensional incompressible Navier-Stokes equations with gravity with respective to the stationary solution $(\bar{u}_j, 0, \bar{p}_j)$) are written as

$$\begin{cases} \left(\frac{\partial}{\partial t} + \bar{u}_j \frac{\partial}{\partial x}\right) u_j' = -\frac{1}{\rho_j}\frac{\partial p_j'}{\partial x}, \\ \left(\frac{\partial}{\partial t} + \bar{u}_j \frac{\partial}{\partial x}\right) \omega_j' = -\frac{1}{\rho_j}\frac{\partial p_j'}{\partial z}, \\ \frac{\partial u_j'}{\partial x} + \frac{\partial \omega_j'}{\partial z} = 0. \end{cases} \tag{5.1.3}$$

Thus, we derive from (5.1.3)

$$\left(\frac{\partial}{\partial t} + \bar{u}_j \frac{\partial}{\partial x}\right)\left(\frac{\partial^2}{\partial x^2} + \frac{\partial^2}{\partial z^2}\right) \omega_j' = 0. \tag{5.1.4}$$

Suppose the two layers fluid are between the rigid boundaries $z = 0$ and $z = H$. Thus, the boundary conditions of the upper and lower surfaces are

$$\begin{aligned} \omega_1'|_{z=0} &= 0, \\ \omega_2'|_{z=H} &= 0. \end{aligned} \tag{5.1.5}$$

By the hydrostatic equilibrium equation, the boundary condition at the interface is given by

$$\left[\left(\frac{\partial}{\partial t} + \bar{u}_j \frac{\partial}{\partial x}\right)\left(p_1' - p_2'\right) - g\left(\rho_1 - \rho_2\right)\omega_j'\right]_{z=H_1} = 0. \tag{5.1.6}$$

Now, under the boundary conditions (5.1.5) and (5.1.6), we solve the eigenvalue problem corresponding to equation (5.1.4). Suppose that the form of wave solutions of equation (5.1.4) is

$$\omega_j' = W_j(z)\mathrm{e}^{ik(x-ct)}. \tag{5.1.7}$$

Substituting (5.1.7) into equation (5.1.4), when $c \neq \bar{u}_j$, we get

$$\frac{\mathrm{d}^2 W_j}{\mathrm{d}z^2} - k^2 W_j = 0 \ (j = 1, 2). \tag{5.1.8}$$

According to (5.1.5), solution of equation (5.1.8) can be written as

$$\begin{cases} W_1(z) = A \sinh kz, \\ W_2(z) = B \sinh k(H - z), \end{cases} \tag{5.1.9}$$

where A and B are constants.

Substituting (5.1.9) into equation (5.1.7), we get ω_j'. Then substituting w_j' into (5.1.3), we obtain u_j', p_j'. Thus, we obtain

$$\begin{cases} u_1' = \mathrm{i}A \cosh kz \mathrm{e}^{\mathrm{i}k(x-ct)}, \\ \omega_1' = A \sinh kz \mathrm{e}^{\mathrm{i}k(x-ct)}, \\ p_1' = \mathrm{i}\rho_1(c - \bar{u}_1)A \cosh kz \mathrm{e}^{\mathrm{i}k(x-ct)}, \end{cases} \tag{5.1.10}$$

$$\begin{cases} u_2' = -\mathrm{i}B \cosh k(H - z)\mathrm{e}^{\mathrm{i}k(x-ct)}, \\ \omega_2' = B \sinh k(H - z)\mathrm{e}^{\mathrm{i}k(x-ct)}, \\ p_2' = -\mathrm{i}\rho_2(c - \bar{u}_2)B \cosh k(H - z)\mathrm{e}^{\mathrm{i}k(x-ct)}. \end{cases} \tag{5.1.11}$$

Substituting $p_1', p_2', \omega_1', \omega_2'$ into the interface condition (5.1.6), we have

$$\begin{aligned} &\left[k(c - \bar{u}_1)^2 \rho_1 \cosh kH_1 - g(\rho_1 - \rho_2) \sinh kH_1\right] A \\ &+ \left[k(c - \bar{u}_1)(c - \bar{u}_2)\rho_2 \cosh kH_2\right] B = 0, \end{aligned} \tag{5.1.12}$$

$$\begin{aligned} &\left[k(c - \bar{u}_1)(c - \bar{u}_2)\rho_1 \cosh kH_1\right] A \\ &+ \left[k(c - \bar{u}_2)^2 \rho_2 \cosh kH_2 - g(\rho_1 - \rho_2) \sinh kH_2\right] B = 0. \end{aligned} \tag{5.1.13}$$

The above two equations are homogeneous linear algebra equations of A, B, of which the necessary and sufficient condition for the existence of nonzero solutions are that the coefficient determinant of A, B is zero. Expanding this coefficient determinant, we have

$$k(c-\bar{u}_1)^2\rho_1\lambda_1 + k(c-\bar{u}_2)^2\rho_2\lambda_2 - g(\rho_1-\rho_2) = 0, \tag{5.1.14}$$

or

$$(\rho_1\lambda_1+\rho_2\lambda_2)c^2 - 2(\rho_1\lambda_1\bar{u}_1+\rho_2\lambda_2\bar{u}_2)c + \left[\rho_1\lambda_1\bar{u}_1^2+\rho_2\lambda_2\bar{u}_2^2 - \frac{g}{k}(\rho_1-\rho_2)\right] = 0, \tag{5.1.15}$$

where

$$\lambda_1 = \coth kH_1, \quad \lambda_2 = \coth kH_2.$$

According to equation (5.1.15), we get

$$c = \frac{\rho_1\lambda_1\bar{u}_1+\rho_2\lambda_2\bar{u}_2}{\rho_1\lambda_1+\rho_2\lambda_2} \pm \sqrt{\frac{g(\rho_1-\rho_2)}{k(\rho_1\lambda_1+\rho_2\lambda_2)} - \frac{\lambda_1\lambda_2\rho_1\rho_2(\bar{u}_2-\bar{u}_1)^2}{(\rho_1\lambda_1+\rho_2\lambda_2)^2}}. \tag{5.1.16}$$

Let

$$\bar{u}_1 = \bar{u} - \hat{u}, \quad \bar{u}_2 = \bar{u} + \hat{u},$$

and

$$\bar{u} = (\bar{u}_1+\bar{u}_2)/2, \quad \hat{u} = (\bar{u}_2-\bar{u}_1)/2,$$

where $\bar{u}$ denotes the average of the basic flow, and $\hat{u}$ denotes the vertical shear of the basic flow. Let

$$\rho = \frac{\rho_1+\rho_2}{2}, \quad \Delta\rho = \rho_1 - \rho_2.$$

Since $\Delta\rho \ll \rho_1, \rho_2$ in general conditions, we take the approxination $\rho \approx \rho_1 \approx \rho_2$. Thus, (5.1.16) is rewritten as

$$\begin{aligned} c &= \frac{\lambda_1\bar{u}_1+\lambda_2\bar{u}_2}{\lambda_1+\lambda_2} \pm \sqrt{\frac{g(\rho_1-\rho_2)}{k(\lambda_1+\lambda_2)\rho} - \frac{\lambda_1\lambda_2(\bar{u}_2-\bar{u}_1)^2}{(\lambda_1+\lambda_2)^2}} \\ &= \bar{u} + \frac{\lambda_2-\lambda_1}{\lambda_2+\lambda_1}\hat{u} \pm \sqrt{\frac{g(\rho_1-\rho_2)}{k(\lambda_1+\lambda_2)\rho} - \frac{4\lambda_1\lambda_2\hat{u}^2}{(\lambda_1+\lambda_2)^2}}. \end{aligned} \tag{5.1.17}$$

Thus we know that, c is related with $\bar{u}, \hat{u}, g, k, \rho_1-\rho_2$ and so on. Moreover, if the quantity $\dfrac{g(\rho_1-\rho_2)}{k(\lambda_1+\lambda_2)\rho} - \dfrac{4\lambda_1\lambda_2\hat{u}^2}{(\lambda_1+\lambda_2)^2}$ is zero or positive, that is, c is real, then a wave with (5.1.7) is neutral wave, and the stratified flow is neutral stable. If the quantity is negative, that is, c is imaginary, then the stratified flow is neutral instable. So, a necessary and sufficient condition of the neutral stability of internal gravity waves in the stratified flow is

$$\frac{g(\rho_1-\rho_2)}{k(\lambda_1+\lambda_2)\rho}-\frac{4\lambda_1\lambda_2\hat{u}^2}{(\lambda_1+\lambda_2)^2}\begin{cases}\geq 0, & \text{neutral stable,}\\ <0, & \text{neutral unstable.}\end{cases} \tag{5.1.18}$$

We know from the above relationship that gravity has an important effect on stability when $\rho_1<\rho_2, \bar{u}_1=\bar{u}_2=0$, and the vertical shear of the wind speed has an effect on instability. Notice that in this problem there has

$$\begin{cases} N^2\equiv -\dfrac{g}{\rho}\dfrac{\partial\rho}{\partial z}\approx\dfrac{g(\rho_1-\rho_2)}{\rho H},\\ \left(\dfrac{\partial\bar{u}}{\partial z}\right)^2\approx\left(\dfrac{\bar{u}_2-\bar{u}_1}{H}\right)^2=\dfrac{4\hat{u}^2}{H^2}.\end{cases}$$

Thus, Richardson number can be defined by

$$Ri=(\rho_1-\rho_2)gH/4\rho\hat{u}^2.$$

If

$$H_1=H_2=H/2,$$

$$\lambda_1=\lambda_2=\lambda=\coth\frac{kH}{2}.$$

Then, (5.1.18) is rewritten as

$$Ri\begin{cases}\geq\dfrac{kH}{2}\lambda=\dfrac{kH}{2}\coth\dfrac{kH}{2}, & \text{neutral stable,}\\ <\dfrac{kH}{2}\lambda=\dfrac{kH}{2}\coth\dfrac{kH}{2}, & \text{neutral unstable.}\end{cases} \tag{5.1.19}$$

Now we discuss two special situations.

(1) When H_1, H_2 are very small, $kH_1\ll 1, kH_2\ll 1$,

$$\lambda_1\approx 1/kH_1, \lambda_2\approx 1/kH_2.$$

(5.1.17) is rewritten as

$$c=\bar{u}+\frac{H_1-H_2}{H_1+H_2}\hat{u}\pm\sqrt{\frac{g(\rho_1-\rho_2)}{\rho}\cdot\frac{H_1H_2}{H_1+H_2}-\frac{4H_1H_2\hat{u}^2}{(H_1+H_2)^2}}. \tag{5.1.20}$$

If $H_1=H_2=H/2$, then we derive from the above relationship that

$$Ri\begin{cases}\geq 1, & \text{neutral stable,}\\ <1, & \text{neutral unstable,}\end{cases} \tag{5.1.21}$$

which is known as Rayleigh Theorem. It's actually the limit situation of (5.1.19) as $kH\to 0$.

(2) When H_1, H_2 are very large, $kH_1 \gg 1, kH_2 \gg 1$,

$$\lambda_1 \approx 1, \lambda_2 \approx 1,$$

(5.1.17) is rewritten as

$$c = \bar{u} \pm \sqrt{\frac{g(\rho_1 - \rho_2)}{2k\rho} - \hat{u}^2}.$$

Then we get from the above relationship that

$$Ri \begin{cases} \geq kH/2, & \text{neutral stable,} \\ < kH/2, & \text{neutral unstable,} \end{cases} \tag{5.1.22}$$

which is the limit situation of (5.1.19) as $kH \to \infty$.

5.1.2 *Stability of General Internal Gravity Wave*

In this subsection, we discuss the neutral stability of internal gravity waves of the three-dimensional Boussinesq equations. Applying the Boussinesq approximation. We obtain the linear equations of a stationary solution $(\bar{u}, 0, 0, \bar{p})$ written as

$$\begin{cases} \left(\dfrac{\partial}{\partial t} + \bar{u}\dfrac{\partial}{\partial x}\right) u' + \dfrac{\partial \bar{u}}{\partial z}\omega' = -\dfrac{1}{\rho_0}\dfrac{\partial p'}{\partial x}, \\ \left(\dfrac{\partial}{\partial t} + \bar{u}\dfrac{\partial}{\partial x}\right) v' = -\dfrac{1}{\rho_0}\dfrac{\partial p'}{\partial y}, \\ \left(\dfrac{\partial}{\partial t} + \bar{u}\dfrac{\partial}{\partial x}\right) \omega' = -\dfrac{1}{\rho_0}\dfrac{\partial p'}{\partial z} - g\dfrac{\rho'}{\rho_0}, \\ \dfrac{\partial u'}{\partial x} + \dfrac{\partial v'}{\partial y} + \dfrac{\partial \omega'}{\partial z} = 0, \\ \left(\dfrac{\partial}{\partial t} + \bar{u}\dfrac{\partial}{\partial x}\right) \rho' - \dfrac{N^2}{g}\rho_0\omega' = 0, \end{cases} \tag{5.1.23}$$

where the basic flow $(\bar{u}, 0, 0)$ and the basic density are only dependent on z. The above equations are the linearization of the three-dimensional Boussinesq equations based on the stationary solution $(\bar{u}(z), 0, 0, \rho_0(z))$.

Taking the derivative of the first and second equations of equations (5.1.23) with respect to x and y respectively, and making use of the fourth equation in (5.1.23), we get

$$\left(\frac{\partial}{\partial t} + \bar{u}\frac{\partial}{\partial x}\right)\left(\frac{\partial \omega'}{\partial z}\right) - \frac{\partial \bar{u}}{\partial z}\frac{\partial \omega'}{\partial x} = \frac{1}{\rho_0}\nabla_h^2 p'.$$

Taking the derivative of the above equation with respect to z, and making use of the third equation in (5.1.23), we have

$$\left(\frac{\partial}{\partial t}+\bar{u}\frac{\partial}{\partial x}\right)\nabla^2\omega' - \frac{\partial^2\bar{u}}{\partial z^2}\frac{\partial\omega'}{\partial x} = -\frac{g}{\rho_0}\nabla_h^2\rho'.$$

Combining the above equation with the fifth equation in (5.1.23), we obtain

$$\left[\left(\frac{\partial}{\partial t}+\bar{u}\frac{\partial}{\partial x}\right)^2\nabla^2 - \frac{\partial^2\bar{u}}{\partial z^2}\left(\frac{\partial}{\partial t}+\bar{u}\frac{\partial}{\partial x}\right)\frac{\partial}{\partial x} + N^2\nabla_h^2\right]\omega' = 0, \quad (5.1.24)$$

where

$$\nabla^2 = \frac{\partial^2}{\partial x^2}+\frac{\partial^2}{\partial y^2}+\frac{\partial^2}{\partial z^2}, \quad \nabla_h^2 = \frac{\partial^2}{\partial x^2}+\frac{\partial^2}{\partial y^2}.$$

The boundary conditions of equation (5.1.24) are given by

$$\begin{cases} \dfrac{\partial\omega'}{\partial y}|_{y=y_1} = 0, \ \dfrac{\partial\omega'}{\partial y}|_{y=y_2} = 0, \\ \omega'|_{z=0} = 0, \quad \omega'|_{z=H} = 0. \end{cases} \quad (5.1.25)$$

When $\bar{u}=0$, (5.1.24) is written as

$$\left(\frac{\partial^2}{\partial t^2}\nabla^2 + N^2\nabla_h^2\right)\omega' = 0. \quad (5.1.26)$$

Let $\omega' = A\mathrm{e}^{\mathrm{i}(k_1x+k_2y+k_3z-\omega t)}$. According to equation (5.1.26), we get

$$\omega^2 = \frac{K_h^2}{K^2}N^2,$$

where $K^2 = k_1^2+k_2^2+k_3^2$, $K_h^2 = k_1^2+k_2^2$. The above relationship shows that in the case of no basic flow, the stability of internal gravity waves is completely dependent on the stratified stability, which is simple but reflects the effect of stratified stability in the stability of internal gravity waves.

When $\bar{u}\neq 0$, according to the boundary condition (5.1.25), we suppose that the solution of equation (5.1.24) is given by

$$\omega' = W(z)\cos l(y-y_1)\mathrm{e}^{\mathrm{i}k(x-ct)}, \quad (5.1.27)$$

where

$$l = 2\pi/L_y, \ L_y = (y_2-y_1),$$

and L_y is the wavelength in north-south direction.

Substituting (5.1.27) into the equation (5.1.24), we have

$$(\bar{u}-c)^2\frac{\mathrm{d}^2W}{\mathrm{d}z^2} + \left[\frac{K_h^2N^2}{k^2} - (\bar{u}-c)\frac{\partial^2\bar{u}}{\partial z^2} - K_h^2(\bar{u}-c)^2\right]W = 0, \quad (5.1.28)$$

where

$$K_h = l^2 + k^2.$$

It is a second-order ordinary differential equation with variable coefficient, which can be solved with the boundary condition when $N(z), c$ and $\bar{u}(z)$ are given. But we consider some necessary conditions of stability here, and should make some exchange. Firstly, let

$$F \equiv W/(\bar{u}-c),$$

thus

$$\frac{\mathrm{d}W}{\mathrm{d}z} = (\bar{u}-c)\frac{\mathrm{d}F}{\mathrm{d}z} + F\frac{\partial \bar{u}}{\partial z},$$

$$\frac{\mathrm{d}^2W}{\mathrm{d}z^2} = (\bar{u}-c)\frac{\mathrm{d}^2F}{\mathrm{d}z^2} + 2\frac{\partial \bar{u}}{\partial z}\frac{\mathrm{d}F}{\mathrm{d}z} + F\frac{\mathrm{d}^2\bar{u}}{\mathrm{d}z^2},$$

$$(\bar{u}-c)\frac{\mathrm{d}^2W}{\mathrm{d}z^2} = \frac{\mathrm{d}}{\mathrm{d}z}\left[(\bar{u}-c)^2\frac{\mathrm{d}F}{\mathrm{d}z}\right] + (\bar{u}-c)F\frac{\partial^2\bar{u}}{\partial z^2}.$$

Substituting the above formulas into (5.1.28), we get

$$\frac{\mathrm{d}}{\mathrm{d}z}\left[(\bar{u}-c)^2\frac{\mathrm{d}F}{\mathrm{d}z}\right] + \left[\frac{K_h^2N^2}{k^2} - K_h^2(\bar{u}-c)^2\right]F = 0. \tag{5.1.29}$$

Let

$$G \equiv (\bar{u}-c)^{\frac{1}{2}}F,$$

then

$$\frac{\mathrm{d}F}{\mathrm{d}z} = (\bar{u}-c)^{-\frac{1}{2}}\frac{\mathrm{d}G}{\mathrm{d}z} - \frac{1}{2}(\bar{u}-c)^{-\frac{3}{2}}G\frac{\partial \bar{u}}{\partial z},$$

$$(\bar{u}-c)^{\frac{3}{2}}\frac{\mathrm{d}F}{\mathrm{d}z} = (\bar{u}-c)\frac{\mathrm{d}G}{\mathrm{d}z} - \frac{1}{2}G\frac{\partial \bar{u}}{\partial z},$$

$$\frac{\mathrm{d}}{\mathrm{d}z}\left[(\bar{u}-c)\frac{\mathrm{d}G}{\mathrm{d}z}\right] = (\bar{u}-c)^{-\frac{1}{2}}\frac{\mathrm{d}}{\mathrm{d}z}\left[(\bar{u}-c)^2\frac{\mathrm{d}F}{\mathrm{d}z}\right] + \frac{1}{4}(\bar{u}-c)^{-1}G\left(\frac{\partial \bar{u}}{\partial z}\right)^2 + \frac{1}{2}G\frac{\partial^2\bar{u}}{\partial z^2}.$$

Substituting the above formulas into (5.1.29), we have

$$\frac{\mathrm{d}}{\mathrm{d}z}\left[(\bar{u}-c)\frac{\mathrm{d}G}{\mathrm{d}z}\right] - \left[\frac{1}{2}\frac{\partial^2\bar{u}}{\partial z^2} + K_h^2(\bar{u}-c) + \frac{\frac{1}{4}\left(\frac{\partial \bar{u}}{\partial z}\right)^2 - \frac{K_h^2N^2}{k^2}}{\bar{u}-c}\right]G = 0. \tag{5.1.30}$$

According to the boundary condition of ω' in the direction of z, we obtain

$$W|_{z=0,H} = F|_{z=0,H} = G|_{z=0,H} = 0. \tag{5.1.31}$$

Now we apply equations (5.1.28)-(5.1.30) and the boundary condition (5.1.31) to discuss the neutral instability of the general internal gravity waves, and obtain the growth rate of instable waves, then obtain Haward semi-circle theorem about the growth rate of instable waves.

5.1.2.1 *Necessary Conditions of Instability*

Taking complex conjugation of (5.1.28), we get

$$(\bar{u}-c^*)^2\frac{\mathrm{d}^2W^*}{\mathrm{d}z^2}+\left[\frac{K_h^2N^2}{k^2}-(\bar{u}-c^*)\frac{\partial^2\bar{u}}{\partial z^2}-K_h^2(\bar{u}-c^*)^2\right]W^*=0, \quad (5.1.32)$$

where W^* is the complex conjugate of W, and c^* is the complex conjugate of c. Taking L^2 inner product of W^* with equation (5.1.28), and taking L^2 inner product of W with equation (5.1.32), then applying

$$W^*\frac{\mathrm{d}^2W}{\mathrm{d}z^2}-W\frac{\mathrm{d}^2W^*}{\mathrm{d}z^2}=\frac{\mathrm{d}}{\mathrm{d}z}\left(W^*\frac{\mathrm{d}W}{\mathrm{d}z}-W\frac{\mathrm{d}W^*}{\mathrm{d}z}\right)$$

and the boundary condition (5.1.31), we have

$$\int_0^H\left\{\frac{K_h^2N^2}{k^2}\left[\frac{1}{(\bar{u}-c)^2}-\frac{1}{(\bar{u}-c^*)^2}\right]-\left(\frac{1}{\bar{u}-c}-\frac{1}{\bar{u}-c^*}\right)\frac{\partial^2\bar{u}}{\partial z^2}\right\}WW^*\mathrm{d}z$$
$$=0. \quad (5.1.33)$$

Noting that

$$WW^*=|W|^2,$$

$$\frac{1}{\bar{u}-c}-\frac{1}{\bar{u}-c^*}=\frac{c-c^*}{|\bar{u}-c|^2}=\frac{2\mathrm{i}c_i}{|\bar{u}-c|^2},$$

$$\frac{1}{(\bar{u}-c)^2}-\frac{1}{(\bar{u}-c^*)^2}=\frac{[2\bar{u}-(c+c^*)]\,(c-c^*)}{|\bar{u}-c|^4}=\frac{4\mathrm{i}(\bar{u}-c_r)c_i}{|\bar{u}-c|^4},$$

we rewrite (5.1.33) as

$$c_i\int_0^H\left[\frac{2K_h^2N^2}{k^2}(\bar{u}-c_r)-|\bar{u}-c|^2\frac{\partial^2\bar{u}}{\partial z^2}\right]\frac{|W|^2}{|\bar{u}-c|^4}\mathrm{d}z=0. \quad (5.1.34)$$

Since $|W|^2/|\bar{u}-c|^4>0$, and the necessary and sufficient condition of the neutral instability of internal gravity waves is $c_i\neq 0$, if (5.1.34) is satisfied, then

$$\frac{2K_h^2N^2}{k^2}(\bar{u}-c_r)-|\bar{u}-c|^2\frac{\partial^2\bar{u}}{\partial z^2}\ \text{changes the symbol in}\ (0,H), \quad (5.1.35)$$

or there is at least one point $z=z_c$ in $(0,H)$ such that

$$\frac{2K_h^2N^2}{k^2}(\bar{u}-c_r)-|\bar{u}-c|^2\frac{\partial^2\bar{u}}{\partial z^2}=0,\ \text{at}\ z=z_c\in(0,H). \quad (5.1.36)$$

(5.1.35) or (5.1.36) is a necessary condition of the neutral instability of the internal gravity waves $(\bar{u},0,0,\rho_0)$.

Making use of equation (5.1.30), we get another stability condition of internal gravity waves. Making L^2 inner product of G^*, with equation (5.1.30), and applying

$$G^* \frac{\mathrm{d}}{\mathrm{d}z}\left[(\bar{u}-c)\frac{\mathrm{d}G}{\mathrm{d}z}\right] = \frac{\mathrm{d}}{\mathrm{d}z}\left[(\bar{u}-c)G^*\frac{\mathrm{d}G}{\mathrm{d}z}\right] - (\bar{u}-c)\left|\frac{\mathrm{d}G}{\mathrm{d}z}\right|^2$$

and boundary condition (5.1.31), we get

$$\int_0^H \left\{(\bar{u}-c)\left(\left|\frac{\mathrm{d}G}{\mathrm{d}z}\right|^2 + K_h^2|G|^2\right)\right.$$
$$\left. - \frac{1}{2}\frac{\partial^2 \bar{u}}{\partial z^2}|G|^2 + \left[\frac{K_h^2 N^2}{k^2} - \frac{1}{4}\left(\frac{\partial \bar{u}}{\partial z}\right)^2\right]\frac{|G|^2}{\bar{u}-c}\right\}\mathrm{d}z = 0.$$

Noting that $1/(\bar{u}-c) = (\bar{u}-c^*)/|\bar{u}-c|^2$, and taking the imaginary part of the above equality, we have

$$c_i \int_0^H \left\{\left|\frac{\mathrm{d}G}{\mathrm{d}z}\right|^2 + K_h^2|G|^2 + \left[\frac{K_h^2 N^2}{k^2} - \frac{1}{4}\left(\frac{\partial \bar{u}}{\partial z}\right)^2\right]\frac{|G|^2}{|\bar{u}-c|^2}\right\}\mathrm{d}z = 0. \tag{5.1.37}$$

If the first two terms in the brace of (5.1.37) are positive, and the third one is nonnegative, that is,

$$\frac{K_h^2 N^2}{k^2} - \frac{1}{4}\left(\frac{\partial \bar{u}}{\partial z}\right)^2 \geq 0, \tag{5.1.38}$$

then $c_i = 0$. Thus the internal gravity wave is stable. (5.1.38) is a sufficient condition of the stability of internal gravity waves. By the definition of Richardson number $Ri = \dfrac{N^2}{\left(\dfrac{\partial \bar{u}}{\partial z}\right)^2}$, (5.1.38) is rewritten as

$$Ri \geq k^2/4K_h^2. \tag{5.1.39}$$

If $l = 0$, then $K_h^2 = k^2$. Thus, (5.1.39) is written as

$$Ri \geq \frac{1}{4}. \tag{5.1.40}$$

That is the so-called Miles Theorem (see [61]). If $Ri \geq \frac{1}{4}$, then the internal gravity wave is neutral stable.

On the other hand, if the internal gravity wave is unstable ($c_i \neq 0$), then, according to (5.1.39),

$$Ri < \frac{k^2}{4K_h^2} = \frac{1}{4}\cdot\frac{1}{1+(l/k)^2} = \frac{1}{4}\cdot\frac{1}{1+(L_x/L_y)^2} < \frac{1}{4}. \tag{5.1.41}$$

When (5.1.41) is satisfied, maybe the internal gravity waves are instable.

Remark 5.1.1. In reference [1], Abarbanel et al. applied the Energy-Casimir method (sometimes also called the Arnold method) to consider the formal stability (its definition can be seen in [104]) of a parallel shear flow (which belongs to the internal gravity waves) in the three-dimensional stratified fluid. Here, we recall their proof briefly. Let $(\boldsymbol{u}_e(\boldsymbol{x}), \rho_e(\boldsymbol{x}))$ $(\boldsymbol{u}_e(\boldsymbol{x}) = (u(y,z),0,0),\ \rho_e(\boldsymbol{x}) = \rho(z),\ u(y,z) = f(y) + U(z),\ f(y) = f_0(y/L)^2, f_0 \ll U(z))$ be a solution of the following equation,

$$\frac{\partial}{\partial t}\boldsymbol{u} + (\boldsymbol{u}\cdot\boldsymbol{\nabla})\boldsymbol{u} = -\boldsymbol{\nabla}p - \rho g\hat{z}, \text{ in } \Omega,$$

$$\frac{\partial}{\partial t}\rho + \boldsymbol{u}\cdot\boldsymbol{\nabla}\rho = 0, \text{ in } \Omega,$$

$$\boldsymbol{\nabla}\cdot\boldsymbol{u} = 0, \text{ in } \Omega,$$

where Ω is a bounded domain, the normal direction part of $\boldsymbol{u}$ at the boundary is zero, and ρ is constant at the boundary. As the energy of the above equations

$$\int d^3\boldsymbol{x}\left[\frac{1}{2}|\boldsymbol{u}|^2 + \rho g z\right]$$

is a conserved quantity,

$$q = (\boldsymbol{\nabla}\times\boldsymbol{u})\cdot\boldsymbol{\nabla}\rho$$

and ρ are conserved along orbits of fluid elements, we get the following Energy-Casimir functional

$$A(\vec{\mathbf{u}},\rho) = \int d^3\boldsymbol{x}\left[\frac{1}{2}|\boldsymbol{u}|^2 + \rho g z + G(q,\rho) + \lambda q\right],$$

which is a conserved quantity of the above equations. We find the first order and the second order variation about the steady solution $(\boldsymbol{u}_e, \rho_e)$ of the above functional respectively as

$$\delta A(\boldsymbol{u}_e,\rho_e) = \int d^3\boldsymbol{x}\{\delta\boldsymbol{u}\left[\boldsymbol{u}_e - G_{qq}\boldsymbol{\nabla}\rho_e\times\boldsymbol{\nabla}q_e\right] + \delta\rho\left[gz + G_\rho(\boldsymbol{\nabla}\times\boldsymbol{u}_e)\cdot\boldsymbol{\nabla}G_q\right]\}$$
$$+(\lambda + G_q)|_s\int \mathrm{d}s\hat{n}\cdot\{\delta\rho\boldsymbol{\nabla}\times\boldsymbol{u}_e - \boldsymbol{\nabla}\rho_e\times\boldsymbol{\nabla}\boldsymbol{u}_e\},$$

and

$$\delta^2 A(\boldsymbol{u}_e,\rho_e) = \int d^3\boldsymbol{x}\left\{|\delta\boldsymbol{u}|^2 + (\delta q,\delta p)\begin{pmatrix}G_{qq} & G_{q\rho}\\ G_{q\rho} & G_{\rho\rho}\end{pmatrix}\begin{pmatrix}\delta q\\ \delta\rho\end{pmatrix}\right\}.$$

If $G_{qq} > 0$ and $G_{qq}G_{\rho\rho} - G_{q\rho}^2 > 0$, then the steady solution $(\boldsymbol{u}_e, \rho_e)$ is formally stable. Thus, Abarbanel et al. [1] find a necessary and sufficient condition for formal stability of the steady solution $(\boldsymbol{u}_e, \rho_e)$, which is,

$$N_{Ri}(z) > 1,$$

where

$$N_{Ri}(z) = N(z)^2 / \left\{ \rho_z^2 \partial^2 \left[\frac{1}{2} U^2(z) \right] / \partial \rho^2 \right\}.$$

5.1.2.2 *The Growth Rate of Unstable Waves*

For neutral unstable waves, $c_i \neq 0$, then (5.1.37) is written as

$$\int_0^H \left\{ \left| \frac{\mathrm{d}G}{\mathrm{d}z} \right|^2 + K_h^2 |G|^2 + \left[\frac{K_h^2 N^2}{k^2} - \frac{1}{4} \left(\frac{\partial \bar{u}}{\partial z} \right)^2 \right] \frac{|G|^2}{|\bar{u} - c|^2} \right\} \mathrm{d}z = 0.$$

Applying the definition of Ri, we rewrite the above equality as

$$\begin{aligned} & k^2 \int_0^H |G|^2 \mathrm{d}z \\ &= \int_0^H \left(\frac{\partial \bar{u}}{\partial z} \right)^2 \cdot \frac{K_h^2}{k^2} \left(\frac{k^2}{4K_h^2} - Ri \right) \frac{|G|^2}{|\bar{u} - c|^2} \mathrm{d}z - \int_0^H \left(\left| \frac{\mathrm{d}G}{\mathrm{d}z} \right|^2 + l^2 |G|^2 \right) \mathrm{d}z. \end{aligned} \tag{5.1.42}$$

Noting

$$\frac{1}{|\bar{u} - c|^2} = \frac{1}{(\bar{u} - c_r)^2 + c_i^2} \leq \frac{1}{c_i^2},$$

we derive from (5.1.42) that

$$0 < k^2 \int_0^H |G|^2 \mathrm{d}z \leq \frac{1}{c_i^2} \int_0^H \left(\frac{\partial \bar{u}}{\partial z} \right)^2 \frac{K_h^2}{k^2} \left(\frac{k^2}{4K_h^2} - Ri \right) |G|^2 \mathrm{d}z.$$

Thus

$$k^2 c_i^2 \leq \max_{z \in (0,H)} \left(\frac{\partial \bar{u}}{\partial z} \right)^2 \frac{K_h^2}{k^2} \left(\frac{k^2}{4K_h^2} - Ri \right).$$

Using (5.1.42) and the above inequality, we know that the growth rate of the unstable internal gravity waves satisfies

$$|kc_i| \leq \max_{z \in (0,H)} \left| \frac{\partial \bar{u}}{\partial z} \right| \frac{K_h}{k} \sqrt{\frac{k^2}{4K_h^2} - Ri}. \tag{5.1.43}$$

5.1.2.3 *Haward Semi-Circle Theorem*

Making use of (5.1.29), we obtain other necessary conditions of the unstable internal gravity wave. Taking L^2 inner product of F^* with equation (5.1.29), and applying

$$F^* \frac{\mathrm{d}}{\mathrm{d}z}\left[(\bar{u}-c)^2 \frac{\mathrm{d}F}{\mathrm{d}z}\right] = \frac{\mathrm{d}}{\mathrm{d}z}\left[(\bar{u}-c)^2 F^* \frac{\mathrm{d}F}{\mathrm{d}z}\right] - (\bar{u}-c)^2 \left|\frac{\mathrm{d}F}{\mathrm{d}z}\right|^2$$

and the boundary condition, we get

$$\int_0^H \left\{(\bar{u}-c)^2 \left(\left|\frac{\mathrm{d}F}{\mathrm{d}z}\right|^2 + K_h^2|F|^2\right) - \frac{K_h^2 N^2}{k^2}|F|^2\right\} \mathrm{d}z = 0.$$

The real and imaginary parts of the above equality are respectively

$$\int_0^H \left[(\bar{u}-c_r)^2 - c_i^2\right] \left(\left|\frac{\mathrm{d}F}{\mathrm{d}z}\right|^2 + K_h^2|F|^2\right) \delta z - \int_0^H \frac{K_h^2 N^2}{k^2}|F|^2 \mathrm{d}z = 0, \tag{5.1.44}$$

$$2c_i \int_0^H (\bar{u}-c_r) \left(\left|\frac{\mathrm{d}F}{\mathrm{d}z}\right|^2 + K_h^2|F|^2\right) \mathrm{d}z = 0. \tag{5.1.45}$$

We know from (5.1.45) that for unstable waves, $c_i \neq 0$,

$$\bar{u} - c_r, \quad \text{changes the symbol in} \quad (0, H) \tag{5.1.46}$$

or there exists at least one point $z = z_c$ in $(0, H)$ such that

$$\bar{u}(z_c) - c_r = 0, \text{ in } z = z_c \in (0, H). \tag{5.1.47}$$

(5.1.46) or (5.1.47) is another necessary condition of the neutral instability of the internal gravity wave, which implies that the neutral unstable internal gravity waves must have a critical layer at $\bar{u} = c_r$.

Let $\bar{u}_m$ and $\bar{u}_M$ respectively be the minimum and maximum of $\bar{u}$ in $(0, H)$. The necessary conditions (5.1.46) or (5.1.47) for instability is rewritten as

$$\bar{u}_m < c_r < \bar{u}_M. \tag{5.1.48}$$

Let

$$Q \equiv \left|\frac{\mathrm{d}F}{\mathrm{d}z}\right|^2 + K_h^2|F|^2 \geq 0.$$

Then (5.1.44) and (5.1.45) can be respectively written as

$$\int_0^H \bar{u}^2 Q \mathrm{d}z = (c_i^2 - c_r^2) \int_0^H Q \mathrm{d}z + 2 \int_0^H \bar{u} c_r Q \mathrm{d}z + \int_0^H \frac{K_h^2 N^2}{k^2}|F|^2 \mathrm{d}z,$$

$$\int_0^H \bar{u}Q\mathrm{d}z = \int_0^H c_r Q\mathrm{d}z.$$

According to the above two equalities, we get

$$\int_0^H \bar{u}^2 Q\mathrm{d}z = (c_i^2 + c_r^2)\int_0^H Q\mathrm{d}z + \int_0^H \frac{K_h^2 N^2}{k^2}|F|^2\mathrm{d}z. \tag{5.1.49}$$

According to $(\bar{u} - \bar{u}_m)(\bar{u} - \bar{u}_M) < 0$, we have

$$\int_0^H \{\bar{u}^2 - (\bar{u}_m + \bar{u}_M)\bar{u} + \bar{u}_m\bar{u}_M\}Q\mathrm{d}z < 0.$$

Thus, we obtain

$$(c_i^2 + c_r^2)\int_0^H Q\mathrm{d}z + \int_0^H \frac{K_h^2 N^2}{k^2}|F|^2\mathrm{d}z - \int_0^H (\bar{u}_m + \bar{u}_M)c_r Q\mathrm{d}z + \int_0^H \bar{u}_m\bar{u}_M Q\mathrm{d}z < 0,$$

that is,

$$\int_0^H \left\{\left[c_r - \frac{1}{2}(\bar{u}_m + \bar{u}_M)\right]^2 + c_i^2 - \frac{1}{4}(\bar{u}_m - \bar{u}_M)^2\right\}Q\mathrm{d}z + \int_0^H \frac{K_h^2 N^2}{k^2}|F|^2\mathrm{d}z < 0. \tag{5.1.50}$$

Thus, for the stable stratified fluids ($N^2 > 0$), we have

$$(c_r - U)^2 + c_i^2 < c_R^2, \tag{5.1.51}$$

where

$$U = \frac{1}{2}(\bar{u}_m + \bar{u}_M),\ c_R^2 = \hat{u}^2 - \frac{\dfrac{K_h^2}{k^2}\displaystyle\int_0^H N^2|F|^2\mathrm{d}z}{\displaystyle\int_0^H Q\mathrm{d}z},\ \hat{u} = \frac{1}{2}(\bar{u}_M - \bar{u}_m).$$

(5.1.51) shows that in the complex plane of phase velocity c, c of the unstable internal gravity waves must be in the upper circle centered at $(U, 0)$ within a radius of c_R (because we can assume $k > 0$, and for the unstable internal gravity waves, $c_i > 0$). For the stable stratified fluids, $N^2 > 0$,

$$c_R^2 < \hat{u}^2.$$

Thus (5.1.51) is written as

$$(c_r - U)^2 + c_i^2 < \hat{u}^2, \tag{5.1.52}$$

which is called Howard semi-circle theorem.

Remark 5.1.2. The Howard semi-circle theorem quantitatively describes the growth rate of unstable waves, which can be seen in [106]. We give an example of application of Howard semi-circle theorem. The linerization equation of two-dimensional Euler equation

$$\partial_t \Delta\psi + \frac{\partial \psi}{\partial y}\frac{\partial \Delta\psi}{\partial x} - \frac{\partial \psi}{\partial x}\frac{\partial \Delta\psi}{\partial y} = 0$$

about a steady solution $\left(\frac{\partial \psi}{\partial y}, -\frac{\partial \psi}{\partial x}\right) = (U(y), 0)$ is

$$\partial_t \Delta\psi' + U\frac{\partial \Delta\psi'}{\partial x} - \frac{\mathrm{d}^2 U}{\mathrm{d}y^2}\frac{\partial \psi'}{\partial x} = 0.$$

Let $\psi' = \phi(y)\mathrm{e}^{\mathrm{i}k(x-ct)}$. The above equation is written as Rayleigh equation

$$(U-c)\left(\frac{\mathrm{d}^2}{\mathrm{d}y^2} - k^2\right)\phi - \frac{\mathrm{d}^2 U}{\mathrm{d}y^2}\phi = 0.$$

In [106], Howard obtained a necessary condition of linear instability of stable solution $(U(y), 0)$

$$\left[c_r - \frac{1}{2}(\bar{U}_m + \bar{U}_M)\right]^2 + c_i^2 < \frac{1}{4}(\bar{U}_M - \bar{U}_m)^2.$$

Lin obtained a sufficient condition of linear instability of the steady solution $(U(y),\ 0)$ by applying Howard semi-circle theorem in [135].

5.1.3 *Stability of General Inertial Internal Gravity Wave*

In this subsection, we shall discuss the stability of inertial internal gravity wave of the three-dimensional Boussinesq equations driven by Coriolis force. Applying Boussinesq approximation, the linearization equations of the inertial internal gravity waves for basic density $\rho_0(z)$ and basic flow $(\bar{u}, 0, 0)$ can be written as

$$\begin{cases} \left(\dfrac{\partial}{\partial t} + \bar{u}\dfrac{\partial}{\partial x}\right)u' - f_0 v' = -\dfrac{1}{\rho_0}\dfrac{\partial p'}{\partial x}, \\ \left(\dfrac{\partial}{\partial t} + \bar{u}\dfrac{\partial}{\partial x}\right)v' + f_0 u' = -\dfrac{1}{\rho_0}\dfrac{\partial p'}{\partial y}, \\ \left(\dfrac{\partial}{\partial t} + \bar{u}\dfrac{\partial}{\partial x}\right)\omega' = -\dfrac{1}{\rho_0}\dfrac{\partial p'}{\partial z} - g\dfrac{\rho'}{\rho_0}, \\ \dfrac{\partial u'}{\partial x} + \dfrac{\partial v'}{\partial y} + \dfrac{\partial \omega'}{\partial z} = 0, \\ \left(\dfrac{\partial}{\partial t} + \bar{u}\dfrac{\partial}{\partial x}\right)\rho' - \dfrac{N^2}{g}\rho_0\omega' = 0, \end{cases} \tag{5.1.53}$$

where the Coriolis parameter f is considered as constant f_0, $\bar{u}$ is a constant and $\rho_0(z)$ is continuous. (5.1.53) is the linearization of the three-dimensional Boussinesq equations driven by Coriolis force on the basis of the steady solution $(\bar{u}, 0, 0, \rho_0(z))$.

Eliminating $v^{'}$ and $u^{'}$ from the first two equations of (5.1.53) respectively, we get

$$\begin{cases} \left\{\left(\dfrac{\partial}{\partial t}+\bar{u}\dfrac{\partial}{\partial x}\right)^2+I^2\right\}u^{'}=-\left(\dfrac{\partial}{\partial t}+\bar{u}\dfrac{\partial}{\partial x}\right)\dfrac{1}{\rho_0}\dfrac{\partial p^{'}}{\partial x}-f_0\dfrac{1}{\rho_0}\dfrac{\partial p^{'}}{\partial y}, \\ \left\{\left(\dfrac{\partial}{\partial t}+\bar{u}\dfrac{\partial}{\partial x}\right)^2+I^2\right\}v^{'}=-\left(\dfrac{\partial}{\partial t}+\bar{u}\dfrac{\partial}{\partial x}\right)\dfrac{1}{\rho_0}\dfrac{\partial p^{'}}{\partial y}+f_0\dfrac{1}{\rho_0}\dfrac{\partial p^{'}}{\partial x}, \end{cases} \tag{5.1.54}$$

where

$$I^2 = f_0^2.$$

Taking the derivative of the first equation of (5.1.54) with respect to x, the second with respect to y, and making use of the fourth equation of (5.1.53), we have

$$\left[\left(\frac{\partial}{\partial t}+\bar{u}\frac{\partial}{\partial x}\right)^2+I^2\right]\frac{\partial \omega^{'}}{\partial z}=\left(\frac{\partial}{\partial t}+\bar{u}\frac{\partial}{\partial x}\right)\frac{1}{\rho_0}\nabla_h^2 p^{'}.$$

Eliminating $\rho^{'}$ from the third and fifth equations of (5.1.53), we obtain

$$\left[\left(\frac{\partial}{\partial t}+\bar{u}\frac{\partial}{\partial x}\right)^2+N^2\right]\omega^{'}=-\left(\frac{\partial}{\partial t}+\bar{u}\frac{\partial}{\partial x}\right)\frac{1}{\rho_0}\frac{\partial p^{'}}{\partial z}.$$

Eliminating p' from the above two equalities, we get

$$\left\{\left[\left(\frac{\partial}{\partial t}+\bar{u}\frac{\partial}{\partial x}\right)^2+N^2\right]\nabla_h^2+\left[\left(\frac{\partial}{\partial t}+\bar{u}\frac{\partial}{\partial x}\right)^2+I^2\right]\frac{\partial^2}{\partial z^2}\right\}\omega^{'}=0. \tag{5.1.55}$$

The boundary conditions of equation (5.1.55) are still taken as (5.1.25).

When $\bar{u}=0$, (5.1.55) becomes

$$\left[\left(\frac{\partial^2}{\partial t^2}+N^2\right)\nabla_h^2+\left(\frac{\partial^2}{\partial t^2}+f_0^2\right)\frac{\partial^2}{\partial z^2}\right]\omega^{'}=0. \tag{5.1.56}$$

Let $\omega^{'}=Ae^{i(k_1x+k_2y+k_3z-\omega t)}$. According to equation (5.1.56), we have

$$\omega^2=\frac{K_h^2N^2+k_3^2f_0^2}{K^2}, \tag{5.1.57}$$

where $K^2 = k_1^2 + k_2^2 + k_3^2$, and $K_h^2 = k_1^2 + k_2^2$. (5.1.57) shows that when there is no basic flow, a necessary condition of instability of inertial internal gravity waves is unstable stratification, that is,

$$\text{unstable} \Rightarrow N^2 < 0,$$

and a sufficient condition of instability is

$$K_h^2 N^2 + k_3^2 f_0^2 < 0 \ (N^2 < 0).$$

We introduce horizontal characteristic scale L and baroclinic Rossby deformed radius L_1, which respectively satisfy

$$L^2 = \frac{k_3^2 H^2}{K_h^2}, L_1^2 = -\frac{N^2 H^2}{f^2}(N^2 < 0).$$

If $k_3 H = 2\pi$, then L is the horizontal wavelength. Thus, (5.1.57) can be rewritten as

$$\omega^2 = \frac{K_h^2 f^2}{K^2 H^2}(L^2 - L_1^2).$$

So, an instability criterion can be rewritten as

$$L < L_1.$$

It shows that when the horizontal scale L is smaller than the Rossby deformed radius L_1, inertial internal gravity waves are unstable; otherwise, when

$$L \geq L_1$$

inertial internal gravity waves must be stable.

When $\bar{u} \neq 0$, by (5.1.25), substituting $\omega' = A e^{i(k_1 x + k_2 y + k_3 z - \omega t)}$ into equation (5.1.57), we get

$$\omega_D^2 = \frac{K_h^2 N^2 + k_3^2 I^2}{K^2}, \tag{5.1.58}$$

where

$$\omega_D = \omega - k_1 \bar{u}$$

is Doppler frequency.

We know from (5.1.58) that a sufficient condition of stability of inertial internal gravity waves is stratified stable, that is,

$$N^2 > 0,$$

and a necessary condition of instability of inertial internal gravity waves is stratified unstable, that is,

$$N^2 < 0.$$

The sufficient condition of instability of inertial internal gravity waves is

$$K_h^2 N^2 + k_3^2 I^2 < 0. \tag{5.1.59}$$

Introduce L and L_2, which respectively satisfy

$$L^2 = k_3^2 H^2 / K_h^2, \ L_2^2 = -N^2 H^2 / I^2.$$

Thus, (5.1.58) can be rewritten as

$$\omega_D^2 = \frac{K_h^2 I^2}{K^2 H^2}(L^2 - L_2^2).$$

So, the instability criterion can be rewritten as

$$N^2 < 0, \quad L < L_2.$$

5.2 Instability of Rossby Waves

In 1939, Rossby studied the atmospheric long waves, and obtained a long-wave formula from the two-dimensional non-divergence vorticity equation, cf. [185]. Later, his students and partners introduced barotropic instable theory and baroclinic instable theory in succession, intensively studied the mechanism of the long-wave production, and well explained the development of the atmospheric circulation. The theory of the atmospheric wave dynamics established by Rossby and his students and partners is one of the most important results in atmospheric science in 20th century.

Afterward, researchers call the long wave generated by the horizontal atmospheric disturbance under the effect of Rossby parameter β ($\beta = \frac{\mathrm{d}f}{\mathrm{d}y}$) Rossby wave, which is also called planetary wave, and is a main wave of the large scale motion and affects the large scale weather. In this section we mainly discuss the instability of Rossby waves. In subsection 5.2.1, we give some necessary conditions of the instability of Rossby waves respectively with energy method and normal mode method. In subsection 5.2.2, we discuss the growth rate of the barotropic instable Rossby waves. In subsection 5.2.3, we discuss the instability of the simplest baroclinic Rossby waves. This section mainly refer to chapter 11 of [145] and Chapter 7 of [172].

5.2.1 *Necessary Conditions of Linear Instability*

Without loss of generality, we discuss the instability of the stationary solutions of the three-dimensional quasi-geostropic equation. Let $\bar{\psi}(y, z)$ be a stationary solution of the following three-dimensional quasi-geostropic e-quation

$$\frac{\partial q}{\partial t}+\frac{\partial \psi}{\partial x}\frac{\partial q}{\partial y}-\frac{\partial \psi}{\partial y}\frac{\partial q}{\partial x}=0,$$

where $q=\left(\frac{\partial^2}{\partial x^2}+\frac{\partial^2}{\partial y^2}\right)\psi+\frac{1}{\rho_0}\frac{\partial}{\partial z}\left(\frac{f_0^2}{N^2}\rho_0\frac{\partial \psi}{\partial z}\right)+\beta_0 y$. Linearizing the above equation about the stationary solution $\bar{\psi}(y,z)$, we get

$$\left(\frac{\partial}{\partial t}+\bar{u}\frac{\partial}{\partial x}\right)q^{'}+\frac{\partial \bar{q}}{\partial y}\frac{\partial \psi^{'}}{\partial x}=0, \tag{5.2.1}$$

where

$$\begin{cases}\bar{u}=-\dfrac{\partial \bar{\psi}}{\partial y}, \quad q^{'}=\nabla_h^2\psi^{'}+\dfrac{1}{\rho_0}\dfrac{\partial}{\partial z}\left(\dfrac{f_0^2}{N^2}\rho_0\dfrac{\partial \psi^{'}}{\partial z}\right),\\ \dfrac{\partial \bar{q}}{\partial y}=\beta_0-\dfrac{\partial^2\bar{u}}{\partial y^2}-\dfrac{1}{\rho_0}\dfrac{\partial}{\partial z}\left(\dfrac{f_0^2}{N^2}\rho_0\dfrac{\partial \bar{u}}{\partial z}\right),\end{cases}$$

$\psi^{'}$ is the stream function of the disturbance, and $q^{'}$ is the potential vorticity of the disturbance. Consider the domain: $0\le x\le 2\pi, y_1\le y\le y_2, 0\le z\le H$. The boundary conditions are given by

$$\begin{cases}v^{'}|_{y=y_1,y_2}=0,\\ \omega^{'}|_{z=0,H}=0,\end{cases}$$

where $\omega^{'}$ is the vertical disturbance velocity, and $\omega^{'}=-\frac{f_0^2}{N^2}\left(\frac{\partial}{\partial t}+u^{'}\frac{\partial}{\partial x}+v^{'}\frac{\partial}{\partial y}\right)\left(\frac{\partial \psi^{'}}{\partial z}\right)$. According to $v^{'}=\frac{\partial \psi^{'}}{\partial x}$ and $u^{'}=-\frac{\partial \psi^{'}}{\partial y}$, the above relationship can be rewritten as

$$\begin{cases}\dfrac{\partial \psi^{'}}{\partial x}|_{y=y_1,y_2}=0,\\ \left\{\left(\dfrac{\partial}{\partial t}+\bar{u}\dfrac{\partial}{\partial x}\right)\dfrac{\partial \psi^{'}}{\partial z}-\dfrac{\partial \bar{u}}{\partial z}\dfrac{\partial \psi^{'}}{\partial x}\right\}\Bigg|_{z=0,H}=0.\end{cases} \tag{5.2.2}$$

Taking L^2 inner product of (5.2.1) with $\rho_0\psi^{'}$, and applying (5.2.2), we have

$$\frac{\partial}{\partial t}\int_0^H\int_{y_1}^{y_2}\mathrm{d}y\mathrm{d}z\frac{\rho_0}{2}\overline{\left[\left(\frac{\partial \psi^{'}}{\partial x}\right)^2+\left(\frac{\partial \psi^{'}}{\partial y}\right)^2+\frac{f_0^2}{N^2}\left(\frac{\partial \psi^{'}}{\partial z}\right)^2\right]}$$

$$= \int_0^H \int_{y_1}^{y_2} \mathrm{d}y\mathrm{d}z \left[\rho_0 \overline{\frac{\partial \psi'}{\partial x} \frac{\partial \psi'}{\partial y} \frac{\partial \bar{u}}{\partial y}} + \rho_0 \frac{f_0^2}{N^2} \overline{\frac{\partial \psi'}{\partial x} \frac{\partial \psi'}{\partial z} \frac{\partial \bar{u}}{\partial z}} \right], \tag{5.2.3}$$

where $\overline{(\ )} = \dfrac{1}{2L_x} \displaystyle\int_{-L_x}^{L_x} (\ )\mathrm{d}x$, and L_x is the period of the disturbance quantity ψ' in x direction. The left-hand side of (5.2.3) stands for the variance ratio of the summation of the kinetic energy and available potential energy of the disturbance field. This kind of increasing or decreasing of the disturbance energy is given by the right side of (5.2.3). When the right side is positive, the disturbance energy is increasing, otherwise, it is decreasing. It depends on the unstable process of the horizontal shear of the basic flow, that is,

$$\overline{\frac{\partial \psi'}{\partial x} \frac{\partial \psi'}{\partial y} \frac{\partial \bar{u}}{\partial y}} > 0,$$

which is called barotropic instability, because it appears in the homogeneous fluid without vertical shear. Another kind of unstable process depends on the existence of the vertical shear of the basic flow, that is,

$$\overline{\frac{\partial \psi'}{\partial x} \frac{\partial \psi'}{\partial z} \frac{\partial \bar{u}}{\partial z}} > 0.$$

Because the vertical shear means the horizontal temperature gradient, this process is called baroclinic instability.

Since

$$\int_0^H \int_{y_1}^{y_2} \left[\rho_0 \overline{\frac{\partial \psi'}{\partial x} \frac{\partial \psi'}{\partial y} \frac{\partial \bar{u}}{\partial y}} \right] \mathrm{d}y\mathrm{d}z = - \int_0^H \int_{y_1}^{y_2} \left[\rho_0 \overline{v_0 u_0} \frac{\partial \bar{u}}{\partial y} \right] \mathrm{d}y\mathrm{d}z,$$

$$\int_0^H \int_{y_1}^{y_2} \left[\rho_0 \frac{f_0^2}{N^2} \overline{\frac{\partial \psi'}{\partial x} \frac{\partial \psi'}{\partial z} \frac{\partial \bar{u}}{\partial z}} \right] \mathrm{d}y\mathrm{d}z = - \int_0^H \int_{y_1}^{y_2} \left[\rho_0 \frac{f_0^2}{N^2} \overline{v_0 \theta_0} \frac{\partial \bar{\theta}}{\partial y} \right] \mathrm{d}y\mathrm{d}z,$$

where $u_0 - \bar{u} = -\dfrac{\partial \psi'}{\partial y}$, $v_0 = \dfrac{\partial \psi'}{\partial x}$, $\theta_0 - \bar{\theta} = \dfrac{\partial \psi'}{\partial z}$, $\dfrac{\partial \bar{u}}{\partial z} = -\dfrac{\partial \bar{\theta}}{\partial y}$, by integration by parts, we have

$$\frac{\partial}{\partial t} E(\psi') = \int_0^H \int_{y_1}^{y_2} \mathrm{d}y\mathrm{d}z \left\{ \bar{u} \left[\frac{\partial}{\partial y} \left(\rho_0 \overline{v_0 u_0} \right) - \frac{\partial}{\partial z} \left(\rho_0 \frac{f_0^2}{N^2} \overline{v_0 \theta_0} \right) \right] \right\}$$
$$+ \int_{y_1}^{y_2} \mathrm{d}y \left[\frac{f_0^2}{N^2} \bar{u} \rho_0 \overline{v_0 \theta_0} \right] \Bigg|_{z=0}^{z=H}, \tag{5.2.4}$$

where

$$E(\psi') = \int_0^H \int_{y_1}^{y_2} \mathrm{d}y\mathrm{d}z \frac{\rho_0}{2} \overline{\left[\left(\frac{\partial \psi'}{\partial x} \right)^2 + \left(\frac{\partial \psi'}{\partial y} \right)^2 + \frac{f_0^2}{N^2} \left(\frac{\partial \psi'}{\partial z} \right)^2 \right]}.$$

Since

$$\rho_0\overline{v_0 q'} = -\left(\frac{\partial}{\partial y}\left(\rho_0\overline{v_0 u_0}\right) - \frac{\partial}{\partial z}\left(\rho_0\frac{f_0^2}{N^2}\overline{v_0\theta_0}\right)\right),$$

(5.2.4) can be rewritten as

$$\frac{\partial}{\partial t}E(\psi') = -\int_0^H\int_{y_1}^{y_2}\mathrm{d}y\mathrm{d}z\left(\bar{u}\rho_0\overline{v_0 q'}\right) + \int_{y_1}^{y_2}\mathrm{d}y\left(\frac{f_0^2}{N^2}\bar{u}\rho_0\overline{v_0\theta_0}\right)\Big|_0^H. \tag{5.2.5}$$

Suppose that function $\eta(x, y, z, t)$ satisfies

$$\frac{\partial\eta}{\partial t} + \bar{u}\frac{\partial\eta}{\partial x} = v_0,$$

we derive from (5.2.1)

$$\left(\frac{\partial}{\partial t} + \bar{u}\frac{\partial}{\partial x}\right)q' = -\left(\frac{\partial}{\partial t} + \bar{u}\frac{\partial}{\partial x}\right)\eta\frac{\partial\bar{q}}{\partial y}, \tag{5.2.6}$$

one solution of this equation is

$$q' = -\eta\frac{\partial\bar{q}}{\partial y}. \tag{5.2.7}$$

According to (5.2.7) and the definition of η, we have

$$\overline{v_0 q'} = -\left(\frac{\partial\overline{\frac{\eta^2}{2}}}{\partial t}\right)\frac{\partial\bar{q}}{\partial y}.$$

By (5.2.2) and the definition of η, we obtain

$$\overline{v_0\theta_0}|_{z=0} = \left[\frac{\partial\bar{u}}{\partial z}\left(\frac{\partial\overline{\frac{\eta^2}{2}}}{\partial t}\right)\right]\Bigg|_{z=0}, \quad \overline{v_0\theta_0}|_{z=H} = \left[\frac{\partial\bar{u}}{\partial z}\left(\frac{\partial\overline{\frac{\eta^2}{2}}}{\partial t}\right)\right]\Bigg|_{z=H}.$$

Applying the above two relationships, (5.2.5) is rewritten as

$$\frac{\partial}{\partial t}\left\{E(\psi') - \int_0^H\int_{y_1}^{y_2}\mathrm{d}y\mathrm{d}z\left[\rho_0\frac{\overline{\eta^2}}{2}\left(\bar{u}\frac{\partial\bar{q}}{\partial y}\right)\right] - \int_{y_1}^{y_2}\mathrm{d}y\left(\frac{f_0^2}{N^2}\bar{u}\frac{\partial\bar{u}}{\partial z}\rho_0\frac{\overline{\eta^2}}{2}\right)\Bigg|_{z=H}\right.$$
$$\left. + \int_{y_1}^{y_2}\mathrm{d}y\left(\frac{f_0^2}{N^2}\bar{u}\frac{\partial\bar{u}}{\partial z}\rho_0\frac{\overline{\eta^2}}{2}\right)\Bigg|_{z=0}\right\} = 0. \tag{5.2.8}$$

We know from (5.2.8) that if

$$\bar{u}\frac{\partial\bar{q}}{\partial y} \le 0,$$

$$\frac{f_0^2}{N^2}\bar{u}\frac{\partial \bar{u}}{\partial z}|_{z=H} \le 0,$$

$$\frac{f_0^2}{N^2}\bar{u}\frac{\partial \bar{u}}{\partial z}|_{z=0} \ge 0,$$

then $\bar{\psi}$ is stable to the small enough disturbance, which means that this stationary solution is linearly stable. Thus any destruction of the former three conditions is a necessary condition of its linear instability.

Now, we use normal mode method to find necessary conditions of instability of the stationary solution. For simplicity, assume N^2 = constant, then

$$\begin{cases} q' = \nabla_h^2 \psi' + \dfrac{f_0^2}{N^2}\dfrac{1}{\rho_0}\dfrac{\partial}{\partial z}\left(\rho_0 \dfrac{\partial \psi'}{\partial z}\right) = \nabla_h^2 \psi' + \dfrac{f_0^2}{N^2}\left(\dfrac{\partial^2 \psi'}{\partial z^2} - \sigma_0 \dfrac{\partial \psi'}{\partial z}\right), \\ \dfrac{\partial \bar{q}}{\partial y} = \beta_0 - \dfrac{\partial^2 \bar{u}}{\partial y^2} - \dfrac{f_0^2}{N^2}\left(\dfrac{\partial^2 \bar{u}}{\partial z^2} - \sigma_0 \dfrac{\partial \bar{u}}{\partial z}\right), \quad \sigma_0 = -\dfrac{\partial ln \rho_0}{\partial z}. \end{cases}$$

Considering the form of q', letting the solution of equation (5.2.1) be

$$\psi' = \psi(y,z)e^{ik(z-ct)+\sigma_0 \frac{z}{2}}, \tag{5.2.9}$$

substituting it into equation (5.2.1), we get

$$(\bar{u}-c)\left[\frac{\partial^2 \psi}{\partial y^2} - k^2\psi + \frac{f_0^2}{N^2}\left(\frac{\partial^2 \psi}{\partial z^2} - \frac{\sigma_0^2}{4}\psi\right)\right] + \frac{\partial \bar{q}}{\partial y}\psi = 0. \tag{5.2.10}$$

Substituting (5.2.9) into (5.2.2), we have

$$\begin{cases} \psi|_{y=y_1,y_2} = 0, \\ \left[(\bar{u}-c)\dfrac{\partial \psi}{\partial z} - \dfrac{\partial \bar{u}}{\partial z}\psi\right]\Bigg|_{z=0,H} = 0. \end{cases} \tag{5.2.11}$$

When $\bar{u} - c \neq 0$, taking L^2 inner product of ψ^*, the conjugate of ψ, with (5.2.10), and using boundary conditions and

$$\begin{cases} \psi^* \dfrac{\partial^2 \psi}{\partial y^2} = \dfrac{\partial}{\partial y}\left(\psi^* \dfrac{\partial \psi}{\partial y}\right) - \left|\dfrac{\partial \psi}{\partial y}\right|^2, \\ \psi^* \dfrac{\partial^2 \psi}{\partial z^2} = \dfrac{\partial}{\partial z}\left(\psi^* \dfrac{\partial \psi}{\partial z}\right) - \left|\dfrac{\partial \psi}{\partial z}\right|^2, \end{cases}$$

we obtain

$$\int_0^H \int_{y_1}^{y_2} \left[\left|\frac{\partial \psi}{\partial y}\right|^2 + \frac{f_0^2}{N^2}\left|\frac{\partial \psi}{\partial z}\right|^2 + \left(k^2 + \frac{\sigma_0 f_0^2}{4N^2}\right)|\psi|^2\right] dy dz$$

$$= \frac{f_0^2}{N^2}\int_{y_1}^{y_2}\left[\frac{(\bar{u}-c^*)\partial \bar{u}/\partial z}{|\bar{u}-c|^2}|\psi|^2\right]\Bigg|_{z=0}^{z=H} dy$$

$$+\int_0^H\int_{y_1}^{y_2}\frac{(\bar{u}-c^*)\partial\bar{q}/\partial y}{|\bar{u}-c|^2}|\psi|^2\mathrm{d}y\mathrm{d}z, \tag{5.2.12}$$

where c^* is the conjugate of c. Dividing the above equality into the real and imaginary parts, we get

$$\int_0^H\int_{y_1}^{y_2}\left[2E_p-\frac{(\bar{u}-c_r)\partial q/\partial y}{|\bar{u}-c|^2}|\psi|^2\right]\mathrm{d}y\mathrm{d}z$$
$$=\frac{f_0^2}{N^2}\int_{y_1}^{y_2}\left[\frac{(\bar{u}-c_r)\partial\bar{u}/\partial z}{|\bar{u}-c|^2}|\psi|^2\right]\Bigg|_{z=0}^{z=H}\mathrm{d}y, \tag{5.2.13}$$

$$c_i\left\{\int_0^H\int_{y_1}^{y_2}\frac{\partial\bar{q}/\partial y}{|\bar{u}-c|^2}|\psi|^2\mathrm{d}y\mathrm{d}z+\frac{f_0^2}{N^2}\int_{y_1}^{y_2}\left(\frac{\partial\bar{u}/\partial z}{|\bar{u}-c|^2}|\psi|^2\right)\Bigg|_{z=0}^{z=H}\mathrm{d}y\right\}=0, \tag{5.2.14}$$

where

$$E_p=\frac{1}{2}\left[\left|\frac{\partial\psi}{\partial y}\right|+\frac{f_0^2}{N^2}\left|\frac{\partial\psi}{\partial z}\right|^2+\left(k^2+\frac{\sigma_0^2f_0^2}{4N^2}\right)|\psi|^2\right]>0.$$

In the quasi-geotropic case, $u'=-\dfrac{\partial\psi'}{\partial y}, v'=-\dfrac{\partial\psi'}{\partial x}, \dfrac{\theta'}{\theta_0}=\dfrac{f_0}{g}\dfrac{\partial\psi'}{\partial z}$, then from (5.2.9) we know that the real values of the disturbance kinetic energy and available potential energy are

$$\begin{cases} K_p=\dfrac{1}{2}((u')^2+(v')^2)=\dfrac{1}{2}\left(\left|\dfrac{\partial\psi}{\partial y}\right|^2+k^2|\psi|^2\right), \\ A_p=\dfrac{g^2}{2N^2}(\dfrac{\theta'}{\theta_0})^2=\dfrac{f_0^2}{2N^2}\left(\left|\dfrac{\partial\psi}{\partial z}\right|^2+\dfrac{\sigma_0^2}{4}|\psi|^2\right). \end{cases}$$

Thus

$$E_p=K_p+A_p,$$

where E_p is the total energy of the disturbance. For the unstable Rossby waves, $c_i\neq 0$, we derive from (5.2.14)

$$\int_0^H\int_{y_1}^{y_2}\frac{\partial\bar{q}/\partial y}{|\bar{u}-c|^2}|\psi|^2\mathrm{d}y\mathrm{d}z+\frac{f_0^2}{N^2}\int_{y_1}^{y_2}\left(\frac{\partial\bar{u}/\partial z}{|\bar{u}-c|^2}|\psi|^2\right)\Bigg|_{z=0}^{z=H}\mathrm{d}y=0. \tag{5.2.15}$$

This is the first necessary condition of the linear instability of Rossby waves. Substituting (5.2.15) into (5.2.13), we get

$$\int_0^H\int_{y_1}^{y_2}2E_p\mathrm{d}y\mathrm{d}z$$

$$= \int_0^H \int_{y_1}^{y_2} \frac{\bar{u}\partial\bar{q}/\partial y}{|\bar{u}-c|^2}|\psi|^2 \mathrm{d}y\mathrm{d}z + \frac{f_0^2}{N^2}\int_{y_1}^{y_2} \left(\frac{\bar{u}\partial\bar{u}/\partial z}{|\bar{u}-c|^2}|\psi|^2 \right)\Bigg|_{z=0}^{z=H} \mathrm{d}y > 0. \tag{5.2.16}$$

This is the second necessary condition of the linear instability of Rossby waves. And we illuminate them in the barotropic and baroclinic situations respectively.

5.2.2 *Linear Instability of Barotropic Rossby Waves*

In the barotropic case, the basic flow is

$$\bar{u} = \bar{u}(y), \tag{5.2.17}$$

thus $\dfrac{\partial\bar{u}}{\partial z} = 0$, $\dfrac{\partial^2\bar{u}}{\partial z^2} = 0$. If the disturbance is independent of z, then

$$q' = \nabla_h^2\psi', \frac{\partial\bar{q}}{\partial y} = \beta_0 - \frac{\partial^2\bar{u}}{\partial y^2}.$$

Here we consider the linear instability of the stationary solution $\bar{u}(y)$ of the two-dimensional quasi-geotropic equation.

5.2.2.1 *Necessary Conditions of the Linear Instability*

In the barotropic situation, (5.2.15) and (5.2.16) are written respectively as

$$\int_{y_1}^{y_2} \frac{\partial\bar{q}/\partial y}{|\bar{u}-c|^2}|\psi|^2 \mathrm{d}y = 0,$$

$$\int_{y_1}^{y_2} \left(\left|\frac{\partial\psi}{\partial y}\right|^2 + k^2|\psi|^2 \right) \mathrm{d}y = \int_{y_1}^{y_2} \frac{\bar{u}\partial\bar{q}/\partial y}{|\bar{u}-c|^2}|\psi|^2 \mathrm{d}y > 0.$$

Due to $|\psi|^2 > 0, |\bar{u}-c|^2 > 0$, the above two relationships respectively require

$$\frac{\partial\bar{q}}{\partial y} = \beta_0 - \frac{\partial^2\bar{u}}{\partial y^2} \text{ must change the symbol in } (y_1, y_2), \text{ and} \tag{5.2.18}$$

$$\bar{u}\frac{\partial\bar{q}}{\partial y} = \bar{u}\left(\beta_0 - \frac{\partial^2\bar{u}}{\partial y^2}\right) \text{ is positive in some domain of } (y_1, y_2). \tag{5.2.19}$$

(5.2.18) is known as Kuo H. L. Theorem (see [125, 126]), which illustrates that the barotropic unstable disturbance requires $\bar{u}(y)$ satisfying $\beta_0 - \dfrac{\partial^2\bar{u}}{\partial y^2}$ to be zero at some point of (y_1, y_2). (5.2.19) is called Fjörtoft Theorem, which illustrates that the increasing of the barotropic unstable disturbance energy requires $\bar{u}(\beta_0 - \frac{\partial^2 \bar{u}}{\partial y^2})$ to be positive at least in some domain of (y_1, y_2). If the Fjörtoft Theorem does not satisfy even if Kuo H. L. Theorem is satisfied, then the barotropic disturbance is stable, because the disturbance energy is decreasing.

5.2.2.2 *The Growth Rate of the Unstable Waves*

In the barotropic situation, equation (5.2.10) and boundary condition (5.2.11) of the disturbance can be respectively written as

$$(\bar{u} - c)\left(\frac{\mathrm{d}^2\psi}{dy^2} - k^2\psi\right) + \frac{\partial \bar{q}}{\partial y}\psi = 0, \tag{5.2.20}$$

$$\psi|_{y=y_1} = 0, \psi|_{y=y_2} = 0. \tag{5.2.21}$$

Thus, (5.2.13) can be rewritten as

$$\int_{y_1}^{y_2} \frac{(\bar{u} - c_r)\partial\bar{q}/\partial y}{|\bar{u} - c|^2}|\psi|^2 \mathrm{d}y = \int_{y_1}^{y_2}\left(\left|\frac{\mathrm{d}\psi}{dy}\right|^2 + k^2|\psi|^2\right)\mathrm{d}y. \tag{5.2.22}$$

Suppose that the north-south width of the disturbance is

$$d = y_2 - y_1.$$

According to the boundary condition (5.2.21), we expand $\psi(y)$ into the following Fourier series

$$\psi(y) = \sum_{n=1}^{\infty} b_n \sin\frac{n\pi(y - y_1)}{d}.$$

Thus

$$\frac{\mathrm{d}\psi}{dy} = \sum_{n=1}^{\infty}\frac{n\pi}{d} b_n \cos\frac{n\pi(y - y_1)}{d}.$$

According to the above two inequalities, we have

$$\int_{y_1}^{y_2} |\psi|^2 \mathrm{d}y = \sum_{n=1}^{\infty} b_n^2 \int_{y_1}^{y_2} \sin^2\frac{n\pi(y - y_1)}{d}\mathrm{d}y = \frac{\mathrm{d}}{2}\sum_{n=1}^{\infty} b_n^2,$$

$$\begin{aligned}\int_{y_1}^{y_2}\left|\frac{\mathrm{d}\psi}{dy}\right|^2 \mathrm{d}y &= \sum_{n=1}^{\infty}\left(\frac{n\pi}{d}\right)^2 b_n^2 \int_{y_1}^{y_2}\cos^2\frac{n\pi(y - y_1)}{d}\mathrm{d}y \\ &= \frac{d}{2}\sum_{n=1}^{\infty}\left(\frac{n\pi}{d}\right)^2 b_n^2 \geq \left(\frac{\pi}{d}\right)^2 \cdot \frac{d}{2}\sum_{n=1}^{\infty} b_n^2.\end{aligned}$$

Thus, (5.2.22) is rewritten as

$$\int_{y_1}^{y_2} \frac{(\bar{u}-c_r)\dfrac{\partial \bar{q}}{\partial y}}{|\bar{u}-c|^2}|\psi|^2 \mathrm{d}y \geq \left(k^2+\frac{\pi^2}{d^2}\right)\int_{y_1}^{y_2} |\psi|^2 \mathrm{d}y. \tag{5.2.23}$$

Since $|\bar{u}-c|^2 = (\bar{u}-c_r)^2 + c_i \geq 2(\bar{u}-c_r)c_i$, the left-hand side of the above inequality satisfies

$$\int_{y_1}^{y_2} \frac{(\bar{u}-c_r)\dfrac{\partial \bar{q}}{\partial y}}{|\bar{u}-c|^2}|\psi|^2 \mathrm{d}y \leq \int_{y_1}^{y_2} \frac{(\bar{u}-c_r)\dfrac{\partial \bar{q}}{\partial y}}{|\bar{u}-c_r|^2}|\psi|^2 \mathrm{d}y$$

$$\leq \int_{y_1}^{y_2} \frac{|\dfrac{\partial \bar{q}}{\partial y}|}{2|c_i|}|\psi|^2 \mathrm{d}y \leq \frac{\max\limits_{(y_1,y_2)}|\dfrac{\partial \bar{q}}{\partial y}|}{2|c_i|}\int_{y_1}^{y_2} |\psi|^2 \mathrm{d}y. \tag{5.2.24}$$

According to (5.2.23) and (5.2.24), we get

$$\left(k^2+\frac{\pi^2}{d^2}\right)\int_{y_1}^{y_2} |\psi|^2 \mathrm{d}y \leq \frac{\max\limits_{(y_1,y_2)}|\dfrac{\partial \bar{q}}{\partial y}|}{2|c_i|}\int_{y_1}^{y_2} |\psi|^2 \mathrm{d}y. \tag{5.2.25}$$

So

$$|c_i| \leq \max_{(y_1,y_2)}|\frac{\partial \bar{q}}{\partial y}|/2\left(k^2+\frac{\pi^2}{d^2}\right).$$

The above inequality gives an upper bound of c_i. Thus we know that the growth rate satisfies

$$|kc_i| \leq k \max_{(y_1,y_2)}|\frac{\partial \bar{q}}{\partial y}|/2\left(k^2+\frac{\pi^2}{d^2}\right). \tag{5.2.26}$$

The right-hand side of the above relationship tends to zero as $k \to 0$ and $=$ $k \to \infty$, which shows that the most unstable wavelengths are not too long or too short.

5.2.2.3 *Semi-Circle Theorem*

Here, we mainly estimate the range of c_r of the unstable Rossby waves. Let

$$\psi(y) = (\bar{u}-c)F(y).$$

Similar to (5.1.29), equation (5.2.20) is rewritten as

$$\frac{\mathrm{d}}{\mathrm{d}y}\left[(\bar{u}-c)^2\frac{\mathrm{d}F}{\mathrm{d}y}\right] + \left[\beta_0(\bar{u}-c) - k^2(\bar{u}-c)\right]F = 0. \tag{5.2.27}$$

Moreover,

$$F|_{y=y_1} = 0, F|_{y=y_2} = 0. \tag{5.2.28}$$

Taking L^2 inner product of F^* with (5.2.27), and by (5.2.28), we get

$$\int_{y_1}^{y_2} (\bar{u} - c)^2 \left(\left| \frac{\mathrm{d}F}{\mathrm{d}y} \right|^2 + k^2 |F|^2 \right) \mathrm{d}y = \int_{y_1}^{y_2} \beta_0 (\bar{u} - c)|F|^2 \mathrm{d}y. \tag{5.2.29}$$

The real and imaginary parts of the above relationship are respectively

$$\int_{y_1}^{y_2} \left[(\bar{u} - c_r)^2 - c_i^2\right] \left(\left| \frac{\mathrm{d}F}{\mathrm{d}y} \right|^2 + k^2 |F|^2 \right) \mathrm{d}y = \int_{y_1}^{y_2} \beta_0 (\bar{u} - c_r)|F|^2 \mathrm{d}y, \tag{5.2.30}$$

$$c_i \left\{ \int_{y_1}^{y_2} (\bar{u} - c_r) \left(\left| \frac{\mathrm{d}F}{\mathrm{d}y} \right|^2 + k^2 |F|^2 \right) \mathrm{d}y - \frac{1}{2} \int_{y_1}^{y_2} \beta_0 |F|^2 \mathrm{d}y \right\} = 0. \tag{5.2.31}$$

For the unstable disturbance, $c_i \neq 0$, we derive from (5.2.31)

$$\int_{y_1}^{y_2} (\bar{u} - c_r) \left(\left| \frac{\mathrm{d}F}{\mathrm{d}y} \right|^2 + k^2 |F|^2 \right) \mathrm{d}y = \frac{1}{2} \int_{y_1}^{y_2} \beta_0 |F|^2 \mathrm{d}y. \tag{5.2.32}$$

Thus

$$c_r = \left(\int_{y_1}^{y_2} \bar{u} Q \mathrm{d}y \Big/ \int_{y_1}^{y_2} Q \mathrm{d}y \right) - \left(\frac{\beta_0}{2} \int_{y_1}^{y_2} |F|^2 \mathrm{d}y \Big/ \int_{y_1}^{y_2} Q \mathrm{d}y \right), \tag{5.2.33}$$

where

$$Q \equiv \left| \frac{\mathrm{d}F}{\mathrm{d}y} \right|^2 + k^2 |F|^2 \geq 0.$$

Let $\bar{u}_m$ and $\bar{u}_M$ be minimum and maximum of $\bar{u}$ in (y_1, y_2), respectively. Noticing $\frac{\beta_0}{2} \int_{y_1}^{y_2} |\psi|^2 d_y > 0$, we derive from (5.2.33)

$$\bar{u}_m - \frac{\beta_0}{2(k^2 + d^2)} < c_r < \bar{u}_M, \tag{5.2.34}$$

which is the restriction of c_r of the barotropic unstable Rossby waves.

According to (5.2.32), (5.2.30) is rewritten as

$$\int_{y_1}^{y_2} \bar{u}^2 Q \mathrm{d}y = (c_r^2 + c_i^2) \int_{y_1}^{y_2} Q \mathrm{d}y + \int_{y_1}^{y_2} \beta_0 \bar{u} |F|^2 \mathrm{d}y. \tag{5.2.35}$$

Similar to (5.1.49),

$$\int_{y_1}^{y_2} [\bar{u}^2 - (\bar{u}_m + \bar{u}_M)\bar{u} + \bar{u}_m \bar{u}_M] Q \mathrm{d}y \leq 0.$$

Substituting (5.2.32) and (5.2.35) into the above inequality, we get

$$(c_r^2+c_i^2)\int_{y_1}^{y_2} Q\mathrm{d}y+\int_{y_1}^{y_2}\beta_0\bar{u}|F|^2\mathrm{d}y-\int_{y_1}^{y_2}(\bar{u}_m+\bar{u}_M)c_rQ\mathrm{d}y$$
$$-\frac{\bar{u}_m+\bar{u}_M}{2}\int_{y_1}^{y_2}\beta_0|F|^2\mathrm{d}y\le 0. \tag{5.2.36}$$

Noting that $\bar{u}\ge\bar{u}_m$, we know that the second term of the left of the above inequality satisfies

$$\int_{y_1}^{y_2}\beta_0\left(\bar{u}-\frac{\bar{u}_m+\bar{u}_M}{2}\right)|F|^2\mathrm{d}y\ge\int_{y_1}^{y_2}\beta_0\left(\bar{u}_m-\frac{\bar{u}_m+\bar{u}_M}{2}\right)|F|^2\mathrm{d}y$$
$$=-\frac{\beta_0}{2}(\bar{u}_M-\bar{u}_m)\int_{y_1}^{y_2}|F|^2\mathrm{d}y\ge-\frac{\beta_0\left(\bar{u}_M-\bar{u}_m\right)/2}{k^2+\dfrac{\pi^2}{d^2}}\int_{y_1}^{y_2}Q\mathrm{d}y.$$

Thus, (5.2.36) becomes to

$$\int_{y_1}^{y_2}\left\{\left[c_r-\frac{1}{2}(\bar{u}_M+\bar{u}_m)\right]^2+c_i^2-\frac{1}{4}(\bar{u}_m+\bar{u}_M)^2-\frac{\beta_0(\bar{u}_M-\bar{u}_m)/2}{k^2+\dfrac{\pi^2}{d^2}}\right\}Q\mathrm{d}y$$
$$\le 0.$$

Therefore, we obtain

$$(c_r-U)^2+c_i^2\le c_R^2, \tag{5.2.37}$$

where

$$U=\frac{1}{2}(\bar{u}_M+\bar{u}_m),c_R^2=U^2+\frac{\beta_0}{k^2+\dfrac{\pi^2}{d^2}}\hat{u},\hat{u}=\frac{1}{2}(\bar{u}_M-\bar{u}_m),$$

(5.2.37) is the semi-circle theorem of the barotropic unstable Rossby waves.

5.2.3 *Linear Instability of Baroclinic Rossby Waves*

In the baroclinic situation, let the basic flow be

$$\bar{u}=\bar{u}(z),$$

then $\dfrac{\partial\bar{u}}{\partial y}=0,\dfrac{\partial^2\bar{u}}{\partial y^2}=0$. Thus

$$\frac{\partial\bar{q}}{\partial y}=\beta_0-\frac{f_0}{N^2}\left(\frac{\partial^2\bar{u}}{\partial z^2}-\sigma_0\frac{\partial\bar{u}}{\partial z}\right).$$

Similarly to the analysis of subsection 5.2.2, we obtain necessary conditions and a semi-circle theorem of the linear in stability of Rossby waves in baroclinic situation. For more details of the basic mechanism of the baroclinic linear instability, see [32, 68, 126, 172].

In the following, we shall discuss the baroclinic neutral stability of Rossby waves by the two-layer quasi-geotrophic model reserving some baroclinic features. The following content can also be found in [178]. The two-layer quasi-geotrophic model (here we assume that the upper and lower layers have the same thickness and the lower layer is without topology, friction and viscosity) is

$$\begin{cases} \left(\frac{\partial}{\partial t} - \frac{\partial \psi_1}{\partial y}\frac{\partial}{\partial x} + \frac{\partial \psi_1}{\partial x}\frac{\partial}{\partial y}\right)\left[\nabla_h^2 \psi_1 + F_1(\psi_2 - \psi_1)\right] + \beta_0 \frac{\partial \psi_1}{\partial x} = 0, \\ \left(\frac{\partial}{\partial t} - \frac{\partial \psi_2}{\partial y}\frac{\partial}{\partial x} + \frac{\partial \psi_2}{\partial x}\frac{\partial}{\partial y}\right)\left[\nabla_h^2 \psi_2 - F_1(\psi_2 - \psi_1)\right] + \beta_0 \frac{\partial \psi_2}{\partial x} = 0, \end{cases} \tag{5.2.38}$$

where ψ_1, ψ_2 are the quasi-geotrophic stream functions of the upper and lower air respectively. Suppose that $(\bar{\psi}_1, \bar{\psi}_2) = (-\bar{u}_1 y, -\bar{u}_2 y)$ is a steady solution of (5.2.38), where $\bar{u}_1$ and $\bar{u}_2$ are constants, and $\bar{u}_2 - \bar{u}_1 \neq 0$. Let

$$\psi_1 = -\bar{u}_1 y + \psi_1', \psi_2 = -\bar{u}_2 y + \psi_2'.$$

Substituting it into equation (5.2.38), we have

$$\begin{cases} \left(\frac{\partial}{\partial t} + \bar{u}_1 \frac{\partial}{\partial x}\right) q_1' + \left[\beta_0 - F_1\left(\bar{u}_2 - \bar{u}_1\right)\right] \frac{\partial \psi_1'}{\partial x} = -J\left(\psi_1', q_1'\right), \\ \left(\frac{\partial}{\partial t} + \bar{u}_2 \frac{\partial}{\partial x}\right) q_2' + \left[\beta_0 + F_1\left(\bar{u}_2 - \bar{u}_1\right)\right] \frac{\partial \psi_2'}{\partial x} = -J\left(\psi_2', q_2'\right), \end{cases}$$

where

$$q_1' = \nabla_h^2 \psi_1' + F_1(\psi_2' - \psi_1'), q_2' = \nabla_h^2 \psi_2' - F_1(\psi_2' - \psi_1').$$

Getting rid of the nonlinear terms of the right side of the above equation, we get

$$\begin{cases} \left(\frac{\partial}{\partial t} + \bar{u}_1 \frac{\partial}{\partial x}\right)(\nabla_h^2 \psi_1' + F_1 \psi_2') - F_1 \left(\frac{\partial}{\partial t} + \bar{u}_2 \frac{\partial}{\partial x}\right)\psi_1' + \beta_0 \frac{\partial \psi_1'}{\partial x} = 0, \\ \left(\frac{\partial}{\partial t} + \bar{u}_1 \frac{\partial}{\partial x}\right)(\nabla_h^2 \psi_2' + F_1 \psi_1') - F_1 \left(\frac{\partial}{\partial t} + \bar{u}_1 \frac{\partial}{\partial x}\right)\psi_2' + \beta_0 \frac{\partial \psi_2'}{\partial x} = 0. \end{cases} \tag{5.2.39}$$

Let (ψ_1', ψ_2') be periodic in the direction of x, and the boundary condition in y direction be homogeneous, that is,

$$\psi_j'|_{y=y_1} = 0, \psi_j'|_{y=y_2} = 0 \ (j = 1, 2).$$

Then we define the solutions of equation (5.2.39) by

$$(\psi_1', \psi_2') = (A, B) \sin l(y - y_1) \mathrm{e}^{\mathrm{i}k(x-ct)}, \tag{5.2.40}$$

where

$$l = 2\pi/d = 2\pi/(y_2 - y_1).$$

Substituting (5.2.40) into equation (5.2.39), we obtain

$$\begin{cases} \left[K_h^2(c - \bar{u}_1) + F_1(c - \bar{u}_2) + \beta_0\right] A - F_1(c - \bar{u}_1)B = 0, \\ -F_1(c - \bar{u}_2)A + \left[K_h^2(c - \bar{u}_2) + F_1(c - \bar{u}_1) + \beta_0\right] B = 0, \end{cases} \tag{5.2.41}$$

where $K_h^2 = l^2 + k^2$. (5.2.41) is linear about A, B. If (5.2.41) has non-zero solutions, then the coefficient determinant of (5.2.41) must be zero. Thus we get the following two-order algebraic equation of c

$$ac^2 + bc + d = 0, \tag{5.2.42}$$

where

$$\begin{cases} a = K_h^2(K_h^2 + 2F_1), b = 2\beta_0(K_h^2 + F_1) - K_h^2(K_h^2 + 2F_1)(\bar{u}_1 + \bar{u}_2), \\ d = \beta_0^2 - \beta_0(K_h^2 + F_1)(\bar{u}_1 + \bar{u}_2) + K_h^2 F_1(\bar{u}_1^2 + \bar{u}_2^2) + K_h^4 \bar{u}_1 \bar{u}_2. \end{cases}$$

Let

$$\bar{u} = (\bar{u}_1 + \bar{u}_2)/2, \hat{u} = (\bar{u}_1 - \bar{u}_2)/2.$$

Then, equation (5.2.42) is written as

$$(c - \bar{u})^2 + \frac{2\beta_0(K_h^2 + F_1)}{K_h^2(K_h^2 + 2F_1)}(c - \bar{u}) + \frac{F_1 - K_h^2 \hat{u}^2(K_h^2 - 2F_1)}{K_h^2(K_h^2 + 2F_1)} = 0. \tag{5.2.43}$$

So

$$c = \bar{u} - \frac{\beta_0}{K_h^2} \cdot \frac{K_h^2 + F_1}{K_h^2 + 2F_1} \pm \frac{\sqrt{\beta_0^2 F_1^2 - K_h^4 \hat{u}^2(4F_1^2 - K_h^4)}}{K_h^2(K_h^2 + 2F_1)}. \tag{5.2.44}$$

We know from (5.2.44) that in a two-layer fluid, a sufficient and necessary condition of the baroclinic neutral stability of Rossby waves is

$$\beta_0^2 F_1^2 - K_h^4 \hat{u}^2(4F_1^2 - K_h^4) \begin{cases} \geq 0, & \text{neutral stable}, \\ < 0, & \text{neutral unstable}. \end{cases} \tag{5.2.45}$$

So β_0 has the effect of stability, and the shear $\hat{u}$ of vertical wind velocity has the effect of instability. If

$$K_h^2 \geq 2F_1,$$

then the wave is stable. If

$$K_h^2 < 2F_1 \tag{5.2.46}$$

then the wave is neutral unstable. When a stationary solution is neutral unstable, it is also linear unstable. According to $K_h^2 = 2F_1$, if $K_h = 2\pi/L$, then the critical wavelength is

$$L_c = \sqrt{2}\pi/\sqrt{F_1}.$$

Thus, (5.2.46) is rewritten as

$$L > L_c, \tag{5.2.47}$$

which means that the wave is neutral unstable only when $L > L_c$. (5.2.46) or (5.2.47) is a necessary condition of the neutral instability (also linear instability) of the two-layer baroclinic Rossby waves. When this condition is satisfied, (5.2.45) is written as

$$\hat{u}\begin{cases} \leq \hat{u}_c, & \text{stable}, \\ > \hat{u}_c, & \text{unstable}, \end{cases} \tag{5.2.48}$$

where

$$\hat{u}_c = \beta_0 F_1/K_h^2\sqrt{4F_1^2 - K_h^4}$$

is the critical value of $\hat{u}$. According to the analysis of actual data, $\hat{u}_c = 4\ \text{m}\cdot\text{s}^{-1}$. Thus $(\bar{u}_2 - \bar{u}_1)_c \approx 8\ \text{m}\cdot\text{s}^{-1}$. Therefore, when the shear of the vertical wind velocity $(\bar{u}_2 - \bar{u}_1) > 8\ \text{m}\cdot\text{s}^{-1}$, there exist neutral instable waves.

Remark 5.2.1. For stationary solutions of the two-dimensional quasi-geotrophic equation, Lin obtained in [136] that for stationary solutions satisfying some conditions, when they are linear unstable, they are also nonlinear unstable.

5.3 Stability of Rossby Waves

5.3.1 *Stability of Bidimensional Quasi-Geostrophic Flow*

The two-dimensional quasi-geotrophic equation in a bounded domain Ω is

$$\left(\partial_t + \frac{\partial\psi}{\partial x}\frac{\partial}{\partial y} - \frac{\partial\psi}{\partial y}\frac{\partial}{\partial x}\right)\omega = 0, \tag{5.3.1}$$

where $\omega = \Delta\psi - F\psi + f(x,y)$, and $f(x,y)$ contains the basin bottom topology and βy. The boundary condition is given by

$$\psi|_{\partial\Omega} = 0. \tag{5.3.2}$$

In the above section, we have recalled some necessary conditions of the linear instability of some stationary solutions of (5.3.1) in a periodic channel. Here we mainly discuss the nonlinear stability of the steady solution $\bar{\psi}$ of (5.3.1) and (5.3.2) satisfying the following conditions,

$$\bar{\psi} = Q(\bar{\omega}) \text{ in } \Omega, \quad \bar{\psi}|_{\partial\Omega} = 0, \tag{5.3.3}$$

where $\bar{\omega} = \Delta\bar{\psi} - F\bar{\psi} + f(x, y)$, and Q is a monotonic continuous differentiable function.

As the two-dimensional quasi-geotrophic equation (5.3.1) with boundary condition (5.3.2) has two conservation laws:

(1) energy conservation law

$$\frac{\mathrm{d}}{\mathrm{d}t}\int_\Omega (|\nabla\psi|^2 + F|\psi|^2)\mathrm{d}x\mathrm{d}y = 0, \quad t \geq 0, \tag{5.3.4}$$

(2) conservation law of the generalized vorticity enstrophy

$$\frac{\mathrm{d}}{\mathrm{d}t}\int_\Omega C(\omega)\mathrm{d}x\mathrm{d}y = 0, \quad t \geq 0, \tag{5.3.5}$$

where C is a two-ordered continuous differentiable function in $(-\infty, +\infty)$. Inspired by the method of Arnold studying the nonlinear stability of the plane ideal incompressible fluid (see [5, 6]), one can use the Energy-Casimir method (see [104]) to study the nonlinear stability of steady solution $\bar{\psi}$, that is, one finds some sufficient conditions of the nonlinear stability of $\bar{\psi}$ by studying Energy-Casimir functional

$$EC(\psi) = \frac{1}{2}\int (|\nabla\psi|^2 + F|\psi|^2) + \int C(\omega), \tag{5.3.6}$$

where C is a second order continuous differentiable function, and $C' = Q$. According to (5.3.3) and (5.3.6), we know that $\bar{\psi}$ is a critical point of $EC(\psi)$.

Suppose that F is a positive constant, and f is a given function. If

$$Q' > 0, \tag{5.3.7}$$

then

$$EC(\psi) - EC(\bar{\psi}) \geq \frac{1}{2}\int_\Omega (|\nabla(\psi - \bar{\psi})|^2 + F|\psi - \bar{\psi}|^2),$$

which implies that $\bar{\psi}$ is the minimizer value point of Energy-Casimir functional. Thus, $\bar{\psi}$ is nonlinearly stable in the following sense, $\forall\varepsilon > 0$ there exists $\delta > 0$, such that, if $\psi(t) \in C([0, T), H^2(\Omega) \cap H_0^1(\Omega))$ is the solution of (5.3.1) and (5.3.2) with initial condition $\psi(t)|_{t=0} = \psi_0$ and $\|\psi_0 - \bar{\psi}\|_{H^2\cap H_0^1} < \delta$, then for any $t \in [0, T)$, $\|\psi(t) - \bar{\psi}\|_1 < \varepsilon$, where $\|\cdot\|_1$

is the norm of $H_0^1(\Omega)$, $\|\psi\|_1^2 = \int_\Omega |\nabla\psi|^2$, $\|\psi\|_{H^2\cap H_0^1}^2 = \int_\Omega |\Delta\psi|^2$, $T > 0$. Similarly, if there exists a positive constant c big enough, such that

$$Q' < -c, \tag{5.3.8}$$

then

$$EC(\psi) - EC(\bar{\psi}) \le \frac{1}{2}\int_\Omega (|\nabla(\psi - \bar{\psi})|^2 + F|\psi - \bar{\psi}|^2) - c\int_\Omega |\Delta(\psi - \bar{\psi})|^2,$$

$$\le -c'\int_\Omega |\Delta(\psi - \bar{\psi})|^2,$$

where c' is a positive constant. The above relationship implies that $\bar{\psi}$ is the maximum value point of EC. Thus, $\bar{\psi}$ is nonlinearly stable in the following sense, $\forall\varepsilon > 0$, there exist $\delta > 0$, such that if $\psi(t) \in C([0,T), H^2(\Omega) \cap H_0^1(\Omega))$ is the solution of (5.3.1) and (5,3,2) with initial condition $\psi(t)|_{t=0} = \psi_0$ and $\|\psi_0 - \bar{\psi}\|_{H^2\cap H_0^1} < \delta$, then, for $t \in [0,T)$, $\|\psi(t) - \bar{\psi}\|_{H^2\cap H_0^1} < \varepsilon$, where $T > 0$.

The above two situations (5.3.7) and (5.3.8) are respectively corresponding to Arnold's first and second stability theorems, and many works (such as [161, 167, 184, 215, 221]) generalize the two situations. In [167], Mu obtained some criteria of the nonlinear stability of the two-dimensional quasi-geotrophic equation in the disturbance of the initial data and parameters, and also illustrated some examples of the nonlinearly stable stationary solutions. Now, we simply recall two nonlinear stability criteria in [167]. For simplicity, we only consider boundary condition (5.3.2). For other conclusions under general boundary conditions, see [167].

Let $(\bar{\psi}, \bar{\omega})$ be a stationary solution of (5.3.1) and (5.3.2), where $F = \bar{F}$, $f = \bar{f}(x,y)$, F is a parameter, and f is a function, that is,

$$J(\bar{\psi}, \bar{\omega}) = 0, \quad \bar{\omega} = \nabla^2\bar{\psi} - \bar{F}\bar{\psi} + \bar{f}(x,y) \text{ in } \Omega,\ \bar{\psi}|_{\partial\Omega} = 0.$$

Moreover, assume that there exists a $Q \in C^1[\alpha, \beta]$ such that

$$\bar{\psi}(x,y) = Q(\bar{\omega}(x,y)), (x,y) \in \Omega, \tag{5.3.9}$$

where $\alpha = \min_{(x,y)\in\Omega} \bar{\omega}(x,y)$, $\beta = \max_{(x,y)\in\Omega} \bar{\omega}(x,y)$, Q is monotonic, and there exist positive constants C_1 and C_2 such that $C_1 \le \mathrm{d}Q/\mathrm{d}\xi \le C_2$ or $C_1 \le -\mathrm{d}Q/\mathrm{d}\xi \le C_2$, $\xi \in [\alpha, \beta]$. For generality, suppose that Q can be extended as a function of $C^1(-\infty, +\infty)$, and satisfies one of the following two conditions

$$0 < C_1 \le \mathrm{d}Q/\mathrm{d}\xi \le C_2 < \infty,\ \xi \in (-\infty, +\infty), \tag{5.3.10}$$

$$0 < C_1 \le -\mathrm{d}Q/\mathrm{d}\xi \le C_2 < \infty,\ \xi \in (-\infty, +\infty). \tag{5.3.11}$$

Because of $\bar{\psi} = Q(\bar{\omega})$, the above two conditions are rewritten as

$$C_1 = \min(\nabla\bar{\psi}/\nabla\bar{\omega}) > 0, C_2 = \max(\nabla\bar{\psi}/\nabla\bar{\omega}) < \infty, \tag{5.3.10'}$$

$$C_1 = \min(-\nabla\bar{\psi}/\nabla\bar{\omega}) > 0, C_2 = \max(-\nabla\bar{\psi}/\nabla\bar{\omega}) < \infty, \tag{5.3.11'}$$

where

$$\nabla\bar{\psi}/\nabla\bar{\omega} = \frac{\partial\bar{\psi}}{\partial x}/\frac{\partial\bar{\omega}}{\partial x} = \frac{\partial\bar{\psi}}{\partial y}/\frac{\partial\bar{\omega}}{\partial y}.$$

Define initial condition disturbance $\delta\bar{\psi}$ and parameter disturbance $\delta\bar{F}$, $\delta\bar{f}$ by

$$\psi_0 \equiv \bar{\psi} + \delta\bar{\psi}, F \equiv \bar{F} + \delta\bar{F}, f(x,y) \equiv \bar{f}(x,y) + \delta\bar{f}, \tag{5.3.12}$$

when $\bar{F} = 0$, suppose $\delta\bar{F} \equiv 0$. ψ is a solution of (5.3.1) and (5.3.2) with parameters F, f and initial data $\psi(0) = \psi_0(\psi_0 \in H^2(\Omega) \cap H_0^1(\Omega))$ (the proof of the global existence of smooth solutions of the two- and three-dimensional quasi-geotrophic equations can be seen in [165, 167]). Under the assumptions (5.3.9) and (5.3.10), Mu obtain the following nonlinear stability theorem.

Theorem 5.3.1 (cf. [152, Theorem 3.1]). If the stationary solution $(\bar{\psi}, \bar{\omega})$ of (5.3.1) and (5.3.2) satisfies (5.3.9) and (5.3.10), then the stationary solution $\bar{\psi}$ is nonlinearly stable with respect to the initial condition and parameter disturbance (5.3.12), that is, for any $\varepsilon_1 > 0, \varepsilon_2 > 0$, there exists $\delta > 0$ such that if

$$\int_\Omega [|\psi_0 - \bar{\psi}|^2 + |\nabla(\psi_0 - \bar{\psi})|^2 + |\omega_0 - \bar{\omega}|^2 + |f - \bar{f}|^2] + |F - \bar{F}|^2 < \delta,$$

then

$$\max_{(x,y)\in\Omega} |\psi - \bar{\psi}| < \varepsilon_1,$$

$$\int_\Omega |\nabla(\psi - \bar{\psi})|^2 + \int_\Omega \bar{F}|\psi - \bar{\psi}|^2 \mathrm{d}x\mathrm{d}y + \int_\Omega |\omega - \bar{\omega}|^2 < \varepsilon_2, \quad t \geq 0. \tag{5.3.13}$$

Remark 5.3.2. Since

$$\omega - \bar{\omega} = \nabla^2(\psi - \bar{\psi}) - \bar{F}(\psi - \bar{\psi}) - (F - \bar{F})\psi - (f - \bar{f}), \tag{5.3.14}$$

the second inequality of (5.3.13) implies that for any $\varepsilon > 0$, there exists $\delta > 0$ such that if

$$\int_\Omega [|\psi_0 - \bar{\psi}|^2 + |\nabla(\psi_0 - \bar{\psi})|^2 + |\omega_0 - \bar{\omega}|^2 + |f - \bar{f}|^2] + |F - \bar{F}|^2 < \delta,$$

then

$$\|\psi(t) - \bar{\psi}\|_{H^2(\Omega)} < \varepsilon, \ t \geq 0.$$

Proof of Theorem 5.3.1. Let

$$H(t) = EC(\psi) - EC(\bar{\psi})$$

$$= \frac{1}{2}\int_\Omega (|\nabla\psi|^2 + F|\psi|^2) - \frac{1}{2}\int_\Omega (|\nabla\bar{\psi}|^2 + \bar{F}|\bar{\psi}|^2) + \int_\Omega [C(\omega) - C(\bar{\omega})],$$

where $C' = Q$. In order to prove the nonlinear stability of the stationary solution with the Energy-Casimir method, the key step is to take second variation of Energy-Casimir functional and derive a convexity estimate from $H(t)$. Here we only need to get a convexity estimate from $H(t)$. Applying (5.3.9), (5.3.10), (5.3.14) and Taylor formula, we have

$$\begin{aligned}
&\frac{1}{2}\int_\Omega (|\nabla\psi|^2 + F|\psi|^2) - \frac{1}{2}\int_\Omega (|\nabla\bar{\psi}|^2 + \bar{F}|\bar{\psi}|^2) + \int_\Omega [C(\omega) - C(\bar{\omega})] \\
=&\frac{1}{2}\int_\Omega [(|\nabla\psi|^2 + F|\psi|^2) - (|\nabla\bar{\psi}|^2 + \bar{F}|\bar{\psi}|^2)] \\
&+ \int_\Omega \left[C'(\bar{\omega})(\omega - \bar{\omega}) + \frac{C''(\omega^*)}{2}(\omega - \bar{\omega})^2\right] \\
\geq&\frac{1}{2}\int_\Omega [|\nabla(\psi - \bar{\psi})|^2 + \bar{F}|\psi - \bar{\psi}|^2] + \frac{1}{2}\int_\Omega (F - \bar{F})|\psi|^2 \\
&- \int_\Omega \bar{\psi}\,[(F - \bar{F})\psi + (\bar{f} - f)] + \frac{1}{3}C_1 \int_\Omega |\omega - \bar{\omega}|^2,
\end{aligned} \tag{5.3.15}$$

where ω^* is between ω and $\bar{\omega}$. According to (5.3.4) and (5.3.5), for any $t > 0$,

$$H(t) = H(0). \tag{5.3.16}$$

Since C is second order continuously differentiable, we know through direct calculation that for any $\varepsilon > 0$, there exists $\delta > 0$ such that if

$$\int_\Omega [|\psi_0 - \bar{\psi}|^2 + |\nabla(\psi_0 - \bar{\psi})|^2 + |\omega_0 - \bar{\omega}|^2 + |f - \bar{f}|^2] + |F - \bar{F}|^2 < \delta,$$

then

$$H(0) < \varepsilon. \tag{5.3.17}$$

Applying (5.3.15), (5.3.16) and (5.3.17), we get the second inequality of (5.3.13).

According to Sobolev Embedding Theorem and (5.3.14), we have

$$\begin{aligned}
&\max_{(x,y)\in\Omega} |\psi - \bar{\psi}|^2 \\
\leq & c\int_\Omega |\nabla^2(\psi - \bar{\psi}) - \bar{F}(\psi - \bar{\psi})|^2 \\
\leq & c\int_\Omega (|\omega - \bar{\omega}|^2 + |F - \bar{F}|^2|\psi|^2 + |f - \bar{f}|^2), \quad t \geq 0,
\end{aligned}$$

where c is a positive constant dependent on Ω. Thus, we obtain the first inequality of (5.3.13). The proof of Theorem 5.3.1 is complete.

Similarly, under the assumptions (5.3.9) and (5.3.11), where C_1 is big enough, Mu obtained the following nonlinear stability theorem.

Theorem 5.3.3 (cf. [167, Theorem 3.2]). If $(\bar{\psi}, \bar{\omega})$ satisfies condition (5.3.9) and (5.3.11), and

$$C_1\bar{\lambda}_1 > 1, \tag{5.3.18}$$

then $\bar{\psi}$ is nonlinearly stable, where $\bar{\lambda}_1$ is the first eigenvalue of the following eigenvalue problem,

$$-\nabla^2\bar{\psi} + \bar{F}\bar{\psi} = \bar{\lambda}\bar{\psi} \text{ in } \Omega, \quad \bar{\psi}|_{\partial\Omega} = 0.$$

Proof. Applying (5.3.9), (5.3.11), (5.3.14) and Taylor formula, we have

$$\begin{aligned}
&\frac{1}{2}\int_\Omega (|\nabla\psi|^2 + F|\psi|^2) - \frac{1}{2}\int_\Omega (|\nabla\bar{\psi}|^2 + \bar{F}|\bar{\psi}|^2) + \int_\Omega [C(\omega) - C(\bar{\omega})] \\
=&\frac{1}{2}\int_\Omega [(|\nabla\psi|^2 + F|\psi|^2) - (|\nabla\bar{\psi}|^2 + \bar{F}|\bar{\psi}|^2)] \\
&+ \int_\Omega \left[C'(\bar{\omega})(\omega - \bar{\omega}) + \frac{C''(\omega^*)}{2}(\omega - \bar{\omega})^2\right] \\
\leq&\frac{1}{2}\int_\Omega [|\nabla(\psi - \bar{\psi})|^2 + \bar{F}|\psi - \bar{\psi}|^2] + \frac{1}{2}\int_\Omega (F - \bar{F})|\psi|^2 \\
&- \int_\Omega \bar{\psi}\left[(F - \bar{F})\psi + (\bar{f} - f)\right] - \frac{C_1 - \varepsilon}{2}\int_\Omega |\omega - \bar{\omega}|^2,
\end{aligned} \tag{5.3.19}$$

where ω^* is between ω and $\bar{\omega}$. According to the definition of $\bar{\lambda}_1$, we get

$$\int_\Omega |\nabla^2(\psi - \bar{\psi}) - \bar{F}(\psi - \bar{\psi})|^2 \geq \bar{\lambda}_1^2 \int_\Omega |\psi - \bar{\psi}|^2. \tag{5.3.20}$$

By integration by parts and Young Inequality, we obtain

$$\int_\Omega (|\nabla(\psi-\bar{\psi})|^2+\bar{F}|(\psi-\bar{\psi})|^2) \leq \frac{\mu}{2}\int_\Omega |\psi-\bar{\psi}|^2+\frac{1}{2\mu}\int_\Omega |\nabla^2(\psi-\bar{\psi})-\bar{F}(\psi-\bar{\psi})|^2, \tag{5.3.21}$$

where μ is a positive constant. Combining (5.3.20) with (5.3.21), and letting $\mu = \bar{\lambda}_1$, we get

$$\frac{1}{2}\int_{\Omega}(|\nabla(\psi-\bar{\psi})|^2+\bar{F}|(\psi-\bar{\psi})|^2) \leq \frac{1}{2\bar{\lambda}_1}\int_{\Omega}|\nabla^2(\psi-\bar{\psi})-\bar{F}(\psi-\bar{\psi})|^2. \quad (5.3.22)$$

According to (5.3.14), (5.3.19) and (5.3.22), we prove Theorem 5.3.3.

Remark 5.3.4. If $EC(\psi)$ is second order continuously differentiable at $\bar{\psi}$, from the proof of Theorem 5.3.1 and Theorem 5.3.3, we know that if $C_2 = +\infty$, then Theorem 5.3.1 and Theorem 5.3.3 are still valid.

Now, we give an example of nonlinear stable stationary solution $\bar{\psi}$, which appears in [63].

Example 5.3.5. Let

$$Q(\xi) = \tan\xi, \quad -\frac{\pi}{2} < \xi < \frac{\pi}{2}.$$

If $f(x, y)$ is a bounded function, then, according to [200, Theorem 1.2] and [204, Corollary 26.13], we prove that the boundary value problem of the ellipse equation

$$\Delta\psi - F\psi + f(x, y) = \arctan(\psi), \ \psi|_{\Omega} = 0,$$

has a unique $\bar{\psi}$, cf. [63]. If $\bar{\omega} = \Delta\bar{\psi} - F\bar{\psi} + f(x, y)$ belongs to a closed interval of $(-\frac{\pi}{2}, \frac{\pi}{2})$, then, $\bar{\psi}$ is nonlinearly stable according to Theorem 5.3.1.

5.3.2 *Stability of Saddle-Point-Type Two-Dimensional Quasi-Geostrophic Flows*

In this subsection, we mainly discuss the stability of the saddle-point-type steady solutions of the two-dimensional quasi-geostrophic equation, which appears in [215]. For non-minimal or non-maximal critical points of Energy-Casimir functional of Hamiltonian systems. Arnold has noticed that saddle-point-type steady solutions are not always unstable (see [7]). In [215], Wolansky and Ghil considered the linear stability of a kind of the saddle-point-type stationary two-dimensional quasi-geostrophic flows.

The two-dimensional quasi-geostrophic equation in a periodic channel is

$$\left(\partial_t + \frac{\partial\psi}{\partial x}\frac{\partial}{\partial y} - \frac{\partial\psi}{\partial y}\frac{\partial}{\partial x}\right)\omega = 0, \quad (5.3.23)$$

where $\omega = \Delta\psi - F\psi + h(y)$, $0 \leq y \leq 1$,

$$\psi|_{y=0} = 0, \psi|_{y=1} = c, \quad (5.3.24a)$$

$$\psi|_{x=0} - \psi|_{x=L} = \partial_x\psi|_{x=0} - \partial_x\psi|_{x=L} = 0, \tag{5.3.24b}$$

and $h(y)$ is a given function describing bottom topology and the effect of the earth rotation βy. Here we mainly study the steady solution $\bar{\psi}(y)$ of (5.3.23) and (5.3.24) satisfying

$$\bar{\psi} = Q(\bar{\omega}), \tag{5.3.25}$$

where $\bar{\omega} = \Delta\bar{\psi} - F\bar{\psi} + h(y)$, and Q is a monotonic continuous differential function.

As the quasi-geotrophic equation (5.3.23) with (5.3.24) has two conservation laws. We know from the above section that researchers can study the nonlinear stability of the barotropic quasi-geotrophic flows with Energy-Casimir method, that is, one can obtain some sufficient conditions of the nonlinear stability of the steady solutions of (5.3.23) and (5.3.24) by studying the following Energy-Casimir functional

$$EC(\psi) = \frac{1}{2}\int_0^1\int_0^L (|\nabla\psi|^2 + F|\psi|^2) + \int_0^1\int_0^L C(\omega), \tag{5.3.26}$$

where C is a second order continuous differential function, and $C^{'} = Q$. According to (5.3.26), we know that $\bar{\psi}(y)$ is a critical point of $EC(\psi)$. If $Q^{'} > 0$ ($Q^{'} < -c$, c is a big enough positive number), then $\bar{\psi}$ is the minimum (maximum) point of EC. According to Arnold first (second) stability theorem, $\bar{\psi}$ is nonlinearly stable. However, if $Q^{'} < 0$, and $\bar{\psi}$ is a saddle-point-type critical point of EC (there does not exist c big enough such that $Q^{'} < -c$), then we can't use Arnold first or second stability theorem to judge the stability of $\bar{\psi}$.

Let $\alpha(y) = -g^{'}(\bar{\psi}) > 0$, where $g = Q^{-1}$. Then

$$E(\phi) = \langle EC_{\bar{\psi}}^{''}(\phi), \phi\rangle = -\int_0^1\int_0^L \frac{1}{\alpha(y)}(\Delta\phi - F\phi)^2 + \int_0^1\int_0^L (|\nabla\phi|^2 + F|\phi|^2), \tag{5.3.27}$$

where $E(\phi)$ is the second order variation of $EC(\psi)$ at point $\bar{\psi}$. Because $Q^{'} < 0$, and there does not exist c big enough such that the second-ordered variation of the above relationship is negative definite, we can't make sure whether $\bar{\psi}$ is stable or not. Now, let's analyze the functional

$$E(\phi) = -\int_0^1\int_0^L \frac{1}{\alpha(y)}(\Delta\phi - F\phi + \alpha(y)\phi)(\Delta\phi - F\phi),$$

and find some conditions satisfied by $\bar{\psi}$ of its stability in some sense.

Let $\phi_{k,j}(x, y) = \theta_j^k(y)e^{\pm 2\pi i k\frac{x}{L}}$ be the eigenfunctions of the eigenvalue problem

$$-(\partial_{xx} + \partial_{yy})\phi + F\phi = \mu\alpha(y)\phi, \quad \phi|_{y=0,1} = 0, \tag{5.3.28}$$

ϕ is a function with period of L in the direction of x, corresponding to eigenvalues μ_j^k. Then $\phi_{k,j}$ is a complete orthogonal basis of space L_p^2, where L_p^2 is the set of L-periodic in x and square-integrable functions with the inner product

$$\langle \phi_1, \phi_2 \rangle = \int_0^1 \alpha(y) \int_0^L \phi_1 \phi_2 \mathrm{d}x\mathrm{d}y.$$

Thus, $\theta_j^k(y)$ are eigenfunctions of the eigenvalue problem

$$-\theta_{yy} + \left[F - \mu\alpha(y) + \left(\frac{2\pi k}{L}\right)^2\right]\theta = 0, \ \ \theta|_{y=0,1} = 0,$$

corresponding to eigenvalues μ_j^k, and θ_j^k are orthogonal. According to Sturm-Lionville Theorem, for $k \geq 0$ fixed, there exists $0 < \mu_0^k < \mu_1^k < \cdots < \mu_j^k \to \infty$. By a comparison argument, we have

$$\mu_j^k > \mu_j^l, \forall j \geq 0, k > l, \text{ and when } k \to \infty, \ \ \mu_0^k \to \infty.$$

If $\mu_0^0 > 1$, then $\mu_j^k > 1$, $\forall k,\ j \geq 0$,

$$E(\phi_{k,j}) = -\int_0^1 \int_0^L (\mu_j^k)^2 \alpha(y) \phi_{k,j}^2 + \int_0^1 \int_0^L \mu_j^k \alpha(y) \phi_{k,j}^2 < 0.$$

Thus, $\forall \phi \in X$,

$$E(\phi) < 0 = E(0), \tag{5.3.29}$$

which implies that 0 is a maximum value point of $E(\phi)$ in X, where $X = H_{0,p}^1 \cap H_p^2$, $H_{0,p}^1$ is a completion space of $C_{0,p}^\infty$ in H^1 norm, and $H_p^2 \subset L_p^2$ is the set of functions admitting square-integrable second derivative in the channel domain. Since $E(\phi)$ is the second order variation of functional $EC(\psi)$ at point $\bar{\psi}$, according to (5.3.29), the second order variation of $EC(\psi)$ at point $\bar{\psi}$ is negative definite. Thus $\bar{\psi}$ is formally stable (the definition of formal stability is seen in [104]), which implies that $\bar{\psi}$ is linearly stable. Indeed, according to the definitions of μ_j^k, we get

$$\int_0^1 \int_0^L \frac{1}{a(y)} |\nabla^2 \phi - \bar{F}\phi|^2 \geq (\mu_0^0)^2 \int_0^1 \int_0^L a(y)|\phi|^2.$$

By integration by parts and Young Inequality, we have

$$\int_0^1 \int_0^L (|\nabla \phi|^2 + \bar{F}|\phi|^2) \leq \frac{\mu}{2} \int_0^1 \int_0^L a(y)|\phi|^2 + \frac{1}{2\mu} \int_0^1 \int_0^L \frac{1}{a(y)} |\nabla^2 \phi - \bar{F}\phi|^2,$$

where μ is a positive constant. By the above two inequalities, and taking $\mu = \mu_0^0$, we get

$$\int_0^1 \int_0^L (|\nabla\phi|^2 + \bar{F}|\phi|^2) \le \frac{1}{\mu_0^0} \int_0^1 \int_0^L \frac{1}{a(y)} |\nabla^2\phi - \bar{F}\phi|^2.$$

Thus

$$E(\phi) - E(0) \le -(1 - \frac{1}{\mu_0^0}) \int_0^1 \int_0^L \frac{1}{\alpha(y)} |\Delta\phi - F\phi|^2. \tag{5.3.30}$$

However, in general, $0 < \mu_0^0 < 1$. Now we suppose

$$0 < \mu_0^0 < \mu_1^0 < \mu_2^0 < \cdots < \mu_n^0 \le 1 < \mu_{n+1}^0 < \cdots,$$

and $\exists L_0 > 0$ such that if $L \le L_0$, then

$$\mu_0^k > 1, k \ge 1.$$

Thus, for $0 \le j \le n$,

$$E(\phi_{0,j}) = -\int_0^1 \int_0^L (\mu_j^0)^2 \alpha(y)\phi_{0,j}^2 + \int_0^1 \int_0^L \mu_j^0 \alpha(y)\phi_{0,j}^2 \ge 0,$$

moreover, for $j \ge n+1, k \ge 0$,

$$E(\phi_{k,j}) = -\int_0^1 \int_0^L (\mu_j^k)^2 \alpha(y)\phi_{k,j}^2 + \int_0^1 \int_0^L \mu_j^k \alpha(y)\phi_{k,j}^2 < 0.$$

So, 0 is not the maximum point of E, and ψ_0 is a saddle-point-type critical point of $EC(\psi)$.

Although 0 is not the maximum point of E, we use the dual variation principle to find a support functional D of E (in which we used Legendre transform) such that 0 is the maximum point of D in space Y (now we take $Y = H_{0,p}^1$). Since

$$E(\phi) = \int_0^1 \int_0^L \left(-\frac{1}{\alpha(y)} |\xi_\phi|^2 - 2\phi\xi_\phi \right) - \int_0^1 \int_0^L (|\nabla\phi|^2 + F\phi^2),$$

let

$$D(\phi) = -\int_0^1 \int_0^L (|\nabla\phi|^2 + F\phi^2) + \sup_{\xi \in L_2^p \cap X_0} \int_0^1 \int_0^L \left(-\frac{1}{\alpha(y)} |\xi|^2 - 2\phi\xi \right),$$

where $\xi_\phi = \Delta\phi - F\phi$,

$$\phi \in X_0 = X \cap \left\{ \phi; \int_0^1 \int_0^L \theta_j^0(y)(\Delta\phi - F\phi) = 0, 1 \le j \le n \right\}.$$

$\int_0^1 \int_0^L \theta_j^0(y)(\Delta\phi - F\phi) = 0$ is a conservation quantity of the following linearized equation. Now let's give the expression of $D(\phi)$. For all $\xi \in L_p^2$, $\xi - \sum_{j=0}^{n}\langle \xi, \theta_j^0\rangle\theta_j^0 \in L_2^p \cap X_0$. Thus,

$$D(\phi) = -\int_0^1 \int_0^L (|\nabla\phi|^2 + F\phi^2)$$

$$+ \sup_{\xi\in L_p^2} \int_0^1 \int_0^L \left[-\frac{1}{\alpha(y)} \left(\xi - \sum_{j=0}^{n}\langle \xi, \theta_j^0\rangle\theta_j^0 \right)^2 - 2\phi \left(\xi - \sum_{j=0}^{n}\langle \xi, \theta_j^0\rangle\theta_j^0 \right) \right].$$

For functional

$$\int_0^1 \int_0^L \left[-\frac{1}{\alpha(y)} \left(\xi - \sum_{j=0}^{n}\langle \xi, \theta_j^0\rangle\theta_j^0 \right)^2 - 2\phi \left(\xi - \sum_{j=0}^{n}\langle \xi, \theta_j^0\rangle\theta_j^0 \right) \right], \quad (5.3.31)$$

taking Frenchet derivative of it with respect to ξ, we know that $\xi = \sum_{j=0}^{n}\langle \xi, \theta_j^0\rangle\theta_j^0 - \alpha(y)\phi + \sum_{j=0}^{n}\langle \phi, \theta_j^0\rangle\alpha\theta_j^0$ is the maximum point of functional (5.3.31).

If $\phi \in X_0$, then $\xi - \sum_{j=0}^{n}\langle \xi, \theta_j^0\rangle\theta_j^0 = -\alpha(y)\phi$. Thus,

$$D(\phi) = -\int_0^1 \int_0^L (|\nabla\phi|^2 + F\phi^2 - \alpha(y)|\phi|^2) - \sum_{j=0}^{n}\langle \phi, \theta_j^0\rangle^2,$$

$D(\phi) \geq E(\phi)$, $\forall\phi \in X_0$, $D(0) = E(0) = 0$, and $D(\phi) < 0$. Indeed, in the orthogonal complement space of $\mathrm{Span}\{\theta_j^0\}_{j=0}^{j=n}$ in Y, D is negative definite. For $\phi \in \mathrm{Span}\{\theta_j^0\}_{j=0}^{j=n}$, let

$$\phi = \beta_0\theta_0^0 + \cdots + \beta_n\theta_n^0,$$

then

$$D(\phi) = \sum_{j=0}^{n}(1 - \mu_j^0)\beta_j^2 - \sum_{j=0}^{n}\beta_j^2 < 0.$$

From the above analysis, we know that if there exists $L_0 > 0$, when $L \leq L_0$, $0 < \mu_0^0 < \mu_1^0 < \mu_2^0 < \cdots < \mu_n^0 < 1 < \mu_{n+1}^0 < \cdots$, then, for $\phi \in X_0$, there exists

$$E(\phi) - E(0) \leq D(\phi) - D(0)$$

$$\leq -\int_0^1\int_0^L (|\nabla\phi|^2 + F|\phi|^2 + \int_0^1\int_0^L \alpha(y)\phi^2$$
$$\leq -\int_0^1\int_0^L (|\nabla\phi|^2 + F|\phi|^2) + \frac{1}{\mu_{n+1}^0}\int_0^1\int_0^L (|\nabla\phi|^2 + F|\phi|^2)$$
$$\leq \left(-1 + \frac{1}{\mu_{n+1}^0}\right)\int_0^1\int_0^L (|\nabla\phi|^2 + F|\phi|^2).$$

Since $E(\phi)$ is a conservation quantity of linearized equation

$$\partial_t(\Delta\phi - F\phi) - \bar{\psi}'(y)\partial_x\left[\Delta\phi - F\phi - g'(\bar{\psi})\phi\right] = 0 \qquad (5.3.32)$$

and $\phi|_{y=0,1} = 0$, ϕ is the function with period of L in the direction of x, $\bar{\phi}$ is linearly stable with respect to disturbance $\psi - \bar{\psi} \in X$ ($\psi - \bar{\psi}$ is a solution of linearized equation (5.3.32)), that is,

$$\forall\varepsilon > 0, \exists\delta > 0, \text{ if } \|\psi(0) - \bar{\psi}\|_X < \delta \text{ and } \psi - \bar{\psi} \in X_0,$$
$$\text{then } \|\psi(t) - \bar{\psi}\|_Y < \varepsilon. \qquad (5.3.33)$$

In particular, if $\mu_0^0 > 1$, then $X_0 = X$.

Remark 5.3.6. In fact, we can also get a relationship similar to (5.3.30). Thus, $\|\psi(t) - \bar{\psi}\|_Y < \varepsilon$ in (5.3.33) is replaced by $\|\psi(t) - \bar{\psi}\|_X < \varepsilon$, and then $\bar{\psi}$ is linearly stable in general sense.

That is, we get the following conclusion.

Theorem 5.3.7. If $\bar{\psi}(y)$ is a stationary solution of (5.3.23) and (5.3.24), and there exists a monotonic decreasing continuous differential function Q such that

$$\bar{\psi} = Q(\bar{\omega}).$$

Furthermore, assume that there exists a $L_0 > 0$, such that for $L \leq L_0$, the eigenvalue μ_j^k of problem

$$-\theta_{yy} + \left[F - \mu\alpha(y) + \left(\frac{2\pi k}{L}\right)^2\right]\theta = 0, \ \theta|_{y=0,1} = 0,$$

satisfies

$$\mu_0^k > 1, k \geq 1,$$

then $\bar{\psi}(y)$ is linearly stable in the sense of (5.3.33).

At last, we give a remark.

Remark 5.3.8. In this section, we only have to introduce the stability of the two-dimensional quasi-geotropic flows by the Energy-Casimir method. Researchers have also used this method to solve the stability of three-dimensional quasi-geotropic flows and some other basic flows of the atmospheric and oceanic science, for example see [104, 146, 166, 221]. In [221], Zeng Qingcun studied nonlinear stability and instability of many important atmospheric motions (including barotropic and baroclinic quasi-geotropic flows and non-geotropic flows, which contain stationary flows and nonstationary) through the nonlinear equations governing the atmosphere motions and variation principle. He obtained some conservation laws with the nonlinear equations governing some kind of atmosphere motions, and constructed some proper Energy-Casimir functionals, through finding the first-order variation and the second order variation of these functionals, he obtained the critical points and some important sufficient conditions of nonlinear stability and necessary conditions of nonlinear instability. For details, see [221].

5.4 Critical Rayleigh Number of Rayleigh-Bénard Convection

Rayleigh-Bénard convection problem is a classical problem of fluid dynamics. In the rigid boundary conditions, Guo Yan and Han Yongqian found the critical Rayleigh number. They proved that when $R_a < R_a^*$, the stationary solution is nonlinearly asymptotically stable, but when $R_a > R_a^*$, the solution is nonlinearly unstable. This result can be seen in [95].

Rayleigh-Bénard convection in a shallow horizontal layer of a fluid heated from below has been widely studied in experiments, numerical simulations and theoretical analysis, cf. [22, 31, 57, 77, 101, 113, 114, 151, 198]. Taking Boussinesq approximation, one can get the initial boundary value problem describing Rayleigh-Bénard convection [31]:

$$\partial_t \boldsymbol{v} + (\boldsymbol{v} \cdot \boldsymbol{\nabla})\boldsymbol{v} + \frac{1}{\rho_0}\boldsymbol{\nabla} p = \nu \Delta \boldsymbol{v} + g\left[\alpha(T - T_0) - 1\right]\mathbf{e}_z, \tag{5.4.1}$$

$$\boldsymbol{\nabla} \cdot \boldsymbol{v} = 0, \tag{5.4.2}$$

$$\partial_t T + (\boldsymbol{v} \cdot \boldsymbol{\nabla})T = \kappa \Delta T, \tag{5.4.3}$$

$$\begin{aligned} &\boldsymbol{v}|_{z=0,h} = 0, T|_{z=0} = T_1,\ T|_{z=h} = T_2, \\ &\boldsymbol{v}, T \text{ have a period of } 2\pi h \text{ in the direction of } x, y \end{aligned} \tag{5.4.4}$$

$$\boldsymbol{v}|_{t=0} = \boldsymbol{v}_0(x,y,z),\ T|_{t=0} = T_0(x,y,z), \tag{5.4.5}$$

where $\boldsymbol{v} = (v_1, v_2, v_3)$ is the velocity field, p is pressure, $\boldsymbol{v}$ is viscosity, α is an positive constant, $\boldsymbol{e}_z = (0,0,1)$ is an upward unit vector, T is temperature,

κ is the thermal diffusive coefficient, T_0 is the reference temperature value, ρ_0 is the density at temperature T_0, and $T_1 > T_2$.

The initial boundary value problem (5.4.1)-(5.4.5) has the following stationary solution

$$\boldsymbol{v}_s \equiv 0,\ p_s = -g\rho_0 + g\alpha\left[(T_s - T_0)z + \frac{T_2 - T_1}{2h}z^2\right],\ T_s = T_1 + \frac{T_2 - T_1}{h}z. \tag{5.4.6}$$

Suppose that the disturbance solution of (5.4.1)-(5.4.5) near the above stationary solution is

$$\boldsymbol{v} = \boldsymbol{v}_s + \boldsymbol{u},\ p = p_s + P,\ T = T_s + \theta.$$

We get

$$\begin{aligned}
&\partial_t \boldsymbol{u} + (\boldsymbol{u}\cdot\boldsymbol{\nabla})\boldsymbol{u} + \frac{1}{\rho_0}\boldsymbol{\nabla} P = \nu\Delta\boldsymbol{u} + g\alpha\theta\mathbf{e}_z,\\
&\boldsymbol{\nabla}\cdot\boldsymbol{u} = 0,\\
&\partial_t\theta + (\boldsymbol{u}\cdot\boldsymbol{\nabla})\theta + \frac{T_2 - T_1}{h}\boldsymbol{u}\cdot\mathbf{e}_z = \kappa\Delta\theta,\\
&\boldsymbol{u}|_{z=0,h} = 0, \theta|_{z=0,h} = 0,\ \text{the period of } \boldsymbol{u},\theta \text{ in } x,y \text{ is } 2\pi h,\\
&\boldsymbol{u}|_{t=0} = \boldsymbol{u}_0(x,y,z),\ \theta|_{t=0} = \theta_0(x,y,z).
\end{aligned}$$

Taking h as the characteristic scale of the vertical length, $(gh)^{1/2}$ as the characteristic scale of the velocity, $(h/g)^{1/2}$ as the characteristic scale of the time, $\rho_0 gh$ as the characteristic scale of the pressure, $T_1 - T_2$ as the characteristic scale of the temperature, we get the non-dimensional form of the initial boundary value problem

$$\partial_t\boldsymbol{u} + (\boldsymbol{u}\cdot\boldsymbol{\nabla})\boldsymbol{u} + \boldsymbol{\nabla} P = \mu_1\Delta\boldsymbol{u} + \mu_1\mu_2 R_a\theta\boldsymbol{e}_z, \qquad \boldsymbol{\nabla}\cdot\boldsymbol{u} = 0, \tag{5.4.7}$$

$$\partial_t\theta + (\boldsymbol{u}\cdot\boldsymbol{\nabla})\theta + \boldsymbol{u}\cdot\boldsymbol{e}_z = \mu_2\Delta\theta, \tag{5.4.8}$$

$$\boldsymbol{u}|_{t=0} = \boldsymbol{u}_0(x,y,z), \quad \theta|_{t=0} = \theta_0(x,y,z), \tag{5.4.9}$$

$$\boldsymbol{u}|_{z=0,h} = 0,\ \theta|_{z=0,h} = 0,\ \text{The period of } \boldsymbol{u},\ \theta \text{ in } x,\ y \text{ is } 2\pi h, \tag{5.4.10}$$

where $\mu_1 = \dfrac{\nu}{g^{1/2}h^{3/2}}$, $\mu_2 = \dfrac{\kappa}{g^{1/2}h^{3/2}}$, and the Rayleigh number is

$$R_a \equiv \frac{\alpha(T_1 - T_2)}{\mu_1\mu_2} > 0. \tag{5.4.11}$$

In order to consider the stability and instability of the stationary solution (5.4.6), we introduce the following critical Rayleigh number R_a^*. For any $k \geq 1$, let $\Theta_k(z)$ be the minimizer of the following variational problem

$$R(k) = \min_{\Theta\in B,\ k^2\int_0^1(|\partial_z\Theta|^2 + k^2|\Theta|^2)\mathrm{d}z = 1,} \int_0^1 |(\partial_z^2 - k^2)^2\Theta|^2 \mathrm{d}z,$$

where B is

$$\left\{ \Theta_k \in H^4,\ \Theta_k\Big|_{z=0,1} = (\partial_z^2 - k^2)\Theta_k\Big|_{z=0,1} = \partial_z(\partial_z^2 - k^2)\Theta_k\Big|_{z=0,1} = 0 \right\}.$$

Define the critical Rayleigh number by

$$R_a^* = \min_{k\neq 0}\{R(k)\}, \tag{5.4.12}$$

the stability of Rayleigh-Bénard convection has been studied by the linear theory method [31, 57, 77, 101] and the nonlinear energy method. Guo Yan and Han Yongqian obtained the critical Rayleigh number of Rayleigh-Bénard convection in [95].

This section is organized as follows: In subsection 5.4.1, we shall prove that the stationary solution (5.4.6) is linearly stable when $R_a < R_a^*$, but linearly unstable when $R_a > R_a^*$. We shall prove by the semigroup method in subsection 5.4.2 that stationary solution (5.4.6) is nonlinearly stable when $R_a < R_a^*$. We prove in subsection 5.4.3 that stationary solution (5.4.6) is nonlinearly unstable when $R_a > R_a^*$. The proof of the nonlinear instability of the stationary solution (5.4.6) is based on the framework given in [94, 96].

We give some notations in this section. Let $(0, 2\pi)^2 = (0, 2\pi) \times (0, 2\pi)$, $(E)^3 = E \times E \times E$, where E is any given Banach space. Suppose that H is

$$\{(u_1, u_2, u_3)|\ u_1, u_2, u_3 \in C_{per}^\infty((0, 2\pi)^2; C_0^\infty(0,1)); \partial_x u_1 + \partial_y u_2 + \partial_z u_3 = 0\},$$

it is about the space completion of the norm of $(L^2(\theta))^3$, where $\theta = (0, 2\pi)^2 \times (0, 1)$.

$$V = \left\{(u_1, u_2, u_3) \in H^1(\theta) \cap L_{per}^2((0, 2\pi)^2; H_0^1(0,1)); \partial_x u_1 + \partial_y u_2 + \partial_z u_3 = 0\right\},$$

its inner product and norm is the same as $(H^1(Q))^3$.

5.4.1 *Linear Stability*

In this subsection, we shall study the linear Boussinesq system disturbed near the stationary solution (5.4.6)

$$\partial_t \boldsymbol{u} + \boldsymbol{\nabla} P = \mu_1 \Delta \boldsymbol{u} + R_a \mu_1 \mu_2 \theta \mathbf{e}_z,\ \boldsymbol{\nabla} \cdot \boldsymbol{u} = 0,\ \partial_t \theta - u_3 = \mu_2 \Delta \theta, \tag{5.4.13}$$

with initial condition (5.4.9) and boundary condition (5.4.10). (5.4.13) is rewritten as

$$\partial_t(\boldsymbol{u}, \theta) = L(\boldsymbol{u}, \theta). \tag{5.4.14}$$

Here we shall make linear analysis of the linear system (5.4.14).

Lemma 5.4.1. The eigenvalue problem

$$-\lambda \boldsymbol{u} + \boldsymbol{\nabla} P = \mu_1 \Delta \boldsymbol{u} + R_a \mu_1 \mu_2 \theta \mathbf{e}_z, \quad \boldsymbol{\nabla} \cdot \boldsymbol{u} = 0, \quad -\lambda \theta - \boldsymbol{u}_3 = \mu_2 \Delta \theta, \tag{5.4.15}$$

has a sequence of countable eigenvalues $\lambda_1 \leq \lambda_2 \leq \lambda_3 \leq \cdots$, and the corresponding eigenfunctions $[\boldsymbol{u}_k, \theta_k]_{k=1}^{\infty}$ constitute an orthogonal basis with respect to the inner product

$$\langle [\boldsymbol{u}, \theta], [\tilde{\boldsymbol{u}}, \tilde{\theta}] \rangle = (\boldsymbol{u}, \tilde{\boldsymbol{u}}) + R_a \mu_1 \mu_2 (\theta, \tilde{\theta}) \tag{5.4.16}$$

where $[\boldsymbol{u}_1, P_1, \theta_1]$ is smooth. Moreover, for the initial condition $[\boldsymbol{u}^0, \theta^0] \in L^2$, if

$$[\boldsymbol{u}^0, \theta^0] = \sum_k \gamma_k [\boldsymbol{u}_k, \theta_k],$$

then the solution of the linear Boussinesq system (5.4.14) can be expressed as

$$\mathrm{e}^{Lt} [\boldsymbol{u}^0, \theta^0] = \sum_k \gamma_k \mathrm{e}^{-\lambda_k t} [\boldsymbol{u}_k, \theta_k]. \tag{5.4.17}$$

In particular, there exists $C > 0$, such that

$$||\mathrm{e}^{Lt} [\boldsymbol{u}^0, \theta^0] || \leq C \mathrm{e}^{-\lambda_1 t} || [\boldsymbol{u}^0, \theta^0] ||. \tag{5.4.18}$$

Proof. Firstly, let's recall the inner product (5.4.16) of $L^2_{R_a}$, the definitions of H and V. Consider the eigenvalue problem

$$-\lambda \boldsymbol{u} = \mu_1 \mathcal{P} \Delta \boldsymbol{u} + R_a \mu_1 \mu_2 \mathcal{P}\{\theta \mathbf{e}_z\}, \quad -\lambda \theta = \mu_2 \Delta \theta + u_3, \tag{5.4.19}$$

where the projection $\mathcal{P}\{L^2(Q)\}^3 \to H$. Obviously, according to the definition of inner product (5.4.16), for λ_0 big enough, the operator

$$\begin{pmatrix} (\mu_1 \Delta - \lambda_0) I & R_a \mu_1 \mu_2 \mathbf{e}_z^t \\ \mathbf{e}_z & (\mu_2 \Delta - \lambda_0) \end{pmatrix}^{-1}$$

is a bounded compact symmetric linear operator mapping $L^2_{R_a}(Q) \cap \{H \times L^2\}$ into itself. According to the compact symmetric operator theory, there exists a sequence of real eigenvalues $\lambda_1 \leq \lambda_2 \leq \cdots \leq \lambda_k \leq \cdots$, with their corresponding eigenfunction $\{(u_k, \theta_k)\}_{k=1}^{\infty}$ constituting a complete orthogonal basis of space $L^2_{R_a}(Q)$. The minimizer $(u_{\lambda_1}, \theta_{\lambda_1})$ of the following variational problem

$$\begin{aligned} &\min_{(U,\Theta) \in A} F(U, \Theta) \\ &= \min_{(U,\Theta) \in A} \int_Q \{\mu_1 |\boldsymbol{\nabla} U|^2 + R_a \mu_1 \mu_2^2 |\boldsymbol{\nabla} \Theta|^2 - 2 R_a \mu_1 \mu_2 U_3 \Theta\} \mathrm{d}x \mathrm{d}y \mathrm{d}z, \end{aligned}$$

is a weak solution of (5.4.19), where the function space A is

$$\Big\{U \in V, \Theta \in H^1(Q) \cap L^2_{per}((0,2\pi)^2; H^1_0(0,1)),\ ||U||^2 + Ra\mu_1\mu_2||\Theta||^2 = 1\Big\}.$$

Since functional $F(U, \Theta)$ is coercive and convex, there exists at least one solution $(u_1, \theta_1) \in A$ of the eigenvalue problem (5.4.19). According to Lemma 1.1 of [76, p. 180], there exists $P_1 \in L^2_{per}((0,2\pi)^2;\ L^2(0,1))$ such that (u_1, P_1, θ_1) is a weak solution of (5.4.15) when $\lambda = \lambda_1$. According to the boundary conditions, (u_1, P_1, θ_1) satisfies the problem (5.4.15) in $\Omega = \{(x,y,z)| -2\pi < x, y < 4\pi, 0 < z < 1\}$ when $\lambda = \lambda_1$. According to Theorem 5.1 in [76, p. 218], we know $u_1 \in (H^3_{per}((0,2\pi)^2; H^3(0,1)))^3 \cap V$, $P_1 \in H^2_{per}((0,2\pi)^2; H^2(0,1))$. With the elliptic regularity theory [82], we have $\theta_1 \in H^3_{per}((0,2\pi)^2; H^3(0,1) \cap H^1_0(0,1))$. Applying the bootstrap method, we get $u_1 \in (H^{m+1}_{per}((0,2\pi)^2; H^{m+1}(0,1)))^3 \cap V$, $P_1 \in H^m_{per}((0,2\pi)^2; H^m(0,1))$, $\theta_1 \in H^{m+1}_{per}((0,2\pi)^2;\ H^{m+1}(0,1) \cap H^1_0(0,1))$, $\forall m \geq 2$.

With the above results, we obtain (5.4.17) and (5.4.18).

Lemma 5.4.2. Suppose that R_a^* is the Rayleigh number defined in (5.4.12). If $R_a < R_a^*$, then $\lambda_1 > 0$. If $R_a > R_a^*$, then $\lambda_1 < 0$.

Proof. For constructing an eigenfunction of (5.4.15), we only have to find u_3 and θ. In fact, taking curl of the first equation of (5.4.13), and letting $\boldsymbol{\omega} = (\omega_1, \omega_2, \omega_3) = \mathbf{curl}\ \boldsymbol{u} = \boldsymbol{\nabla} \times \boldsymbol{u}$, we get

$$\partial_t \boldsymbol{\omega} = \mu_1 \Delta \omega + R_a \mu_1 \mu_2 (\boldsymbol{\nabla} \times \mathbf{e}_z)\theta.$$

Taking curl of the above equation, we have

$$\partial_t(\boldsymbol{\nabla} \times \boldsymbol{\omega}) = -\partial_t \Delta \boldsymbol{u} = -\mu_1 \Delta^2 \boldsymbol{u} + R_a \mu_1 \mu_2 \Big(\boldsymbol{\nabla} \times (\boldsymbol{\nabla} \times \mathbf{e}_z)\Big)\theta.$$

Since the horizontal part (u_1, u_2) of the velocity field can be denoted by u_3 and ω_3 (see [31]), the eigenvalue problem (5.4.15) is equivalent to

$$-\lambda\omega_3 = \mu_1 \Delta\omega_3,\ -\lambda \Delta u_3 = \mu_1 \Delta^2 u_3 + R_a \mu_1 \mu_2 (\partial_x^2 + \partial_y^2)\theta,\ -\lambda\theta = \mu_2 \Delta\theta + u_3. \tag{5.4.20}$$

By studying the following reduced variational problem about $U_3(z)$ and $\Theta(z)$, we get the eigenfunctions of the eigenvalue problem (5.4.15). For $k \geq 1$, define

$$\begin{aligned} F_3(U_3, \Theta, R_a) \equiv \int_0^1 & [\mu_1 |\partial_z^2 U_3 - k^2 U_3|^2 + R_a \mu_1 \mu_2^2 k^2 (|\partial_z \Theta|^2 + k^2 |\Theta^2|) \\ & - 2R_a \mu_1 \mu_2 k^2 U_3 \Theta] \mathrm{d}z. \end{aligned} \tag{5.4.21}$$

Then let's consider the variational problem

$$\lambda(R_a) = \min_{(U,\Theta)\in A_3} F_3(U_3, \Theta, R_a), \tag{5.4.22}$$

where

$$A_3 = \Big\{U_3 \in H_0^2,\ \Theta \in H_0^1,\ \int_0^1 (|\partial_z U_3|^2 + k^2|U_3|^2 + R_a\mu_1\mu_2 k^2|\Theta|^2)\mathrm{d}z = 1\Big\}.$$

With the standard method, we prove that the problem (5.4.22) has a minimizer $[U_3, \Theta]$, which satisfies Euler-Lagrange equation

$$-\lambda(R_a)(\partial_z^2 - k_0^2)U_3 = \mu_1(\partial_z^2 - k^2)^2 U_3 - R_a\mu_1\mu_2 k^2\Theta, \tag{5.4.23}$$

$$-\lambda(R_a)\Theta = \mu_2(\partial_z^2 - k^2)\Theta + U_3, \tag{5.4.24}$$

with boundary condition $U_3\Big|_{z=0,1} = \partial_z U_3\Big|_{z=0,1} = 0,\ \Theta\Big|_{z=0,1} = 0$. From direct calculation, we know that

$$\left(-\frac{1}{k_0}\partial_z U_3(z)\sin(k_0x),\ \ 0,\ \ U_3(z)\cos(k_0x),\ \ \Theta(z)\cos(k_0x)\right)$$

is a solution of the problem (5.4.15) when $\lambda = \lambda(R_a)$.

First we prove that if $R_a > R_a^*$, then $\lambda_1 \le \lambda(R_a) < 0$. According to (5.4.12), there exists $k_0 \ge 1$ such that

$$R(k_0) = \min_{\Theta\in B,\ k_0^2\int_0^1(|\partial_z\Theta|^2 + k_0^2|\Theta|^2)\mathrm{d}z = 1,} \int_0^1 |(\partial_z^2 - k_0^2)^2\Theta|^2\mathrm{d}z < R_a.$$

Since operator $(\partial_z^2 - k_0^2)^{-1} : L^2(0,1) \to L^2(0,1)$ is a bounded compact symmetric linear operator, the eigenvalue $0 < R(k_0)$ is real and finite multiplicity, and the variational problem $R(k_0)$ exists a minimizer $\Theta_{k_0} \in B$. Let

$$U_{3,k_0} \equiv -\mu_2(\partial_z^2 - k_0^2)\Theta_{k_0},$$

substituting $[U_{3,k_0}, \Theta_{k_0}]$ into (5.4.21), we get

$$\begin{aligned}
&\lambda(R_a)\int_0^1 (|\partial_z U_{3,k_0}|^2 + k^2|U_{3,k_0}|^2 + R_a\mu_1\mu_2 k_0^2|\Theta|^2)\mathrm{d}z\\
&\le \int_0^1 \{\mu_1|(\partial_z^2 - k_0^2)U_{3,k_0}|^2 + R_a\mu_1\mu_2[\mu_2 k_0^2(|\partial_z\Theta_{k_0}|^2 + k_0^2|\Theta_{k_0}^2|)\\
&\quad -2k_0^2 U_{3,k_0}\Theta_{k_0}]\}\\
&= \mu_1\mu_2^2\int_0^1 [|(\partial_z^2 - k_0^2)^2\Theta_{k_0}|^2 - R_a k_0^2(|\partial_z\Theta_{k_0}|^2 + k_0^2|\Theta_{k_0}|^2)]\mathrm{d}z
\end{aligned}$$

$$< \mu_1\mu_2^2 \int_0^1 [|(\partial_z^2 - k_0^2)^2 \Theta_{k_0}|^2 - R(k_0)k_0^2(|\partial_z \Theta_{k_0}|^2 + k_0^2|\Theta_{k_0}|^2)]\mathrm{d}z$$
$$= 0.$$

Thus, $\lambda(R_a) < 0$.

In the following, we prove that if $R_a < R_a^*$, then $\lambda_1 > 0$ with the contradiction method. If $\lambda_1 \leq 0$, then the first equation of'(5.4.20) means $\boldsymbol{\omega} = 0$. Thus the corresponding eigenfunction $[u_3, \theta]$ satisfies the second and third equations in (5.4.20), or (5.4.23) and (5.4.24), which implies that for $k \geq 1$, there exists

$$\lambda(R_a) \leq \lambda_1 \leq 0.$$

Letting $\tilde{\Theta} = \sqrt{R_a}\Theta$, we have

$$\lambda(R_a) = \min_{\int_0^1(|\partial_z U_3|^2 + k^2|U_3|^2 + \mu_1\mu_2 k^2|\tilde{\Theta}|^2)\mathrm{d}z = 1} F_3(U_3, \tilde{\Theta}, R_a), \tag{5.4.25}$$

where

$$F_3(U_3, \tilde{\Theta}, R_a) = \int_0^1 [\mu_1|\partial_z^2 U_3 - k^2 U_3|^2 + \mu_1\mu_2^2 k_0^2(|\partial_z \tilde{\Theta}|^2 + k^2|\tilde{\Theta}^2|) - 2\sqrt{R_a}\mu_1\mu_2 k^2 U_3 \tilde{\Theta}]\mathrm{d}z.$$

$\lambda(R_a)$ is a continuous function of R_a. In fact, for any two Rayleigh numbers R_{a_1} and R_{a_2}, we choose two corresponding minimizers $[U_1, \Theta_1]$ and $[U_2, \Theta_2]$. According to the above expression of F_3,

$$|F_3(U_1, \tilde{\Theta}_1, R_{a_1}) - F_3(U_1, \tilde{\Theta}_1, R_{a_2})| \leq C|R_{a_1} - R_{a_2}|,$$

$$|F_3(U_2, \tilde{\Theta}_2, R_{a_2}) - F_3(U_2, \tilde{\Theta}_2, R_{a_1})| \leq C|R_{a_1} - R_{a_2}|.$$

Thus,

$$\lambda(R_{a_1}) = F_3(U_1, \tilde{\Theta}_1, R_{a_1}) < \lambda(R_{a_2}) + C|R_{a_1} - R_{a_2}|.$$

Similarly, we get

$$\lambda(R_{a_2}) = F_3(U_2, \tilde{\Theta}_2, R_{a_2}) < \lambda(R_{a_1}) + C|R_{a_1} - R_{a_2}|.$$

So, we obtain the continuity of $\lambda(R_a)$. According to the expression of F_3, letting $U_3 = \tilde{\Theta}$, we get

$$\lim_{R_a \to \infty} \lambda(R_a) = -\infty,$$

which implies that for any $k \geq 1$, there exists $R_a^0(k) \leq R_a$ such that

$$\lambda(R_a^0(k)) = 0.$$

Denote the minimizer of (5.4.25) by $[U_3^0, \Theta^0]$, which satisfies the latter two equations of (5.4.20) when $\lambda(R_a^0(k)) = 0$, that is,

$$0 = \mu_1(\partial_z^2 - k^2)^2 U_3^0 - R_a^0 \mu_1 \mu_2 k^2 \Theta^0,$$

$$0 = \mu_2(\partial_z^2 - k^2)\Theta^0 + U_3^0.$$

The above two equations are equivalent to

$$-(\partial_z^2 - k^2)^4 \Theta^0 - R_a^0(k) k^2 (\partial_z^2 - k^2)\Theta^0 = 0$$

which means $R_a^0\,(k) \geq R(k)$. Thus, according to (5.4.12),

$$R_a \geq R_a^0\,(k) \geq R_a^*,$$

which is a contradiction. Thus the lemma is proved.

5.4.2 *Nonlinear Stability for $R_a < R_a^*$*

Theorem 5.4.3. If Rayleigh number $R_a < R_a^*$, then the stationary solution $(\boldsymbol{u}_s, P_s, T_s)$ of (5.4.6) is unconditional nonlinearly stable with respect to the norm of $C(0,\infty; L^2(Q))$, and it is nonlinearly stable with respect to norm of $C(0,\infty; H^2(Q)) \cap W_\infty^1(0,\infty; L^2(Q))$.

Proof. The proof of the unconditional nonlinear stability of $(\boldsymbol{u}_s, P_s, T_s)$ with respect to the norm of $C(0,\infty; L^2(Q))$ appears in [77]. In fact, taking L^2 inner product of (5.4.7) with $\boldsymbol{u}$, and taking L^2 inner product of (5.4.8) with θ, we get

$$\|\boldsymbol{u}(t)\|_{L^2}^2 + \|\theta(t)\|_{L^2}^2 \leq (\|\boldsymbol{u}_0\|_{L^2}^2 + \|\theta_0\|_{L^2}^2)\mathrm{e}^{-2\tilde{\lambda}_1 t},$$

where the definition of $\tilde{\lambda}_1$ will be given later.

Now, let's prove the nonlinear stability of $(\boldsymbol{u}_s, P_s, T_s)$ with respect to the norm of $C(0,\infty; H^2(Q)) \cap W_\infty^1(0,\infty; L^2(Q))$. Let

$$\mathcal{A} = \begin{pmatrix} (\mu_1 \mathcal{P}\Delta)I & 0 \\ 0 & \mu_2\Delta \end{pmatrix}, \quad \mathcal{B} = \begin{pmatrix} 0 & \sqrt{R_a\mu_1\mu_2}\,\mathbf{e}_z^t \\ \sqrt{R_a\mu_1\mu_2}\,\mathbf{e}_z & 0 \end{pmatrix},$$

where $\mathbf{e}_z = (0,0,1)$, and I is a 3×3 unit matrix. If $\tilde{\lambda}$ is an eigenvalue of the eigenvalue problem

$$(\mathcal{A} + \mathcal{B})\left[\tilde{\boldsymbol{u}}, \tilde{\theta}\right] = -\tilde{\lambda}\left[\tilde{\boldsymbol{u}}, \tilde{\theta}\right], \tag{5.4.26}$$

and $\left[\tilde{\boldsymbol{u}}, \tilde{\theta}\right]$ is its corresponding eigenfunction, then $\lambda = \tilde{\lambda}$ is an eigenvalue of (5.4.15), and $[\boldsymbol{u}, \theta] = \left[\tilde{\boldsymbol{u}}, \tilde{\theta}/\sqrt{R_a\mu_1\mu_2}\right]$ is its corresponding eigenfunction. Similarly, if λ is an eigenvalue of (5.4.15), and $[\boldsymbol{u}, \theta]$ is its corresponding eigenfunction, then $\tilde{\lambda} = \lambda$ is an eigenvalue of (5.4.26), and $\left[\tilde{\boldsymbol{u}}, \tilde{\theta}\right] = \left[\boldsymbol{u}, \sqrt{R_a\mu_1\mu_2}\,\theta\right]$ is its corresponding eigenfunction. Thus, $\tilde{\lambda}_1 = \lambda_1 > 0$.

$\mathcal{A} + \mathcal{B}$ is the infinitesimal generator of analytic semigroup $\mathrm{e}^{t(\mathcal{A}+\mathcal{B})}$: $H \times L^2(Q) \to H \times L^2(Q)$,

$$\|\mathrm{e}^{t(\mathcal{A}+\mathcal{B})}\| \leq C\mathrm{e}^{-\lambda_1 t}, \quad \forall t > 0, \tag{5.4.27}$$

$$\|(-\mathcal{A} - \mathcal{B})^{1/2}\mathrm{e}^{t(\mathcal{A}+\mathcal{B})}\| \leq Ct^{-1/2}\mathrm{e}^{-\lambda_1 t}, \quad \forall t > 0, \tag{5.4.28}$$

cf. [171, 176]. Since $(-\mathcal{A} - \mathcal{B})^{-1}$: $H \times L^2(Q) \to H \times L^2(Q), \mathcal{A}(-\mathcal{A} - \mathcal{B})^{-1}$: $H \times L^2(Q) \to H \times L^2(Q)$ is a self-adjoint bounded linear operator, $\forall t > 0$, $[\boldsymbol{u}, \theta] \in H \times L^2(Q)$,

$$\begin{aligned}
&\|(-\mathcal{A})^{1/2}\mathrm{e}^{t(\mathcal{A}+\mathcal{B})}[\boldsymbol{u}, \theta]\| \\
=&\Big[(-\mathcal{A})(-\mathcal{A} - \mathcal{B})^{-1}(-\mathcal{A} - \mathcal{B})^{1/2}\mathrm{e}^{t(\mathcal{A}+\mathcal{B})}[\boldsymbol{u}, \theta], \\
&(-\mathcal{A} - \mathcal{B})^{1/2}\mathrm{e}^{t(\mathcal{A}+\mathcal{B})}[\boldsymbol{u}, \theta]\Big]^{1/2} \\
\leq&\|\mathcal{A}(-\mathcal{A} - \mathcal{B})^{-1}\|^{1/2}\, \|(-\mathcal{A} - \mathcal{B})^{1/2}\mathrm{e}^{t(\mathcal{A}+\mathcal{B})}[\boldsymbol{u}, \theta]\| \\
\leq&Ct^{-1/2}\mathrm{e}^{-\lambda_1 t}\|[\boldsymbol{u}, \theta]\|.
\end{aligned} \tag{5.4.29}$$

Equations (5.4.7) and (5.4.8) are rewritten as

$$\begin{aligned}
&\left[\boldsymbol{u}(t),\ \sqrt{R_a\mu_1\mu_2}\,\theta(t)\right] \\
=&\mathrm{e}^{t(\mathcal{A}+\mathcal{B})}(\boldsymbol{u}_0,\ \sqrt{R_a\mu_1\mu_2}\,\theta_0) \\
&- \int_0^t \mathrm{e}^{(t-s)(\mathcal{A}+\mathcal{B})}\left[(\boldsymbol{u}\cdot\boldsymbol{\nabla})\boldsymbol{u}(s),\ (\boldsymbol{u}\cdot\boldsymbol{\nabla})(\sqrt{R_a\mu_1\mu_2}\,\theta(s))\right]\mathrm{d}s.
\end{aligned} \tag{5.4.30}$$

Applying (5.4.27), (5.4.29) and (5.4.30), we get

$$\begin{aligned}
&\|\boldsymbol{u}(t)\|_{L^2} + \|\theta(t)\|_{L^2} \\
\leq&C_1\mathrm{e}^{-\lambda_1 t}\Big(\|\boldsymbol{u}_0\|_{L^2} + \|\theta_0\|_{L^2}\Big) \\
&+ C_2\int_0^t \mathrm{e}^{-\lambda_1(t-s)}\Big(\|(\boldsymbol{u}\cdot\boldsymbol{\nabla})\boldsymbol{u}(s)\|_{L^2} + \|(\boldsymbol{u}\cdot\boldsymbol{\nabla})\theta(s)\|_{L^2}\Big)\mathrm{d}s \\
\leq&C_1\Big(\|\boldsymbol{u}_0\|_{L^2} + \|\theta_0\|_{L^2}\Big) + C_2\sup_{0\leq s\leq T}\Big(\|\boldsymbol{u}(s)\|^2_{H^2} + \|\theta(s)\|^2_{H^2}\Big), \quad \forall T \geq t,
\end{aligned}$$

$$
\begin{aligned}
&\|\Delta \boldsymbol{u}(t)\|_{L^2}+\|\Delta\theta(t)\|_{L^2}\\
\leq &C_1 e^{-\lambda_1 t}\Big(\|\Delta \boldsymbol{u}_0\|_{L^2}+\|\Delta\theta_0\|_{L^2}\Big)\\
&+C_2\int_0^t (t-s)^{-1/2}e^{-\lambda_1(t-s)}\Big(\|\nabla\{(\boldsymbol{u}\cdot\nabla)\boldsymbol{u}(s)\}\|_{L^2}+\|\nabla\{(\boldsymbol{u}\cdot\nabla)\theta(s)\}\|_{L^2}\Big)\mathrm{d}s\\
\leq &C_1\Big(\|\Delta \boldsymbol{u}_0\|_{L^2}+\|\Delta\theta_0\|_{L^2}\Big)+C_2\sup_{0\leq s\leq T}\Big(\|\boldsymbol{u}(s)\|_{H^2}^2+\|\theta(s)\|_{H^2}^2\Big),\quad \forall T\geq t.
\end{aligned}
$$

Let

$$E(T)=\sup_{0\leq s\leq T}\Big(\|\boldsymbol{u}(s)\|_{H^2}+\|\theta(s)\|_{H^2}\Big).$$

Thus, we obtain

$$E(T)\leq C_1\Big(\|\boldsymbol{u}_0\|_{H^2}+\|\theta_0\|_{H^2}\Big)+C_2E^2(T),\quad \forall T\geq 0, \tag{5.4.31}$$

where C_1 and C_2 are positive constants independent of T. Therefore, according to (5.4.31), we know that if $\|\boldsymbol{u}_0\|_{H^2}+\|\theta_0\|_{H^2}$ is small enough, then (5.4.7) and (5.4.8) have a unique solution $(\boldsymbol{u},\theta)\in C\Big([0,\infty);\ \big(H^2(Q)\big)^4\Big)$, such that

$$\|\boldsymbol{u}(t)\|_{H^2}+\|\theta(t)\|_{H^2}\leq 2C_1\Big(\|\boldsymbol{u}_0\|_{H^2}+\|\theta_0\|_{H^2}\Big),\quad \forall t\geq 0.$$

The theorem is proved.

5.4.3 *Nonlinear Instability for $R_a > R_a^*$*

Let's study the nonlinear instability of stationary solution (5.4.6). Firstly, we recall an important lemma.

Lemma 5.4.4 (cf. [94]). Let L be a linear operator in Banach space X with norm $\|\cdot\|$, e^{tL} be a strong continuous semigroup in X, and there exist $C_L>0$ and $\lambda>0$ such that

$$\|e^{tL}\|_{(X,X)}\leq C_L e^{t\lambda}.$$

Suppose that $N(y)$ is a nonlinear operator in X, $|||\cdot|||$ is another norm in X, and there exists a positive constant C_N such that for $y\in X$, $|||y|||<\infty$,

$$\|N(y)\|\leq C_N|||y|||^2.$$

Assume that $y(t)$ is a solution of the following equation,

$$y'=Ly+N(y),$$

with $|||y(t)|||^2\leq\sigma$, and there exists $C_\sigma>0$ such that for any $\varepsilon>0$, there exists a $C_\varepsilon>0$ such that

$$\frac{\mathrm{d}}{\mathrm{d}t}|||y(t)|||\leq\varepsilon|||y(t)|||+C_\sigma|||y(t)|||+C_\varepsilon\|y(t)\|. \tag{5.4.32}$$

If $y^\delta(0) = \delta y_0$, $\|y_0\| = 1$, $|||y_0||| < \infty$, and β_0 is a fixed small enough constant, then there exists a constant $C > 0$ such that if

$$0 \le t \le T^\delta \equiv \frac{1}{\lambda}\log\frac{\beta_0}{\delta},$$

then

$$\|y(t) - \delta \mathrm{e}^{tL} y_0\| \le C\big(|||y_0|||^2 + 1\big)\delta^2 \mathrm{e}^{2\lambda t}.$$

In particular, if there exists a constant C_p such that $\|\delta \mathrm{e}^{tL} y_0\| \ge C_p \delta \mathrm{e}^{\lambda t}$, then there exists $T^{esc} \le T^\delta$ such that

$$\|y(T^{esc})\| \ge \tau_0 > 0,$$

where τ_0 is dependent on $C_L, C_N, C_\sigma, C_p, \lambda, y_0, \sigma$, but independent of δ.

For proving the nonlinear instability of stationary solution (5.4.6) with the above lemma, we must choose a proper Sobolev norm $|||\cdot|||$, then get the estimate (5.4.32). Let $\|\cdot\| = \|\cdot\|_{L^2}$, $D^k_{x,y} = \sum\limits_{k_1+k_2=k} \partial_x^{k_1}\partial_y^{k_2}$,

$$E_0 = \|\boldsymbol{u}(t)\|^2 + \|\theta(t)\|^2,$$

$$E_k = E_0 + \|\boldsymbol{\nabla} \boldsymbol{u}(t)\|^2 + \|\boldsymbol{\nabla}\theta(t)\|^2 + \|D^k_{x,y}\boldsymbol{\nabla}\boldsymbol{u}(t)\|^2 + \|D^k_{x,y}\boldsymbol{\nabla}\theta(t)\|^2.$$

Lemma 5.4.5. Suppose $k \ge 2$. We have

$$\|(\boldsymbol{u}\cdot\boldsymbol{\nabla})\boldsymbol{u}\| + \|(\boldsymbol{u}\cdot\boldsymbol{\nabla})\theta\| \le CE_k.$$

Proof. By Sobolev Embedding Theorem and Hölder Inequality, we get

$$\begin{aligned}\boldsymbol{u}^2(x,y,z,t) =& 2\int_0^z \boldsymbol{u}_z \boldsymbol{u}(x,y,s,t)\mathrm{d}s\\ \le& C\int_0^z \|\boldsymbol{u}_z(\cdot,s,t)\|_{H^2_{x,y}}\|\boldsymbol{u}(\cdot,s,t)\|_{H^2_{x,y}}\mathrm{d}s,\end{aligned}$$

$$\|\boldsymbol{u}(t)\|_{L^\infty} \le C(\|\boldsymbol{u}(t)\| + \|(\partial_x^2+\partial_y^2)\boldsymbol{u}(t)\| + \|\partial_z \boldsymbol{u}(t)\| + \|(\partial_x^2+\partial_y^2)\partial_z\boldsymbol{u}(t)\|).$$

With multiplicative inequality [14, p. 323], we have

$$\|(\boldsymbol{u}\cdot\boldsymbol{\nabla})\boldsymbol{u}\| + \|(\boldsymbol{u}\cdot\boldsymbol{\nabla})\theta\| \le \|\boldsymbol{u}\|_{L^\infty}(\|\boldsymbol{\nabla}\boldsymbol{u}\| + \|\boldsymbol{\nabla}\theta\|) \le CE_k,$$

in the process of proving the above inequality we used the following estimate

$$\|\boldsymbol{\nabla} D^{k-l}_{x,y}\boldsymbol{u}\| \le C\|\boldsymbol{\nabla}\boldsymbol{u}\|^{l/k}\|\boldsymbol{\nabla} D^k_{x,y}\boldsymbol{u}\|^{(k-l)/k},\ \ \forall 1 \le l \le k-1.$$

Lemma 5.4.6. Let $k \ge 3$, we obtain

$$\frac{\mathrm{d}}{\mathrm{d}t}E_k \le \varepsilon E_k + CE_k^2 + C_\varepsilon E_0, \quad \forall \varepsilon > 0.$$

Proof. Taking L^2 inner product of (5.4.7) with $\boldsymbol{u}$, and taking L^2 inner product of (5.4.8) with θ, we have

$$\frac{\mathrm{d}}{\mathrm{d}t}\|\boldsymbol{u}\|^2 + 2\mu_1\|\nabla\boldsymbol{u}\|^2 \le 2R_a\|u_3\|\|\theta\|,$$

$$\frac{\mathrm{d}}{\mathrm{d}t}\|\theta\|^2 + 2\mu_2\|\nabla\theta\|^2 \le 2\|u_3\|\|\theta\|.$$

Taking L^2 inner product of (5.4.7) with $\boldsymbol{u}_t$, and taking L^2 inner product of (5.4.8) with $\Delta\theta$, we obtain

$$\frac{\mathrm{d}}{\mathrm{d}t}\mu_1\|\nabla\boldsymbol{u}\|^2 + 2\|\boldsymbol{u}_t\|^2 \le 2\|(\boldsymbol{u}\cdot\nabla)\boldsymbol{u}\|^2 + \|\partial_t u_3\|^2 + C\|\theta\|^2,$$

$$\frac{\mathrm{d}}{\mathrm{d}t}\|\nabla\theta\|^2 + 2\mu_2\|\Delta\theta\|^2 \le C\|(\boldsymbol{u}\cdot\nabla)\theta\|^2 + \mu_2\|\Delta\theta\|^2 + C\|u_3\|^2.$$

Similarly,

$$\frac{\mathrm{d}}{\mathrm{d}t}\mu_1\|\nabla D^k_{x,y}\boldsymbol{u}\|^2 + 2\|D^k_{x,y}\boldsymbol{u}_t\|^2 \le 2\|D^k_{x,y}\{(\boldsymbol{u}\cdot\nabla)\boldsymbol{u}\}\|^2 + \|D^k_{x,y}\boldsymbol{u}_t\|^2 + C\|D^k_{x,y}\theta\|^2,$$

$$\begin{aligned}\frac{\mathrm{d}}{\mathrm{d}t}\|\nabla D^k_{x,y}\theta\|^2 + 2\mu_2\|\Delta D^k_{x,y}\theta\|^2 \le{}& C\|D^k_{x,y}\{(\boldsymbol{u}\cdot\nabla)\theta\}\|^2 + \mu_2\|\Delta D^k_{x,y}\theta\|^2\\ &+ C\|D^k_{x,y}\boldsymbol{u}\|^2.\end{aligned}$$

Applying Hölder Inequality, Sobolev Embedding Theorem, the multiplicative inequality [15, p. 323] and Lemma 5.4.5, we get

$$\begin{aligned}&\|D^k_{x,y}\{(\boldsymbol{u}\cdot\nabla)\boldsymbol{u}\}\|\\
\le{}& C\sum_{l=0}^{k-2}\|D^l_{x,y}\boldsymbol{u}\|_{L^\infty}\|D^{k-l}_{x,y}\nabla\boldsymbol{u}\|\\
&+ 2\|D^{k-1}_{x,y}\boldsymbol{u}\|_{L^\infty_z(0,1;L^2_{x,y})}\|D_{x,y}\nabla\boldsymbol{u}\|_{L^2_z(0,1;L^\infty_{x,y})}\\
&+ 2\|D^k_{x,y}\boldsymbol{u}\|_{L^\infty_z(0,1;L^2_{x,y})}\|\nabla\boldsymbol{u}\|_{L^2_z(0,1;L^\infty_{x,y})}\\
\le{}& CE_k + C\|D^{k-1}_{x,y}\boldsymbol{u}\|_{H^1_z(0,1;L^2_{x,y})}\|D_{x,y}\nabla\boldsymbol{u}\|_{L^2_z(0,1;H^2_{x,y})}\\
&+ C\|D^k_{x,y}\boldsymbol{u}\|_{H^1_z(0,1;L^2_{x,y})}\|\nabla\boldsymbol{u}\|_{L^2_z(0,1;H^2_{x,y})} \le CE_k,\\
&\|D^k_{x,y}\{(\boldsymbol{u}\cdot\nabla)\theta\}\|\\
\le{}& C\sum_{l=0}^{k-2}\|D^l_{x,y}\boldsymbol{u}\|_{L^\infty}\|\nabla D^{k-l}_{x,y}\theta\|_{L^2}\\
&+ C\|D^{k-1}_{x,y}\boldsymbol{u}\|_{L^\infty_z(0,1;L^2_{x,y})}\|D_{x,y}\nabla\theta\|_{L^2_z(0,1;L^\infty_{x,y})}\end{aligned}$$

$$
\begin{aligned}
&+C\|D^k_{x,y}\boldsymbol{u}\|_{L^\infty_z(0,1;L^2_{x,y})}\|\nabla\theta\|_{L^2_z(0,1;L^\infty_{x,y})}\\
\le & C\|D^{k-1}_{x,y}\boldsymbol{u}\|_{H^1_z(0,1;L^2_{x,y})}\|D_{x,y}\nabla\theta\|_{L^2_z(0,1;H^2_{x,y})}+CE_k\\
&+C\|D^k_{x,y}\boldsymbol{u}\|_{H^1_z(0,1;L^2_{x,y})}\|\nabla\theta\|_{L^2_z(0,1;H^2_{x,y})}\le CE_k,
\end{aligned}
$$

where we use

$$
\|\nabla D^{k-l}_{x,y}\theta\|\le C\|\nabla\theta\|^{l/k}\|\nabla D^k_{x,y}\theta\|^{(k-l)/k},\quad \forall 1\le l\le k-1.
$$

Meanwhile

$$
\begin{aligned}
&C\|D^k_{x,y}\boldsymbol{u}\|^2+C\|D^k_{x,y}\theta\|^2\\
\le & C\|D^{k+1}_{x,y}\boldsymbol{u}\|^{2k/(k+1)}\|\boldsymbol{u}\|^{2/(k+1)}+C\|D^{k+1}_{x,y}\theta\|^{2k/(k+1)}\|\theta\|^{2/(k+1)}\\
\le & \varepsilon\{\|D^{k+1}_{x,y}\boldsymbol{u}\|^2+\|D^{k+1}_{x,y}\theta\|^2\}+C_\varepsilon\{\|\boldsymbol{u}\|^2+\|\theta\|^2\}.
\end{aligned}
$$

Using all the above inequalities, we prove Lemma 5.4.6.

As $\lambda_1<0$ when $R_a>R_a^*$, there exists $k^*\ge 1$ such that $\lambda^*=\lambda(R_a)<0$, where $\lambda(R_a)$ is the minimum value defined in (5.4.22). When $k=k^*$. $-\lambda^*>0$ is the so-called fastest growth rate. Let $[U_k,\Theta_k]$ be the minimizer of the variational problem.

Definition 5.4.7. Define the generic profile of the initial disturbance as

$$
\begin{aligned}
\left[\tilde{\boldsymbol{u}},\tilde{\theta}\right]&=(\tilde{u}_1,\tilde{u}_2,\tilde{u}_3,\tilde{\theta})\\
&=\sum_{k_1^2+k_2^2=k^2}\Big(V_{1,k_1,k_2},V_{2,k_1,k_2},v_{k_1,k_2}U_k(z),\vartheta_{k_1,k_2}\Theta_k(z)\Big)\mathrm{e}^{\mathrm{i}k_1x+\mathrm{i}k_2y},
\end{aligned}
$$

where if $k_1^2+k_2^2=(k^*)^2$. Then at least one of v_{k_1,k_2} and ϑ_{k_1,k_2} is non-zero. $V_{1,0,0}=V_{2,0,0}=0$,

$$
V_{1,k_1,k_2}=\frac{\mathrm{i}k_1}{k^2}v_{k_1,k_2}\partial_zU_k(z),\quad \forall\, k_1^2+k_2^2=k^2\ge 1,
$$

$$
V_{2,k_1,k_2}=\frac{\mathrm{i}k_2}{k^2}v_{k_1,k_2}\partial_zU_k(z),\quad \forall\, k_1^2+k_2^2=k^2\ge 1.
$$

Combining Lemma 5.4.4, Lemma 5.4.5 with Lemma 5.4.7, we get the following theorem.

Theorem 5.4.8. Let Rayleigh number $R_a>R_a^*$, and $\left[\boldsymbol{u}^\delta,\theta^\delta\right]$ be a solution of (5.4.7)-(5.4.10) with initial data

$$
\left[\boldsymbol{u}^\delta(0),\theta^\delta(0)\right]=\delta\left[\tilde{\boldsymbol{u}},\tilde{\theta}\right],
$$

where $[\tilde{\boldsymbol{u}},\tilde{\theta}]$ is the generic profile defined above, $\left\|\left[\tilde{\boldsymbol{u}},\tilde{\theta}\right]\right\|=1$. Then for small enough $\delta\le\delta_0$, $0\le t\le T^\delta$,

$\|\left[\boldsymbol{u}^\delta(t),\ \theta^\delta(t)\right]$

$$- \sum_{k_1^2+k_2^2=(k^*)^2} \delta e^{tL}\Big(V_{1,k_1,k_2},\ V_{2,k_1,k_2},\ v_{k_1,k_2}U_k(z),\ \vartheta_{k_1,k_2}\Theta_k(z)\Big)e^{ik_1x+ik_2y}\|$$

$$\leq C\left\{1+\left(\Big\|\Big[\nabla\tilde{\boldsymbol{u}},\nabla\tilde{\theta}\Big]\Big\|^2+\Big\|\Big[\partial_x^k\nabla\tilde{\boldsymbol{u}},\partial_x^k\nabla\tilde{\theta}\Big]\Big\|^2\right)\right\}\delta^2 e^{-2\lambda^* t},\ k\geq 3,$$

where $\|\,[\boldsymbol{u},\theta]\,\| = \|u_1\| + \|u_2\| + \|u_3\| + \|\theta\|$.

Remark 5.4.9. Notice that $e^{tL}(V_{1,k_1,k_2},\ V_{2,k_1,k_2},\ v_{k_1,k_2}U_k(z),\ \vartheta_{k_1,k_2}\Theta_k(z))e^{ik_1x+ik_2y}$ is given concretely in (5.4.17). Obviously, if $k_1^2+k_2^2 = (k^*)^2$, v_{k_1,k_2} or ϑ_{k_1,k_2} is non-zero, then a positive constant C exists such that

$$\|e^{tL}\Big(V_{1,k_1,k_2},\ V_{2,k_1,k_2},\ v_{k_1,k_2}U_k(z),\ \vartheta_{k_1,k_2}\Theta_k(z)\Big)e^{ik_1x+ik_2y}\| \geq Ce^{-\lambda^* t}.$$

Thus, according to Lemma 5.4.4 and Theorem 5.4.8, stationary solution $(\boldsymbol{u}_s, P_s, T_s)$ is nonlinearly unstable.

According to Lemmas 5.4.4-5.4.6, choosing $|||\cdot||| = E_k^{1/2}$ and $\|\cdot\| = E_0^{1/2}$, we prove Theorem 5.4.8.

References

[1] ABARBANEL H D I, HOLM D D, MARSDEN J E, et al. Richardson number criterion for the nonlinear stability of the three-dimensional stratified flow. Phys. Rev. Lett., 1984, 52(26): 2352–2355.

[2] ABIDI H, HMIDI T. On the global well-posedness of the critical quasi-geostrophic equation. SIAM J. Math. Anal., 2008, 40(1): 167–185.

[3] ADAMS R A. Sobolev Space. New York: Academic Press, 1975.

[4] ARNOLD L. Random Dynamical System. Springer Monographs in Mathematics. Berlin: Springer-Verlag, 1998.

[5] ARNOLD V I. Conditions for nonlinear stability of the stationary plane curvilinear flows of an ideal fluid. Dokl. Mat. Nauk., 1965, 162(5): 773–777.

[6] ARNOLD V I. On an a priori estimate in the theory of hydrodynamic stability. English Transl: Am. Math. Soc. Transl., 1969, 19: 267–269.

[7] ARNOLD V I. Mathematical Methods of Classical Mechanics. Berlin: Springer, 1989.

[8] BABIN A V, VISHIK M I. Attractors of Evolution Equations. Amsterdam: North Holland, 1992.

[9] BARCILON V. Stability of a non-divergent Ekman layer. Tellus, 1965, 17: 53–68.

[10] BATES P W, LU K N, WANG B X. Random attractors for stochastic reaction-diffusion equations on unbounded domains, J. Differential Equations, 2009, (246): 845–869.

[11] BEALE J, KATO T, MAJDA A. Remarks on breakdown of smooth solutions for the 3-D Euler equations. Comm. Math. Phys., 1984, 94: 61–66.

[12] BENNETT A F, KLOEDEN P E. The simplified quasi-geostrophic equations: existence and uniqueness of strong solutions. Mathematika, 1980, 27: 287–311.

[13] BENNETT A F, KLOEDEN P E. The dissipative quasi-geostrophic equations. Mathematika, 1981, 28: 265–285.

[14] BENZI R, PARISI G, SUTERA A, et al. Stochastic resonance in climatic change. Tellus, 1982, 34: 10–16.

[15] BESOV O V, IL'IN V P, NIKOL'SKII S M. Integral Representations of Functions and Imbedding Theorems. Vol. I, J. Wiley, New York, 1978.

[16] BJERKNES V. Das Problem von der Wettervorhersage, betrachtet vom Standpunkt der Mechanik und der Physik. Meteor. Z., 1904, 21: 1–7.

[17] BLÖMKER D, DUAN J Q, WANNER T. Enstrophy dynamics of stochastically forced large-scale geophysical flows. J. Math. Phys., 2002, 43(5): 2616–2626.

[18] BLUMEN W. Theory of wave interactions and two-dimensional turbulence. Uniform potential vorticity flow. Part I. J. Atmos. Sci., 1978, 35: 774–783.

[19] BOURGEOIS A J, BEALE J T. Validity of the quasi-geostrophic model for large-scale flow in the atmosphere and ocean. SIAM J. Math. Anal., 1994, 25: 1023–1068.

[20] BRANNAN J R, DUAN J Q, WANNER T. Dissipative quasi-geostrophic dynamics under random forcing. J. Math. Anal. Appl., 1998, 228: 221–233.

[21] BRYAN K. A numerical method for the study of the circulation of the world ocean. J. Comput. Phys., 1969, 4(3): 347–376.

[22] BUSSE F H. Transition to turbulence in Rayleigh-Bénard convection, in Hydrodynamic Instabilities and the Transition to Turbulence. 2nd ed. H. L. Swinney and J. P. Gollub, eds., Berlin: Springer-Verlag, 1985.

[23] CAFFARELLI L, VASSEUR A. Drift diffusion equations with fractional diffusion and the quasi-geostroophic equation. Ann. math., 2010, 171(3): 1903–1930.

[24] CAO C S, TITI E S. Global well-posedness and finite-dimensional global attractor for a 3-D planetary geostrophic viscous model. Comm. Pure Appl. Math, 2003, 56: 198–233.

[25] CAO C S, TITI E S, ZIANE M. A "horizontal" hyper-diffusion 3-D thermocline planetary geostrophic model: well-posedness and long-time behavior. Nonlinearity, 2004, 17: 1749–1776.

[26] CAO C S, TITI E S. Global well-posedness of the three-dimensional viscous primitive equations of large-scale ocean and atmosphere dynamics. Ann. Math., 2007, 166: 245–267.

[27] CAO C S, TITI E S. Global well-posedness of the 3D primitive equations with partial vertical turbulence mixing heat diffusion. Comm. Math. Phys., 2012, 310(2): 537-568.

[28] CESSI P, IERLEY G R. Symmetry-breaking multiple equilibria in quasi-geostrophic wind-driven flows. J. Phys. Ocean., 1995, 25: 1196–1205.

[29] CHAE D. On the regularity conditions for the dissipative quasi-geostrophic equations. SIAM J. Math. Anal., 2006, 37(5): 1649–1656.

[30] CHAE D, LEE J. Global well-posedness in the super-critical dissipative quasi-geostrophic equations. Comm. Math. Phys., 2003, 233(2): 297–311.

[31] CHANDRASEKHAR S. Hydrodynamic and Hydromagnetic Stability. Oxford: Oxford University Press, 1961.

[32] CHARNEY J G. The dynamics of long waves in a baroclinic westerly current. J. Meteor., 1947, 4: 135–163.

[33] CHARNEY J G, FJÖRTOFT R, VON NEUMANN J. Numerical integration of the barotropic vorticity equation. Tellus, 1950, 2: 237–254.

[34] CHARNEY J G, PHILLIPS N A. Numerical integration of the quasi-geostrophic equations for barotropic simple baroclinic flows. J. Meteor., 1953, 10: 71–99.

[35] CHARVE F. Global well posedness and asymptotics for a geophysical fluid system. Comm. PDE, 2005, 29(11/12), 1919–1940.

[36] CHEN Q L, MIAO C X, ZHANG Z F. A new Bernstein's inequality and the 2D dissipative quasi-geostrophic Equation. Comm. Math. Phys.,

2007, 271(3): 821–838.
[37] CHEPYZHOV V V, VISHIK M I. Evolution equations and their trajectory attractors. J. Math. Pures Appl., 1997, 2: 913–964.
[38] CHOU J F. Long Term Weather Prediction (in Chinese). Beijing: Meteorology Press, 1986.
[39] COLIN L. The Cauchy problem and the continuous limit for the multilayer model in geophysical fluid dynamics. SIAM J. Math. Anal., 1997, 28(3): 516–529.
[40] CONSTANTIN P, CORDOBA D, WU J H. On the critical dissipative quasi-geostrophic equation. Indiana Univ. Math. J., 2001, 50(Special Issue): 97–107.
[41] CONSTANTIN P, MAJDA A, TABAK E. Formation of strong fronts in the 2-D quasi-geostrophic thermal active scalar. Nonlinearity, 1994, 7: 1495–1533.
[42] CONSTANTIN P, MAJDA A, TABAK E. Singular front formation in a model for quasi-geostrophic flow. Phys. Fluids, 1994, 6: 9–11.
[43] CONTANTIN P, NIE Q, SCHORGHOFER N. Nonsingular surface quasi-geostrophic flows. Phys Lett. A, 1998, 241(3): 168–172.
[44] CONTANTIN P, NIE Q, SCHORGHOFER N. Front formation in atctive scalar. Phys. Rev. E., 1999, 60(3): 2858–2863.
[45] CONSTANTIN P, WU J H. Behavior of solutions of 2D quasi-geostrophic equations. SIAM J. Math. Anal., 1999, 30: 937–948.
[46] CONSTANTIN P, WU J H. Regularity of Hölder continuous solutions of the supercritical quasi-geostrophic equation. Ann. I. H. Poincaré-AN, 2008, 25: 1103–1110.
[47] CORDOBA D. Nonexistence of simple hyperbolic blow-up for the quasi-geostrophic equation. Ann. Math., 1998, 148: 1135–1152.
[48] CORDOBA A, CORDOBA D. A maximum principle applied to quasi-geostrophic equations. Comm. Math. Phys., 2004, 249: 511–528.
[49] CORDOBA D, FEFFERMAN C. Growth of solutions for QG and 2D Euler equations. J. Am. Math. Soc., 2002, 15(3): 665–670.
[50] COX M D. A Primitive Equation, Three-Dimensional Model of the O-cean. (GFDL ocean group, technical report) Princeton: Geophysical Fluid Dynamics Laboratory, 1984.
[51] CRAUEL H, DEBUSSCHE A, FLANDOLI F. Random attractors. J. Dyn. Diff. Eq., 1997, 29(2): 307–341.
[52] CRAUEL H, FLANDOLI F. Attractors of random dynamics systems. Prob. Th. Rel. Fields, 1994, 100: 365–393.
[53] DA PRATO G, DEBUSSCHE A, TEMAM R. Stochastic Burger's equations. Nonl. Diff. Eq. Appl., 1994, 1: 389–402.
[54] DA PRATO G, ZABCZYK J. Stochastic Equations in Infinite Dimensions, Encyclopedia of Mathematics and its Application. Cambridge: Cambridge University Press, 1993.
[55] DA PRATO G, ZABCZYK J. Ergodicity for Infinite Dimensional Systems. London Mathematical Society Lecture Note Series 229. Cambridge: Cambridge Univ. Press, 1996.

[56] DA PRATO G, ZABCZYK J. Evolution Equations with white-noise boundary conditions. Stocha. and Stocha. Reports, 1993, 42(3/4): 167–182.

[57] DAVIS S H. On the principle of exchange of stabilities. Proc. Roy. Soc. London A, 1969, 310: 341–358.

[58] DEBUSSCHE A, GLATT-HOLTZ N, TEMAM R, et al. Global existence and regularity for the 3D stochastic primitive equations of the ocean and atmosphere with multiplicative white noise. Nonlinearity, 2012, 25(7): 2093-2118.

[59] DESJARDINS B, GRENIER E. Linear instability implies nonlinear instability for various types of viscous boundary layers. Ann. I. H. Poincare-AN, 2003, 20(1): 87–106.

[60] DIJKSTRA A HENK. Nonlinear Physical Oceanography: A Dynamical Systems Approach to the Large Scale Ocean Circulation and ElNiño. 2nd Revised and Enlarged Edition. Berlin/ New York: Springer, 2005.

[61] DRAZIN P G, REID W H. Hydrodynamic Stability. Cambridge: Cambridge Univ. Press, 1981.

[62] DUAN J Q, KLOEDEN P E, SCHMALFUSS B. Exponential stability of the quasigeostrophic equation under random perturbations. Prog. Probability, 2001, 49: 241–256.

[63] DUAN J Q, HOLM D D, LI K T. Variational methods and nonlinear quasigeostrophic waves. Phys. Fluid, 1999, 11(14): 875–879.

[64] DUDIS J J, DAVIS S H. Energy stability of the buoyancy boundary layer. J. Fluid Mech., 1971, 47(2): 381–403.

[65] DUDIS J J, DAVIS S H. Energy stability of the Ekman boundary layer. J. Fluid Mech., 1971, 47(2): 405–413.

[66] DUTTON J A. The nonlinear quasi-geostrophic equation: existence and uniqueness of solutions on a bounded domain. J. Atmos. Sci., 1974, 31: 422–433.

[67] EADY E T. Long waves and cyclone waves. Tellus, 1949, 1: 33–52.

[68] EKMAN V W. On the influence of the earth's rotation on ocean currents. Arkiv Matem. Astr. Fysik (Stockholm), 1962, 1–52.

[69] EMBID P F, MAJDA A J. Averaging over fast gravity waves for geophysical flows with arbitrary potential vorticity. Comm. in PDE., 1996, 21: 619–658.

[70] EPSTEIN E S. Stochastic dynamic prediction. Tellus, 1969, 21(66), 739–759.

[71] EVANS L C, GASTLER R. Some results for the primitive equations with physical boundary conditions. Zeitschrift für angewandte Mathematik und Physik, 2013, DOI: 10.1007/s00033-013-0320-6.

[72] FLANDOLI F. Dissipativity and invariant measures for stochastic Navier-Stokes equations. Nonl. Diff. Eq. Appl., 1994, 1(4): 403–423.

[73] FLANDOLI F, SCHMALFUSS B. Random attractors for the 3D stochastic Navier-Stokes equation with multiplicative noise, Stochastics and Stochastic Reports, 59 (1996) 21–45.

[74] FRANKIGNOUL C, HASSELMANN K.Stochastic climate models II:

Application to sea-surface temperature anomalies and thermocline variability. Tellus, 1977, 29: 289–305.
[75] FRANZKE C, MAJDA A J, VANDEN-EIJNDEN E. Low-order stochastic mode reduction for a realistic barotropic model climate. J. Atmos. Sci., 2005, 62: 1722–1745.
[76] GALDI G P. An Introduction to the Mathematical Theory of the Navier-Stokes Equations. Vol. I. New York: Springer-Verlag, 1994.
[77] GALDI G P, STRAUGHAN B. Exchange of stabilities, symmetry and nonlinear stability. Arch. Rational Mech. Anal. 1985, 89(3): 211–228.
[78] GAO H J, SUN C F. Random attractor for the 3D viscous stochastic primitive equations with additive noise. Stoch. Dyn., 2009, 9: 293-313.
[79] GAO H J, SUN C F. Hausdorff dimension of random attractor for stochastic Navier-Stokes-Voight equations and primitive equations. Dyn. PDE, 2010, 7(4): 307-326.
[80] GAO H J, SUN C F. Large deviations for the stochastic primitive equations in two space dimensions. Comm. Math. Sci., 2012, 10: 575-593.
[81] GIGA Y, INUI K, MAHALOV A, et al. Rotating Navier-Stokes equations in $\mathbb{R}^3_+$ with initial data non-decreasing at infinity: the Ekman boundary layer problem. Arch. Rational Mech. Anal., 2007, 186(2): 177–224.
[82] GILBARG D, TRUDINGER N S. Elliptic Partial Differential Equations of the Second Order. 2nd Ed., Berlin/New York: Springer-Verlag, 2001.
[83] GILL A E. The boundary-layer regime for convection in a rectangular cavity. J. Fluid Mech., 1966, 26(3): 515–536.
[84] GILL A E. Atmosphere-Ocean Dynamics. Canifornia: Academic Press, 1982.
[85] GLATT-HOLTZ N, TEMAM R. Pathwise solutions of the 2-D stochastic primitive equations. Appl. Math. Opti., 2011, 63 (3): 401–433.
[86] GRIFFIES S M, TZIPERMAN E. A linear thermohaline oscillator driven by stochastic atmospheric forcing. J. Climate, 1995, 8(10): 2440–2453.
[87] GUILLÉN-GONZÁLEZ F, MASMOUDI N, RODRÍGUEZ-BELLIDO M A. Anisotropic estimates and strong solutions for the primitive equations. Diff. Int. Equ., 2001, 14: 1381–1408.
[88] GUO B L, HUANG D W. Existence of weak solutions and trajectory attractors for the moist atmospheric equations in geophysics. J. Math. Phys., 2006, 47(8): 083508.
[89] GUO B L, HUANG D W. 3D stochastic primitive equations of the large-scale ocean: global well-posedness and attractors. Comm. Math. Phys., 2009, 286(2): 697–723.
[90] GUO B L, HUANG D W. On the 3D viscous primmitive equations of the large-scale atmosphere. Acta Math. Sci, 2009, 29B(4): 846–866.
[91] GUO B L, HUANG D W. Existence and stability of steady waves for the Hasegawa-Mima equation. Bound. Val. Prob., 2009, doi:10.1155/2009/509801.
[92] GUO B L, HUANG D W. On the primitive equations of the large-scale ocean with stochastic boundary. Diff. Int. Equ., 2010, 23(3,4): 373–398.

[93] GUO B L, HUANG D W. Existence of the universal attractor for the 3-D viscous primitive equations of large-scale moist atmosphere. J. Diff. Equ., 2011, 251 (3): 457-491.
[94] GUO Y, HALLSTROM C, SPIRN D. Dynamics near unstable, interfacial fluids. Comm. Math. Phys., 2007, 270(3): 635–689.
[95] GUO Y, HAN Y Q. Critical Rayleigh Number in Rayleigh-Bénard Convection. Quart. Appl. Math., 2010, 68: 149–160.
[96] GUO Y, STRAUSS W A. Instability of periodic BGK equilibria. Comm. Pure Appl. Math. 2006, 48(8): 861–894.
[97] HALE J K. Asymptotic Behavior of Dissipative Systems. Math, Surveys Monographs, Amer. Math. Soc., Vol. 25, Providence, R. I., 1988.
[98] HALTINER G J, WILLIAMS R T. Numerical Prediction and Dynamic Meteorology. 2nd edition. New York: John Wiley and Sons, 1980.
[99] HASSELMANN K. Stochastic climate models. Part I: Theory. Tellus, 1976, 28: 473–485.
[100] HELD I M, PIERREHUMBERT R T, GERNER S, et al. Surface quasi-geostrophic dynamics. J. Fluid Mech., 1995, 282: 1–20.
[101] HERRON I H. On the principle of exchange of stabilities in Rayleigh-Bénard convection. SIAM J. Appl. Math., 2000, 61(4): 1362–1368.
[102] HESS M, HIEBER M, MAHALOV A, et al. Nonlinear stability of Ekman boundary layers. Konstanzer Schriften in Mathematic and Informatik, Nr. 242, Feb, 2008.
[103] HMIDI T, KERAANI S. Global solutions of the super-critical 2D quasi-geostrophic equations in Besov spaces. Adv, Math., 2007, 214: 618–638.
[104] HOLM D D, MARSDEN R T, RATIU T, et al. Nonlinear stability of fluid and plasma equilibria. Phys. Rep., 1985, 123(1/2): 1–116.
[105] HOLTON J R. An Introduction to Dynamic Meteorology. Third Edition. Elsevier Academic Press, 1992.
[106] HOWARD LN. Note on a paper of John W. Miles. J. Fluid Mech., 1961, 10(4): 509–512.
[107] HU C B, TEMAM R, ZIANE M. The primitive equations on the large scale ocean under the small depth hypothesis. Disc. Cont. Dyn. Sys., 2003, 9(1): 97–131.
[108] HUANG D W, GUO B L. On two-dimensional large-scale primitive equations in oceanic dynamics (I), (II). Applied Mathematics and Mechanics, 2007, 28(5): 521–538.
[109] HUANG D W, GUO B L, HAN Y Q. Random attractors for a quasi-geostrophic dynamical system under stochastic forcing. Int. J. Dyn. Syst. Differ. Equ., 2008, 3(1): 147–154.
[110] HUANG D W, GUO B L. On the existence of atmospheric attractors. Science in China Series D: Earth Science, 2008, 51(3): 469–480.
[111] HUANG H Y, GUO B L. Existence of the solutions and the attractors for the large-scale atmospheric equations. Science in China: Series D Earth Science (in Chinese), 2006, 49(6): 650–660.
[112] ITOO H, KURIHARA Y, ASAI T, et al. Numerical test of finite-difference form of primitive equations for barotropic case. J. Meteo.

Soc. Japan, 1962, 40(2).
[113] JEFFREYS H. The stability of a layer of fluid heated from below. Phil. Mag. 1926, 2(10): 833–844.
[114] JOSEPH D D. On the stability of the Boussinesq equations. Arch. Rational Mech. Anal. 1965, 20(1): 59–71.
[115] JOSEPH D D. Nonlinear stability of the Boussinesq equations by the method of energy. Arch. Rational Mech. Anal., 1966, 22(3): 163–184.
[116] JU N. Existence and uniqueness of the solution to the dissipative 2D quasi-geostrophic equations in the Sobolev Space. Comm. Math. Phys., 2004, 251(2): 365–376.
[117] JU N. The maximum principle and the global attractor for the dissipative 2D quasi-geostrophic equations. Comm. Math. Phys., 2005, 255(1): 161–181.
[118] JU N. Global solutions to the two dimensional quasi-geostrophic equation with critical or super-critical dissipation. Math. Ann., 2006, 334(3): 627–642.
[119] KASAHARA A. Computational aspects of numerical models for weather prediction and Climate simulation. Methods Comput. Phys., 1977, 17: 1–66.
[120] KASAHARA A, WASHINGTON W M. NCAR global general circulation model of the atmosphere. Mon. Wea. Rec., 1967, 95(7): 389–402.
[121] KHOUIDER B, TITI E S. An inviscid regularization for the surface quasi-geostrophic equation. Comm. Pure Appl. Math., 2008, 61(10): 1331–1346.
[122] KISELEV A, NAZAROV F, VOLBERG A. Global well-posedness for the critical 2D dissipative quasi-geostrophic equation. Invent. Math., 2007, 167(3): 445–453.
[123] KLEEMAN R, MOORE A. A theory for the limitation of ENSO predictability due to stochastic atmospheric transients. J. Atmos. Sci., 1997, 54(6): 753–767.
[124] KUKAVICA I, ZIANE M. On the regularity of the primitive equations of the ocean. Nonlinearity, 2007, 20(12): 2739–2753.
[125] KUO H L. Dynamics instability of two-dimensional non-divergent flow in a barotropic atmosphere. J. Meteor., 1949, 6: 105–122.
[126] KUO H L. Dynamics of quasi-geostrophic flows and instability theory. Adv. Appl. Mech., 1973, 13: 247–330.
[127] KURIHARA Y, TULEYA R E. Structure of a tropical cyclone development in a three-dimensional numerical simulation model. J. Atmos. Sci., 1974, 31(4): 893–919.
[128] LEITH C E. Climate response and fluctuation dissipation. J. Atmos. Sci., 1975, 32: 2022–2026.
[129] LEITH C E. Nonlinear normal mode initialization and quasi-geostrophic theory. J. Atmos. Sci., 1980, 37: 958–968.
[130] LI J P, CHOU J F. Existence of atmosphere attractor. Science in China (Series D), 1997, 40(2): 215–224.
[131] LI J P, CHOU J F. Asymptotic behavior of solutions of the moist atmo-

spheric equations. Acta Meteor. Sinica, 1998, 56(2): 61–72 (in Chinese).
[132] LI J P, CHOU J F. The qualitative theory on the dynamical equations of atmospheric motion and its applications. Chinese J. of Atmos. Sci., 1998, 22(4): 348–360.
[133] LI J P, CHOU J F. The global analysis theory of climate system and its applications. Chinese Sci. Bull., 2003, 48(10): 1034–1039.
[134] LI M C, HUANG J Y. A stochastic climatic model on the quasi three-yearly and half-yearly oscillation of the sea surface temperature. Acta Meteorologica Sinica, 1984, 42(2): 168–176.
[135] LIN Z W. Instability of some ideal plane flows. SIAM J. Math. Anal., 2003, 35(2): 318–356.
[136] LIN Z W. Nonlinear instability of some ideal plane flows. Int. Math. Res. Not., 2004, 41: 2147–2178.
[137] LIONS J L. Quelques méthodes de résolutions des problèmes aux limites nonlinéaires. Paris: Dunod, 1969.
[138] LIONS J L, MAGENES E. Problèmes aux limites non homogènes et applications. Paris: Dunod, 1968.
[139] LIONS J L, MANLEY O, TEMAN R, et al. Physical interpretation of the attractor for a simple model of atmospheric circulation. J. Atmos. Sci., 1997, 54(9): 1137–1143.
[140] LIONS J L, TEMAM R, WANG S. New formulations of the primitive equations of atmosphere and applications. Nonlinearity, 1992, 5: 237–288.
[141] LIONS J L, TEMAM R, WANG S. On the equations of the large-scale ocean. Nonlinearity, 1992, 5: 1007–1053.
[142] LIONS J L, TEMAM R, WANG S. Models of the coupled atmosphere and ocean (CAO I). Computational Mechanics Advance, 1993, 1: 5–54.
[143] LIONS J L, TEMAM R, WANG S. Numerical analysis of the coupled models of atmosphere and ocean (CAO II). Computational Mechanics Advance, 1993, 1(11): 55–119.
[144] LIONS J L, TEMAM R, WANG S. Mathematical theory for the coupled atmosphere-ocean models(CAO III). J. Math. Pures Appl., 1995, 74: 105–163.
[145] LIU S S, LIU S D. Atmospheric Dynamics. Beijing: Peking University Press, 1991.
[146] LIU Y M, MU M, SHEPHERD T G. Nonlinear stability of continuously stratified quasi-geostrophic flow. J. Fluid Mech., 1996, 325: 419–439.
[147] LORENZ E N. Energy and numerical weather prediction. Tellus, 1960, 12(4): 364–373.
[148] LORENZ E N. Dimension of weather and climate attractors. Nature, 1991, 353: 241–244.
[149] LORENZ E N. An attractor embedded in the atmosphere. Tellus, 2006, 58A: 425–429.
[150] LU H, LÜ S, XIN J, et al. A random attractor for the stochastic quasi-geostrophic dynamical system on unbounded domains. Nonl. Anal. T-MA, 2013, 90: 96–112.

[151] LUNARDI A. Analytic Semigroups and Optimal Regularity in Parabolic Problems. Progress in Nonlinear Differential Equations and Their Applications 16. Basel, Boston, Berlin: Birkhäuser, 1995.

[152] MAJDA A. Compressible Fluid Flow and Systems of Conservation Laws in Several Space Variables. Appl. Math. Sci. Vol. 53, New York: Springer, 1984.

[153] MAJDA A. Introduction to PDEs and Waves for the Atmosphere and Ocean. Courant Lecture Notes Mathematics, Vol. 9, 2003.

[154] MAJDA A, TIMOFEYEV I, VANDEN-EIJNDEN E. Models for stochastic climate prediction. Proc. Nati. Acad. Sci. U.S.A., 1999, 96: 14687–14691.

[155] MAJDA A, TIMOFEYEV I, VANDEN-EIJNDEN E. A mathematical framework for stochastic climate models. Comm. Pure Appl. Math., 2001, 54(8): 891–974.

[156] MAJDA A, TIMOFEYEV I, VANDEN-EIJNDEN E. A priori tests of a stochastic model reduction strategy. Physica D., 2002, 170: 206–252.

[157] MAJDA A, TIMOFEYEV I, VANDEN-EIJNDEN E. Systematic strategies for stochastic mode reduction in climate. J. Atmos. Sci., 2003, 60: 1705–1722.

[158] MAJDA A, TABAK E G. A two-dimensional model for quasigeostrophic flow: comparison with the two-dimensional euler flow. Physical D: Nonlinear Phenomena, 1996, 98: 515–522.

[159] MAJDA A, WANG X M. Validity of the one and one-half layer quasigeostrophic model and effective topography. Comm. PDE, 2005, 30: 1305–1314.

[160] MAJDA A, WANG X M. The emergence of large-scale coherent structure under small-scale random bombardments. Comm. Pure Appl. Math., 2001, 59: 467–500.

[161] MCINTYRE M E, SHEPHERD T G. An exact local conservation theorem for finite-amplitude disturbances to non-parallel shear flows with remarks on Hamiltonian structure and on Arnold's stability theorems. J. Fluid Mech., 1987, 181: 527–565.

[162] MIKOLAJEWICZ U, MAIER-REIMER E. Internal secular variability in an OGCM. Climate Dyn., 1990, 4: 145–156.

[163] MÜLLER P. Stochastic forcing of quasi-geostrophic eddies. Stochastic Modelling in Physical Oceanography, R. J. Adler, P. Müller and B. L. Rozovskii, Eds, Boston: Birkhäuser, 1996.

[164] MU M. Global classical solutions of initial-boundary value problems for nonlinear vorticity equation and its applications. Acta Math. Scientia, 1986, 6: 201–218.

[165] MU M. Global classical solutions of initial-boundary value problems for generalized vorticity equations. Scientia Sinica (Ser. A), 1987, 30: 359–371.

[166] MU M. Nonlinear stability criteria for motions of multilayer quasi-geostrophic flow. Scientia Sinica (Ser. B), 1991, 34(12): 1516–1528.

[167] MU M. Nonlinear stability of two-dimensional quasigeostrophic motions.

Geophys. Astrophys. Fluid Dyn., 1992, 65: 57–76.

[168] NICOLIS C, NICOLIS G. Is there a climatic attractor. Nature, 1984, 311: 529–532.

[169] OHKITAMI K, YAMADA M. Inviscid and inviscid-limit behavior of a surface quasi-geostrophic flow. Phys. Fluids, 1997, 9(4): 876–882.

[170] PAZY A. Semigroup of Linear Operators and Applications to Partial Differential Equations. Applied Mathematical Sciences 44, Berlin/New York: Springer-Verlag, 1983.

[171] PEDLOSKY J. Geophysical Fluid Dynamics. 1st Edition. Berlin/New York: Springer-Verlag, 1979.

[172] PEDLOSKY J. Geophysical Fluid Dynamics. 2nd Edition. Berlin/New York: Springer-Verlag, 1987.

[173] PEDLOSKY J. Ocean Circulation Theory. Berlin/New York: Springer-Verlag, 1996.

[174] PEIXOTO J P, OORT A H. Physics of Climate. New York/Berlin/Herdelberg: Springer-Verlag, 1992.

[175] PELLEW M A, SOUTHWELL R V. On maintained convection motion in a fluid heated from below. Proc. Roy. Soc. London A, 1940, 176(966): 312–343.

[176] PENLAND C, MATROSOVA L. A balance condition for stochastic numerical models with applications to El Niño-Southern oscillation. J. Climate, 1994, 7(9): 1352–1372.

[177] PHILLIPS N A. Energy transformations and meridional circulations associated with simple baroclinic waves in a two-level quasi-geostrophic model. Tellus, 1954, 6(3): 273–286.

[178] PHILLIPS O M. On the generation of waves by turbulent winds. J. Fluid Mech., 1957, 2(5): 417–445.

[179] PIERREHUMBERT R T, HELD I M, SWANSON K L. Spectra of local and nonlocal two-dimensional turbulence. Chaos, Solitons, and Fractals, 1994, 4(6): 1111–1116.

[180] RAYLEIGH LORD. On the stability, or instability, of certain fluid motions. Proc. London Math. Soc., 1879, 11(1): 57–72.

[181] REISER H. Baroclinic forecasting with the primitive equation. Proc. Int. Symp. on Num. Weather Pred. in Tokyo, Tokyo, 1962.

[182] RESNICK S G. Dynamical problems in nonlinear advective partial differential equation. Ph.D. Thesis, University of Chicago, 1995.

[183] RICHARDSON L F. Weather Prediction by Numerical Process. Cambridge: Cambridge University Press, 1922 (reprinted Dover, New York, 1965).

[184] RIPA R. Arnold's second stability theorem for the equivalent barotropic model. J. Fluid Mech., 1993, 257: 597–605.

[185] ROSSBY C G. Relation between variations in the intensity of the zonal circulation of the atmosphere and the displacements of the semi-permanent centers of action. J. Mar. Res., 1939, 2: 38–55.

[186] RUBENSTEIN D. A spectral model of wind-forced internal waves. J. Phys. Oceanogr., 1994, 24(4): 819–831.

[187] SALTZMAN B. Stochastically-driven climatic fluctuation in the sea-ice, ocean temperature, CO_2 feedback system. Tellus, 1981, 34(2): 97–112.
[188] SAMELSON R, TEMAM R, S. WANG. Some mathematical properties of the planetary geostrophic equations for large-scale ocean circulation. Appl. Anal., 1998, 70(1/2): 147–173.
[189] SAMELSON R, TEMAM R, C. WANG. et al. Surface pressure Poisson equation formulation of the primitive equations: numerical schemes. SIAM J. Numer Anal., 2003, 41(3): 1163–1194.
[190] SCHOCHET S. Singular limits in bounded domains for quasilinear symmetric hyperbolic systems having a vorticity equation. J. Diff. Eqns., 1987, 68(3): 400–428.
[191] SERRIN J. On the stability of viscous fluid motions. Arch. Rational Mech. Anal., 1959, 3(1): 1–13.
[192] SHEN J, WANG S H. A fast and accurate numerical scheme for the primitive equations of the atmosphere. SIAM J. Numer. Anal., 1999, 36(3): 719–737.
[193] SHEPHERD T G. Rigorous bounds on the nonlinear saturation of instabilities to parallel shear flows. J. Fluid Mech., 1988, 196: 291–322.
[194] SHUMAN F G, HOVERMALE J B. A Six-level primitive equation model. J. Appl. Meteor, 1968, 7: 525–531.
[195] SIMONNET E, TACHIM T MEDJO, TEMAM R. Barotropic-baroclinic formulation of the primitive equations of the ocean. Appl. Anal., 2003, 82(5): 439–456.
[196] SMAGORINSKY J. General circulation experiments with the primitive equations. I. the basic experiment, Mon. Wea. Rev., 1963, 91(3): 99–164.
[197] STEIN E M. Singular Integrals and Differentiability Properties of Functions. Princeton, NJ: Princeton University Press, 1970.
[198] STRAUGHAN B. The Energy Method, Stability, and Nonlinear Convection. Appl. Math. Sci. Vol. 91, New York: Springer-Velag, 1992.
[199] STRUWE M. Variational Methods: Applications to Nonlinear Partial Differential Equations and Hamiltonian Systems. New York: Springer-Velag, 1990.
[200] TANABE H. Functional Analytic Methods for Partial Differential Equations. Monographs and textbook in pure and applied mathematics: Vol. 204. Marcel Dekker, INC., New York, 1997.
[201] TEMAM R. Infinite-Dimensional Dynamical Systems in Mechanics and Physics. 2nd ed. Appl. Math. Ser., Vol. 68. New York: Springer-Verlag, 1997.
[202] TEMAM R. Navier-Stokes Equation: Theory and Numerical Analysis. Revised Edition. North-Holland, 1984.
[203] TEMAM R, ZIANE M. Some mathematical problems in geophysical fluid dynamics. Handbook of Mathematical Fluid Dynamics, 2005, 3: 535–658.
[204] TRIEBEL H. Theory of Function Spaces. Basel-Boston: Birkhäuser, 1983.

[205] VALLIS G K. Atmospheric and Oceanic Fluid Dynamics: Fundamental and Large-Scale Circulation. Cambridge: Cambridge Univ. Press, 2006.
[206] VISHIK M I, CHEPYZHOV V V. Trajectory and global attractors of three-dimensional Navier-Stokes systems. Math. Notes, 2002, 71(1/2): 177–193.
[207] VISHIK M I, CHEPYZHOV V V. Averaging of trajectory attractors of evolution equations with rapidly oscillating terms. Russian Acad. Sci. Sb. Math., 2001, 192(1): 16–53.
[208] WANG B X. Random attractors for stochastic FitzHugh-Nagumo system on unbounded domains, Nonlinear Analysis, 2009, (71): 2811–2828.
[209] WANG S H. On the 2-D model of large-scale atmospheric motion: well-posedness and attractors. Nonlinear Anal. Theo. Math. Appl., 1992, 18(1): 17–60.
[210] WANG S H. Attractors for the 3-D baroclinic quasi-geostrophic equations of large-scale atmosphere. J. Math. Anal. Appl., 1992, 165(1): 266–283.
[211] WASHINGTON W M, PARKINSON C L. An Introduction to Three-Dimensional Climate Modelling. Oxford: Oxford Univ. Press, 1986.
[212] WASHINGTON W M, SEMTNER A J, et al. A general circulation experiment with a coupled atmosphere. ocean and sea ice model. J. Phys. Oceanogr., 1980, 10(12): 1887–1908.
[213] WOLANSKY G. Existence, uniqueness, and stability of stationary barotropic flow with forcing and dissipation. Comm. Pure Appl. Math., 1988, 41(1): 19–46.
[214] WOLANSKY G. The barotropic vorticity equation under forcing and dissipation: bifurcations of nonsymmetric responses and multiplicity of solutions. SIAM J. Appl. Math., 1989, 49(6): 1585–1607.
[215] WOLANSKY G, GHIL M. Stability of quasi-geostrophic flow in a periodic channel. Phys. Letters A, 1995, 202(1): 111–116.
[216] WU J H. The two-dimensional quasi-geostrophic equation with critical or supercritical dissipation. Nonlinearity, 2005, 18(1): 139–154.
[217] WU J H. Global solutions of the 2D dissipative quasi-geostrophic equations in Besov spaces. SIAM J. Math. Anal., 2005, 36(3): 1014–1030.
[218] YAMASAKI M. Numerical Simulation of tropical cyclone development with the use of primitive equations. J. Meteor. Soc. Japan, 1968, 46(3): 178–201.
[219] ZEIDLER E. Nonlinear Functional Analysis and its Application. Vol. II. Berlin: Springer, 1990.
[220] ZENG Q C. Mathematic and Physical Base of Numerical Weather Prediction. The first volume. Beijing: Science Press, 1979.
[221] ZENG Q C. Variational principle of instability of atmospheric motions. Adv. Atmos. Sci., 1989, 6(2): 137–172.
[222] ZHOU X J. Atmospheric stochastic dynamics and predictability. Acta Meteorologica Sinica, 2005, 63(5): 806–811.